LINEAR INTEGRATED NETWORKS

LINEAR INTEGRATED NETWORKS
Fundamentals

GEORGE. S. MOSCHYTZ

Bell Telephone Laboratories
presently professor,
Swiss Federal Institute of Technology
Zurich, Switzerland

VAN NOSTRAND REINHOLD COMPANY
New York/Cincinnati/Toronto/London/Melbourne

Van Nostrand Reinhold Company Regional Offices:
New York   Cincinnati   Chicago   Millbrae   Dallas

Van Nostrand Reinhold Company International Offices:
London   Toronto   Melbourne

Manufactured in the United States of America

Published by Van Nostrand Reinhold Company
450 West 33rd Street, New York, N.Y. 10001

Published simultaneously in Canada by Van Nostrand Reinhold Ltd.

15   14   13   12   11   10   9   8   7   6   5   4   3   2   1

Library of Congress Cataloging in Publication Data

Moschytz, George S.
    Linear integrated networks: fundamentals.

    (Bell Telephone Laboratories series)
    Includes bibliographical references.
    1. Linear integrated circuits.   2.   Hybrid inte-
grated circuits.   I.   Title.
TK7874.M534 5   621.381'71   74-7250
ISBN 0-442-25581-0

To My Wife Doris
For Her Constant Encouragement and Understanding

FOREWORD

It is quite timely for a work to appear on the theory and design of linear integrated networks which emphasizes the growing recognition and importance of hybrid integrated circuits based on silicon semiconductor technology and precision passive film technology. Dr. Moschytz has undertaken a project which provides a sound theoretical basis for understanding linear active networks while stressing the practical active network realizations which were achieved during recent years, particularly at Bell Laboratories where he supervised much of the advanced development and design effort on active linear networks. Many of the examples and illustrations in the work are based on these designs, with careful consideration given to the numerous design and performance trade offs involved in developing high performance active networks using a hybrid technology.

This work will improve the communications and understanding necessary among the specialists in semiconductor device, component, and network technologies, and it effectively complements other recent books by providing practical illustrations of active networks designed to meet special requirements. The author gives special consideration to the unique features of hybrid integrated circuits with regard to small size, low power dissipation, batch fabrication, parasitic effects, network and device sensitivity, functional adjustment, and network performance.

The two volumes, devoted respectively to fundamentals and design, should become a valuable addition to the literature, and will be of considerable benefit to students and to practicing industry engineers and scientists. The volumes are the latest in the Bell Telephone Laboratories series, in which we attempt to bring the engineering and scientific community up to date on subjects of major current interest.

David Feldman
Director
Film and Hybrid Technology Laboratory
Bell Telephone Laboratories

PREFACE

The elimination of inductors from linear microcircuits is a particularly severe restriction in conventional LCR filter design, which is based on the duality between capacitor and inductor network characteristics. If frequency selective networks are to be integrated, methods must be found of functionally replacing their conventional LCR predecessors by a combination of passive-RC and active integrated circuits. As in previous cases when new technologies replaced old ones (the vacuum-tube transistor change-over comes to mind), this calls for a completely new and unbiased approach to network design. Such a new approach must take into consideration both the new concepts emerging in linear active network theory and the completely new circuit design techniques amenable to integrated circuit fabrication.

Contrary to previous technology transitions, the present one is still more far reaching in that it requires a complete realignment of the disciplines involved and the way they interact with one another. Where electronic systems previously were the result of a step-by-step process in which materials were developed, these materials then incorporated into components, components into circuits, and circuits into functional systems, these processes are rapidly converging into two separate activities. The first activity entails the fundamental study of phenomena as well as the control of physical, chemical, and electronic properties of materials for their subsequent fabrication into components and devices; the second involves the synthesis of devices, circuits, and subsystems into a unified and mutually compatible integrated design. Whereas the former must be left to separate specialized groups, the latter is replacing the activity previously assumed by the circuit and network designer.

This book has been written to provide a link between the specialized areas developing materials and integrated devices and those concerned primarily with the theoretical concepts of linear active network analysis and synthesis. I am familiar with excellent texts in the field of linear active network theory and others in the field of linear integrated circuit design. However, in order to design linear integrated active networks successfully and efficiently, the network theorist must know more about integrated-circuit fabrication techniques and device characterization and, conversely, the linear integrated-circuit designer must be more familiar with the emerging concepts of active network theory. Thus, this book is directed at the designer of linear integrated devices, at the linear active network theorist, and, in particular, at the "man-in-between" who is attempting to combine the newly appearing linear integrated devices into usable structures as covered in the companion volume *Linear Integrated Networks: Design*.

I am well aware that in setting myself the goal of dealing only with those topics in device technology and network theory that I consider relevant to the

ix

relatively new discipline of linear integrated network design, I am running the risk of doing full justice to neither. As a result, the specialist in either field may feel tempted to accuse me of superficiality. On the other hand, it may be that this book is one of an emerging kind that is less concerned with theory for its own sake than with the application of theory, blended with practical experience, for the sake of developing a useful product: linear integrated networks (LINs). Besides, having found myself in the position of the man-in-between, who has to glean the relevant information from the detailed literature of the various areas involved, I feel that I know which fundamentals of each discipline it is useful for the designer of linear integrated networks to find in a single volume. At the same time, for the specialists on either side of linear integrated networks, this book describes the fundamentals of their respective sciences in mutually comprehensible terms, thereby contributing to improved communications between them. Finally, for the graduate student who will one day embark on the adventure of engineering in the real world of nonideal devices, parasitic effects, and the application of scientific principles to the production of economic systems, a course on the material in this book combined with the material covered in *LINs: Design* should provide some valuable insight into the challenges and problems that he will encounter when designing linear integrated networks.

The material in this book has been divided into two parts. Part 1, comprising the first five chapters, deals with the fundamentals of linear active networks. The material is limited to those aspects of active network theory that are considered necessary for the practical design of linear integrated networks. The topics covered in Chapter 1 include the characterization of general linear networks and the characteristics of passive-RC networks. Pole-zero plots are introduced and the effects on the poles of passive-RC networks, due to the incorporation of active devices, are examined. Chapter 2 deals with feedback techniques in linear active networks, such as root loci and signal-flow graphs. Methods of analyzing networks are discussed in Chapter 3 .These include the indefinite admittance matrix and various other schemes for the analysis of active networks. In Chapter 4, those concepts of sensitivity are discussed that are most important in active network design, and in Chapter 5, basic active network elements are introduced.

Part 2 approaches the design of linear integrated networks from the point of view of integrated circuit technology and fabrication. The three chapters in this section deal with the fundamentals of hybrid-integrated circuit characterization and design. The emphasis is on the combination of film-integrated circuits and silicon-integrated circuits (SICs) into hybrid integrated subsystems, this combination being ideally suited for linear integrated network design. Chapter 6 deals with the basic processes involved in film and silicon integrated circuit fabrication, and with the characteristics of the resulting components and devices. The material in Chapters 7 and 8 represents the first step in applying the material described in Chapter 6 to hybrid integrated network design. Here are described some basic integrated circuit modules or building blocks that have proven to be important cornerstones in linear integrated network design. SIC operational amplifiers and gyrators (Chapter 7) as well as highly stable null networks using film resistors and capacitors (Chapter 8) are discussed in detail. From these three chapters, guidelines for hybrid-integrated circuit design evolve, which differ markedly from those with which the designer of conventional circuits has been familiar.

As in Part 1, the material in Part 2 is introductory in nature and is limited to those topics considered necessary for the design of linear integrated networks. At the end of the book, references are given for more detailed and in-depth treatment of the relevant topics.

The material covered in this book sets the stage for the companion volume *Linear Integrated Networks: Design* by establishing all the necessary concepts, terms and definitions and by introducing the network theory and device technology on which the material in the companion volume is based. The book is course-oriented and problems with worked out solutions are available separately. Part 1 lends itself to relatively short, point-by-point summaries at the end of each chapter which serve to emphasize those items that are of particular importance, amd that are repeatedly referred to in the companion volume. Because the book is directed at the engineer in the field at least as much as at the student, it is written in a form suitable for self study. For this reason numerous examples are worked out step-by step and liberal use is made of figures in order to ensure comprehension by the reader.

Much of the material covered in this book and the companion volume (*LINs: Design*) has been "class-room tested" as a two-semester course with two rather different types of students. First given as an out-of-hours course at Bell Telephone Laboratories, Holmdel, N.J., the students were exclusively engineers who, in some way or another, had use for certain aspects, if not for outright design, of hybrid-integrated networks in the course of their daily work. The second time the material was taught as a first year graduate course at the Swiss Federal Institute of Technology, Zurich, Switzerland*. Before going to the material covered in *Design* it was found necessary to go through the material covered in this book (*Fundamentals*) with both groups. With the former this material represented a refresher course for those engineers who had recently graduated; for most of the others it was new material. The second group had been taught much of the material in various other non-application oriented courses but had, to some extent, lost sight of the forest for the trees. Thus, since this book emphasises the most important results of more comprehensive topics (without going through the proofs), and associates them directly to the design of linear integrated networks wherever possible, it was found to provide a useful exercise in the application of previously somewhat obscure theory to practice.

---

* While on leave of absence from Bell Telephone Laboratories as a visiting professor at the Institute of Telecommunications.

In teaching the material covered in this and the companion book in a two-semester course, the following sequence has been found useful. Chapters 1 to 4 (*Fundamentals*) as well as Chapter 2 (*Design*) are covered in the first semester in order to provide a somewhat rounded off treatment of linear active networks in general. Chapter 5 *Fundamentals* and selected topics of Chapters 1, 3, 4, 6 and 7 *Design* are then dealt with in the second semester, whereby Chapters 6, 7 and 8 *Fundamentals* are suggested as additional reading material. In this way the second semester covers the integrated circuit aspects of linear network design. Depending on how much of *Fundamentals* it is deemed necessary to teach, this program can, of course, be drastically shortened, if need be, to a one-semester course.

GEORGE S. MOSCHYTZ

ACKNOWLEDGMENTS

I am greatly indebted to T. F. Epley, formerly of the Van Nostrand Reinhold Co., whose enthusiasm and encouragement provided the necessary motivation for this book and the companion volume *Linear Integrated Networks: Design*, and whose guidance through the early stages was invaluable; and to his successor B. R. Nathan and his colleagues J. Stirbis and Mrs. A. W. Gordon whose much appreciated advice assisted me up to the completion of both books.

I am indebted to many friends, colleagues and students for their valuable comments and suggestions during the preparation of this book. I am grateful for the constructive criticisms of the reviewers at Bell Telephone Laboratories (BTL), in particular to R. K. Even, D. Hilberman, E. Lueder (now professor at the Stuttgart University) D. G. Marsh, D. R. Means, W. H. Orr, and B. A. Unger. I am especially indebted to R. K. Even for his very thorough review of the entire manuscript, and for his numerous constructive criticisms and helpful advice regarding organization and presentation of the book. I also greatly appreciate the numerous helpful comments from my students at the Swiss Federal Institute of Technology (SFIT) in Zurich in particular from I. E. Berkovics who reviewed most of the manuscript and from R. Hammer, P. Horn and M. Kuenzli. I am most grateful for the kind cooperation of D. Hirsch and E. R. Kretzmer of BTL and of Professor H. Weber of the SFIT for making it possible for the manuscript to progress smoothly while I was in Zurich on a leave of absence from Bell Telephone Laboratories. My deep appreciation also goes to the extensive secretarial assistance of Miss P. Jobes at BTL and Mrs. A. Huebscher and Mrs. L. Hubmann at the SFIT, as well as to the excellent typing skills of Mrs. M. Taolise. Finally, my thanks go to my wife Doris and my daughters Joy, Helen and Miriam for the interest and understanding they showed during this undertaking.

G. S. MOSCHYTZ

LINEAR
INTEGRATED
NETWORKS

CONTENTS

LINEAR
INTEGRATED
NETWORKS

PART 1:
FUNDAMENTALS OF LINEAR ACTIVE NETWORKS

1

THE CHARACTERIZATION OF LINEAR NETWORKS

INTRODUCTION

The main incentive for the recent interest in linear active networks has come from the rapidly progressing field of microminiaturization. There are numerous reasons why microcircuits look so promising to the electronics industry as a whole. The most important ones can be summarized as follows:

1. Increased equipment density, and with it a reduction in equipment size and weight.

2. Increased system reliability, obtained by reducing the number of metallic interfaces (e.g., thermal-compression-bonded leads). Thus "interconnection degradation" caused by open metallic interfaces within electronic systems containing otherwise faultless components is greatly reduced.

3. Cost reduction, due to the batch processing production methods available and to economy in installation and maintenance.

4. Increased operating speeds, due to the elimination of stray capacitance and inductance and to the decrease in propagation delay that goes with reduction in size.

On the other hand, the following limitations, imposed both on the designers, manufacturers and users of microcircuits, go with the anticipated benefits:

1. For appreciable cost reduction, production quantities must be high. Only in this way can the initial high tooling costs be effectively absorbed. To

achieve this, general purpose, off-the-shelf functional building blocks have been developed for digital systems and, to an ever increasing extent, this same course is being followed in the field of linear networks.

2. For compatibility with the two major microminiaturization technologies (film and semiconductor integrated circuits) there are certain limitations on the passive components available to the designer of microcircuits. Most severe among these is the elimination of magnetic elements such as inductors and transformers. The reason for this is that the quality factor of inductors deteriorates considerably when reduced in size to an extent comparable with other miniaturized components.[1] By contrast the quality factor of capacitors remains constant in the same process. However, both resistors and capacitors are limited to medium values.

3. Finally, the inherent high packaging density of microcircuits limits their power handling capability. This has led to an interest in "micropower" integrated circuits for use particularly in military, space, and medical electronic systems.

Because of the anticipated benefits—foremost of which are the decrease in manufacturing cost and the increase in system reliability—concerted efforts in research and development have been devoted to overcoming these limitations and to making microcircuits a practical reality. To a great extent these endeavors have been immensely successful, and today's proliferation of electronic systems with highly sophisticated microcircuits both in the industrial, military, and commercial fields presumably exceeds the most optimistic predictions made a decade ago.

Only one limitation has so far resisted a wholly satisfying and general remedy: it is the absence of any microminiaturized equivalent or substitute for inductors. This is a particularly severe restriction in conventional filter design, which is based on the duality between capacitor and inductor network characteristics. If systems employing linear frequency-selective networks are to be miniaturized all the same, methods must be found of functionally replacing passive *LCR* networks. This becomes all the more important at low frequencies (say, up to 100 kHz), where inductors are particularly cumbersome. The methods most frequently investigated to circumvent this limitation consist of various combinations of active and passive *RC* networks; these constitute the newly evolving and vigorously pursued field of active network synthesis.

It is with the functional replacement of passive *LCR* networks by active *RC* networks that this book* is concerned. In this chapter we shall first take a brief look at the general transmission functions realizable by linear lumped-parameter finite networks. Some important concepts will be considered that

1. A. Rand, Inductor size vs. *Q*: A dimensional analysis, *IEEE Trans. Component Parts*, **CP–10**, 31–35 (1963).
*We refer here both to this book and its companion: *Linear Integrated Networks: Design.*

are necessary to characterize the overall behavior of these functions. The characteristics of passive RC networks will then be briefly reviewed in light of the concepts derived. Finally, we shall show, in a brief and cursory manner, how the inherent limitations of passive RC networks can be overcome by combining them with active elements.

1.1 THE TRANSMISSION FUNCTION OF LINEAR LUMPED-PARAMETER FINITE NETWORKS

The output signal of an nth-order linear lumped-parameter finite (LLF) network[2] can be found in terms of the input signal by solving a linear nth-order differential equation as follows:

$$a_n \frac{d^n y}{dt^n} + a_{n-1} \frac{d^{n-1} y}{dt^{n-1}} + \cdots + a_1 \frac{dy}{dt} + a_0 y$$

$$= b_m \frac{d^m x}{dt^m} + b_{m-1} \frac{d^{m-1} x}{dt^{m-1}} + \cdots + b_1 \frac{dx}{dt} + b_0 x \quad (1\text{-}1)$$

where

$x(t)$ is the input signal,
$y(t)$ is the output signal, and
$n \geq m$,

and the coefficients a_i and b_i are real.

Applying the Laplace transform to this equation, the required transmission function $T(s) = Y(s)/X(s)$ results as the ratio of two polynomials with real coefficients in the complex frequency variable s:

$$T(s) = \frac{Y(s)}{X(s)} = \frac{b_m s^m + b_{m-1} s^{m-1} + \cdots + b_1 s + b_0}{a_n s^n + a_{n-1} s^{n-1} + \cdots + a_1 s + a_0} \qquad [1\text{-}2][3]$$

where

$$s = \sigma + j\omega.$$

Expressing both numerator and denominator polynomials in their factored form we obtain the poles and zeros of the transmission function:

$$T(s) = K \frac{(s - z_1)(s - z_2) \cdots (s - z_m)}{(s - p_1)(s - p_2) \cdots (s - p_n)} = K \frac{\prod\limits_{i=1}^{m} (s - z_i)}{\prod\limits_{j=1}^{n} (s - p_j)} \qquad [1\text{-}3]$$

The poles p_j and zeros z_i need not necessarily be simple; furthermore, they can be either real or complex numbers. However, since the coefficients in the

2. Only LLF networks will be considered in this book.
3. Square brackets are used to emphasize important definitions or results.

transmission function are always real, complex poles and zeros must always occur in conjugate pairs. Therefore, the numerator and denominator can be factored into first- and second-order terms, each having real coefficients. The transmission function then has the form

$$T(s) = K \frac{\displaystyle\prod_{i=1}^{m/2} (s - z_{i_1})(s - z_{i_2})}{\displaystyle\prod_{j=1}^{n/2} (s - p_j)(s - p_j^*)} = K \frac{\displaystyle\prod_{i=1}^{m/2} (s^2 + 2\sigma_{z_i} s + \omega_{z_i}^2)}{\displaystyle\prod_{j=1}^{n/2} (s^2 + 2\sigma_{p_j} s + \omega_{p_j}^2)} \qquad [1\text{-}4]$$

for m and n even, and

$$T(s) = K \frac{(s - z_m) \displaystyle\prod_{i=1}^{(m-1)/2} (s - z_{i_1})(s - z_{i_2})}{(s - p_n) \displaystyle\prod_{j=1}^{(n-1)/2} (s - p_j)(s - p_j^*)}$$

$$= K \frac{(s - z_m) \displaystyle\prod_{i=1}^{(m-1)/2} (s^2 + 2\sigma_{z_i} s + \omega_{z_i}^2)}{(s - p_n) \displaystyle\prod_{j=1}^{(n-1)/2} (s^2 + 2\sigma_{p_j} s + \omega_{p_j}^2)} \qquad [1\text{-}5]$$

for m and n odd. Complex conjugate pole pairs have been assumed since, as we shall soon see, this is the case of most interest to us. In contrast, the zeros z_{i_1}, z_{i_2} may be either real or complex.

By comparing the coefficients in the two expressions in (1-4) and (1-5) it follows directly that the real coefficients σ_{z_i}, ω_{z_i}, σ_{p_j} and ω_{p_j} can be related to the corresponding zeros z_i and poles p_j by the expressions

$$\sigma_{z_i} = -\frac{z_{i_1} + z_{i_2}}{2} \qquad (1\text{-}6a)$$

$$\omega_{z_i}^2 = z_{i_1} z_{i_2} \qquad (1\text{-}6b)$$

$$\sigma_{p_j} = -\frac{p_j + p_j^*}{2} \qquad (1\text{-}7a)$$

$$\omega_{p_j}^2 = p_j p_j^* \qquad (1\text{-}7b)$$

Clearly, whether the zeros z_i and poles p_j are real or complex conjugate the coefficients σ_{z_i}, ω_{z_i}, σ_{p_j}, and ω_{p_j} must be real. In particular, if we let

$$\operatorname{Re} p_j = \operatorname{Re} p_j^* = -\sigma_{p_j} \qquad (1\text{-}8a)$$

and

$$\operatorname{Im} p_j = -\operatorname{Im} p_j^* = \tilde{\omega}_{p_j} \qquad (1\text{-}8b)$$

then it follows from (1-7) that

$$\omega_{p_j}^2 = \sigma_{p_j}^2 + \tilde{\omega}_{p_j}^2 \qquad (1\text{-}9)$$

A similar expression holds for $\omega_{z_i}^2$ if the zero pair z_{i_1}, z_{i_2} are complex conjugate.

1.2 POLE–ZERO DIAGRAMS

The poles and zeros of any transmission function can be represented graphically by a so-called pole–zero diagram.[4] This is the geometric diagram of the
factored form of the transmission function as given by (1-3), (1-4), or (1-5) in
the complex frequency, or s plane
 Pole–zero diagrams are very useful, since the poles and zeros of a transmission function contain all transmission characteristics of a network except
the scale factor K. This is demonstrated by the transmission functions and
corresponding poles and zeros of some well-known network functions in
Figs. 1-1 to 1-3. The poles of the fourth-order[5] maximally flat, or Butterworth,
low-pass function shown in Fig. 1-1a are shown in Fig. 1-1b. They are equispaced on a unit circle in the left half plane (LHP). If the poles fall inside the
Butterworth unit circle, ripple is introduced in the passband of the filter.
More specifically, if the pole contour is an ellipse (Fig. 1-2a), a Chebyshev
or equiripple low-pass function results, as shown in Fig. 1-2b. The pass-band
ripple is determined by the ellipse eccentricity ε. In fact the pole coordinates
can be obtained directly from those of the maximally flat filter of the same
order by multiplying the real parts of those poles by ε. Displacing the poles
still further, so that they are equispaced on a line parallel to the imaginary
axis in the LHP (see Fig. 1-3a), and adding symmetrical zeros in the right-
half plane (RHP), a flat-delay all-pass network can be approximated,[6] as
shown in Fig. 1-3b. Networks of this kind, whose transmission functions have
RHP zeros, are called *nonminimum-phase networks.*
 The pole-zero diagrams of the examples discussed above exhibit the characteristic features associated with the transmission functions of a particular
group of LLF networks, namely those consisting of inductors, capacitors,
and resistors (*LCR* networks). The location of the zeros is anywhere in
the s plane, that of the poles anywhere inside the left half s plane[7] (not
including the $j\omega$ axis in view of the presence of resistive elements in the
network).
 Returning to the aforementioned objective of realizing this type of transmission function without using inductors, we shall now show that this cannot

4. See, e.g., N. Balabanian and T. A. Bickart, *Electrical Network Theory* (New York: John Wiley
  & Sons, 1969), p. 392. For a brief but excellent treatment of this subject, see also E. J. Angelo,
  Jr. and A. Papoulis, *Pole–Zero Patterns in the Analysis of Low-Order Systems* (New York:
  McGraw-Hill, 1964).
5. The order of a network transmission function will be assumed to be the order of its denominator
  polynomial.
6. This network is in a special category, as its transmission function e^{-sT} is an entire function
  with no poles or zeros for any finite s. To realize it by LLF networks, e^{-sT} must be approximated by a rational function in s, of the kind given by (1–2). Methods of doing this, one of
  which results in the pole-zero diagram shown in Fig. 1–3a, have been developed. For example,
  methods by W. E. Thomson, L. Storch, and E. S. Kuh are given in L. Weinberg, *Network
  Analysis and Synthesis* (New York: McGraw-Hill, 1962), pp. 499–506.
7. The restriction on the poles is a result of the fact that to be physically realizable the denominator
  polynomial of the transmission function which determines the natural frequencies of the
  network is required to be a *Hurwitz polynomial*; see A. E. Guillemin, *Modern Methods of
  Network Synthesis*, Advances in Electronics, Vol. 3 (New York: Academic Press, 1951), p. 275.

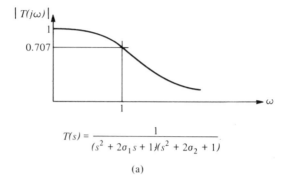

$$T(s) = \frac{1}{(s^2 + 2\sigma_1 s + 1)(s^2 + 2\sigma_2 + 1)}$$

(a)

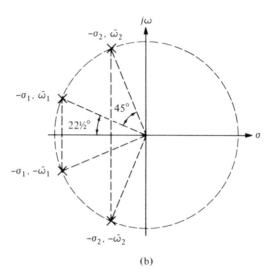

(b)

FIG. 1-1. Fourth-order, maximally flat (Butterworth) low-pass filter. (a) Transmission function in s and amplitude vs. frequency response. (b) Pole–zero diagram consisting of poles equispaced on a unit circle in the LHP.

be accomplished by resistors and capacitors alone. As we shall see in the following section, although the location of zeros is unrestricted with passive RC networks, poles can be generated only on the negative real axis. As the examples above showed, however, complex conjugate poles are required for the realization of commonplace filter networks. It therefore becomes necessary to find some means by which the poles generated by passive RC networks can be "lifted off" the negative real axis and placed within the left half s plane. As will be discussed later, this is accomplished by introducing active networks into the system in the numerous ways suggested by the relatively new field of active network synthesis. First, however, let us become more familiar with the general properties of passive RC networks.

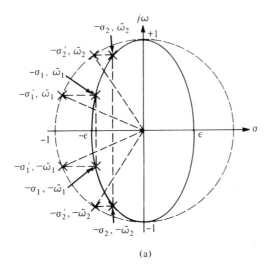

(a)

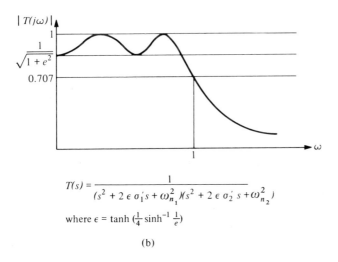

$$T(s) = \frac{1}{(s^2 + 2 \epsilon \sigma_1' s + \omega_{n_1}^2)(s^2 + 2 \epsilon \sigma_2' s + \omega_{n_2}^2)}$$

where $\epsilon = \tanh (\frac{1}{4} \sinh^{-1} \frac{1}{e})$

(b)

FIG. 1-2. Fourth-order equiripple (Chebyshev) low-pass filter. (a) Pole–zero diagram consisting of poles on an ellipse. (b) Transmission function in s and amplitude vs. frequency response. The passband ripple is a function of the ellipse eccentricity ε.

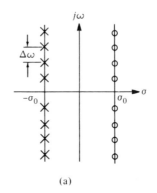

(a)

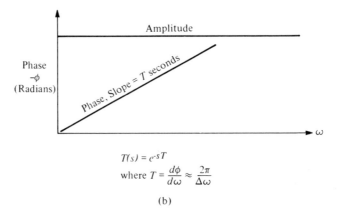

$$T(s) = e^{-sT}$$
$$\text{where } T = \frac{d\phi}{d\omega} \approx \frac{2\pi}{\Delta\omega}$$

(b)

FIG. 1-3. Pole–zero diagram of flat delay (linear phase) all-pass network (nonminimum-phase network). (a) Pole–zero diagram (nonminimum phase because of RHP zeros). (b) Transmission function and frequency response of amplitude and phase.

1.3 POLES AND ZEROS OF *RC* NETWORK FUNCTIONS[8]

If $T(s)$ is the transfer, impedance, or admittance function of a passive *RC* network it can be written in the form

$$T(s) = \frac{N(s)}{D(s)} \tag{1-10}$$

where $N(s)$ and $D(s)$ are polynomials in s with real coefficients. The poles and zeros of $T(s)$ completely determine $T(s)$ except for a constant multiplier, which is only of secondary importance since it can always be adjusted for by an amplifier or an attenuator.

8. This section is based to some extent on J. G. Truxal, *Automatic Feedback Control System Synthesis* (New York: McGraw-Hill 1955), Chapter 3. This chapter is recommended as a concise review of *RC* synthesis techniques.

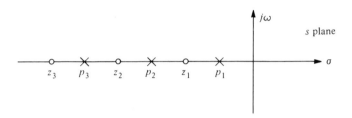

FIG. 1-4. Typical pole–zero configuration for $Z(s)$ given by (1-11).

Characteristics of poles and zeros associated with an RC network depend on the nature of $T(s)$, whether it is a driving point impedance, a voltage ratio, or whatever. We must therefore consider the various possible forms of $T(s)$ separately.

1.3.1 RC Driving Point Impedance Functions

The necessary and sufficient conditions for a rational function of s, $T(s)$, to be a driving point impedance of an RC network are:[9]

1. All poles and zeros are simple and lie on the negative real axis in the s plane.

2. The poles and zeros alternate along the axis.

3. The lowest critical frequency (that nearest the origin) is a pole.

4. The highest critical frequency (that farthest from the origin) is a zero.

Let us consider the following typical impedance function:

$$Z(s) = K \frac{(s - z_1)(s - z_2)(s - z_3)}{(s - p_1)(s - p_2)(s - p_3)} \qquad (1\text{-}11)$$

By condition 1 above, the z_i and p_i are all negative real numbers, and by conditions 3 and 4 these roots may typically be located as shown in Fig. 1-4. Either or both of two special cases are possible for $Z(s)$:

Case 1. The lowest critical frequency, the pole p_1 in Fig. 1-4, may be located at the origin, in which case

$$Z(s) = K \frac{(s - z_1)(s - z_2)(s - z_3)}{s(s - p_2)(s - p_3)} \qquad (1\text{-}12a)$$

and

$$Z(0) = \infty \qquad (1\text{-}12b)$$

9. E. A. Guillemin, *Communication Networks*, Vol. II, (New York: John Wiley & Sons, 1935), pp. 211–212.

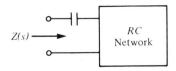

FIG. 1-5. Network with driving point impedance pole at origin; thus $Z(0) = \infty$, or $Y(0) = 0$.

The corresponding network has the form shown in Fig. 1-5.

Case 2. The largest critical frequency, i.e., the zero z_3, may be located at infinity:

$$Z(s) = K \frac{(s - z_1)(s - z_2)}{(s - p_1)(s - p_2)(s - p_3)} \qquad (1\text{-}13a)$$

and

$$Z(\infty) = 0 \qquad (1\text{-}13b)$$

The corresponding network has the form shown in Fig. 1-6. Naturally, neither of these two special cases need occur, i.e., $Z(s)$ may be constant at both frequency extremes.

So far we have sufficient information to define completely the conditions under which a given rational algebraic function in s can be realized as the driving point impedance of an RC network. However, other characteristics of functions meeting these realizability conditions are also useful.

Since all poles and zeros of the impedance function are located on the negative real axis, the characteristics of $Z(s)$ are completely portrayed by a plot of the impedance as a function of the real variable σ. For the function with the pole–zero configuration shown in Fig. 1-4 such a plot takes on the form of Fig. 1-7. This plot demonstrates two additional characteristics of $Z(s)$ which can be listed in general form as follows:

5. $Z(\infty) \le Z(0)$.

For our example, this follows directly from (1-11), i.e.,

$$Z(\infty) = K \qquad (1\text{-}14)$$

$$Z(0) = K\frac{z_1 z_2 z_3}{p_1 p_2 p_3} \qquad (1\text{-}15)$$

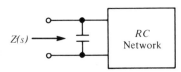

FIG. 1-6. Network with driving point impedance zero at infinity; thus $Z(\infty) = 0$, or $Y(\infty) = \infty$.

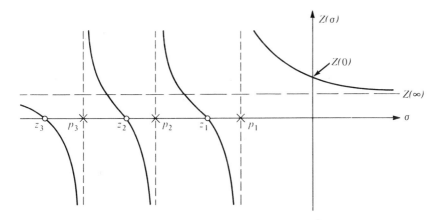

FIG. 1-7. Impedance as function of real variable σ corresponding to pole–zero plot of Fig. 1-4.

From conditions (1–4) above it follows that $z_1 > p_1$, $z_2 > p_2$, $z_3 > p_3$. As a direct consequence, $Z(0) \geq Z(\infty)$. $Z(\infty)$ can only equal $Z(0)$ in the special case that $Z(s)$ is a constant (e.g., resistive network).

6. $\dfrac{dZ(\sigma)}{d\sigma} < 0$ for all σ.

This can be demonstrated for our example in Fig. 1-7 by considering the corresponding partial fraction expansion of $Z(s)$:

$$Z(s) = k + \frac{k_1}{s - p_1} + \frac{k_2}{s - p_2} + \frac{k_3}{s - p_3} \qquad (1\text{-}16)$$

It can be shown that the residues k_i are real and positive (there always being an even number of negative multiplicative factors in their evaluation). Therefore

$$Z(\sigma) = k + \frac{k_1}{\sigma - p_1} + \frac{k_2}{\sigma - p_2} + \frac{k_3}{\sigma - p_3} \qquad (1\text{-}17)$$

and

$$\frac{dZ(\sigma)}{d\sigma} = -\frac{k_1}{(\sigma - p_1)^2} - \frac{k_2}{(\sigma - p_2)^2} - \frac{k_3}{(\sigma - p_3)^2} \qquad (1\text{-}18)$$

This is negative for all σ as stated above. Using similar reasoning, the general proof can be derived.

With the last two properties we can now state the necessary and sufficient conditions for a function $Z(s)$ to be realizable by the driving point impedance of an *RC* network, in terms of the poles and residues (instead of in terms of the poles and zeros):

A. All poles of $Z(s)$ are simple and lie on the negative real axis.
B. All residues are real and positive.
C. $Z(s)$ may have a pole at the origin but not at infinity.

1.3.2 *RC* Driving Point Admittance Functions

Since $Y(s)$ is the reciprocal of $Z(s)$, the rules for the poles and zeros are now interchanged. Another consideration in the derivation of the properties of $Y(s)$ is that the function $Y(s)/s$ has properties identical to those of the driving point *impedance* of an *RC* network. The realizability conditions for *RC* driving point impedance and admittance functions are summarized in Table 1-1.

TABLE 1-1.* REALIZABILITY CONDITIONS FOR *RC* DRIVING POINT FUNCTIONS

	Driving point impedance functions, $Z(s)$	Driving point admittance functions, $Y(s)$
Necessary and sufficient conditions on the rational functions $Z(s)$ or $Y(s)$ for realization as an *RC* network in terms of poles and zeros	1. All poles and zeros simple and on the negative real axis in the *s* plane 2. Poles and zeros alternate along axis 3. Lowest critical frequency a *pole* 4. Highest critical frequency a *zero*	1. All poles and zeros simple and on the negative real axis in the *s* plane 2. Poles and zeros alternate along axis 3. Lowest critical frequency a *zero* 4. Highest critical frequency a *pole*
Necessary and sufficient conditions in terms of poles and residues	1. All poles of $Z(s)$ simple and on negative real axis 2. All residues of $Z(s)$ real and positive 3. $Z(s)$ may have pole at origin, none at infinity	1. All poles of $Y(s)/s$ simple and on negative real axis 2. All residues of $Y(s)/s$ real and positive 3. $Y(s)/s$ may have pole at origin, none at infinity
Form of sketch of $Z(\sigma)$ or $Y(\sigma)$†		
Important characteristics	1. $\dfrac{dZ(\sigma)}{d\sigma}$ negative 2. $Z(\infty) \leq Z(0)$	1. $\dfrac{dY(\sigma)}{d\sigma}$ positive 2. $Y(\infty) \geq Y(0)$

* From *Automatic Feedback Control System Synthesis* by John G. Truxal, copyright 1955 by McGraw-Hill Book Company. Used with permission of McGraw-Hill Book Company.
† As a special case, it is possible that either $Z(\infty)$ or $Y(0)$ may equal zero.

1.3.3 RL Driving Point Functions

Although we are primiarily concerned with RC networks, we shall see later that it is useful to be aware of the close relationship between the properties of RC and RL driving point functions. Since the dual[10] of an RC network is an RL network, the driving point *impedance* $Z(s)$ of an RL network has the same properties as the driving point *admittance* of an RC network. Thus the properties of an RC driving point admittance listed in Table 1-1 are the same as the properties of the driving point impedance of an RL one-port. Conversely, the properties of an RC driving point impedance listed in Table 1-1 are the same as the properties of an RL driving point admittance. Thus the properties of RC driving point functions listed in Table 1-1 can be used to test the realizability of the dual RL driving point functions, where the impedance function is the dual of an admittance function and vice versa.

1.3.4 RC Transfer Immittance Functions

The characteristics of any linear bilateral two-port network can be completely described in terms of either of two sets of three network parameters:

$$z_{11}, z_{22}, \text{ and } z_{12}$$

or

$$y_{11}, y_{22}, \text{ and } y_{12}$$

The relationships between the various types of transfer functions and the corresponding z and y parameters are summarized in Table 1-2.

Let us now briefly consider the realizability conditions for a set of z parameters of an RC network. The parameters z_{11} and z_{22} must satisfy the conditions for an RC driving point impedance already given above. This leaves us with the conditions for the open-circuit transfer impedance z_{12}. These are as follows:

1. If k_{11}, k_{22}, and k_{12} are the residues of z_{11}, z_{22}, and z_{12}, respectively, at any one of the poles of z_{12}, the following residue condition must be satisfied:

$$k_{11}k_{22} - k_{12}^2 > 0 \qquad (1\text{-}19)$$

with

$$k_{11} > 0, k_{22} > 0$$

---

10. Throughout this book we shall dwell only on those concepts of network and other theories necessary for the understanding of active RC networks. Since the dual of an RC network is an RL network, the general network-theoretical principle of duality is of no further interest to us here than to justify the statement made above concerning the identity between the properties of RC driving point impedance and RL driving point admittance functions and vice versa. For anyone interested in network duality in more detail, the following references are recommended: M. E. Van Valkenburg, *Network Analysis*, 2nd edition (Englewood Cliffs, N. J.: Prentice-Hall, 1964), p. 77; S. P. Chan, *Introductory Topological Analysis of Electrical Networks* (New York: Holt, Rhinehart & Winston, 1969), p. 149; S. K. Mitra, *Analysis and Synthesis of Linear Active Networks* (New York: John Wiley & Sons, 1969), p. 90. In contrast to general network duality, a special category, namely, RC:CR duality, will be of considerable interest to us and will be dealt with in detail in Chapter 3.

TABLE 1-2.* RELATIONS BETWEEN TRANSFER FUNCTIONS AND NETWORK PARAMETERS

Source impedance	Load impedance	System schematic	Transfer function in terms of z_{11}, z_{22}, z_{12}	Transfer function in terms of y_{11}, y_{22}, y_{12}
0	∞		$\dfrac{V_2}{V_1} = \dfrac{z_{12}}{z_{11}}$	$\dfrac{V_2}{V_1} = -\dfrac{y_{12}}{y_{22}}$
	Z_L		$\dfrac{V_2}{V_1} = \dfrac{z_{12}Z_L}{\Delta_z + z_{11}Z_L}$	$\dfrac{V_2}{V_1} = -\dfrac{y_{12}}{y_{22}+Y_L}$
∞	∞		$\dfrac{V_2}{I_1} = z_{12}$	$\dfrac{V_2}{I_1} = -\dfrac{y_{12}}{\Delta_y}$
	Z_L		$\dfrac{V_2}{I_1} = \dfrac{z_{12}Z_L}{z_{22}+}$	$\dfrac{V_2}{I_1} = -\dfrac{y_{12}}{\Delta_y + y_{11}Y_L}$
Z_s	∞		$\dfrac{V_2}{V_s} = \dfrac{z_{12}}{z_{11}+Z_s}$	$\dfrac{V_2}{V_s} = -\dfrac{y_{12}}{y_{22}+\Delta_y Z_s}$
	Z_L		$\dfrac{V_2}{V_s} = \dfrac{z_{12}Z_L}{\Delta_z + z_{22}Z_s + z_{11}Z_L + Z_sZ_L}$	$\dfrac{V_2}{V_s} = -\dfrac{y_{12}Y_s}{\Delta_y + y_{11}Y_L + y_{22}Y_s + Y_LY_s}$

* From *Automatic Feedback Control System Synthesis* by John G. Truxal, copyright 1955 by McGraw-Hill Book Company. Used with permission of McGraw-Hill Book Company.

This condition must hold for *all* poles of z_{12}. Consequently:

a. All poles of z_{12} must be simple, lie on the negative real axis and also be poles of z_{11} and z_{22}. (Otherwise k_{11} or k_{22} is zero at one of the poles of z_{12}.) On the other hand z_{11} and z_{22} can have "private" poles which do not appear in z_{12}.

b. Since z_{12} is the ratio of two polynomials with real coefficients,[11] all residues must be real.

c. The residues of z_{12} can be either positive or negative (remember that those of z_{11} and z_{22} must be positive), but they are restricted in absolute value with respect to the corresponding product k_{11}, k_{22}. This condition places a constraint on the magnitude of the multiplying factor which can be realized for $z_{12}(s)$.

2. At infinity $(s \to \infty)$

$$z_{11}z_{22} - z_{12}^2 \geq 0 \tag{1-20}$$

3. No conditions are imposed on the location of the zeros of $z_{12}(s)$ (in contrast to the zeros of driving point impedances) other than the elementary restriction that these zeros must occur in complex conjugate pairs.[12]

The conditions for realizability by an RC network of y_{11}, y_{22}, and y_{12} can be presented in a similar manner. As before, the discussion is simplified if the functions considered are y_{11}/s, y_{22}/s, and y_{12}/s, since these three functions must satisfy the same conditions as the corresponding set of impedance functions.

1.4 REALIZABILITY CONDITIONS FOR RC TRANSFER FUNCTIONS

Once the conditions for realizability by RC networks have been established for the z and y parameters, Table 1-2 can be used to determine appropriate conditions for the various transfer functions (open-circuit voltage ratio, etc.). These have been summarized in Table 1-3. To demonstrate how they were obtained, we shall derive the realizability conditions for the first case of Table 1-3 which consists of a signal source with zero impedance driving an RC network with no load (open circuit at the output) as shown in Fig. 1-8. The corresponding voltage transfer function is

$$\frac{V_2}{V_1} = \frac{z_{12}}{z_{11}} \tag{1-21}$$

We are now interested in finding the permissible locations of the poles and zeros of V_2/V_1, and the behavior of the network at both zero and infinite frequency, if it is to be realizable in passive RC form.

---

11. In addition the denominator must be a Hurwitz polynomial.
12. It will be shown in Section 1.5 that the permissible location of zeros in the s plane for *unbalanced RC* networks is more restricted.

TABLE 1-3.* REALIZABILITY CONDITIONS FOR RC TRANSFER FUNCTIONS

Source impedance	Load impedance	System schematic	Transfer function in terms of z's or y's	Realizability conditions
0	∞		$\dfrac{V_2}{V_1} = \dfrac{z_{12}}{z_{11}} = -\dfrac{y_{12}}{y_{22}}$	No poles at ∞ or 0 Poles simple, on negative real axis
	$Z_L = 1$		$\dfrac{V_2}{V_1} = \dfrac{z_{12}}{z_{11} + \Delta_z} = -\dfrac{y_{12}}{1 + y_{22}}$	No poles at ∞ or 0 Poles simple, on negative real axis
∞	∞		$\dfrac{V_2}{I_1} = z_{12} = -\dfrac{y_{12}}{\Delta_y}$	No pole at ∞ Poles simple, on negative real axis
	$Z_L = 1$		$\dfrac{V_2}{I_1} = \dfrac{z_{12}}{1 + z_{22}} = -\dfrac{y_{12}}{y_{11} + \Delta_y}$	No poles at 0 or ∞ Poles simple, on negative real axis

$Z_s = 1$

∞

$$\frac{V_2}{V_s} = \frac{z_{12}}{1+z_{11}} = -\frac{y_{12}}{y_{22}+\Delta_y}$$

No poles at 0 or ∞

Poles simple, on negative real axis

$Z_s = 1$

$$\frac{V_2}{V_s} = \frac{z_{12}}{1+z_{11}+z_{22}+\Delta_z} = \frac{-y_{12}}{1+y_{11}+y_{22}+\Delta_y}$$

No poles at 0 or ∞

Poles simple, on negative real axis

General conditions

z_{11}, z_{22}, z_{12}

y_{11}, y_{22}, y_{12}

z_{11} and z_{22}:
RC driving point impedances

z_{12}:
Poles simple, on negative real axis, residue condition satisfied

y/s functions same as z

*From *Automatic Feedback Control System Synthesis* by John G. Truxal, copyright 1955 by McGraw-Hill Book Company. Used with permission of McGraw-Hill Book Company.

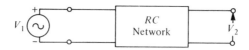

FIG. 1-8. *RC* network driven from zero-impedance source, no output load.

To begin with, we can assume that z_{12} and z_{11} have the forms

$$z_{12}(s) = \frac{m(s)}{q(s)} \qquad (1\text{-}22)$$

and

$$z_{11}(s) = \frac{p(s)}{r(s)q(s)} \qquad (1\text{-}23)$$

respectively. Then the following statements can be made:

1. The poles of $z_{12}(s)$ are the zeros of $q(s)$.
2. All these poles must be present in $z_{11}(s)$.
3. $z_{11}(s)$ may have additional poles, represented by the negative real zeros of the polynomial $r(s)$.
4. $p(s)$ has all zeros on the negative real axis.
5. $m(s)$ may have zeros anywhere in the s plane.

Substituting (1-22) and (1-23) into (1-21) we get

$$\frac{V_2}{V_1} = \frac{m(s)r(s)}{p(s)} \qquad (1\text{-}24)$$

Equation (1-24) contains the required information about the finite nonzero critical frequencies:

1. All finite poles of V_2/V_1 are at the zeros of $p(s)$ and hence must be simple and lie on the negative real axis.
2. The zeros of V_2/V_1 may lie anywhere in the s plane by the nature of $m(s)$.

The infinite-frequency behavior of V_2/V_1 is most readily investigated by consideration of (1-21):

1. Neither z_{12} nor z_{11} may have a pole at infinity.
2. If z_{11} has a simple zero at infinity, z_{12} must have at least a simple zero in order to satisfy (1-20).
3. Consequently, V_2/V_1 cannot have a pole at infinity, but it can have either a constant value or a zero.
4. In other words, the degree of the denominator polynomial of V_2/V_1 must be at least as high as the degree of the numerator, or, in terms of (1-10),

$$\text{Degree of } N(s) \leq \text{Degree of } D(s) \qquad [1\text{-}25]$$

As a last step, the zero-frequency behavior of the voltage transfer function must be considered. The following pairs[13] of possibilities exist for the zero-frequency behavior of z_{11} and z_{12}:

z_{11}	z_{12}	$V_2/V_1 = z_{12}/z_{11}$
constant	constant	constant
constant	zero of any order	zero of any order
simple pole	constant	simple zero
simple pole	simple pole	constant
simple pole	zero of any order	zero of any order

Consequently, V_2/V_1 can have either a constant value or a zero of any order. No pole is permitted at the origin. Thus, to summarize the characteristics of (1-21) when realized in RC form:

A. Finite nonzero critical frequencies:
Poles simple and on negative real axis.
Zeros anywhere in the s plane.
B. $s \to \infty$:
V_2/V_1 constant or zero; no poles.
(Degree of numerator $\leq$ Degree of denominator).
C. $s \to 0$:
V_2/V_1 constant or zero; no poles.

Similar reasoning provides the realizability conditions for the other transfer functions given in Table 1-3.

1.5 TRANSFER FUNCTIONS OF UNBALANCED RC NETWORKS

Up to now we have discussed RC networks in general, not distinguishing between balanced and unbalanced networks. Actually all the conditions listed so far, except those for the zeros of z_{12} (see condition 3 in Section 1.3.4) are the same for both network types. In the following we shall consider the limitations placed on the zeros of unbalanced RC networks, that is three-terminal RC networks with a common ground at input and output. For an unbalanced RC two-port whose transfer function is given by:

$$T(s) = \frac{b_n s^n + b_{n-1} s^{n-1} + \cdots + b_1 s + b_0}{a_n s^n + a_{n-1} s^{n-1} + \cdots + a_1 s + a_0} \tag{1-26}$$

---

13. The case of $z_{11} = $ constant, z_{12} possessing a simple pole, does not, of course, exist, since z_{11} must contain all poles of z_{12}.

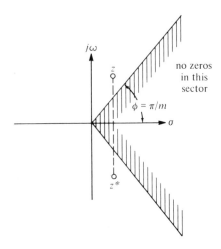

FIG. 1-9. Wedge of angle $2\pi/m$ in the RHP from which zeros of unbalanced network with mth-order numerator are excluded.

Fialkow and Gerst[14] derived the following conditions:

1. $T(s)$ can have no positive real transmission zeros.
2. $b_j \geq 0$ and $b_j \leq a_j$ for $j = 0, 1, \ldots, n$ (provided that no common factors of the numerator and denominator of $T(s)$ have been cancelled).
3. $0 \leq T(\sigma) \leq 1$ for $\sigma > 0$. $T(\sigma)$ can equal unity only at the origin and/or at infinity.

An important consequence of condition 2 relates the permissible location of the finite zeros of $T(s)$ to the order of the numerator polynomial:

4. The arguments of the m finite zeros of an mth order numerator $N(s)$ must be greater than or equal to π/m. Thus, the m finite zeros of an unbalanced network are not only excluded from the positive real axis but from a whole segment of $\pm \pi/m$ in the RHP as shown in Fig. 1-9.

This last property is a very useful one to remember in practice. We will therefore do well to emphasize its significance by going through a simple but no less elegant proof.[15]

Consider the numerator polynomial

$$N(s) = b_m s^m + b_{m-1} s^{m-1} + \cdots + b_1 s + b_0 \qquad (1\text{-}27)$$

where all the b_i are real and

$$b_m > 0$$

$$b_i \geq 0, \, i = 0, 1, \ldots, m - 1$$

14. A. D. Fialkow and I. Gerst, The transfer function of networks without mutual reactance, *Quart. Appl. Math.*, **12**, 117–131 (1954).
15. Brought to the attention of the author by E. Lueder, this proof originates from the French mathematician Jules Henri Poincaré (1854–1912).

Let $N(s)$ have a zero:

$$s = s_0 \, e^{j\phi} \tag{1-28}$$

Then

$$N(s_0 \, e^{j\phi}) = b_m s_0^m \cos m\phi + b_{m-1} s_0^{m-1} \cos (m-1)\phi + \cdots + b_1 s_0 \cos \phi + b_0$$
$$+ j[b_m s_0^m \sin m\phi + b_{m-1} s_0^{m-1} \sin (m-1)\phi + \cdots + b_1 s_0 \sin \phi]$$
$$= 0 \tag{1-29}$$

We must now find the limits of ϕ such that the following conditions can be satisfied:

$$\text{Re } N(s_0 \, e^{j\phi}) = 0$$
$$\text{Im } N(s_0 \, e^{j\phi}) = 0 \tag{1-30}$$

Let us assume that $|\phi| < \pi/m$. Then

$$\sin k\phi > 0 \quad \text{for} \quad \phi > 0$$
$$\sin k\phi < 0 \quad \text{for} \quad \phi < 0$$

where $k = 1, 2, \ldots, m$. Clearly, the imaginary part of $N(s_0 \, e^{j\phi})$ cannot equal zero. Thus $s = s_0 \, e^{j\phi}$ cannot be a zero of $N(s)$ as long as $|\phi| < \pi/m$. We are left with the condition for the argument ϕ of a root of $N(s)$ that

$$|\phi| \geq \pi/m \tag{1-31}$$

Let us consider some examples. Assume that

$$T(s) = \frac{N(s)}{D(s)} = \frac{\displaystyle\sum_{i=0}^{m} b_i s^i}{\displaystyle\sum_{j=0}^{6} a_j s^j} \tag{1-32}$$

is the transfer function of an unbalanced RC network. We are interested in the regions of permissible zeros of $T(s)$ in the s plane as a function of the order of the numerator polynomial m. Let $m = 0$. Then $T(s)$ has the form

$$T(s) = \frac{b_0}{a_6 s^6 + a_5 s^5 + \cdots + a_0} \tag{1-33}$$

The minimum argument $\phi_{\min} = \pi/m$ is here undetermined, since we obtain infinity. A glance at (1-33) explains why: $T(s)$ has no *finite* zeros for which, alone, property 4 is valid. The zeros of (1-33) all occur at infinity.

Assume now that in (1-32) $m = 1$. $T(s)$ then has the form

$$T(s) = \frac{b_1 s + b_0}{a_6 s^6 + a_5 s^5 + \cdots + a_0} \tag{1-34}$$

Here the minimum argument $\phi_{\min}$ equals π, which implies that all zeros of $T(s)$ must lie on the negative real axis as shown in Fig. 1.10a.

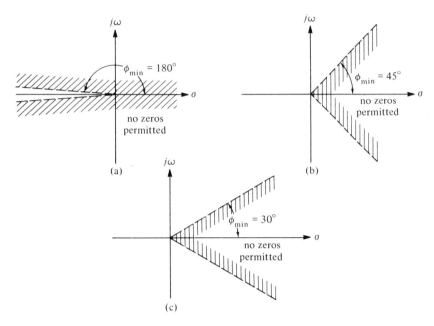

FIG. 1-10. Region of permissible transmission zeros when the order m of numerator polynomial varies. (a) $m = 1$; (b) $m = 4$; (c) $m = 6$.

Let $m = 4$. Without needing to write out $T(s)$, whose numerator now is fourth-order (its denominator, of course, remains sixth-order), we can immediately say that $\phi_{min} = 45°$ and that the permissible region of transmission zeros is bounded as shown in Fig. 1.10b. Finally, if $m = 6$ the nonpermissible region is bounded by $\phi_{min} = \pm 30°$, as shown in Fig. 1.10c. Notice that for $T(s)$ to remain a realizable transfer function, m may not exceed the order n of the denominator, which here is 6. Thus, whereas for any values of m less than n, the order of the denominator is not required to establish the theoretical boundaries on the zero location in the s plane, it is ultimately the degree n of the denominator, or the order of the network, which determines the maximum possible region in the s plane in which zeros may be located.

Another comment seems in order. In discussing the transfer function $T(s)$ and the boundaries on its transmission zeros, we have not made any statements yet as to how the function $T(s)$ itself is to be realized in passive RC form. The numerator of $T(s)$ may well have some order m, but that in itself does not guarantee that the m zeros of $T(s)$ are realizable anywhere in the s plane outside the RHP wedge bounded by $\pm \pi/m$. Indeed it may well be in the nature of the particular network under consideration that the m zeros of $T(s)$ are limited to the negative real axis, no matter how small π/m may be. Thus it is important to realize that there are only certain categories of RC networks with which complex zeros can be obtained at all, while there are others, no matter how complex they may be and how high the order of their numerators, with which zeros are realizable only on the negative real axis. These categories will be discussed next.

1.5.1 Transfer Characteristics of RC Ladder Networks

Since RC ladder networks are very commonly used, it is worth considering their characteristics separately. The general form of a ladder network is shown in Fig. 1-11, where the end branch at either side of the ladder may be in series or shunt. It is assumed here that any arm of the ladder may be an RC two-terminal (i.e., one-port) configuration of any complexity. Thus it may be characterized by the RC driving point conditions listed in Table 1-1. Assuming, for example, that the Z_i are all equal to Z and the Y_i equal to Y, the transfer function of the ladder network shown in Fig. 1-11 can easily be derived. One obtains

$$\frac{V_2}{V_1} = \frac{1}{a + bZY + cZ^2Y^2 + \cdots + Z^nY^n} \qquad [1\text{-}35]$$

The coefficients can be derived from Pascal's triangle by entering the triangle at the number of loops in the network and reading from left to right the coefficients $a, b, c, \ldots, k$. This is demonstrated in Fig. 1-12. If we consider, for example, the three-loop network of Fig. 1-13, the resulting voltage transfer function follows directly from Pascal's triangle:

$$\frac{V_2}{V_1} = \frac{1}{1 + 6sRC + 5s^2R^2C^2 + s^3R^3C^3} \qquad (1\text{-}36)$$

More general methods of analyzing ladder networks will be discussed in Chapter 3. Here we are more concerned with the general network characteristics (poles and zeros) realizable with these networks.

As would be expected, the characteristics of the transfer function given by (1-35) are most readily determined by consideration of the admissible pole and zero locations in the s plane. As with any passive RC network, the poles of the ladder network must lie on the negative real axis, with the exception of infinity. Although Table 1-3 implies that some transfer functions may have a pole at zero, this too is nonrealizable, at least in practice. This can be shown

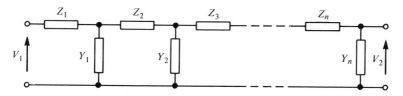

FIG. 1-11. General ladder network.

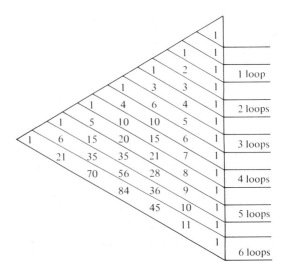

FIG. 1-12. Pascal's triangle.

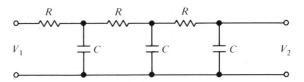

FIG. 1-13. Three-loop network where $Z = R$ and $Y = j\omega C$

to be a direct consequence of the characteristics of the z or y parameters of an RC network.

The location of the zeros of a ladder network is readily obtained by considering that such a configuration has only one path from the input to the output. A transmission zero can therefore be realized in only two ways: either a series arm goes into antiresonance and becomes an open circuit (pole), or a shunt arm goes into resonance and becomes a short circuit (zero). It can happen, of course, that antiresonance and resonance frequencies coincide and cancel each other out, in which case a constant (i.e., frequency-independent) voltage or current attenuation, rather than a transmission zero, results.[16]

The comments above concerning zero realizations are valid for any ladder networks (e.g., also LCR). For an RC ladder the poles of the series impedances and of the shunt admittances must lie on the negative real axis and, therefore, so must the transmission zeros resulting from them. However, the zeros may be of any multiplicity, since more than one antiresonance, resonance, or

16. This scheme is generally used to provide "frequency-independent" probes for electronic equipment such as oscilloscopes.

combination of the two may occur in various branches at the same frequency. Nevertheless, RC ladder networks have the significant limitation of being unable to generate complex conjugate zeros in the s plane.

1.5.2 RC Networks with Complex Transmission Zeros

We have seen that RC ladder networks can realize zeros only on the negative real axis. To generate complex zeros a ladder must also include inductors besides the resistors and capacitors, since only then can the impedance or admittance of one arm become infinite at a complex value of s. If complex zeros are to be generated with R's and C's only, the network configuration required to do so must differ fundamentally from the ladder network, in that it has more than one path from input to output or, at least, a path stretching over more than one series arm of the network. The two most common methods known for the generation of complex zeros with an unbalanced RC network are:

1. Guillemin's procedure,[17] using ladders connected in parallel.
2. Dasher's procedure,[18] using a cascade connection of bridged twin-T structures.

Guillemin's procedure can be used for transfer functions having zeros in the RHP (exclusive of the positive real axis); Dasher's procedure is restricted to functions with zeros on the $j\omega$ axis or in the LHP. Dasher's procedure can be used to realize specified impedance functions (z_{11} and z_{12}) as well as admittance functions, while Guillemin's procedure is restricted to the realization of specified admittance functions (y_{11} and y_{12}). In complicated examples amenable to both procedures, Dasher's method yields fewer elements and often higher gain constants.

As an illustration let us consider two networks that may typically result when using Guillemin's and Dasher's methods to realize the same transfer function. If the function is second-order and of the general form

$$T(s) = \frac{N(s)}{D(s)} = \frac{b_2 s^2 + b_1 s + b_0}{a_2 s^2 + a_1 s + a_0} \qquad (1\text{-}37)$$

then Guillemin's method may result in the parallel ladder network shown in Fig. 1-14, Dasher's in the somewhat simpler bridged twin-T shown in Fig. 1-15. Both networks are rather complicated and their synthesis somewhat lengthy. In fact we shall see in the course of this book, that whenever complex zeros are required, we shall not in general go to the trouble of synthesizing a network by either one of the methods described above, but will use one of

17. E. A. Guillemin, Synthesis of RC Networks, *J. Math. Phys.*, **28**, 22–42 (1949).
18. B. J. Dasher, Synthesis of RC transfer functions as unbalanced two terminal-pair networks, *Trans. IRE Profess. Group Circuit Theory*, **PGCT1**, 20–34 (1952).

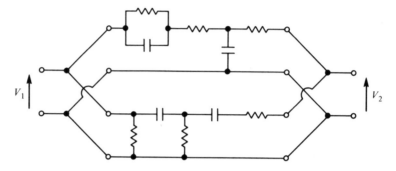

Fig. 1-14. Parallel ladder network (Guillemin's method) to realize complex zeros.

two basic networks that can be interpreted as rudimentary Guillemin or Dasher networks depending on one's point of view.

The first of these networks, the bridged-T shown in Fig. 1-16, has the transfer function

$$T(s) = \frac{Z_1 Z_2 + Z_2'(Z_1 + Z_1' + Z_2)}{Z_1 Z_2' + (Z_1 + Z_2')(Z_1' + Z_2)} \tag{1-38}$$

The two RC versions of the bridged-T shown in Fig. 1.17 are commonly used. Substituting the R's and C's for the Z's in (1-38), a transfer function of the type represented by (1-37) is obtained. For the circuit of Fig. 1-17a the coefficients of (1-37) are

$$a_2 = b_2 = 1 \tag{1-39a}$$

$$a_1 = \frac{1}{R_2' C_1} + \frac{C_1 + C_2}{R_1' C_1 C_2} \tag{1-39b}$$

$$b_1 = \frac{C_1 + C_2}{R_1' C_1 C_2} \tag{1-39c}$$

$$a_0 = b_0 = (R_1' R_2' C_1 C_2)^{-1} \tag{1-39d}$$

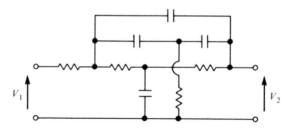

Fig. 1-15. Bridged-T network (Dasher's method) to realize complex zeros.

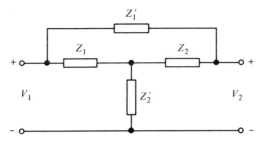

FIG. 1-16. Generalized bridged-T network.

For the circuit of Fig. 1-17b they are

$$a_2 = b_2 = 1 \tag{1-40a}$$

$$a_1 = \frac{1}{R_2 C_1'} + \frac{R_1 + R_2}{C_2' R_1 R_2} \tag{1-40b}$$

$$b_1 = \frac{R_1 + R_2}{C_2' R_1 R_2} \tag{1-40c}$$

$$a_0 = b_0 = (R_1 R_2 C_1' C_2')^{-1} \tag{1-40d}$$

To evaluate over which range of the s plane either of the bridged-T networks of Fig. 1.17 can provide zeros, it is now advantageous to recall property 4 of unbalanced networks, according to which "the arguments of the m zeros of an mth-order numerator $N(s)$ must be greater than or equal to π/m." The order m of $N(s)$ in (1-37) is 2, and the argument $\phi > \pi/2$. As shown in Fig. 1.18 the permissible region for zero realization using an RC bridged-T is therefore restricted to the LHP, with the imaginary axis as the limiting bound. Zeros on the imaginary axis can only be obtained in the limit when the resistors or capacitors of the bridged-T have an infinite spread (e.g., zero and infinite values), at which time the network becomes rudimentary and of no practical use. For all practical purposes, then, the restricted area is bounded

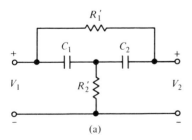

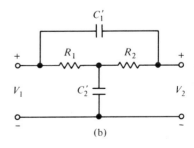

(a)   (b)

FIG. 1-17. Two commonly used RC versions of bridged-T networks. (a) Resistive shunt; (b) capacitive shunt.

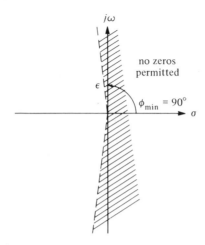

FIG. 1-18. Region of permissible bridged-T zeros in the s plane.

by the two dashed radial lines shown in Fig. 1-18, whose argument is $\pm(90° + \varepsilon)$, where ε depends on the range of component values that can be tolerated in the network.[19] A range of 10^6 (e.g., $R_1 = 1\ M\Omega$, $R_2 = 1\ \Omega$) will for example generally be unacceptable in any practical network.

The second useful circuit for complex zero realization is the twin-T network shown in Fig. 1-19. It has a transfer function of the form

$$T(s) = \frac{N(S)}{D(S)} = \frac{b_3 s^3 + b_2 s^2 + b_1 s + b_0}{a_3 s^3 + a_2 s^2 + a_1 s + a_0} \tag{1-41}$$

where

$$a_3 = b_3 = R_1 R_2 R_3 C_1 C_2 C_3 \tag{1-42a}$$

$$a_2 = R_3[R_1 C_3 (C_1 + C_2) + (R_1 + R_2)C_1 C_2] + R_1 R_2 C_2 C_3 \tag{1-42b}$$

$$b_2 = R_3(R_1 + R_2)C_1 C_2 \tag{1-42c}$$

$$a_1 = R_3(C_1 + C_2) + R_2 C_2 + R_1(C_2 + C_3) \tag{1-42d}$$

$$b_1 = R_3(C_1 + C_2) \tag{1-42e}$$

$$a_0 = b_0 = 1 \tag{1-42f}$$

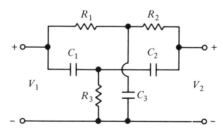

FIG. 1-19. General twin-T network.

19. This is demonstrated in *Linear Integrated Networks: Design* , Chapter 2, Section 2.3.3.

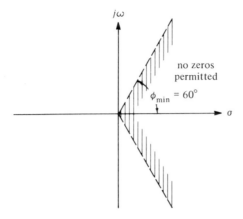

FIG. 1-20. Region of permissible twin-T zeros in the s plane.

The region of permissible zeros follows as before from property 4 of un-balanced networks. Here the order m of $N(s)$ is 3, and the minimum argument is therefore 60°. The resulting permissible area in the s plane is shown in Fig. 1.20. It is bounded by the two radial dashed lines whose argument is $\pm 60°$. Clearly the twin-T network is more powerful than the bridged-T, since it can provide RHP zeros (nonminimum-phase networks), even if only within a restricted area. The cost of this added capability is a third resistor-capacitor combination, which increases the order m of the numerator to 3. Presumably the addition of a fourth RC combination connected appropriately to the twin-T would increase m to 4, thereby decreasing ϕ_{min} to 45°, while a fifth RC combination would decrease ϕ_{min} to 36°, and so on. Clearly, to obtain zeros on the positive real axis ϕ_{min} must equal zero and m infinity. This corresponds to a network with an infinite number of RC combinations which is an impossibility with linear lumped elements (but not with distributed networks); consequently, positive real zeros are an impossibility with un-balanced networks. This reiterates property 1 of the Fialkow–Gerst conditions listed at the beginning of this Section (1.5).

1.6 SECOND-ORDER PASSIVE RC NETWORKS

In progressing through this book, and as we get more involved in the subject of active networks, we shall find over and over again that our interest is centered not so much on the characteristics of general nth-order networks but on the individual second- (or third-) order networks that are cascaded to make up the complex forms. Such cascades correspond to the decompositions given by (1-4) and (1-5). There are numerous reasons for cascading $n/2$ second-order networks instead of designing a specified nth-order network directly. These will be discussed in more detail later. For the present, when

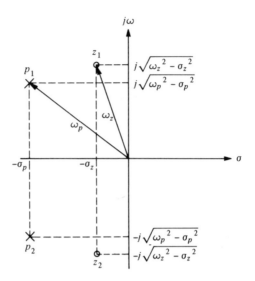

FIG. 1-21. Pole–zero diagram of a general second-order network.

considering any type of network affiliated with active RC realizations, we shall consider the second-order version of that network in particular. In this vein the purely passive RC network is also of interest to us in its second-order form.

From (1-4) and (1-5) the general transfer function of any second-order LLF network can be written as

$$T(s) = \frac{N(s)}{D(s)} = K \frac{s^2 + 2\sigma_z s + \omega_z^2}{s^2 + 2\sigma_p s + \omega_p^2} \qquad [1\text{-}43]$$

where, by definition, the σ's and ω's are real (as is the scaling factor K) and the subscripts z and p designate "zero" and "pole," respectively. The zeros of (1-43) are given by

$$z_{1,2} = -\sigma_z \pm j\sqrt{\omega_z^2 - \sigma_z^2} \qquad (1\text{-}44)$$

the poles by

$$p_{1,2} = -\sigma_p \pm j\sqrt{\omega_p^2 - \sigma_p^2} \qquad (1\text{-}45)$$

The plot of $z_{1,2}$ and $p_{1,2}$ in the s plane shown in Fig. 1-21 explains the choice of 2σ and ω as the coefficients of $T(s)$ in (1-43). It is useful now to define a term q that characterizes the location of a pole or zero with respect to the real or imaginary axis of the s plane. This term is given by

$$q = \frac{\omega}{2\sigma} \qquad [1\text{-}46]$$

and enables us to write (1-43) in the form

$$T(s) = K \frac{s^2 + (\omega_z/q_z)s + \omega_z^2}{s^2 + (\omega_p/q_p)s + \omega_p^2} \qquad [1\text{-}47]$$

By inspection of Fig. 1-21 it can be seen that the q of a root pair anywhere between the real and imaginary axis has the limits

$$0.5 \leq q \leq \infty \qquad (1\text{-}48)$$

where an infinite q corresponds to a pair of conjugate roots on the $j\omega$ axis and a q of 0.5 corresponds to a double root on the real axis. If we recall that the poles of a passive RC transmission function must be simple and on the negative real axis, it follows that

$$(q_p)_{RC} = \hat{q}_p < 0.5 \qquad [1\text{-}49]$$

The location of the zeros of a passive RC network is not in general restricted, and the limits on q_z are therefore

$$0 \leq |q_z| \leq \infty \qquad (1\text{-}50)$$

If the network is unbalanced, the Fialkow–Gerst conditions limit the location of the zeros, or specify a minimum q_z, in the RHP. Thus for a second-order numerator ($m = 2$) the zeros cannot extend into the RHP and are limited by the $j\omega$ axis ($\phi \geq \pi/2$). Thus[20] for RHP zeros

$$q_{z_{\min}}\big|_{m=2} = \infty \qquad (1\text{-}51)$$

This corresponds to an RC bridged-T network. For a third-order (e.g., twin-T) numerator ($m = 3$; $\phi \geq \pi/3$) the zeros in the RHP are limited to

$$q_{z_{\min}}\big|_{m=3} = 1 \qquad (1\text{-}52)$$

and for a fourth-order numerator to

$$q_{z_{\min}}\big|_{m=4} = \sqrt{2}/2 \qquad (1\text{-}53)$$

As we pointed out earlier, the degree m of $N(s)$ must go to infinity (infinite number of discrete resistors and capacitors) in order to obtain a minimum q_z less than or equal to 0.5 in the RHP.[21]

As an illustrative example of a second-order RC network let us find the transfer function and frequency response of the network shown in Fig. 1-22. What kind of general network does it represent?

20. Rather than put a sign on q to indicate whether the root in question is in the LHP (q negative) or the RHP (q positive), we shall refer to a positive q throughout and simply state in which half-plane the root is located. In this way we shall not come into conflict with the conventional term Q defined later on.
21. Note that a distributed RC network can be interpreted as one with an infinite number of infinitesimal resistors and capacitors.

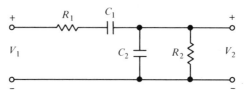

FIG. 1-22. A second-order RC network.

By straightforward analysis we obtain the voltage transfer function

$$\frac{V_2}{V_1}(s) = \frac{1}{R_1 C_2} \cdot \frac{s}{s^2 + \dfrac{R_1 C_1 + R_2(C_1 + C_2)}{R_1 R_2 C_1 C_2} s + \dfrac{1}{R_1 R_2 C_1 C_2}} \tag{1-54}$$

Letting

$$K = \frac{1}{R_1 C_2} \tag{1-55a}$$

$$2\sigma_p = \frac{R_1 C_1 + R_2(C_1 + C_2)}{R_1 R_2 C_1 C_2} \tag{1-55b}$$

$$\omega_p^2 = \frac{1}{R_1 R_2 C_1 C_2} \tag{1-55c}$$

(1-54) becomes:

$$T(s) = \frac{V_2}{V_1} = K \cdot \frac{s}{s^2 + 2\sigma_p s + \omega_p^2} \tag{1-56}$$

Let us compare this with the gain function of the RLC tuned circuit shown in Figure 1-23. The source impedance is considered infinitely large, i.e., the tank circuit is fed from a current source and loaded only with the parallel resistor R. From Table 1-2 we therefore obtain

$$\frac{V_2}{I_1}(s) = Z(s) = \frac{1}{C} \cdot \frac{s}{s^2 + \dfrac{s}{RC} + \dfrac{1}{LC}} \tag{1-57}$$

This has precisely the same form as (1-56), where now

$$K = \frac{1}{C} \tag{1-58a}$$

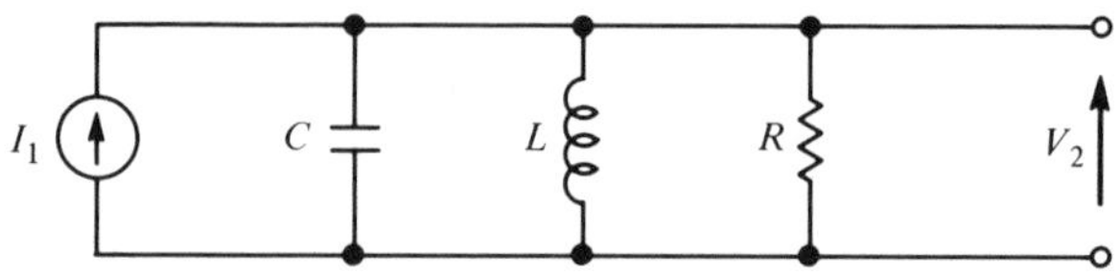

FIG. 1-23. RLC tuned circuit.

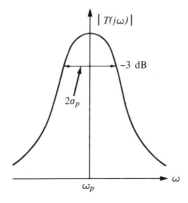

Fig. 1-24. Frequency response of a resonant tank circuit.

$$2\sigma_p = \frac{1}{RC} \qquad\qquad (1\text{-}58b)$$

$$\omega_p^2 = \frac{1}{LC} \qquad\qquad (1\text{-}58c)$$

The frequency response of the tank circuit, obtained by calculating $|T(s)|_{s=j\omega}$, is well known and is given in Fig. 1.24. It represents a simple second-order bandpass function which is characterized by the general expression given by (1-57). Since the circuit of Fig. 1-22 is described by the same function, it must represent a passive RC version of a second-order bandpass network. What then is the fundamental difference between the LCR and the RC network? We know that the -3 dB bandwidth of the tank circuit equals $1/RC$ and the center frequency is $1/\sqrt{LC}$. From (1-58b) and (1-58c) it therefore follows that the -3 dB bandwidth of the general second-order bandpass network is given by

$$BW\big|_{-3\,\text{dB}} = 2\sigma_p \qquad\qquad (1\text{-}59)$$

and the center frequency is equal to ω_p. Since the Q of an LCR tank circuit is defined by:

$$Q \equiv \frac{\text{Center frequency}}{-3\text{ dB Bandwidth}} \qquad\qquad [1\text{-}60]$$

it follows from (1-46) and (1-60) that for the general second-order bandpass function defined by (1-56) we have:[22]

$$Q = q_p = \frac{\omega_p}{2\sigma_p} \qquad\qquad (1\text{-}61)$$

22. We shall in the remainder of this book differentiate between the zero and pole Q, q_z, and q_p, respectively, of a general second-order network function, and Q, the quality factor or pole Q of a second-order bandpass function as given by (1–56). Strictly speaking, only the latter can be defined by (1–60). However, if the dominant pole pair of an nth-order function is close enough to the $j\omega$-axis, its pole Q can also be approximated by (1–60).

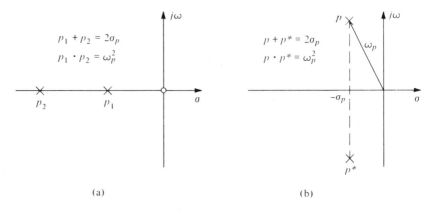

(a)   (b)

FIG. 1-25. Pole–zero plots of second-order bandpass networks. (a) Passive RC network; (b) LCR network.

However, for the RC network of Fig. 1-22 the poles must be negative real (see Fig. 1-25a), and therefore from (1-49) we have [23]

$$Q_{RC} < 0.5 \qquad (1\text{-}62)$$

For the LCR tank of Fig. 1-23 the poles may be anywhere in the LHP (see Fig. 1-25b), and we therefore have, from (1-48) and (1-50),

$$0 < Q_{LCR} < \infty \qquad [1\text{-}63]$$

Thus, the fact that the poles of an RC network are restricted to the negative real axis has the fundamental and in practice very severe limitation that only networks with very poor frequency selectivity or very high damping can be realized with them. As was demonstrated in Section 1.3, this is by no means adequate for any kind of practical filter, which invariably requires complex conjugate poles. It will be clear now that the closer to the $j\omega$ axis the poles are required to be, the higher will be the Q's of the networks involved. It is now our task to investigate methods of combining active devices with passive RC networks in such a way as to overcome the severe limitation on frequency selectivity imposed by the negative real poles of passive RC networks or, in other words, to generate the high-Q complex conjugate poles generally required in filter design.

23. This is also readily obtained for our example. Calculating $Q_{RC} = \omega_p/2\sigma_p$ from (1–55), we obtain $Q_{RC}^{-1} = \sqrt{R_1 C_1/R_2 C_2} + \sqrt{R_2 C_2/R_1 C_1} + \sqrt{R_2 C_1/R_1 C_2}$. This has the form $x + 1/x + \text{const.}$ Since $x + 1/x \geq 2$ for all x, (1-62) follows directly.

1.7 Complex Pole Realization with Active RC Networks

In the previous section we tried to present the case for active RC networks. It is, in a nutshell, that $\hat{q}_p$, which is the pole Q of a passive RC network, is limited to values less than 0.5. This results in greatly overdamped networks providing very poor frequency selectivity. We shall briefly here, and in more detail in later chapters, attempt to show how this severe limitation can be overcome with the help of active devices combined with RC networks. In so doing, we shall restrict ourselves to second-order networks (i.e., complex conjugate pole pairs). As was already pointed out earlier it is with these that active network design is mainly concerned, since all higher-order networks are, in practice, broken down into cascades of second- (or at most third-) order structures.

Let us start out with the two forms of a second-order transfer function describing a passive RC network, namely,

$$\hat{T}(s) = \frac{N(s)}{\hat{D}(s)} = \frac{N(s)}{s^2 + 2\hat{\sigma}_p s + \omega_p^2} \qquad \text{(1-64a)}$$

$$= \frac{N(s)}{s^2 + \dfrac{\omega_p}{\hat{q}_p} s + \omega_p^2} \qquad \text{(1-64b)}$$

$N(s)$ is a polynomial of at most second order, i.e.,

$$0 \leq \text{degree of } N(s) \leq 2 \qquad \text{(1-65)}$$

The roots of $N(s)$ may, for all practical purposes, lie anywhere in the LHP.[24] The overscript "^" on $T(s)$, $D(s)$, σ_p and q_p are to imply that these terms characterize a passive RC network, i.e., that the roots of $D(s)$ are negative real or that

$$0 < \hat{q}_p = \frac{\omega_p}{2\hat{\sigma}_p} < 0.5 \qquad \text{(1-66)}$$

In order to obtain better frequency selectivity than $\hat{T}(s)$ can provide, we must decrease the coefficient of the linear term of $\hat{D}(s)$. In so doing, it will be clear to the reader by now that we shall be substituting complex conjugate poles (whose q_p value is larger than 0.5) for the original negative real ones.

---

24. If $\hat{T}(s)$ is originally of a higher order but reduced to second order due to pole-zero cancellation then $N(s)$ may have zeros as far into the RHP as the original order of $N(s)$ will allow, based on the discussion in Section 1-5.

One way of decreasing the coefficient of the linear term in $\hat{D}(s)$ is to divide it by a quantity

$$m = 1 + \mu \tag{1-67}$$

where

$$0 < \mu < \infty \tag{1-68}$$

Thus

$$1 < m < \infty \tag{1-69}$$

From (1-64b) we then obtain

$$T(s) = \frac{N(s)}{s^2 + \dfrac{\omega_p}{m\hat{q}_p} s + \omega_p^2} \tag{1-70}$$

Clearly $T(s)$ no longer qualifies as a passive RC network since its pole Q, which is now given by

$$q_p = m\hat{q}_p = (1 + \mu)\hat{q}_p \tag{1-71}$$

can be arbitrarily large. Thus

$$T(s) = \frac{N(s)}{s^2 + \dfrac{\omega_p}{(1 + \mu)\hat{q}_p} s + \omega_p^2} \tag{1-72}$$

which may be rewritten as

$$T(s) = (1 + \mu) \cdot \frac{N(s)}{s^2 + \dfrac{\omega_p}{\hat{q}_p} s + \omega_p^2} \cdot \frac{1}{1 + \mu \left[\dfrac{s^2 + \omega_p^2}{s^2 + \dfrac{\omega_p}{\hat{q}_p} s + \omega_p^2} \right]} \tag{1-73}$$

or, with (1-64a),

$$T(s) = \frac{m}{\mu} \hat{T}(s) \cdot \frac{\mu}{1 + \mu \cdot \hat{t}_1(s)} \tag{1-74}$$

where

$$\hat{t}_1(s) = \frac{s^2 + \omega_p^2}{s^2 + \dfrac{\omega_p}{\hat{q}_p} s + \omega_p^2} = \frac{s^2 + \omega_p^2}{\hat{D}(s)} \tag{1-75}$$

Inspection of (1-73) shows that the multiplication of the pole Q by the factor m can be accomplished by a negative feedback configuration, as shown by the block diagram of Fig. 1-26.

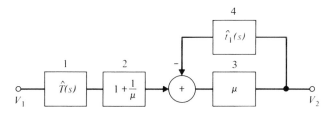

IG. 1-26. Block diagram of negative feedback scheme combining gain blocks with RC
etworks to generate a complex conjugate pole pair. Boxes 2 and 3 represent gain blocks
ith amplification of $1 + (1/\mu)$ and μ, respectively, and boxes 1 and 4 represent passive RC
etworks with identical poles.

Another way of decreasing the coefficient of the linear term of $\hat{D}(s)$ in
1-64) is to subtract a quantity κ from it, where

$$\kappa < 2\hat{\sigma}_p \tag{1-76}$$

Vith (1-64a) we obtain

$$T(s) = \frac{N(s)}{s^2 + (2\hat{\sigma}_p - \kappa)s + \omega_p^2} \tag{1-77}$$

hich can be rewritten as

$$T(s) = \frac{N(s)}{s^2 + 2\hat{\sigma}_p s + \omega_p^2} \cdot \frac{1}{1 - \kappa \dfrac{s}{s^2 + 2\hat{\sigma}_p s + \omega_p^2}} \tag{1-78}$$

r, with (1-64a),

$$T(s) = \frac{\hat{T}(s)}{\kappa} \cdot \frac{\kappa}{1 - \kappa \hat{i}_2(s)} \tag{1-79}$$

here

$$\hat{i}_2(s) = \frac{s}{s^2 + 2\hat{\sigma}_p s + \omega_p^2} = \frac{s}{\hat{D}(s)} \tag{1-80}$$

nspection of (1-79) shows that another way of increasing the pole Q of $T(s)$
s by a positive feedback configuration, as shown by Fig. 1-27. Instead of a
$\mathcal{Q}$ multiplication, as obtained by the negative feedback configuration of Fig.
-26, we obtain an increase in pole Q here by subtracting κ from the band-
vidth term $2\hat{\sigma}_p$. Contrary to the negative feedback case, in which the μ
mplification could take on very large values, the gain of the block κ (box 2)
s here limited by (1-76) to prevent an infinite pole Q, i.e. oscillation.
    The purpose of the two examples shown in Figs. 1-26 and 1-27 is to demon-
trate how the pole Q the passive RC network started out with (see (1-64a)
nd (1-64b)) can be increased (and a conjugate complex pole pair thereby

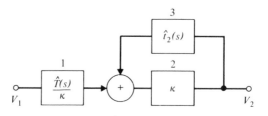

FIG. 1-27. Block diagram of positive feedback scheme combining gain blocks with RC networks to generate a conjugate complex pole pair.

realized) by combining the passive RC network with an active gain block in either a positive or negative feedback loop. Indeed, it has been shown[2] that most of the well known active synthesis methods can be classified into one or another type of feedback configuration. Thus the feedback network is a fundamental structure in active-network design which must be thoroughly understood. Two of the modern methods of analyzing feedback or control systems, namely, root locus theory and signal-flow graph theory, have been found to be extremely useful for this purpose. These will be discussed in the next chapter.

1.8 SUMMARY

1. The transmission function of an LLF (linear, lumped-parameter, finite) network is a real rational function in s, i.e., the ratio of two polynomials in s with real coefficients.

2. Poles are the roots of the denominator polynomial of the transfer function of an LLF network.

3. The poles of a rational transmission function represent its singularities. They are the natural frequencies of the corresponding LLF network (roots of its characteristic equation). These are of fundamental importance since they partially determine the stability of the network and characterize its behavior in general (e.g., selectivity, sensitivity, etc.).

4. Zeros are the roots of the numerator polynomial of the transfer function of an LLF network.

5. Practical (i.e., LCR) filters are networks possessing complex conjugate and real poles in the left half s plane, excluding the $j\omega$ axis and infinity.

6. General LCR networks can have zeros anywhere in the s plane. Unbalanced LCR networks without mutual coupling or ideal transformers

25. I. M. Horowitz and G. R. Branner. A unified survey of active RC synthesis techniques, *Proc. Nat. Electron. Conf. (NEC)*, **23**, 257–261 (1967).

annot have zeros on the positive real axis. In any physical network there is at east one zero at infinity, since in any practical circuit the gain always falls off o zero as the frequency is increased indefinitely (parasitic capacitance!).

7. Transfer functions of general passive RC networks have simple poles on he negative real axis and zeros anywhere in the s plane.

8. If the passive RC networks are unbalanced, no zeros may lie on the positive real axis.

9. The zeros of an unbalanced passive RC network are no more limited han those of an unbalanced LCR network with no mutual coupling. The ocation of zeros in the RHP (nonminimum phase networks) depends on the order m of the polynomial $N(s)$ determining them. The higher the order m, the arger the sector in the RHP in which zeros may lie. Referring to Fig. 1-9 we lave in general

$$|\phi| > \frac{\pi}{m}$$

and since

$$q_z = \frac{1}{2|\cos \phi|} \tag{1-81}$$

ve have

$$q_z > \frac{1}{2|\cos (\pi/m)|}, \qquad m = 1, 2, \ldots \tag{1-82}$$

10. The transfer function of an RC ladder network possesses poles which re simple and restricted to the negative real axis excluding the origin and nfinity.

11. The transmission zeros of an RC ladder network may be of any order but are restricted to the negative real axis, including the origin and infinity.

12. Complex zeros can be realized only with RC networks possessing more han one transmission path from input to output, or at least a path stretching over more than one series branch of the network. One common procedure generates complex zeros by connecting ladder networks in parallel (Guillemin) he other by cascading bridged twin-T structures (Dasher). In complicated ases, the latter procedure results in a considerable saving in the number of lements and often leads to appreciably higher gain constants. However, it uffers from considerable calculation complexity if the polynomials involved re of high degree (e.g., greater than four).

13. The "case for active RC networks" can be said to consist in the fact hat $\hat{q}_p$, which is the Q of a pole pair realized by a passive RC network, is imited to values less than 0.5. In terms of pole locations this means that the poles are confined to the negative real axis.

14. The term $\hat{q}_p$ can be increased by numerous ways in order to provide complex poles. All the methods require a combination of active devices with passive RC networks. Most of them can be reduced to feedback techniques of one kind or another.

15. Because active network techniques can generally be reduced to feedback schemes, feedback and control system concepts are very useful in this field.

16. Some of the important control system concepts in active network design are: root locus and flow-graph techniques, stability and sensitivity, return difference, and null return difference. These will be discussed in the next chapter.

CHAPTER

2

FEEDBACK TECHNIQUES IN LINEAR ACTIVE NETWORKS

INTRODUCTION

At the end of the previous chapter two examples demonstrated how the generation of complex conjugate poles by the combination of passive RC networks with active devices automatically leads to feedback configurations. Whereas the block diagrams shown were by no means optimized, nor even necessarily practical, they did show up the close tie between active network design and feedback or control system concepts. It was pointed out, in fact, that most active synthesis methods can be reduced to a basic feedback structure of one form or another. As we shall see in Chapter 2 of *Linear Integrated Networks: Design* this interpretation is not only enlightening analytically; it can also be very useful practically. It is therefore not surprising that numerous analytical tools that were originally developed in conjunction with control systems have found their way into the field of active network design. One of these tools is root locus analysis; another is signal-flow graph theory. Both topics will be dealt with in this chapter.

2.1 ROOT LOCUS METHODS IN LINEAR ACTIVE NETWORKS[1]

Let us first reconsider the feedback scheme dealt with in the preceding chapter (Fig. 1-26,) which is shown again in Fig. 2-1. The transfer function for

---

1. The root locus method was first presented by W. R. Evans: Graphical analysis of control systems, *Trans. AIEE*, **67**, 547–551 (1948).

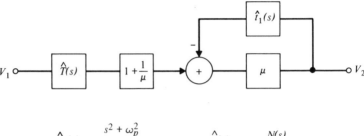

$$\hat{t}_1(s) = \frac{s^2 + \omega_p^2}{s^2 + \dfrac{\omega_p}{\hat{q}_p} s + \omega_p^2} \qquad\qquad \hat{T}(s) = \frac{N(s)}{s^2 + \dfrac{\omega_p}{\hat{q}_p} s + \omega_p^{\,2}}$$

FIG. 2-1. Negative feedback network.

this configuration was shown to be

$$T(s) = \frac{N(s)}{s^2 + \dfrac{\omega_p}{(1 + \mu)\hat{q}_p} s + \omega_p^2} \tag{2-1}$$

Solving for the poles, which are the roots of the denominator, we obtain:

$$p_{1,2} = \frac{-\omega_p}{2(1 + \mu)\hat{q}_p} \pm j\omega_p\sqrt{1 - \frac{1}{4(1 + \mu)^2\hat{q}_p^2}} \tag{2-2}$$

The poles are complex conjugate and their location in the s plane is a function of the gain element μ. In designing active networks it is of interest to know what this function is. Taking the extremes of μ we can tell by inspection of (2-2) that when $\mu = 0$, the poles are negative real (remember that $\hat{q}_p$ is less than 0.5), and when $\mu = \infty$ they lie on the $j\omega$ axis. However, the pole locations which are of particular interest, namely, those in between these two extremes, can only be obtained by a rather lengthy, if straightforward, numerical evaluation of (2-2).

There is nothing general about the extreme pole locations of (2-2). If, for example, we look for the pole locations of the feedback network of Fig. 1-27 of the preceding chapter we find that for this case the poles corresponding to the extremes of the active element are quite different. Here the transfer function is given by

$$T(s) = \frac{N(s)}{s^2 + (2\hat{\sigma}_p - \kappa)s + \omega_p^2} \tag{2-3}$$

and the poles

$$p_{1,2} = -\left(\hat{\sigma}_p - \frac{\kappa}{2}\right) \pm j\omega_p\sqrt{1 - \frac{1}{\omega_p^2}\left(\hat{\sigma}_p - \frac{\kappa}{2}\right)^2} \tag{2-4}$$

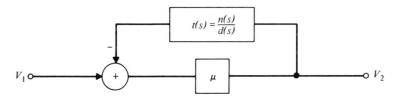

FIG. 2-2. Generalized negative feedback system.

Here again the poles are negative real when the gain element κ is equal to zero. It is not directly apparent where the poles move to when κ becomes very large, although inspection of (2-4) shows that they lie on the $j\omega$ axis when $\kappa = 2\hat{\sigma}_p$. It should be clear from these two examples that the dependence of the pole location on a parameter of an active network (often, but not always, a parameter that characterizes the active device) is of great significance but is, in general, rather laborious to determine. The root locus method provides a useful tool for obtaining precisely this dependence in a very straightforward and simple way.

2.1.1 Negative Feedback: The 180° Root Locus

Let us consider the generalized, closed-loop, negative feedback network shown in Fig. 2-2. μ is the open-loop gain constant. It is independent of both time and the complex frequency s. The feedback path is characterized by the transfer function $t(s)$, which is assumed to consist of the ratio of two polynomials $n(s)$ and $d(s)$. We obtain for the closed-loop transfer function of the feedback network

$$T(s) = \frac{\mu}{1 - g(s)} \qquad (2\text{-}5)$$

where $g(s)$ is the open-loop transfer function. For the negative feedback system of Fig. 2-2 it is given by

$$g(s) = -\mu t(s) \qquad (2\text{-}6)$$

where $0 \le \mu \le \infty$. With

$$t(s) = \frac{n(s)}{d(s)} \qquad (2\text{-}7)$$

(2-5) can be written as

$$T(s) = \frac{N(s)}{D(s)} = \frac{\mu}{1 + \mu t(s)} = \frac{\mu d(s)}{d(s) + \mu n(s)} \qquad (2\text{-}8)$$

It can be shown that the denominator of $T(s)$ is the characteristic polynomial of the network; hence the so-called *characteristic equation* is

$$1 - g(s) = 1 + \mu t(s) = 0 \qquad \text{[2-9a]}$$

This is often also written in the forms

$$1 + \mu \frac{n(s)}{d(s)} = 0 \qquad \text{[2-9b]}$$

or

$$d(s) + \mu n(s) = 0 \qquad \text{[2-9c]}$$

The roots of the characteristic equation are the poles of the closed-loop system; they characterize its transient response (hence the designation "characteristic equation"). As pointed out in Chapter 1, the stability and the frequency response (e.g., ripple in the passband) of a network is largely determined by the s-plane location of these poles with respect to the $j\omega$ axis.

Root locus theory provides a graphical trial-and-error technique for the determination of the roots of the characteristic equation $1 + \mu t(s) = 0$ for all values of μ. Substituting a specific test point into the characteristic equation (e.g., $s = s_i$) will, in general, form a complex number $\mu t(s_i)$. Since two complex numbers can be equal if, and only if, their respective real and imaginary parts are identical, and since $1 + \mu t(s_i) = 0$ if s_i is to be a root of the characteristic equation, it is necessary that

$$\text{Re } \mu t(s_i) = -1 \qquad \text{[2-10a]}$$

and

$$\text{Im } \mu t(s_i) = 0 \qquad \text{[2-10b]}$$

Thus, $\mu t(s_i)$ must be a complex quantity with unity amplitude and a phase angle of 180°.

To investigate more closely the implications of criteria (2-10a) and (2-10b), consider the open-loop transfer function

$$\mu t(s) = \mu \cdot \frac{(s - z_1)(s - z_2) \cdots (s - z_i) \cdots (s - z_m)}{s(s - p_1)(s - p_2) \cdots (s - p_j) \cdots (s - p_n)} \qquad (2\text{-}11)$$

This equation consists of the ratio of simple first-order zeros and poles,[2] where z_i and p_j can be complex. The characteristic equation corresponding to this pole–zero array is

$$1 + \mu \frac{(s - z_1)(s - z_2)(s - z_3) \cdots (s - z_m)}{s(s - p_1)(s - p_2)(s - p_3) \cdots (s - p_n)} = 0 \qquad (2\text{-}12)$$

---

2. For the sake of clarity in the derivation, we assume only first-order poles and zeros here. The resulting root locus is also valid for multiple roots, however.

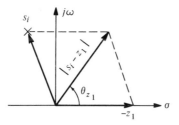

FIG. 2-3. Vector addition of s_i and $-z_1$.

Rearranging, we obtain

$$\frac{\mu(s - z_1)(s - z_2)(s - z_3) \cdots (s - z_m)}{s(s - p_1)(s - p_2)(s - p_3) \cdots (s - p_n)} = -1$$

$$= \exp{(j\,180°)} \qquad (2\text{-}13)$$

Let us now select an arbitrary test point $s = s_i$ and determine whether it respresents a root of (2-13). Substituting into (2-13) we investigate whether the following identity is possible:

$$\frac{\mu(s_i - z_1)(s_i - z_2)(s_i - z_3) \cdots (s_i - z_m)}{s_i(s_i - p_1)(s_i - p_2)(s_i - p_3) \cdots (s_i - p_n)} \equiv \exp{(j\,180°)} \qquad (2\text{-}14)$$

To do so each component of (2-14) must be evaluated in terms of amplitude and phase. For the first factor in the numerator, a phasor diagram, illustrated in Fig. 2-3, is employed to sum the complex quantities s_i and $-z_1$. The resulting phasor has a magnitude of $|s_i - z_1|$ and a phase angle of $\tan^{-1}[\text{Im}\,(s_i - z_1)/\text{Re}\,(s_i - z_1)]$. In abbreviated form, $(s_i - z_1)$ is equivalent to the polar coordinate representation $A_{z1}\exp{j\theta_{z1}}$.

It will prove convenient in further developing the theory to translate each phasor so that it terminates at the test point s_i. Without changing either the magnitude or phase characteristics, $A_{z1}\exp{j\theta_{z1}}$ can be moved horizontally until the arrowhead points to s_i. The phasor is now observed to originate at the associated open-loop zero location, i.e., at $s = z_1$. Thus, the quantity $A_{z1}\exp{j\theta_{z1}}$ can be described by the length and phase angle of a phasor which originates at the corresponding open-loop zero and terminates on the test point, as illustrated in Fig. 2-4. The phasor magnitudes and angles of the remaining open-loop components may be evaluated in an identical manner.

Expressing all components of (2-14) in polar coordinates as suggested in the equivalent s-plane phasor diagram of Fig. 2-5 , we obtain

$$\frac{\mu(A_{z1}\exp{j\theta_{z1}})(A_{z2}\exp{j\theta_{z2}}) \cdots (A_{zm}\exp{j\theta_{zm}})}{(Ap_0\exp{j\theta_{p0}})(A_{p1}\exp{j\theta_{p1}}) \cdots (A_{pn}\exp{j\theta_{pn}})} \equiv \exp{(j\,180°)} \qquad (2\text{-}15)$$

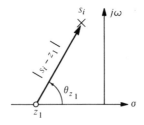

FIG. 2-4. Vector addition of s_i and z_1 after horizontal translation.

Rearranging the above,

$$\frac{\mu A_{z1} A_{z2} \cdots A_{zm} \exp j(\theta_{z1} + \theta_{z2} + \cdots + \theta_{zm})}{A_{p0} A_{p1} \cdots A_{pn} \exp j(\theta_{p1} + \theta_{p2} + \cdots + \theta_{pn})} \equiv \exp(j\,180°) \qquad (2\text{-}16)$$

This can be further reduced to

$$\frac{\mu A_{z1} A_{z2} \cdots A_{zm}}{A_{p0} A_{p1} \cdots A_{pn}} \exp j(\theta_{z1} + \theta_{z2} + \cdots + \theta_{zm} - \theta_{p0} - \theta_{p1} - \cdots - \theta_{pn})$$

$$\equiv \exp(j\,180°) \qquad (2\text{-}17)$$

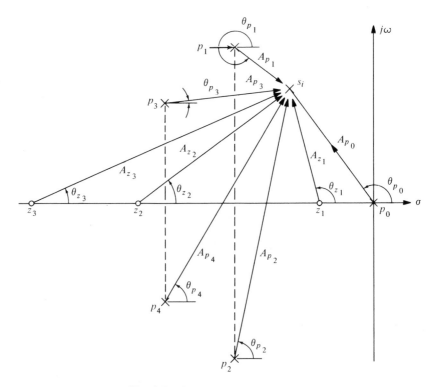

FIG. 2-5. Phasor diagram in the s plane.

Now, for $s = s_i$ to be a root of the characteristic equation, and thus to satisfy (2-17) it is necessary and sufficient that

$$\theta_{z1} + \theta_{z2} + \cdots + \theta_{zm} - \theta_{p0} - \theta_{p1} - \cdots - \theta_{pn} = \pm(2q + 1)180° \quad [2\text{-}18]$$

where $q = 0, 1, 2, \ldots$, and

$$\mu \cdot \frac{A_{z1} A_{z2} \cdots A_{zm}}{A_{p0} A_{p1} \cdots A_{pn}} = 1 \quad [2\text{-}19]$$

Equation (2-18) is referred to as the *angle criterion* of root locus theory, while (2-19) is the *magnitude criterion*. The former states simply that if the test point s_i is to be a root of the characteristic equation of a negative feedback system, the sum of all open-loop zero-phasor angles minus the sum of all open-loop pole-phasor angles must equal 180° (or some odd multiple of 180°). Stated differently, the root loci represent all those points s_i for which the loop phase shift is $(2q + 1)$ times 180°; hence the designation "180° root locus." Equation (2-19) provides the method for explicitly calculating the corresponding value of open-loop gain μ. Thus, at $s = s_i$,

$$\mu_{s_i} = \frac{A_{p0} A_{p1} \cdots A_{pn}}{A_{z1} A_{z2} \cdots A_{zm}} \quad (2\text{-}20)$$

It should be noted that, as stated, (2-20) is valid over the entire s plane. However, only at those positions where the angle criterion is satisfied does the gain calculation become meaningful. Consequently, the loci of all roots are first established through an application of (2-18); they are then calculated for gain by means of (2-20).

2.1.2 Rules for Designing the 180° Root Locus

Based on the angle criterion of (2-18) and on the closed-loop transfer function (2-8) we can now derive some simple rules for root locus design. The most important of these are summarized in the following basic eight rules. For a more detailed discussion, including the proofs of these rules, the reader is referred to two detailed and well written contributions in the literature.[3] For our part, we shall restrict ourselves to the rules of root locus design, and hope that, given sufficient illustrative examples, the reader will be in a position to apply these rules to problems of his own. More does not seem necessary here, since one of the main attractions of systematic analysis tools (such as root locus design) lies in the fact that they can be used directly, by routine observance of design rules, without the necessity of going into the

---

3. J. G. Truxal, *Automatic Feedback Control Systems Synthesis*, (New York: McGraw-Hill, 1955), Chapter 4; and A. W. Langill, Jr., Root locus systems analysis, *Electro-Technology*, October 1963, pp. 79–94.

detailed proofs and justifications for the existence of those rules every time they are applied.

For the purpose of defining terms, let us first return to the characteristic equation given by (2-9a), namely,

$$1 + \mu t(s) = 0$$

where $t(s) = n(s)/d(s)$ and $n(s)$ is a polynomial of mth degree, $d(s)$ a polynomial of nth degree. We then refer to the roots of the numerator of $1 + \mu t(s) = 0$ as the *closed*-loop poles and the roots of $n(s) = 0$ and $d(s) = 0$ as the *open*-loop zeros and poles, respectively. Using this terminology we can now list the eight rules with which the root locus of the characteristic equation of a negative feedback network can be simply derived.

1. Locus end points. As the open-loop gain μ varies from 0 to ∞ the loci (i.e., curves traversed by the *closed*-loop) poles start from the *open*-loop poles (i.e., roots of $d(s) = 0$) and terminate at the open-loop zeros (i.e., roots of $n(s) = 0$). When $t(s)$ is the voltage transfer function of a passive RC network, the loci therefore start on the negative real axis.

2. Number of loci. The number of separate locus branches equals the number of poles or zeros of $t(s)$, where critical frequencies at infinity are included and multiple-order critical frequencies are counted according to their order.

3. Conjugate values. Complex parts of the loci always appear in complex conjugate pairs if, as is customary in the design of linear networks, the coefficients of the polynomials $n(s)$ and $d(s)$ are real.

4. Loci on real axis. The root locus exists at any point on the real axis which lies to the *left* of an *odd* number of singularities.

5. Asymptotes of loci.
a. The loci near infinity (i.e., for large values of s) approach asymptotic lines whose directions are given by

$$\theta_i = \pm \frac{k\,180°}{n - m} \tag{2-21}$$

where k is an odd integer (i.e., 1, 3, 5, ...) and $n - m$ is the difference between the number of finite poles and zeros of $t(s)$.
b. The asymptotes intersect on the real axis at σ_i, which is determined from the formula

$$\sigma_i = \frac{\sum\limits^{n} \text{open-loop poles} - \sum\limits^{m} \text{open-loop zeros}}{n - m} \tag{2-22}$$

σ_i is often referred to as the center of gravity of the roots.

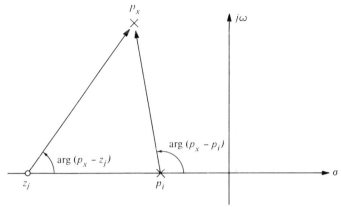

FIG. 2-6. Angles of departure and arrival.

6. Angles of departure and arrival (open-loop poles and zeros).
a. Referring to Fig. 2-6, the angle of departure from a pole p_x is given by

$$\phi_d = \sum_{j=1}^{m} \arg (p_x - z_j) - \sum_{\substack{i=1 \\ i \neq x}}^{n} \arg (p_x - p_i) + 180° \qquad (2\text{-}23)$$

b. The angle of arrival at a zero, z_x, is given by

$$\phi_a = \sum_{i=1}^{n} \arg (z_x - p_i) - \sum_{\substack{j=1 \\ j \neq x}}^{m} \arg (z_x - z_j) - 180° \qquad (2\text{-}24)$$

7. Angles of arrival and departure on the negative real axis. Root locus branches (corresponding to complex conjugate poles) depart from or arrive on the real axis in pairs at angles of $\pm 90°$. Occasionally a root locus involves the coalescence of more than two, namely, λ roots. Arrival or departure then takes place along equal angles of $180°/\lambda$ with respect to the negative real axis.

8. Breakaway point on the real axis. The breakaway points are determined by solving the following equation:

$$\sum \frac{1}{|s_b - z_l|} - \sum \frac{1}{|s_b - p_l|} = \sum \frac{1}{|s_b - z_r|} - \sum \frac{1}{|s_b - p_r|} \qquad (2\text{-}25)$$

where s_b is the position of the breakaway point, subscripts l and r represent singularities to the left and to the right of the breakaway point, respectively, and $|s_b - z_i|$ is the length of the vector from an arbitrary zero located on the real axis to the breakaway point.

The breakaway points are very often more easily obtained by solving the equation:

$$\frac{d\mu}{ds} = \frac{d}{ds} \left[-\frac{1}{t(s)} \right] = 0 \qquad (2\text{-}26)$$

This expresses the fact that the breakaway point from the real axis, which involves a 90° change of direction, corresponds to that value of s which produces a zero net change of angle for a small vertical displacement off the axis.

2.1.3 Examples of Root Locus Construction

To understand the use of the rules listed above we shall go through the step-by-step construction of some typical root loci as they may occur in active network design.

EXAMPLE 1:
As our first example we return to the feedback network shown in Fig. 2-1, which we know has the capability of realizing complex roots (poles) for μ values larger than zero. The loop is given by

$$g(s) = -\mu \hat{\imath}_1(s) = -\mu \frac{s^2 + \omega_p^2}{s^2 + \dfrac{\omega_p}{\hat{q}_p} s + \omega_p^2} \tag{2-27}$$

Since $\hat{q}_p < 0.5$, we shall assume the value $\hat{q}_p = 0.25$. Then

$$\hat{\imath}_1(s) = \frac{(s + j\omega_p)(s - j\omega_p)}{[s + (2 - \sqrt{3})\omega_p][s + (2 + \sqrt{3})\omega_p]} \tag{2-28}$$

Following the same sequence of rules as above we can now proceed to draw the root locus of (2-27) with respect to μ.

1. Locus end points (see Fig. 2-7):
   $\mu = 0$: Open-loop poles (poles of $\hat{\imath}_1(s)$):

   $$p_1 = -(2 - \sqrt{3}) = -0.3$$
   $$p_2 = -(2 + \sqrt{3}) = -3.7$$

   $\mu = \infty$: Open-loop zeros (zeros of $\hat{\imath}_1(s)$]:

   $$z_{1,2} = \pm j\omega_p$$

2. Number of locus branches: Two, since there are two open-loop poles and two finite open-loop zeros.
3. Conjugate values: Since there is a pair of complex conjugate zeros there must be a pair of complex conjugate root locus branches.
4. Loci on real axis: "Left of odd number of singularities" is between the two negative real poles (see Fig. 2-7).
5. Asymptotes of loci: Since $n = m$ (equal number of open loop poles and zeros), there are no asymptotes.

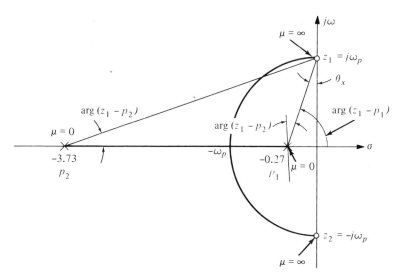

FIG. 2-7. Root locus construction for the loop gain of (2-27). Note that $p_1/z_1 = z_1/p_2$. Therefore $\theta_x = \arg(z_1 - p_2)$ and $\arg(z_1 - p_1) + \arg(z_1 - p_2) = 90°$.

6. Angles of departure and arrival:
   a. Departure from p_1:

$$\phi_d = \arg(p_1 - z_1) + \arg(p_1 - z_2) - \arg(p_1 - p_2) + 180°$$

From Fig. 2-7 we see that

$$\arg(p_1 - z_1) = -\arg(p_1 - z_2)$$
$$\arg(p_1 - p_2) = 0°$$

Therefore $\phi_d = 180°$ (see Fig. 2-8).
   b. Arrival at z_1:

$$\phi_a = \arg(z_1 - p_1) + \arg(z_1 - p_2) - \arg(z_1 - z_2) - 180°$$

From (2-28) it follows that $z_1^2 = p_1 p_2$; as a result, we find from straightforward geometrical analysis (see Fig. 2-7) that

$$\arg(z_1 - p_1) + \arg(z_1 - p_2) = 90°$$

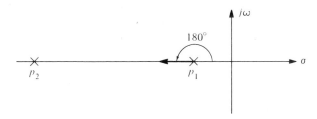

FIG. 2-8. Angle of departure for the loop gain of (2-27).

Furthermore, we see by inspection that

$$\arg(z_1 - z_2) = 90°$$

Thus

$$\phi_a = -180°$$

i.e., the tangent to the root locus at z_1 is horizontal.

7. Angles of departure on negative real axis: $\pm 90°$.
8. Breakaway point on the real axis: Sometimes a simpler method than either one given under rule 8 above can be used to obtain the breakaway point. In the present case we can consider the general pole equation given by (2-2). The breakaway point is determined from (2-2) by the condition that

$$j\omega_p \sqrt{1 - \frac{1}{4(1 + \mu)^2 \hat{q}_p^2}} = 0$$

Thus, with $\hat{q}_p = \tfrac{1}{4}$ we obtain

$$4(1 + \mu)^2 \frac{1}{16} = 1$$

Solving for the μ value, μ_b, corresponding to the breakaway point, we get

$$\mu_b = 1$$

Substituting $\mu_b = 1$ and $\hat{q}_p = \tfrac{1}{4}$ in (2-2) we obtain the breakaway point:

$$s_b = -\frac{\omega_p}{2 \cdot 2 \cdot \tfrac{1}{4}} = -\omega_p$$

The resulting root locus is shown in Fig. 2-9. Notice that it is confined to the LHP for all values of μ, thus practically guaranteeing an unconditionally stable system. Theoretically, for $\mu = \infty$, the closed-loop poles will coincide with the open-loop zeros z_1 and z_2 and lie on the $j\omega$ axis, thus causing instability. (As was pointed out in Chapter 1, the denominator of the transfer function of a realizable, passive network must be a Hurwitz polynomial. The roots of such a polynomial lie *inside* the left half s plane, not including the imaginary axis. Notice that violation of this criterion by an active network does not mean that it is unrealizable; it means that it is unstable.)

In order to guarantee unconditional stability theoretically as well as practically (since $\mu = \infty$ is not practically realizable), the open-loop zeros z_1 and z_2 need only be shifted by an infinitesimal amount to the left of the $j\omega$ axis. Care must be taken, however, that under worst-case conditions the RC network $\hat{t}_1(s)$ in Fig. 2-1 generating these zeros does not drift in such a way as to cause z_1 and z_2 to shift onto, or even to the right of, the $j\omega$ axis. This

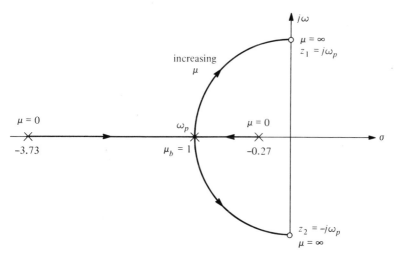

FIG. 2-9. Root locus of the open-loop transfer function of (2-27).

is a technological problem, since it depends on the stability properties (with respect to aging, temperature, humidity, etc.) of the resistors and capacitors making up $\hat{t}_1(s)$. More will be said about these properties for hybrid integrated networks in Chapter 6.

EXAMPLE 2:

As a second example we shall sketch the root locus for the open-loop transfer function

$$g(s) = \frac{-\mu}{(s+1)(s+2)} \tag{2-29}$$

1. Locus end points:
   $\mu = 0$ (open-loop poles): $p_1 = -1$, $p_2 = -2$.
   $\mu = \infty$ (open-loop zeros): $z_{1,2} = \infty$.
2. Two locus branches because of two open-loop poles.
3. Two complex-conjugate locus branches since the open-loop poles converge and take off from the negative real axis.
4. Locus exists between the two open-loop poles p_1, p_2 on the negative real axis.
5. Asymptotes:
   a. Angle:

$$\theta_i = \pm \frac{k \cdot 180°}{2 - 0} = \pm 90°$$

   b. intersection on negative real axis:

$$\sigma_i = \frac{-1 - 2}{2} = -\frac{3}{2}$$

6. Angles of departure and arrival: Trivial, since open-loop poles are on the negative real axis and open-loop zeros at infinity.
7. Angle of departure from real axis: $\pm 90°$.
8. Breakaway point: From the characteristic equation (2-9a) we have

$$\mu = -(s^2 + 3s + 2)$$

and

$$\frac{d\mu}{ds} = -2s - 3 = 0$$

Consequently

$$s_b = -\tfrac{3}{2} \quad \text{and} \quad \mu_b = 0.25$$

Thus the loci break away at the intersection of the asymptotes; in fact the root locus coincides with the section on the negative real axis between the open-loop poles and the asymptotes perpendicular to the real axis, as shown in Fig. 2-10. Notice that the roots of the characteristic equation do not ever cross the $j\omega$ axis no matter how large μ gets. In other words, the root locus of Fig. 2-10 typifies an unconditionally stable system with respect to μ, [since the root locus never crosses the imaginary axis for any possible values of μ (i.e., $0 \le \mu \le \infty$)]. At the same time, and for precisely this reason, it is difficult to realize high-Q poles (poles very close to the imaginary axis) with a system of this kind. Thus, unconditional stability and high-Q poles are very often incompatible requirements for a given system.

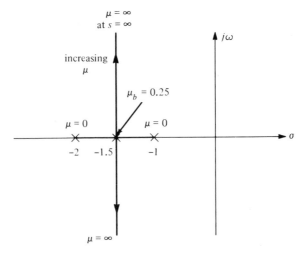

Fig. 2-10. Root locus of the open-loop transfer function $g(s) = -\mu/(s + 1)(s + 2)$.

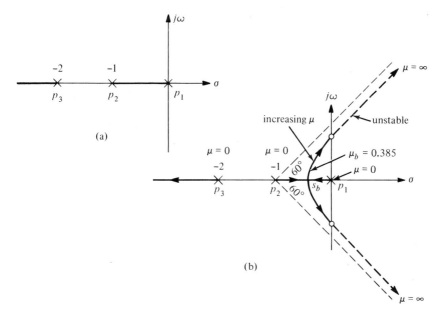

FIG. 2-11. Root locus of the open-loop transfer function $g(s) = -\mu/s(s+1)(s+2)$. (a) Existence of root locus on the negative real axis. (b) Complete root locus with respect to μ.

EXAMPLE 3:

Let us now consider what happens to the root locus of Fig.2-10 if we add a pole at the origin to (2-29), thereby making the open-loop transfer function third-order:

$$\mu t(s) = \frac{\mu}{s(s+1)(s+2)} \tag{2-30}$$

Using the same numbering as above we now have:

1. Locus starting points (open loop poles)

$$p_1 = 0, p_2 = -1, p_3 = -2$$

2. Three locus branches because three open-loop poles.
3. At least one of the locus branches must be on the negative real axis; the other two will be complex conjugate.
4. On the negative real axis, the locus exists between p_1 and p_2 and to the left of p_3 (see Fig. 2-11a).
5. Asymptotes:
   a. Angle:

$$\theta_i = \pm \frac{k \cdot 180°}{3} = \pm 60°, 180°$$

b. Intersection on negative real axis:

$$\sigma_i = \frac{-1-2}{3} = -1$$

6. Angles of departure and arrival: Trivial, since open-loop poles on the negative real axis and open-loop zeros coincide with asymptotes at infinity.
7. Angles of departure from negative real axis: $\pm 90°$.
8. Breakaway point: From the characteristic equation we have:

$$\mu = -(s^3 + 3s^2 + 2s)$$

and

$$\frac{d\mu}{ds} = -(3s^2 + 6s + 2) = 0$$

Therefore

$$s_{b_{1,2}} = -1 \pm \frac{\sqrt{3}}{3}$$

Since the breakaway point must lie between 0 and -1, the only possible solution is

$$s_b = -1 + \frac{\sqrt{3}}{3} = -0.423$$

which occurs when

$$\mu_b = 0.385$$

The resulting root locus is plotted in Fig. 2-11b. A comparison with the root locus in Fig. 2-10, [which corresponds to the open-loop function of (2-29)], shows that the addition of a pole at the origin in (2-30) modifies the unconditionally stable system of Fig. 2-10 into the conditionally stable system of Fig. 2-11b. The latter becomes unstable for the value of μ for which the two root locus branches intersect the imaginary axis. On the other hand if μ (and the open-loop singularities) can be very precisely controlled and maintained during operation of the system, arbitrarily high-Q poles are now attainable.

EXAMPLE 4:
Instead of adding a third pole we here add a negative real zero to (2-29) and observe how this affects the original root locus of Fig. 2-10. Letting the open-loop transfer function have the form

$$g(s) = -\mu \cdot \frac{s+5}{(s+1)(s+2)} \tag{2-31}$$

the corresponding root locus construction proceeds as follows:

1. Locus starting points (open-loop poles):

$$p_1 = -1, p_2 = -2$$

   Locus end points (open-loop zeros)

$$z_1 = -5, z_2 = \infty$$

2. Two locus branches because only two starting points (open-loop poles).
3, 4. Locus exists between p_1 and p_2 and to the left of z_1 (see Fig. 2-12a). Thus the locus must leave the negative real axis, resulting in two complex conjugate branches.
5. Asymptotes:
   a. Angle:

$$\theta_i = \pm \frac{k\,180°}{2-1} = 180°$$

   b. Since the angle θ_i is 180° the asymptote must lie *on* the negative real axis and there can be no intersection with it.
6. Angles of departure and arrival are trivial, since the open-loop poles and zeros are on the negative real axis.
7. Departure from negative real axis is $\pm 90°$.
8. Breakaway point: From the characteristic equation we have

$$\mu = -\frac{s^2 + 3s + 2}{s + 5}$$

Thus

$$\frac{d\mu}{ds} = -\frac{(s+5)(2s+3) - (s^2 + 3s + 2)}{(s+5)^2} = 0$$

Solving for the breakaway points (there are two) we obtain:

$$s_{b_1} = -1.54, \qquad s_{b_2} = -8.46$$

The complete root locus is shown in Fig. 2-12b. Notice that where the addition of a pole to (2-29) bent the root locus toward the imaginary axis, thereby permitting high-Q poles but also risking instability, the addition of a zero has turned the root locus of Fig. 2-10 away from the imaginary axis, making oscillations but also the realization of high-Q poles virtually impossible. Thus the three root loci of Figs. 2-10, 2-11, and 2-12 demonstrate how the addition of a pole or zero can have opposite effects on a second-order system. Manipulation of a root locus by the addition of poles and zeros is common in control systems, where a system may have to be undamped by some amount, or stabilized. In active network design the desired pole-zero pattern is usually specified. Naturally the poles must be specified left of the $j\omega$ axis. Having

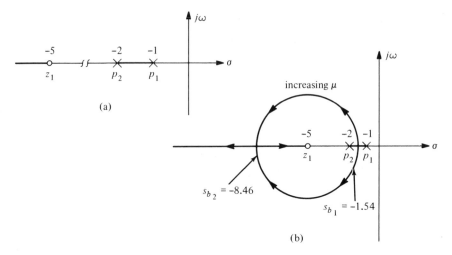

FIG. 2-12. Root locus of the open-loop transfer function $g(s) = -\mu(s + 5)/(s + 1)(s + 2)$. (a) Existence of root locus on the negative real axis. (b) Complete root locus with respect to μ.

designed a network with the desired poles the problem becomes one of guaranteeing that they remain at the specified location in spite of variations in time or temperature of individual network components. The closer a pole pair is required to be to the $j\omega$ axis (i.e., the higher the pole Q) the more critical the problem of network stability is likely to be. A given slight variation, say of the active element, may cause the pole to shift onto, or right of, the $j\omega$ axis. Thus the sensitivity of a pole, and particularly of a high-Q pole, to variations of the network elements generating it, is of greatest importance. We shall devote a later chapter to this question of network sensitivity and to methods of minimizing it. First, however, there are some additional feedback concepts with respect to active networks that must be considered. These include the root locus design of a positive feedback system.

2.1.4 Positive Feedback: The 0° Root Locus

The derivation of root locus theory as it was presented in Section 2.1.2 was based exclusively on a system with negative feedback. Actually two sets of branches constitute a complete system of root loci, one for positive feedback and the other for negative feedback. If we consider the generalized closed-loop positive feedback network shown in Fig. 2-13 we obtain the transfer function

$$T(s) = \frac{\mu}{1 - g(s)} = \frac{\mu}{1 - \mu t(s)} \tag{2-32}$$

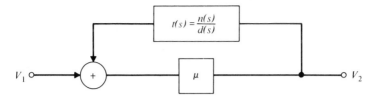

FIG. 2-13. Generalized positive feedback system.

where the open-loop transfer function is

$$g(s) = \mu t(s) \tag{2-33}$$

and $0 \le \mu \le \infty$. Actually, if μ is the gain element, its polarity will generally determine whether the feedback is positive or negative (μ being positive for positive feedback, negative for negative feedback). Thus, if only the magnitude of μ is considered, the equations describing the root loci for positive and negative feedback are:

$$\text{Positive feedback:} \quad |\mu| t(s) = +1 \tag{2-34}$$

$$\text{Negative feedback:} \quad |\mu| t(s) = -1 \tag{2-35}$$

For positive feedback, the root loci constitute all those values of s for which the loop phase shift is $2q \cdot 180°$, where q is any integer including zero. This is why we refer here to a "$0°$ root locus." (By contrast, recall that for negative feedback, as dealt with above and considered most frequently, the root loci constitute all those values of s for which the loop phase shift is $(2q + 1) \cdot 180°$, for which reason it is called the "$180°$ root locus.")

Most of the rules for the construction of a $180°$ root locus that were discussed above apply also to a $0°$ root locus. One significant difference is in the parts of the real axis that may be occupied in the two cases:

Positive Feedback ($0°$ Root Locus): The loci may occupy only portions of the real axis lying to the left of any *even* number of critical frequencies (zeros or poles) and the entire portion of the real axis to the right of the rightmost critical frequency.

Negative Feedback ($180°$ Root Locus): The loci may occupy only portions of the real axis lying to the left of any *odd* number of critical frequencies.

Another difference between the two root loci is in the angles of the asymptotes with the real axis:

Asymptotes: The root loci tend to infinity when the loop transmission $t(s)$ has one or more zeros at infinity. The number of asymptotes v is determined by the number of infinite zeros of $t(s)$. For the angle requirement to be met, the angles which the asymptotes make with the real axis are 0, $\pm 2 \cdot 180°/v$, $\pm 4 \cdot 180°/v$, ... for positive feedback and $\pm 180°/v$, $\pm 3 \cdot 180°/v$, $\pm 5 \cdot 180°/v$,

... for negative feedback. The term v is the difference between the number of finite poles and zeros of $t(s)$, or, in terms of the order n of its denominator and m of its numerator, $v = n - m$.

When compared with the 180° root locus, the 0° root locus has some other unique characteristics. For one thing its direction of depature from the open loop poles is always in the direction opposite (i.e., turned by an additional 180°) to that of the corresponding 180° root locus. This is because the expressions for the angles of departure and arrival are the same as those pertaining to the 180° locus, (2-23) and (2-24), except that the 180° term in both expressions is now replaced by 360°. Furthermore, the positive real axis is always an asymptote of the 0° root locus. Nevertheless, the two locus types (i.e., 180° and 0°) always approach the same points as μ approaches infinity.

The differences between the 0° and 180° root loci can, of course, be directly derived from the respective angle criterion that has to be fulfilled. Two examples of 0° and 180° root loci, shown in Figs. 2-14 and 2-15, should illustrate clearly the differences between the locus types. Notice, that although the angles of the asymptotes differ as mentioned above, the intercept point on the negative real axis is the same for both.

EXAMPLE:

To go briefly through an example of positive feedback root locus design we shall consider the open-loop transfer function

$$g(s) = \kappa \cdot \frac{s}{s^2 + 2\hat{\sigma}_p s + \omega_p^2} \qquad (2\text{-}36)$$

This will be recognized as the open-loop transfer function of the positive feedback scheme in Fig. 1-27 of the preceding chapter. Since $\hat{t}_2(s)$ describes a passive RC network, $\hat{\sigma}_p$ and ω_p are constrained by the condition that

$$\hat{\sigma}_p > \omega_p \qquad (2\text{-}37)$$

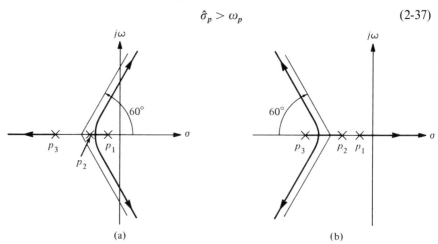

(a)   (b)

FIG. 2-14. Root locus of the open-loop transfer function $g(s) = \mu/(s - p_1)(s - p_2)(s - p_3)$. (a) Negative feedback. (b) Positive feedback.

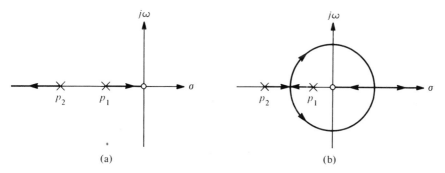

(a)   (b)

FIG. 2-15. Root locus of the open-loop transfer function $g(s) = \mu s/(s - p_1)(s - p_2)$. (a) Negative feedback. (b) Positive feedback.

Assuming that $\hat{\sigma}_p = 2\omega_p$, we obtain the open-loop poles

$$p_1 = \left(-2 + \sqrt{3}\right)\omega_p, p_2 = -\left(2 + \sqrt{3}\right)\omega_p$$

and, with the open-loop zero at the origin, we can immediately fill in the sections of the real axis along which the locus exists (see Fig. 2-16a). There is only one asymptote, since (2-36) has only one infinite zero; its angle is $0°$. Only the positive real axis meets this requirement, as indeed it must, since the

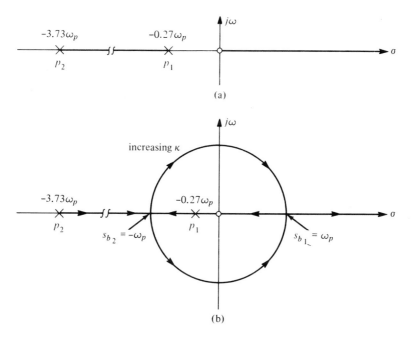

FIG. 2-16. Root locus of the open-loop transfer function $g(s) = \kappa s/(s^2 + 4\omega_p + \omega_p^2)$. (a) Existence of root locus on the negative real axis. (b) Complete root locus with respect to κ.

positive real axis is *always* an asymptote of the $0°$ root locus. (Thus, there is always at least one asymptote with the $0°$ root locus, namely, the positive real axis.)

Calculating the breakaway points we start out from the characteristic equation (2-34) and obtain, with (2-36),

$$\kappa = \frac{s^2 + 4\omega_p s + \omega^2}{s}$$

Differentiating according to (2-26) we have

$$\frac{d\kappa}{ds} = \frac{s(2s + 4\omega_p) - (s^2 + 4\omega_p s + \omega_p^2)}{s^2} = 0$$

Solving for the breakaway points, we get

$$s_{b_1} = \omega_p, \, s_{b_2} = -\omega_p$$

They occur for

$$\kappa_1 = 6\omega_p, \, \kappa_2 = 2\omega_p$$

(These breakaway points could have been obtained directly from (2-4) by setting the imaginary term equal to zero and solving for the κ values corresponding to s_{b_1} and s_{b_2}.) We now have sufficient information to draw the root locus as shown in Fig. 2-16b. It corresponds to the representative $0°$ locus shown in Fig. 2-15b. Notice that a sign change in (2-36) produces the corresponding $180°$ root locus shown in Fig. 2-15a.

2.2 SIGNAL–FLOW GRAPH TECHNIQUES IN LINEAR ACTIVE NETWORKS

Signal-flow graphs, introduced by S. J. Mason in 1953[4] have been used since then to graphically solve sets of linear equations characterizing both passive and active networks. Their usefulness stems from the fact that, once a network has been reduced to a flow graph, well defined rules allow network transfer functions to be evaluated by inspection.

We shall restrict ourselves here to those aspects of signal-flow graph theory deemed necessary or useful for the analysis and understanding of linear active network design. Indeed, although we shall spend some time on the derivation of signal-flow graphs from the corresponding physical networks for analysis purposes, this is of secondary interest to us in the present context, since there is a more suitable circuit analysis tool, namely, the indefinite admittance matrix, which is discussed in Chapter 3. Thus our main interest in signal-flow graph theory here is with respect to basic concepts in feedback theory, which

4. S. J. Mason, Feedback theory—Some properties of signal flow graphs, *Proc. IRE*, **41**, 1144–1156 (1953). S. J. Mason and H. J. Zimmerman, *Electronic Circuits, Signals and Systems*, (New York; John Wiley & Sons, 1965).

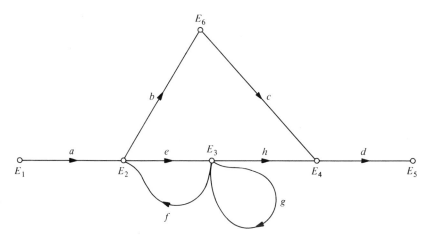

Fɪɢ. 2-17. General signal-flow graph.

are greatly clarified by its introduction. For additional treatment of the general subject of signal-flow graph theory the reader is referred to the numerous expositions on the subject in the literature.[5]

A signal-flow graph is a graphical representation of a set of linear equations in terms of nodes and branches. Before discussing the rules required for the construction of signal-flow graphs, some basic terms must first be explained. Referring to the general signal-flow graph in Fig. 2-17 we shall use the following terms in the manner explained below:

Node: Represents a specified quantity (e.g., signal) or a variable.

Branch: Represents the functional dependence between the variables (i.e., nodes). The value of a variable represented by a node is obtained by adding the branch outputs entering that node. Signals travel along branches only in the direction described by the arrows of the branches (i.e., in the branch direction).

Source node: A node with no branches entering (e.g., node E_1 in Fig. 2-17). It represents an independent variable.

5. An excellent treatment of signal-flow graph theory as applied to feedback systems can be found in: J. G. Truxal, op. cit., Chapter 2. Other detailed treatments of the subject for linear network and system analysis are: Y Chow and E. Cassignol, *Linear Signal-Flow Graphs and Applications* (New York: John Wiley & Sons, 1962); S. C. Lorens, *Flow Graphs for the Modeling and Analysis of Linear Systems* (New York: McGraw-Hill, 1964); J. R. Ward and R. D. Strum, *The Signal-Flow Graph in Linear System Analysis* (Englewood Cliffs, N. J.: Prentice-Hall, 1968); J. R. Abrahams and G. P. Coverly, *Signal-Flow Analysis* (New York and London: Pergamon Press, 1965); S. P. Chan, *Introductory Topological Analysis of Electrical Networks* (New York: Holt, Rhinehart and Winston, 1969), Chapter 7. Concise expositions on signal-flow graph analysis have frequently been published in engineering magazines. Among those to be recommended are: J. Mittelman, Signal-flow graphs, *EEE*, Nov. 1962, p. 62, Feb. 1963, p. 38, March 1963, p. 63; and D. Perti and J. Ricci, Flow graphs simplify circuit analysis, *EDN Mag.*, October 1967 and June 10, 1968.

Sink node: A node with no branches leaving (node E_5). It represents a variable that no other variable in the system is dependent upon.

Intermediate node: A node with branches entering and leaving (nodes E_2, E_3, E_4, and E_6).

Path: A continuous unidirectional succession of branches, all of which are traveling in the same direction.

Open path: Any path along which a node is encountered only once (*abcd* or *aeh*; *aef* is not open).

Forward path: An open path between a source node and a sink node (*abcd* or *aehd*).

Loop: A closed path which returns to the starting node and in which no node is encountered more than once, excepting the starting node (loops *g* and *ef*, but not *egf*).

Self loop: A feedback loop consisting of a single branch and node (loop *g*).

Branch transmission: The linear quantity, regardless of its dimension, relating one node of a branch to the other. Thus, for example, a signal x_k, traveling along a branch between x_k and x_j, is multiplied by the transmission (e.g., gain) t_{kj} of the branch, so that a signal of $t_{kj} x_k$ is delivered at node x_j.

Loop transmission: The product of the transmission of branches in the closed loop.

2.2.1 Construction of Signal–Flow Graphs from Physical Linear Networks

Much as we did in the section on root locus design, we can summarize the construction of a signal-flow graph by a few basic rules:

1. The variables (e.g., signals) at the nodes of a signal-flow graph are associated with the variables of a set of linear equations, while transmittances of the branches represent the relating constants or coefficients.
2. Signals travel along branches only in the direction of the arrows.
3. A signal traveling along any branch is multiplied by the transmittance of that branch.
4. The value of the variable represented by any node is the sum of all signals entering the node.
5. The value of the variable represented by any node is transmitted on all branches leaving that node.

The signal-flow graph of a physical linear network is nothing more than a graphical representation of the set of linear equations characterizing that network; indeed it is generally derived directly from those equations. Since there are many different forms in which the equations of a network can be written (e.g., mesh or node equations) there are equally many different forms of signal-flow graphs that can be used to describe that same network. Naturally some sets of equations produce simpler signal-flow graphs for a given network than others. Generalizations are difficult, however, because the form of network equations providing the simplest flow graph depends on the network in question, that is, on its topology (e.g., ladder or lattice), on its complexity, and on the number of independent variables contained in it. Thus the dexterity with which a complex network can be converted into a tractable signal-flow graph depends to some extent on experience. On the other hand, as we shall see shortly, one of the big advantages of signal-flow graphs is that, even if equations are chosen that result in an unnecessarily cumbersome graph, the easily implemented rules for reducing the graph to its basic and simplest possible form make its usefulness independent of the elegance with which the initial graph was derived.

EXAMPLES:

To illustrate the construction of a signal-flow graph from a physical network we shall go through some representative examples. Consider first the resistive network in Fig. 2-18. We can write the following network equations

$$V_3 = R_4 I_3 \qquad (2\text{-}38a)$$

$$I_3 = V_2 \frac{1}{R_3 + R_4} \qquad (2\text{-}38b)$$

$$V_2 = I_2 R_2 \qquad (2\text{-}38c)$$

$$I_2 = I_1 - I_3 \qquad (2\text{-}38d)$$

$$I_1 = \frac{V_1}{R_1} - \frac{V_2}{R_1} \qquad (2\text{-}38e)$$

The variables of our network which will be the nodes of our flow graph are $V_1, V_2, V_3, I_1, I_2, I_3$. Notice that we have a ladder type of network here, that we worked our way from the output (sink) to the input (source),

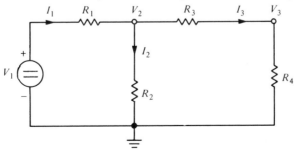

FIG. 2-18. Resistive network.

and that we alternated voltage and current equations. This enables us to d aw the corresponding flow graph directly from the equations, proceeding from the sink to the source as prescribed by the sequence of equations in (2-38), as shown in Fig. 2-19.

As another example, consider the RC ladder network shown in Fig. 2-20a. The network equations can be written by inspection:

$$V_3 = R_3 I_3 \tag{2-39a}$$

$$I_3 = sC_3(V_2 - V_3) \tag{2-39b}$$

$$V_2 = R_2(I_2 - I_3) \tag{2-39c}$$

$$I_2 = sC_2(V_1 - V_2) \tag{2-39d}$$

$$V_1 = R_1(I_1 - I_2) \tag{2-39e}$$

$$I_1 = sC_1(V_s - V_1) \tag{2.39f}$$

Notice that the equations progress from the sink to the source. They interrelate the network voltages and currents which are the variables of the equ-

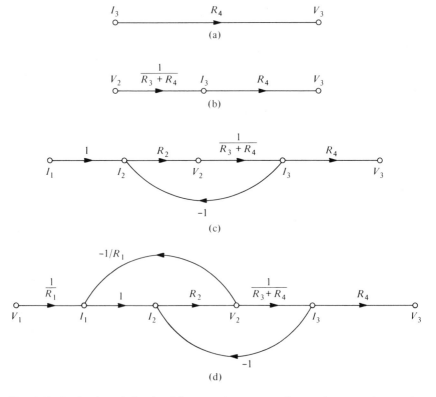

FIG. 2-19. Derivation of the signal-flow graph corresponding to the network equations (2-38a–d).

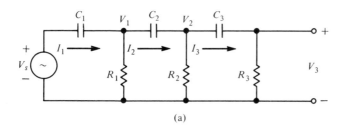

(a)

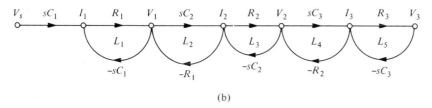

(b)

FIG. 2-20. (a) RC ladder network. (b) Corresponding signal-flow graph.

ations (nodes of the signal-flow graph). Because voltage and current equations alternate, the corresponding signal-flow graph can be drawn directly, proceeding in the same sequence as the network equations, from sink to source. The resulting flow graph is shown in Fig. 2-20b. It would be the same if the sequence of equations had proceeded in the opposite direction, from source to sink. It would not be the same if the describing equations had not progressed continuously in one direction or if, for example, only node or mesh equations had been used instead of alternating between the two. For ladder networks, alternating equations derived in the manner described above result in signal-flow graphs with the regular pattern shown in Fig. 2-20b.

Let us now derive the signal-flow graph of an active network. Consider, for example, the transistor amplifier shown in Fig. 2-21a. Using the transistor equivalent-T circuit shown in Fig. 2-21b we obtain the equivalent circuit of the transistor amplifier in Fig. 2-21c. Notice that, in spite of the fact that the equivalent circuit contains a dependent voltage source, we have essentially a ladder network, as in the previous examples, and therefore proceed accordingly. The circuit equations are

$$v_0 = -R_3 i_c \tag{2-40a}$$

$$i_c = \frac{v_0 + r_m i_b - v_1}{r_c(1 - \alpha)} \tag{2-40b}$$

$$v_1 = (r_e + R_2)(i_b + i_c) \tag{2-40c}$$

$$i_b = \frac{v_s - v_1}{R_1 + r_b} \tag{2-40d}$$

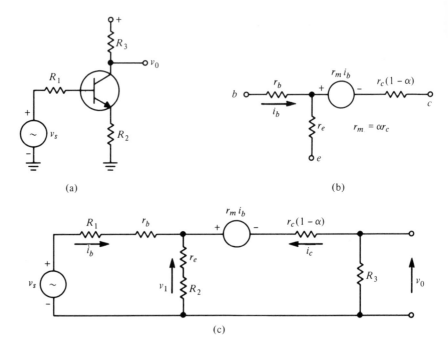

(a)   (b)

(c)

FIG. 2-21. (a) Transistor amplifier. (b) Transistor equivalent-T circuit with controlled voltage source. (c) Equivalent circuit of the transistor amplifier.

and the corresponding signal-flow graph follows directly, as shown in Fig. 2.22. Notice that the presence of the dependent voltage source $r_m i_b$ adds a branch to the typical flow-graph topology of a ladder network.

Finally, we shall derive the signal-flow graph of the emitter-follower shown in Fig. 2-23a and, this time, use the transistor equivalent-T circuit shown in Fig. 2-23b, which uses a dependent current source. The resulting equivalent circuit of the emitter follower is shown in Fig. 2-23c. As drawn it does not resemble a ladder network (although it could be redrawn to do so), and the

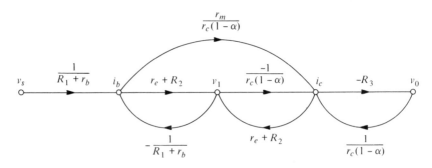

FIG. 2-22. Signal-flow graph of transistor amplifier in Fig. 2-21a.

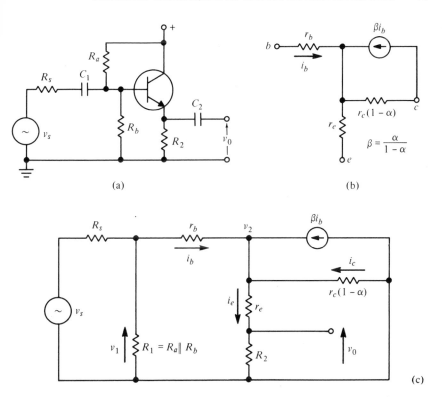

FIG. 2-23. (a) Emitter follower. (b) Transistor equivalent-T circuit with controlled current source. (c) Equivalent circuit of the emitter follower.

set of circuit equations will not follow the pattern of the preceding examples nor result in a signal-flow graph of the same kind. A set of suitable equations is

$$v_0 = i_e R_2 \qquad (2\text{-}41a)$$

$$i_e = (\beta + 1)i_b + i_c \qquad (2\text{-}41b)$$

$$i_c = -\frac{v_2}{r_c(1 - \alpha)} \qquad (2\text{-}41c)$$

$$v_2 = i_e(r_e + R_2) \qquad (2\text{-}41d)$$

$$i_b = \frac{v_1 - v_2}{r_b} \qquad (2\text{-}41e)$$

Constructing the signal-flow graph in the order of the equations given above we obtain the configuration shown in Fig. 2-24. As expected, it has no resemblance to a ladder-type signal-flow graph.

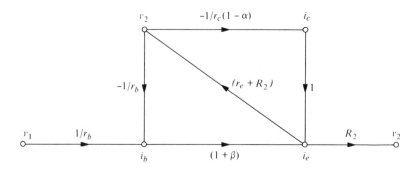

FIG. 2-24. Signal-flow graph of the emitter follower in Fig. 2-23a.

2.2.2 Rules for the Reduction of Signal-Flow Graphs

We have seen in the previous section that the signal-flow graph of a network does not contain any more information than the describing equations from which it is derived. Nevertheless, there may be numerous good reasons for converting a set of network equations into an equivalent graphical form. For one thing it permits the visualization of the signal transmission paths in a network; for another, it indicates the interaction between variables within the network and points out existing feedback loops. Finally, and perhaps most important of all, having obtained a given graph of arbitrary complexity (depending on the equivalent circuit and network equations from which it is derived) a set of simple rules permit its reduction to simpler and more tractable forms by successive elimination of dependent variables, or, in other words, by elimination of graph nodes. Thus the reduction of signal-flow graphs is equivalent to solving a corresponding set of network equations but, due to the existence of certain systematic rules, the reduction of a graph is often more convenient. As demonstrated in what follows, almost every one of the reduction rules corresponds directly to an equivalent manipulation of linear equations in which a dependent variable is eliminated.

Rule 1.   Cascade transformation
   The total transmission of a series of branches in cascade (i.e., path) equals the product of the individual branch transmissions. An example is given in Fig. 2-25 together with the equivalent equations.

Rule 2.   Parallel transformation
   The total transmission of parallel branches equals the sum of the individual branch transmissions, as shown in Fig. 2-26.

Rule 3.   Removal of a node
   The node of origin or termination of a transmission may be removed or shifted, as long as the transmittance between nodes of interest in the system

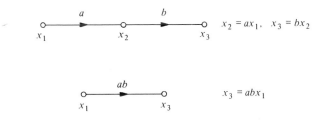

$$x_2 = ax_1, \quad x_3 = bx_2$$

$$x_3 = abx_1$$

FIG. 2-25. Reduction rule 1: cascade transformation.

remains unchanged. (Also see rule 4 below.) A node may be removed as shown in Fig. 2-27.

Rule 4. Shifting a transmittance

Starting out with the signal-flow graph shown in Fig. 2-28a, the starting point of a branch may be shifted as shown in Fig. 2-28b, the termination of an internal branch shifted as shown in Fig. 2-28c, and the termination of a branch containing a source node (x_4) shifted as shown in Fig. 2-28d. Notice that a new variable x'_3 must be introduced when the termination of the internal branch is shifted (see Fig. 2-28c). A consequence of the transmittance shifting rules is the "Y" transformation shown in Fig. 2-29a and the star-to-mesh transformation shown in Fig. 2-29b.

Rule 5. Path inversion

In order to invert a path originating at a *source* node x_i and terminating at a node x_j whose branch transmission is μ, we replace it by a path from x_j to x_i having a branch transmission $1/\mu$. Furthermore, all other branches originally terminating at x_j are shifted to terminate at x_i and their branch transmissions are multiplied by $-1/\mu$. This is demonstrated in Fig. 2-30, where the path a from x_2 to x_1 (Fig. 2-30a) is inverted (Fig. 2-30b). Notice that the inversion of a path (whose starting node must, by definition, be a source) has the effect of shifting the source from one end of the path to the other. Thus a succession of inversions in a flow graph will successively transfer the corresponding

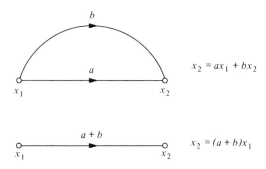

$$x_2 = ax_1 + bx_2$$

$$x_2 = (a + b)x_1$$

FIG. 2-26. Reduction rule 2: parallel transformation.

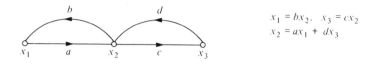

$$x_1 = bx_2, \quad x_3 = cx_2$$
$$x_2 = ax_1 + dx_3$$

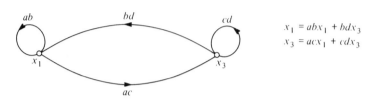

$$x_1 = abx_1 + bdx_3$$
$$x_3 = acx_1 + cdx_3$$

Fig. 2-27. Reduction rule 3: removal of a node.

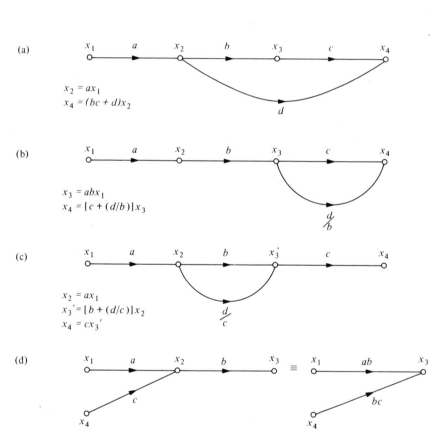

(a)

$$x_2 = ax_1$$
$$x_4 = (bc + d)x_2$$

(b)

$$x_3 = abx_1$$
$$x_4 = [c + (d/b)]x_3$$

(c)

$$x_2 = ax_1$$
$$x_3' = [b + (d/c)]x_2$$
$$x_4 = cx_3'$$

(d)

FIG. 2-28. Reduction rule 4: shifting a transmittance. (a) Original signal-flow graph. (b) Shifting the starting point of a branch. (c) Shifting the termination point of an internal branch. (d) Shifting the termination point of a branch containing a source.

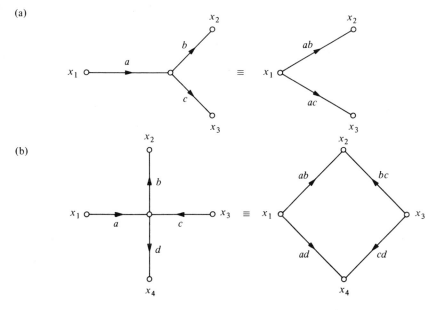

FIG. 2-29. (a) "Y" transformation. (b) Star-to-mesh transformation.

source accordingly. This is demonstrated in Fig. 2-31, where the path from x_1 to x_5 is inverted in four steps and it is shown how the source consequently travels from x_1 to x_5.

Rule 6.   Removal of a self loop

A self loop whose loop transmission is equal to L is removed from a node by dividing the transmission of all other branches entering that node by $(1 - L)$. An example is shown in Fig. 2-32.

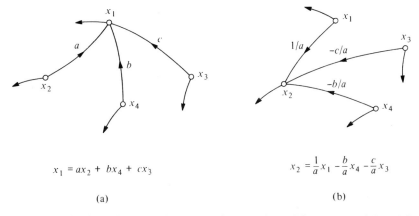

$$x_1 = ax_2 + bx_4 + cx_3$$

(a)

$$x_2 = \frac{1}{a}x_1 - \frac{b}{a}x_4 - \frac{c}{a}x_3$$

(b)

FIG. 2-30. Reduction rule 5: path inversion. (a) Original signal-flow graph. (b) Signal-flow graph with path a inverted.

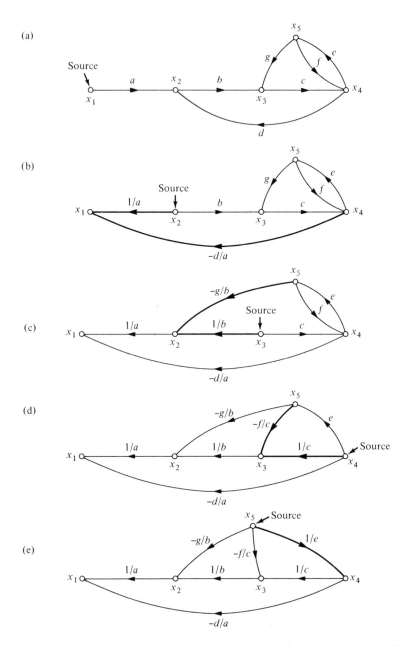

FIG. 2-31. Stepwise inversion of a path. (a) Original graph; (b–e) successive inversions. Note the shifting of the source through each node.

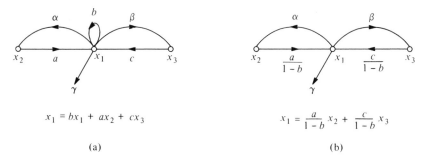

$$x_1 = bx_1 + ax_2 + cx_3$$

(a)

$$x_1 = \frac{a}{1-b}x_2 + \frac{c}{1-b}x_3$$

(b)

FIG. 2-32. Rule 6: Removal of a self loop. (a) Original signal-flow graph with self loop b. (b) Signal-flow graph with self loop b removed.

Rule 7.   Loop reduction

a. Single loop.   The transmission from an independent variable x_i (i.e., source node) to a dependent variable x_j (i.e., internal or sink node) in a signal-flow graph containing only one loop and one path equals

$$T_{ij} = \frac{P_{ij}}{1 - L} \qquad (2\text{-}42)$$

where P_{ij} is the forward path transmission from x_i to x_j, and L is the transmission of the loop. An example is shown in Fig. 2-33.

b. Multiloop with nontouching loops.   With a cascade of nontouching loops (i.e., loops possessing no common nodes) the overall transmission is the product of the individual transmissions as given by (2-42), namely,

$$T_{in} = \frac{P_{ij}}{1 - L_j} \cdot \frac{P_{jk}}{1 - L_k} \cdots \frac{P_{(n-1)n}}{1 - L_n} \qquad (2\text{-}43)$$

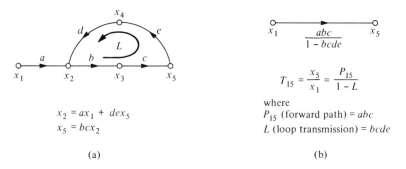

$$x_2 = ax_1 + dex_5$$
$$x_5 = bcx_2$$

(a)

$$T_{15} = \frac{x_5}{x_1} = \frac{P_{15}}{1 - L}$$

where
P_{15} (forward path) = abc
L (loop transmission) = $bcde$

(b)

FIG. 2-33. Rule 7a: Reduction of a single loop. (a) Original signal-flow graph. (b) Flow graph with loop reduced.

(a)

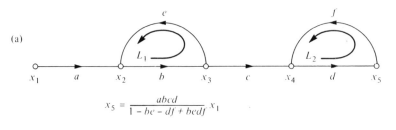

$$x_5 = \frac{abcd}{1 - be - df + bedf} x_1$$

(b)

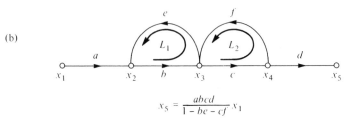

$$x_5 = \frac{abcd}{1 - be - cf} x_1$$

FIG. 2-34. Multiloops (a) with nontouching loops, (b) with two touching loops.

Applying (2-43) to the example shown in Fig. 2-34a we obtain

$$T_{15} = \frac{x_5}{x_1} = \frac{P_{13}}{1 - L_1} \cdot \frac{P_{35}}{1 - L_2} = \frac{P_{15}}{1 - L_1 - L_2 + L_1 L_2}$$

$$= \frac{abcd}{1 - be - df + bedf} \qquad (2\text{-}44)$$

It is important to note here, that the two loops do *not* share a node. If they did we would obtain another result, as shown in the next case.

c. Multiloop with touching loops. For the case that the two loops do share at least one node, as shown in Fig. 2-34b, the overall transmission is given by:

$$T_{15} = \frac{x_5}{x_1} = \frac{abcd}{1 - be - cf} = \frac{P_{15}}{1 - L_1 - L_2} \qquad (2\text{-}45)$$

Comparison of this with (2-44) shows that the cross term $L_1 L_2$ is missing here because the two loops of the signal-flow graph are touching at node x_3. In fact the question whether the loops of a multiloop do or do not have common nodes is very important in the evaluation of the corresponding signal-flow graphs. Fortunately these and all other cases can be summarized in the two general rules discussed next.

Rule 8.   The general multiloop single-path reduction rule
We consider here the multiloop case where there is only one path available from a source node x_i to a dependent node x_j, whereby this path touches every loop in the signal-flow graph, (i.e., has at least one node in common with

each). Stated briefly, the rule for the reduction of a single-path multiloop says that

$$T_{ij} = \frac{P_{ij}}{\Delta} \qquad [2\text{-}46]$$

The quantity Δ is known as the graph or network determinant and is evaluated as follows:

$\Delta = 1 - $ (sum of all loops[6] taken one at a time) $+$ (sum of products of all nontouching loops taken two at a time) $-$ (sum of products of all nontouching loops taken three at a time) $+ \cdots$

For a graph containing n loops, Δ may be expressed mathematically as

$$\Delta = [(1 - L_1)(1 - L_2)(1 - L_3) \cdots (1 - L_n)]' \qquad [2\text{-}47]$$

where the prime indicates that the terms containing products of touching loops shall be equated to zero. The quantity P_{ij} is defined as the path from the independent variable[7] x_i (source) to the dependent variable x_j.

Although the definition of the graph determinant seems rather complicated, a few examples will show how easily it can be evaluated.

EXAMPLES:

As our first example we return to the resistive network of Fig. 2-18. The corresponding signal-flow graph is redrawn in Fig. 2-35a; this can be reduced to the graph in Fig. 2-35b using rule 4.

Clearly the signal-flow graph in Fig. 2-35b could be simplified more, e.g., by eliminating the loops or at least by combining the two into one. It was pointed out earlier, however, and should be evident in this example, that in order to apply the general multiloop reduction formula as given by (2-46) the signal-flow graph need not by any means be reduced to its least redundant or most efficient form. Thus, using (2-46), the representation in Fig. 2.35b permits us to obtain any desired network transmission function virtually by inspection. Herein lies one of the main advantages of signal-flow graph design. By inspection of Fig. 2-35b we can see that the overall voltage transmission is

$$T_{13} = \frac{V_3}{V_1} = \frac{(R_2/R_1)R_4/(R_3 + R_4)}{1 + (R_2/R_1) + R_2/(R_3 + R_4)}$$

$$= \frac{R_2 R_4}{R_1(R_3 + R_4) + R_2(R_3 + R_4) + R_1 R_2} \qquad (2\text{-}48)$$

Likewise, the voltage transmission $T_{12} = V_2/V_1$ is

$$T_{12} = \frac{V_2}{V_1} = \frac{R_2/R_1}{1 + R_2/R_1 + R_2/(R_3 + R_4)} = \frac{R_2(R_3 + R_4)}{(R_1 + R_2)(R_3 + R_4) + R_1 R_2}$$

$$(2\text{-}49)$$

6. We refer here, for brevity, to loops, where loop transmissions are, of course, intended.
7. If x_i is not a source it can either be transformed into one by inversion and other manipulations, or else the transmission from the real source (e.g., x_s) to x_i (T_{si}) and to x_j (T_{sj}) can be obtained separately. Then $T_{ij} = T_{sj}/T_{si}$.

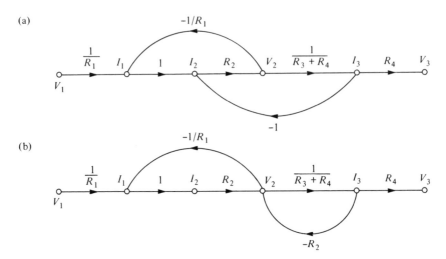

FIG. 2-35. (a) Signal-flow graph of the resistive network in Fig. 2-18. (b) Graph reduction using rule 4.

Notice that the transmission paths $P_{13} = (R_2/R_1)[R_4/(R_3 + R_4)]$ in T_{13} and $P_{12} = R_2/R_1$ in T_{12} fulfill the condition necessary to satisfy (2-46), i.e., that they "touch every loop in the flow-graph." If, however, we wish to obtain the input conductance $G_{11} = I_1/V_1$ this is no longer true, since the transmission path $P_{11} = 1/R_1$ does not touch the loop between V_2 and I_3. Thus (2-46) does not apply and we shall have to wait until we have discussed the still more general multiloop multipath reduction rule (rule 9), before this conductance can be obtained by inspection of the signal-flow graph.

First, however, let us examine some more representative single-path signal-flow graphs and the application of (2-46) to them. Consider, for example, the signal-flow graph in Fig. 2-36. Applying (2-46) to obtain the overall transmission we have

$$T_{17} = \frac{x_7}{x_1} = \frac{P_{17}}{\Delta}$$

$$= \frac{abcdef}{1 - (L_1 + L_2 + L_3 + L_4 + L_5) + (L_1 L_3 + L_1 L_4 + L_1 L_5 + L_2 L_4}$$
$$+ L_2 L_5 + L_3 L_5) - L_1 L_3 L_5$$

$$(2\text{-}50a)$$

$$= \frac{abcdef}{1 - (bg + ch + di + ej + fk) + (bgdi + bgej + bgfk + chej + chfk}$$
$$+ difk) - bgdifk$$

$$(2\text{-}50b)$$

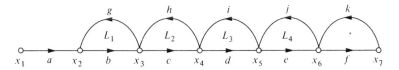

FIG. 2-36. Multiloop single-path signal-flow graph with touching loops.

Since the signal-flow graph in Fig. 2-36 is of the type obtained for ladder networks, we can apply (2-50a) directly to obtain the overall transmission of the RC ladder network in Fig. 2-20a. By inspection of the corresponding signal-flow graph in Fig. 2-20b we have

$$P_{s3} = R_1 R_2 R_3 C_1 C_2 C_3 s^3 \qquad (2\text{-}51a)$$

and

$$L_1 = -R_1 C_1 s; \quad L_2 = -R_1 C_2 s; \quad L_3 = -R_2 C_2 s;$$
$$L_4 = R_2 C_3 s; \quad L_5 = -R_3 C_3 s \qquad (2\text{-}51b)$$

Substituting into (2-50a) we obtain

$$\frac{V_3}{V_s} = \frac{P_{s3}}{\Delta} \qquad (2\text{-}52a)$$

where

$$\Delta = 1 + (R_1 C_1 + R_1 C_2 + R_2 C_2 + R_2 C_3 + R_3 C_3)s$$
$$+ (R_1 R_2 C_1 C_2 + R_1 R_2 C_1 C_3 + R_1 R_3 C_1 C_3 + R_1 R_2 C_2 C_3$$
$$+ R_1 R_3 C_2 C_3 + R_2 R_3 C_2 C_3)s^2 + R_1 R_2 R_3 C_1 C_2 C_3 s^3 \qquad (2\text{-}52b)$$

Even though the expressions for ladder networks rapidly get long and cumbersome, it should be evident from this example that the systematic method of analysis prescribed by the general reduction rule (2-46) makes the calculation basically very simple.

If we consider three nontouching loops, as shown in Fig. 2-37, the general reduction rule (2-46) supplies the overall transmission directly by inspection:

$$T_{16} = \frac{x_6}{x_1} = \frac{P_{16}}{\Delta}$$

$$= \frac{abcde}{1 - (L_1 + L_2 + L_3) + (L_1 L_2 + L_1 L_3 + L_2 L_3) - L_1 L_2 L_3}$$

$$= \frac{abcde}{1 - af - cg - eh + afcg + afeh + cgeh - afcgeh} \qquad (2\text{-}53)$$

instead of by cascading the individual loops as was done in (2-44).

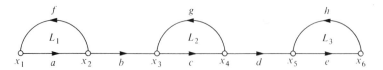

FIG. 2-37. Multiloop with three nontouching loops.

As a final example we consider the multiloop signal-flow graph in Fig. 2-38. Application of (2-46) provides the overall transmission by inspection:

$$T_{16} = \frac{x_6}{x_1} = \frac{P_{16}}{\Delta} = \frac{abcde}{1 - (L_1 + L_2 + L_3) + (L_2 L_3)}$$

$$= \frac{abcde}{1 - bcdg - cf - eh + cfeh} \tag{2-54}$$

Notice that no significance should be given to the fact that a branch of loop L_2 is contained within loop L_1.

Rule 9. The general multiloop multipath reduction rule[8]

If a signal-flow graph contains more than one path from source to sink (or intermediate node), as shown in Fig. 2-39 then the transmission may be evaluated as follows:

$$T_{16} = \frac{x_6}{x_1} = T_1 + T_2 = \frac{P_1}{1 - L_1} + \frac{P_2}{1 - L_2} = \frac{P_1}{\Delta_2} + \frac{P_2}{\Delta_1}$$

$$= \frac{P_1 \Delta_1 + P_2 \Delta_2}{\Delta} = \frac{abc(1 - eh) + def(1 - bg)}{1 - bg - eh + bgeh} \tag{2-55}$$

where

$$\Delta = \Delta_1 \Delta_2$$

$$\Delta_1 = 1 - L_2$$

$$\Delta_2 = 1 - L_1$$

$$T_1 = \frac{P_1}{1 - L_1}$$

$$T_2 = \frac{P_2}{1 - L_2} .$$

Note that each path P_k is modified by Δ_k, i.e., by the determinant of that part of the graph not touching the kth path. This modification of P_k permits us to drop the provision required in rule 8, that the path considered "touches

8. This rule is often called Mason's gain formula, after S. J. Mason, who first introduced it.

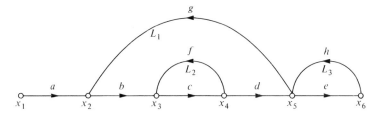

FIG. 2-38. Multiloop single-path signal-flow graph.

every loop in the signal-flow graph." We may therefore generalize the previous results (rule 8) to obtain the following formula for the transmission of an arbitrary signal-flow graph:

$$T_{ij} = \frac{\sum_k P_k \Delta_k}{\Delta}$$ [2-56]

where

P_k is the transmission of the kth path between x_i and x_j, where x_i must be a source, x_j however not necessarily being a sink (i.e. intermediate nodes permitted);

Δ is the graph or network determinant; and

Δ_k is the cofactor of the kth path, i.e., the determinant of the part of the graph that does not touch, i.e., does not have any common nodes with, the kth path.

Equation (2-56) above represents the most general relationship defining the transmission of a signal-flow graph. It can be applied directly to any flow graph without first going through the reduction processes described earlier.

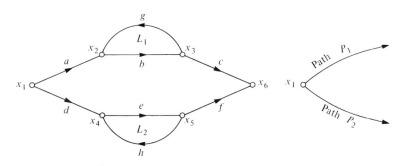

FIG. 2-39. Multipath multiloop signal-flow graph.

Often this is the most expedient course to take. If, however, sensitivity or stability studies are to be undertaken, then it will be more useful to reduce a general flow graph into one of the essential signal-flow graphs described later. First, however, we shall illustrate the application of (2-56) to some typical examples.

EXAMPLES:

Returning once again to the resistive network of Fig. 2-18 and its partially reduced signal-flow graph in Fig. 2-35b, we are now in a position to calculate the input conductance $G_{11} = I_1/V_1$. It will be recalled that we cannot use (2-46) of rule 8 for this purpose, since the corresponding path $P_{11} = 1/R_1$ does not touch both loops of the graph. With (2-56) this is no longer required and we have, by inspection of Fig. 2.35b,

$$G_{11} = \frac{I_1}{V_1} = \frac{P_{11} \cdot \Delta_{11}}{\Delta} \qquad (2\text{-}57)$$

With

$$P_{11} = 1/R_1 \qquad (2\text{-}58a)$$

$$\Delta_{11} = 1 + \frac{R_2}{R_3 + R_4} \qquad (2\text{-}58b)$$

and

$$\Delta = 1 + \frac{R_2}{R_1} + \frac{R_2}{R_3 + R_4} \qquad (2\text{-}58c)$$

(2-57) becomes

$$G_{11} = \frac{I_1}{V_1} = \frac{R_2 + R_3 + R_4}{(R_1 + R_2)(R_3 + R_4) + R_1 R_2} \qquad (2.59)$$

As a second example, consider the signal-flow graph shown in Fig. 2-40. Applying (2-56) to calculate the overall transmission we obtain

$$T_{18} = \frac{x_8}{x_1} = \frac{P_{18} \Delta_{18}}{\Delta} \qquad (2\text{-}60)$$

where

$$\Delta = 1 - \sum_i L_i + \sum_{i,j} L'_i L'_j - \sum_{i,j,k} L''_i L''_j L''_k + \cdots \qquad (2\text{-}61)$$

and

$\sum_i L_i$ is the sum of all loop transmissions;
$L'_i L'_j$ is the product of the loop transmissions of any two loops that do not have a node or a branch in common;
$\sum_{i,j} L'_i L'_j$ is the sum of all such products;
$L''_i L''_j L''_k$ is the product of the loop transmission of any three loops that do not have any branches or nodes in common; and so on.

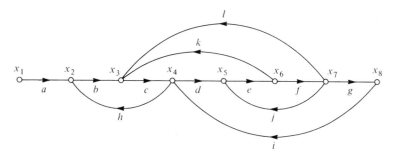

FIG. 2-40. Multiloop signal-flow graph.

For a complex graph such as that shown in Fig. 2-40 it is best to find the sums and products of Δ in a systematic way. Thus to find $\sum_i L_i$ we start out with branch a and list all loops containing a, then go on to branch b and take all loops containing b that have not already been listed, and so on. Thus

$$\sum_i L_i = bch + cdek + cdefl + defgi + efj \tag{2-62}$$

Next we find $\sum_{i,j} L_i' L_j'$, which is the sum of the products of the loop transmissions taken two at a time, omitting all those that hav any node or any branch in common. To evaluate $\sum_{i,j} L_i' L_j'$, the list in $\sum_i L_i$ is examined. The first term in $\sum_i L_i$ is bch. The other terms are scanned to detect those that do not contain b, c, or h. The first term satisfying this condition is $defgi$, but on examining the graph we find that it has a node in common with bch and therefore cannot be included. However, the last term efj has no branches or nodes in common with bch, and therefore $bchefj$ appears in $\sum_{ij} L_i' L_j'$. Next we note the second term in $\sum_i L_i$, i.e., $cdek$, and examine as above. In this way we find that

$$\sum_{i,j} L_i' L_j' = (bch)(efj) \tag{2-63}$$

Since $\sum_{ij} L_i' L_j'$ contains only one term it is impossible for $\sum_{i,j,k} L_i'' L_j'' L_k''$ to exist. Thus

$$\Delta = 1 - \sum_i L_i + \sum_{i,j} L_i' L_j'$$

$$= 1 - (bch + cdek + cdefl + defgi + efj) + bchefj \tag{2-64}$$

Continuing with the evaluation of P_{18} and Δ_{18} we obtain

$$P_{18} = abcdefg \tag{2-65}$$

Since each loop in $\sum_i L_i$ has at least one branch present in P_{18} we have

$$\Delta_{18} = 1 \tag{2-66}$$

and

$$P_{18} \Delta_{18} = abcdefg. \tag{2-67}$$

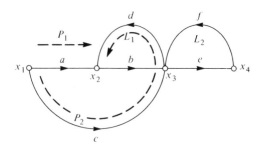

FIG. 2-41. Multipath multiloop.

The fact that $\Delta_{18} = 1$ is clear from inspection of the signal-flow graph, which is actually a multiloop graph with a single path. Naturally (2-56) of rule 9 is general enough to include the analysis of single-path multiloops as well. Substituting the expressions obtained above into (2-60) the overall transmission T_{18} is obtained.

To illustrate the application of (2-56) to genuine multipath multiloops let us now consider the signal-flow graph shown in Fig. 2-41. With (2-56) we have

$$T_{12} = \frac{x_2}{x_1} = \frac{\sum_k P_k \Delta_k}{\Delta} = \frac{P_1 \Delta_1 + P_2 \Delta_2}{\Delta}$$

$$= \frac{P_1(1 - L_2) + P_2}{1 - L_1 - L_2} = \frac{a(1 - ef) + cd}{1 - db - ef} \tag{2-68}$$

Note that Δ_2 equals unity, since $P_2 = cd$ touches every loop in the graph.

Another multipath multiloop is shown in Fig. 2-42. Applying (2-56) the following transmissions are obtained by inspection:

$$T_{12} = \frac{x_2}{x_1} = \frac{P_1 \Delta_1 + P_2 \Delta_2 + P_3 \Delta_3}{\Delta} = \frac{a(1 - cg) + de + dcf}{1 - be - bcf - cg} \tag{2-69a}$$

$$T_{13} = \frac{x_3}{x_1} = \frac{P_4 \Delta_4 + P_5 \Delta_5}{\Delta} = \frac{ab + d}{1 - be - bcf - cg} \tag{2-69b}$$

$$T_{14} = \frac{x_4}{x_1} = \frac{P_6 \Delta_6 + P_7 \Delta_7}{\Delta} = \frac{abc + dc}{1 - be - bcf - cg} \tag{2-69c}$$

Notice that Δ (the determinant of the network) is defined for a given graph, irrespective of the transmission required, and therefore need be evaluated only once.

It should be evident from the examples above that once the general reduction procedure contained in rule 9 has been thoroughly mastered, the transmission of any signal-flow graph can be obtained by inspection with relative

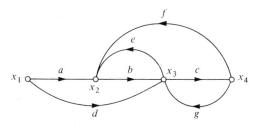

FIG. 2-42. Multipath multiloop.

ease, without actually reducing the graph to a simple form by using any of the other reduction rules. It should be pointed out however, that it is sometimes simpler to use some of the partial reduction rules preceding rule 9 than to use rule 9 itself, as the following example will show. Consider the signal-flow graph shown in Fig. 2-43. Using rules 2 and 7a we can immediately write, by inspection,

$$T_{18} = \frac{P_1}{1 - L_1} + \frac{P_2}{1 - L_2} + \frac{P_3}{1 - L_3}$$

$$= \frac{abc}{1 - bj} + \frac{def}{1 - ek} + \frac{ghi}{1 - lh} \tag{2-70}$$

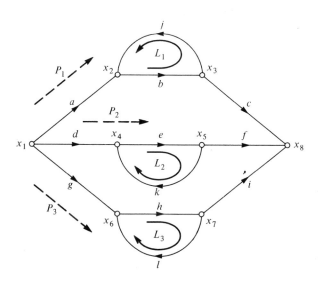

FIG. 2-43. Multipath multiloop.

Multiplying out, we obtain

$$T_{18} = \frac{P_1(1 - L_2)(1 - L_3) + P_2(1 - L_1)(1 - L_3) + P_3(1 - L_1)(1 - L_2)}{(1 - L_1)(1 - L_2)(1 - L_3)} \qquad (2\text{-}71)$$

Expanding (2-71), T_{18} is of course exactly in the form prescribed by (2-56):

$$T_{18} = \frac{P_1[1 - (L_2 + L_3) + L_2 L_3] + P_2[1 - (L_1 + L_3) + L_1 L_3]}{1 - (L_1 + L_2 + L_3) + (L_1 L_2 + L_1 L_3 + L_2 L_3) - L_1 L_2 L_3}$$

$$+ \frac{P_3[1 - (L_1 + L_2) + L_1 L_2]}{1 - (L_1 + L_2 + L_3) + (L_1 L_2 + L_1 L_3 + L_2 L_3) - L_1 L_2 L_3} \qquad (2\text{-}72a)$$

$$= \frac{P_1 \Delta_1 + P_2 \Delta_2 + P_3 \Delta_3}{\Delta} \qquad (2\text{-}72b)$$

Clearly the derivation of T_{18} from (2-70) is more direct and simpler in this case than the derivation from (2-56). Thus it depends very much on the topology of the signal-flow graph which method of transmission derivation should be used. In general, though, the more complex and untractable the signal-flow graph is, the less likely it is that any but the general reduction rule (rule 9) can provide a desired transmission directly, and it is without question the most powerful and important of the rules discussed. As a final demonstration of this, Fig. 2-44 contains a variety of signal-flow graphs, none of which, at first glance, seems trivial to analyze. However, as the reader may readily verify, the transmissions may be obtained by inspection, using the general reduction rule given by (2-56).

Finally, to end this section on a practical note, let us return to the common emitter transistor amplifier of Fig. 2-21a and its corresponding signal-flow graph in Fig. 2-22. Calculating the voltage gain of this amplifier we have:

$$T_{so} = \frac{v_o}{v_s} = \frac{\sum\limits_k P_k \Delta_k}{\Delta}$$

$$= \frac{P_{so}}{\Delta} \qquad (2\text{-}73)$$

since $\Delta_{so} = 1$. By inspection of Fig. 2-22 we have

$$P_{so} = \frac{1}{R_1 + r_b} \left\{ (r_e + R_2) \cdot \left[-\frac{1}{r_c(1 - \alpha)} \right] + \frac{r_m}{r_c(1 - \alpha)} \right\} (-R_3)$$

$$= \frac{R_3}{R_1 + r_b} \left\{ \frac{r_e + R_2 - r_m}{r_c(1 - \alpha)} \right\} \qquad (2\text{-}74)$$

and

$$\Delta = 1 + \frac{r_e + R_2}{R_1 + r_b} + \frac{r_m}{r_c(1 - \alpha)} \frac{(r_e + R_2)}{R_1 + r_b} + \frac{r_e + R_2}{r_c(1 - \alpha)}$$

$$+ \frac{R_3}{r_c(1 - \alpha)} + \left(-\frac{r_e + R_2}{R_1 + r_b} \right) \left[-\frac{R_3}{r_c(1 - \alpha)} \right] \qquad (2\text{-}75)$$

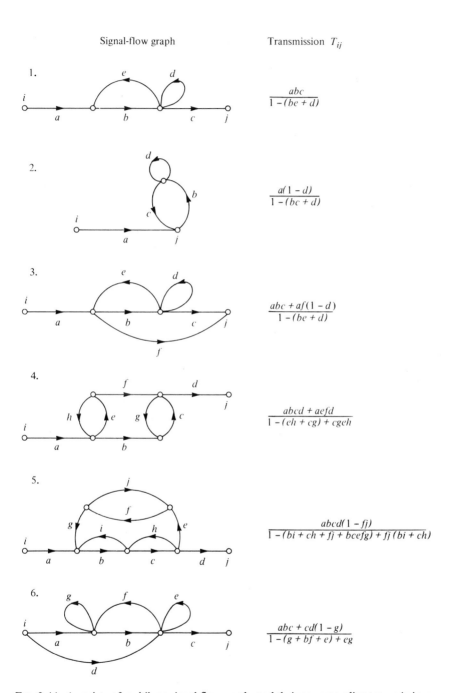

FIG. 2-44. A variety of multiloop signal-flow graphs and their corresponding transmissions.

Expanding (2-75) we obtain

$$\Delta = [r_c(1 - \alpha)(R_1 + r_b + r_e + R_2) + (R_1 + r_b)(r_e + R_2 + R_3)$$

$$+ (r_e + R_2)(r_m + R_3)] \frac{1}{r_c(1 - \alpha)(R_1 + r_b)} \tag{2-76}$$

Assuming that $\alpha \approx 1$ we obtain, with (2-74) and (2-76),

$$\frac{v_o}{v_s} = \frac{R_3(r_e + R_2 - r_m)}{(r_e + R_2)(R_1 + r_b + R_3 + r_m) + R_3(R_1 + r_b)} \tag{2-77}$$

In general $r_m = \alpha r_c$ is much larger than either the emitter resistance $(R_2 + r_e)$, the base resistance $(R_1 + r_b)$, the load resistance R_3, or the combinations of the three appearing in (2-77). Thus the voltage gain can be approximated as

$$\frac{v_o}{v_s} \approx \frac{-r_m R_3}{r_m(r_e + R_2) + R_3(R_1 + r_b)} \approx \frac{R_3}{r_e + R_2} \tag{2-78}$$

2.2.3 Signal–Flow Graph Representation of Two–Port Parameters

A useful application of signal-flow graphs is in the representation of two-port parameters. This will be shown by two examples. First we shall find the input impedance and voltage transfer function of the two-port shown in Fig. 2-45. It is characterized by its h parameters as follows:

$$V_1 = h_{11}I_1 + h_{12}V_2 \tag{2-79a}$$

$$I_2 = h_{21}I_1 + h_{22}V_2 \tag{2-79b}$$

with the constraint that

$$V_2 = -I_2 Z_L \tag{2-79c}$$

Note that the constraint $I_2 = -(1/Z_L)V_2$ cannot be used, since I_2 is already defined by (2-79b) and each node can be defined by only one equation. In general a constraint can only define a source variable (i.e., independent variable).

The signal-flow graph corresponding to Fig. 2-45 is given in Fig. 2-46. Notice that although V_2 started out as an independent variable (i.e., source)

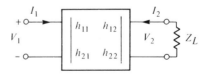

FIG. 2-45. Two-port characterized by its h parameters.

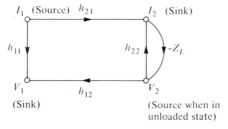

FIG. 2-46. Signal-flow graph corresponding to the two-port in Fig. 2-45.

for the case of the unloaded two-port, the branch $-Z_L$ entering it (which is a result of the loading constraint given by (2-79c)) transforms it into an intermediate node.

To obtain the input impedance $V_1 = Z_{in}I_1$ the nodes V_2 and I_2 must be eliminated. Removing I_2 we obtain the reduced graph shown in Fig. 2-47a. The self-loop at V_2 is removed next, resulting in 2-47b. Finally, removing node V_2 we obtain the input impedance by inspection:

$$Z_{in} = \frac{h_{11} + Z_L \Delta h}{1 + Z_L h_{22}}$$ (2-80)

where

$$\Delta h = h_{11}h_{22} - h_{12}h_{21}$$

It may be of interest to the reader to derive (2-80) directly by inspection of Fig. 2-46, using the general reduction formula (2-56). The remarkable analytic simplicity afforded by this formula is thereby well demonstrated.

We recall now that rule 9 (Mason's formula) is applicable only to the computation of the graph transmission from a source node to any other node. Thus we cannot directly calculate the voltage transfer function V_2/V_1 for the graph of Fig. 2-46, since V_1 is a sink, not a source. One way of overcoming this difficulty is first to calculate V_1/V_2 and then to take the reciprocal of the result. This is relatively simple here, since V_2 becomes a source after the

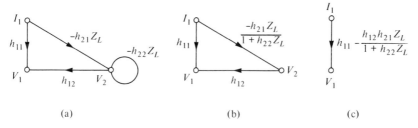

(a)   (b)   (c)

FIG. 2-47. Reduction of the signal-flow graph in Fig. 2-46. (a) I_2 removed. (b) Self-loop at V_2 removed. (c) V_2 removed.

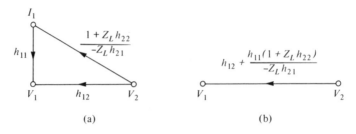

(a)   (b)

FIG. 2-48. (a) Signal-flow graph of Fig. 2-47b with the branch between I_1 and V_2 inverted. (b) Resulting inverse voltage transfer ratio.

inversion of only one branch, namely $-Z_L$ (Fig. 2-48a). Adding the two resulting paths (see Fig. 2-48b) we obtain

$$\frac{V_1}{V_2} = \frac{h_{11} + \Delta h Z_L}{-Z_L h_{21}}$$ (2-81a)

Taking the reciprocal of this value, we have

$$\frac{V_2}{V_1} = -\frac{Z_L h_{21}}{h_{11} + \Delta h Z_L}$$ (2-81b)

Another common method of overcoming the problem that arises when neither of the nodes between which the transmission is to be calculated is a source, is first to compute the graph transmissions from the source node to each of the two nodes in question and then to take the ratio of the two. For our example this means calculating V_2/I_1 and V_1/I_1 (since I_1 is a source) and then taking the ratio of the two. Thus, we have by inspection

$$\frac{V_2}{I_1} = -\frac{Z_L h_{21}}{1 + h_{22} Z_L}$$ (2-82)

and $V_1/I_1 = Z_{in}$, as already calculated and given by (2-80). Dividing (2-82) by (2-80) we obtain the voltage transfer function (2-81b). Notice that the second method was particularly easy here, since we already had calculated Z_{in} and required no additional inversion.

As a second example we consider the cascade of two two-ports N and N' shown in Fig. 2-49a; the latter are characterized by the h and y parameters, respectively. In order to cascade N and N' the following constraints must be satisfied:

$$I_2 = -I_2'$$ (2-83a)

$$V_2 = V_2'$$ (2-83b)

$$V_3 = -I_3 Z_L$$ (2-83c)

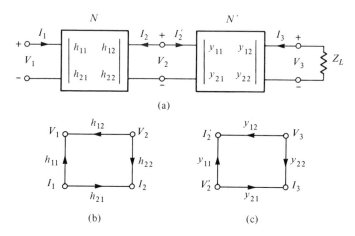

(a)

(b)   (c)

FIG. 2-49. Cascading two-ports. (a) Cascade of networks N and N'. (b) Signal-flow graph of N. (c) Signal-flow graph of N'.

The first two constraints cannot be met by the graphs in their present form, since I_2 and I_2' are both sink nodes, and V_2 and V_2' are both source nodes. In order to change this we can invert path h_{22} or y_{11}. Inverting y_{11} we obtain the signal-flow graph of Fig. 2-50a, which then simplifies as in Fig. 2-50b. From here we can proceed with rule 9 to calculate any desired network functions directly. By inspection of Fig. 2-50b we obtain the network determinant

$$\Delta = 1 - (L_1 + L_2 + L_3) + L_1 L_3$$

$$= 1 + \frac{1}{y_{11}} [h_{22} + (\Delta y + h_{22} y_{22}) Z_L] \tag{2-84}$$

where

$$\Delta y = y_{11} y_{22} - y_{12} y_{21}$$

To calculate the input impedance, for example, we obtain

$$Z_{in} = \frac{V_1}{I_1} = \frac{P_1 \Delta_1 + P_2 \Delta_2}{\Delta}$$

$$= \frac{h_{11} \Delta - \dfrac{h_{12} h_{21}}{y_{11}} (1 - L_3)}{\Delta}$$

$$= h_{11} - \frac{h_{12} h_{21} (1 + y_{22} Z_L)}{y_{11} + h_{22} + (\Delta y + h_{22} y_{22}) Z_L} \tag{2-85}$$

(a)

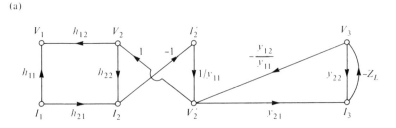

(b)

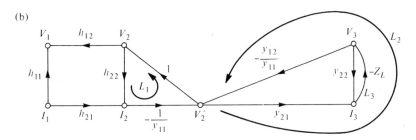

FIG. 2-50. Two steps in the signal-flow graph derivation of the two-port cascade shown in Fig. 2-49. (a) Inversion of y_{11}. (b) Elimination of node I_2'.

To calculate the voltage transfer function V_3/V_1 we can invert either h_{11} or h_{12} in order to obtain a forward path from V_1 to V_3 with V_1 as a source. An alternative is to calculate the inverse transfer function V_1/V_3 as we did in (2-81). However, this requires the inversion of h_{21} in Fig. 2-50, so that no advantage results. (Another alternative is to calculate V_3/I_1 divided by $V_1/I_1 = Z_{in}$, which is valid, since I_1 is a source.) Inverting h_{11} (see Fig. 2-51) we have

$$\frac{V_3}{V_1} = \frac{P_{31}}{1 - (L_1 + L_2 + L_3 + L_4) + (L_1 L_3 + L_3 L_4)}$$

$$= \frac{\dfrac{1}{h_{11}} \cdot \dfrac{h_{21}}{y_{11}} \cdot y_{21} Z_L}{\Delta + L_4 (L_3 - 1)}$$

$$= \frac{h_{21} y_{21} Z_L}{h_{11} y_{11} + \Delta h(1 + y_{22} Z_L) + h_{11} \Delta y Z_L} \tag{2-86}$$

Notice that the inversion of h_{11} resulted in an additional loop L_4 in Fig. 2-51 and, with that, in a modified network determinant $\Delta' = \Delta + L_4(L_3 - 1)$.

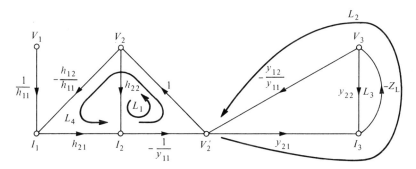

FIG. 2-51. Inversion of h_{11} in the signal-flow graph of Fig. 2-50.

2.2.4 The Essential Signal-Flow Graph

It will be apparent from the preceding sections that the complexity of a signal-flow graph is the result of the complexity of the particular set of equations used to describe a given system rather than an indication of the system complexity itself. We need only consider the two flow graphs shown in Fig. 2-52 to see that the number of feedback loops is no indication of the complexity of a system. The two graphs represent the same system and an equivalent set of equations, yet Fig. 2-52a exhibits two feedback loops, Fig. 2-52b only one. The *order* (or essential complexity) of a system depends on the number of *independent* feedback loops, where the term "independent" means that the loops cannot be combined by simple addition.

The concept of order may be further clarified by defining *essential nodes* of a signal-flow graph, i.e., those nodes which must be removed in order to open all feedback loops. The *removal* of a node must not be confused with the elimination of a node according to rule 7. It must be interpreted here as an equating of the value of the node variable to zero or, equivalently, as a deletion of all branches leaving the node. Then we can state that *the order of a signal-flow graph corresponds to the number of its essential nodes*, i.e., to the number of nodes which must be removed in order to open all feedback loops.

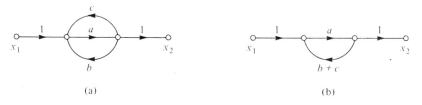

FIG. 2-52. Two signal-flow graphs of different complexity representing the equation $x_2/x_1 = a/[1 - a(b + c)]$. (a) Two feedback loops. (b) Single feedback loop.

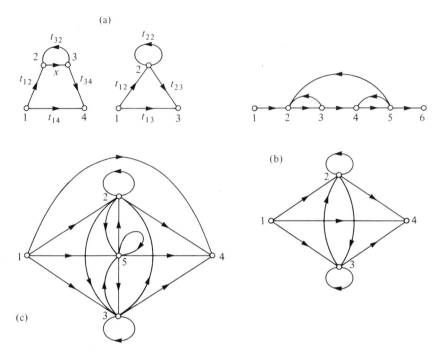

FIG. 2-53. Essential signal-flow graphs of varying order. (a) Order one. (b) Order two. (c) Order three.

Examples of signal-flow graphs of orders one, two, and three are shown in Fig. 2-53.

In a complex signal-flow graph with order greater than 2, evaluation of the order is often a difficult task, requiring a trial-and-error analysis to determine which nodes form the essential set. However, in general it is not too important to be able to ascertain the exact order of a system. If the order determined is inaccurately high, the analysis is not basically affected, although it may be more complicated than it would have been if it had been recognized that the graph possesses a lower order. Fortunately, in the great majority of active networks and feedback systems encountered in practice, the order is less than or equal to 2.

A fundamental characteristic of an essential node is that its elimination (in contrast to its removal) involves division of all branches entering it by the one-minus-loop-gain term. All other nodes can be eliminated by a process of addition and multiplication of transmittances of the original graph.[9] As a

9. In terms of network equations, the reduction to an essential signal-flow graph corresponds to the elimination of variables in the original equations by direct substitution. No division of coefficients is necessary.

consequence, if the transmittances in the original graph are stable (i.e., without poles in the RHP), then all essential transmittances are also stable.[10] For the purposes of stability and sensitivity analysis, the simpler essential flow graph should be used in preference to the more complex multiloop signal-flow graph drawn directly from a given network.

The advantages associated with signal-flow graphs are basically the advantages resulting from a pictorial representation of the network equations. As has been pointed out before, nothing can be obtained with signal-flow graphs that cannot be obtained with the corresponding network equations. Arguments for the use of signal-flow graphs must therefore be based on the greater ease with which analysis can be carried out. Although such an argument can be convincingly presented in terms of the ideas presented in the preceding sections, the advantages of signal-flow graphs are most pronounced when some of the more specialized aspects of feedback theory are presented. To demonstrate this we shall first have a closer look at the most commonly occurring signal-flow graph in feedback networks, namely, that of order one.

2.2.5 The Essential Signal-Flow Graph of Order One

In using signal-flow graphs for the analysis of active networks we should remember that active networks generally occur as individual, or cascades of, second-order networks. As a result, the active networks themselves do not become unduly complex and it is rare to find an equivalent essential flow graph of higher order than one. Furthermore, as has already been pointed out, the signal-flow graph is particularly useful when used in sensitivity and stability studies, and these are usually conducted on graphs of order one—at least when active networks are involved.[11] Thus we see that the essential signal-flow graph of order one has special significance when used in the context of active networks.

In what follows, we shall first show that the four basic transmittances of an essential signal-flow graph of order one can be derived from a multiloop graph in one step rather than in the many steps required by the reduction rules described earlier. Consider the essential signal-flow graph shown in Fig. 2-54. Besides the input and output nodes an essential signal-flow graph contains essential nodes equal in number to the order of the graph. In this case the essential signal-flow graph has the single essential node E_x besides the input and output nodes E_i and E_o. Furthermore, in an essential diagram of order one, there are only four possible transmittances t_{io}, t_{ix}, t_{xo} and t_{xx}. The determination of an essential diagram therefore reduces to an evaluation of these

10. This is no longer true if, say, a second-order signal-flow graph is reduced to one of first order; see, e.g., the reduction in Fig. 2–57, which is discussed below.
11. Even if there is more than one active device in the network, sensitivity studies usually consider one device at a time, or lump the effect of numerous devices into the effect of an equivalent single device.

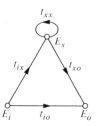

FIG. 2-54. Essential signal-flow graph of order one.

four transmittances. That these can be obtained directly by inspection of the original signal-flow graph will now be illustrated by an example.

Consider the flow graph shown in Fig. 2-55. Notice that we can immediately identify the node x_7 as what must be the essential node E_x. Our assignment is now to find the essential transmittances of Fig. 2-54 in terms of the branch transmittances of Fig. 2-55.

The essential transmission t_{io} measures the total transmission from input to output without passing through node E_x. By inspection of Fig. 2-55 there is only one such path, namely the horizontal one from input to output; any downward digression immediately leads to E_x. Hence

$$t_{io} = abcde \qquad (2\text{-}87)$$

The total transmission t_{ix} from the input x_1 to E_x, without departing from E_x, involves two parallel paths. By inspection of Fig. 2-55 and comparison with Fig. 2-54 we obtain

$$t_{ix} = ab(h + cdk) \qquad (2\text{-}88)$$

The transmittance t_{xo} from the essential node E_x to the output without entering the node E_x is the sum of the four separate transmittances representing the four possible paths of departure from E_x:

$$t_{xo} = fbcde + gcde + ide + je \qquad (2\text{-}89)$$

Finally t_{xx} can be calculated by considering each of the four paths leaving E_x:

$$t_{xx} = fb(h + cdk) + g(h + cdk) + idk + jk \qquad (2\text{-}90)$$

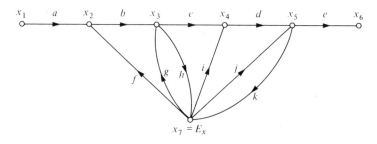

FIG. 2-55. Signal-flow graph of order one.

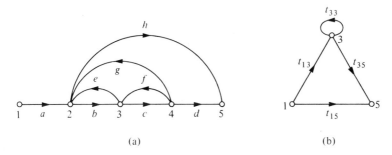

(a)   (b)

FIG. 2-56. Signal-flow graph of order one. (a) Original graph. (b) Equivalent essential graph.

Notice that a number of terms in each of the last two equations can be combined. As a general procedure, however, it is convenient first to consider the paths leaving each node individually.

Two additional examples should help to illustrate the reduction of a general signal-flow graph to its essential form. Consider first the graph in Fig. 2-56a. Despite the presence of three feedback loops in this graph, its order is only unity, since the removal of node 3 opens all the feedback loops. The essential node of the essential signal-flow graph must therefore be node 3, and it will have the form of the essential graph in Fig. 2-56b. By inspection of Fig. 2-56a the four basic transmittances are:

t_{15} (leakage path, i.e., all paths from input to output not entering 3) $= ah$
t_{13} (all paths from input which enter node 3) $= ab$
t_{35} (all paths to output, leaving node 3) $= cd + eh + cgh$
t_{33} (all self loops of node 3) $= eb + cf + cgb$

As a second example, consider the signal-flow graph in Fig. 2-57a. Clearly it is a second-order graph, since x_k and x_j are both essential nodes. We can

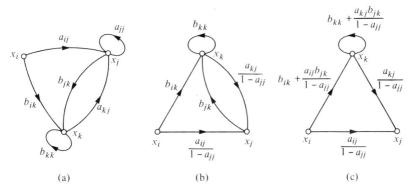

(a)   (b)   (c)

FIG. 2-57. Reduction of second-order to first-order essential flow graph. (a) Original graph. (b) Reduction of a_{jj}. (c) Resulting first-order graph.

reduce it to a first-order flow graph, but this involves division by a $(1 - L)$-term. Nevertheless the reduction to a lower-order essential signal-flow graph is sometimes desirable, particularly when feedback aspects of the system are to be examined. Let us then, as an illustrative example, find the first-order essential signal-flow graph of Fig. 2-57a with respect to the essential node x_k. Eliminating the self loop a_{jj} by dividing all branches entering node x_j by $1 - a_{jj}$ we obtain the graph in Fig. 2-57b. The resulting first-order essential flow graph (Fig. 2-57c) is obtained by inspection.

Once the transmittances of any essential signal-flow graph of order one have been obtained, the transfer function follows directly from inspection of Fig. 2-54:

$$T(s) = \frac{E_o}{E_i} = t_{io} + \frac{t_{ix} t_{xo}}{1 - t_{xx}} \qquad [2\text{-}91]$$

Multiplying out we obtain

$$T(s) = \frac{N(s)}{D(s)} = \frac{(t_{io} + t_{ix} t_{xo}) - t_{io} t_{xx}}{1 - t_{xx}} \qquad [2\text{-}92]$$

Equations (2-91) or (2-92) give the transfer function of a typical active network with feedback. In general, the network will contain one or more active devices with which the desired characteristics (in particular complex poles, for which, it will be remembered, the active devices are combined with passive RC components in the first place) are obtained. It is of particular interest—and will be discussed in more detail in a later chapter—what effect parameter variations of any particular device will have on the network characteristics. A useful method of establishing the effect of device gain on pole location was found to be the root locus. However, this still leaves unanswered the question of the effects of the device parameters on other characteristics of the network, such as input and output impedance, stability, etc., and gives no indication of the effects of feedback on the characteristics of a general network. To provide a deeper insight into these questions we shall introduce the concepts of return difference and null return difference.

2.2.6 Return Difference and Null Return Difference

In an active network, it is not the active device as such but the part it plays in the feedback loop of the network that is decisive in how effectively a desired change in network characteristics can be achieved. An active buffer stage between passive RC networks will, after all, not affect the basic characteristics of the passive networks at all. Thus it is important to have some measure of the effect that feedback has on the basic characteristics of a network.

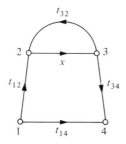

FIG. 2-58. First-order signal-flow graph with parameter x appearing in only one branch.

The effect of feedback can be measured by a quantity called the return difference. The return difference referred to a particular parameter x is most readily defined in terms of a signal-flow graph. The graph is drawn in such a way that x appears in only one branch and as the transmittance of that branch. One suitable form of the graph is shown in Fig. 2-58. Note that its form corresponds to one step in reduction prior to the essential diagram of order one (shown in Fig. 2-54). The essential diagram itself is not suitable, since no network parameter is separately available in it; that is, the parameter x is implicit in the branch t_{xx} of Fig. 2-54, but its effect on characteristics of the feedback loop cannot be explicitly identified.

If we consider feedback with respect to one particular network element x it should be clear from the preceding discussions on signal-flow graph manipulations, that we can always obtain a graph of the form shown in Fig. 2-58, in which x appears as the transmittance of only one branch. The element x will in general characterize an active device in terms of its gain, transconductance or the like. It can however also refer to the characteristics of a passive component (e.g., resistance, conductance, etc.). The branch transmittances other than x are generally functions of s. The transmittance from the input to the origin of the x branch is given by $t_{12}(s)$ and from the end terminal of the x branch to the output terminal by $t_{34}(s)$. The transmittance of the feedback branch around the x branch is given by $t_{32}(s)$. In general there will also be a leakage transmittance directly from the input to the output terminal. It is given by $t_{14}(s)$ and is equal to the overall system transmittance when $x = 0$. For an arbitrary value of x the overall transmission function is given by

$$T(s, x) = t_{14}(s) + x \frac{t_{12}(s)t_{34}(s)}{1 - xt_{32}(s)} \qquad \text{[2-93]}$$

The return difference referred to the network element x is then defined by

$$F_x(s) = 1 - xt_{32}(s) \qquad \text{[2-94]}$$

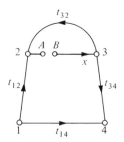

FIG. 2-59. First-order signal-flow graph illustrating the meaning of return difference.

Physically, the return difference can be directly measured. In terms of the corresponding flow graph, which is shown again in Fig. 2-59, the measurement entails the following steps:

1. Open the system at the origin of the branch with transmittance x (see terminals A, B in Fig. 2-59).
2. Set the system input (terminal 1) equal to zero.
3. Apply a unit signal at terminal B and determine the signal returning at terminal A.
4. The difference between the transmitted signal (terminal B) and the returned signal (terminal A) is th return difference given by (2-94).

Very often it is not necessary to manipulate the signal-flow graph into the form of Fig. 2-58, since the return difference can be obtained by inspection of the original signal-flow graph derived directly from the given network.

The return difference with respect to a given element is a quantitative measure of the feedback around the element. In the single-loop system shown in Fig. 2-60, for example, the return difference $F_\mu(s)$ with respect to μ is $1 - \mu t(s)$. From (2-93) and (2-94) it follows that the zeros of $F_x(s)$ are the poles of $T(s, x)$ (assuming there is no cancellation of terms in (2-93)). Thus, as a quantitative measure of feedback, the return difference not only indicates the effect of feedback in controlling network impedances and sensitivity, but also serves as the basis for stability analysis. In the single-loop case of Fig. 2-60, for example, the location of the zeros of $F_\mu(s)$ determines the stability of the system.

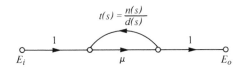

FIG. 2-60. Single-loop feedback system.

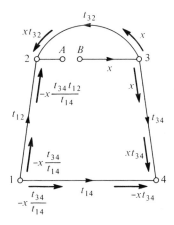

FIG. 2-61. First-order signal-flow graph illustrating the meaning of null return difference.

Another term that is useful in characterizing feedback networks, especially with respect to stability and sensitivity studies, is the null return difference $F_x^0(s)$. It is defined as the return difference with respect to an element x, evaluated under the condition that the input is adjusted to give zero output. In terms of the signal-flow graph of Fig. 2-61, for example, the null return difference is determined as follows:

1. The graph is broken at terminals A and B and a unit signal is transmitted from B.
2. The signal reaching terminal 3 is x. This is transmitted back to 2 and also on to the output node 4.
3. At node 4, the signal xt_{34} arrives along the branch from 3. If the input is adjusted to make the output zero, the signal $-xt_{34}$ must arrive at 4 along the branch from 1. Hence the signal at 1 is $(-x\, t_{34}/t_{14})$.
4. The input from 1 is also transmitted along the branch from 1 to 2, and the signal returning to A is the sum of the two signals arriving at 2.
5. The null return difference with reference to x is therefore given by the expression

$$F_x^0(s) = 1 - \left(xt_{32} - x\frac{t_{34}\,t_{12}}{t_{14}} \right) \tag{2-95}$$

Substituting (2-94) into (2-95) we have

$$F_x^0(s) = F_x(s) + x\frac{t_{34}\,t_{12}}{t_{14}} \tag{2-96}$$

With (2-93) we can also express $t_{34}\,t_{12}$ in terms of T, t_{14}, and F_x as follows:

$$F_x^0(s) = F_x + \frac{T - t_{14}}{t_{14}}\, F_x \tag{2-97}$$

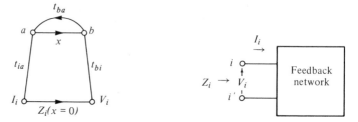

FIG. 2-62. Signal-flow graph characterizing   FIG. 2-63. Network corresponding to the
the driving point impedance of a network.   signal-flow graph in Fig. 2-62.

Dividing (2-97) by F_x we obtain an expression relating the overall gain, the leakage transmittance and the two return differences:

$$\frac{T}{t_{14}} = \frac{F_x^0}{F_x}$$ [2-98]

It is clear from (2-96) that in the case of a network in which the leakage transmission t_{14} equals zero, the inverse null return difference $1/F_x^0$ also equals zero. The significance of this fact will become clearer in the discussion on network sensitivity.

The null return difference can be evaluated directly from the original signal-flow diagram in much the same way as the return difference. The significant difference between the two is that the return difference is measured with zero input; the null return difference is measured with the input adjusted to make the output zero. Both return differences are functions of s and both depend on the particular variables chosen as the input and output signals. For example, we have tacitly assumed so far that the return differences in (2-98) and in preceding expressions correspond to transmittances T and t_{ij}, which are the ratios of output to input and internal node voltages, respectively. This need by no means be the case, however, and other variables, such as currents or a combination of currents and voltages, may be selected as the input and output signals of the flow graph. It must be remembered, however, that for every choice of input and output variable there will be a corresponding pair of return differences.

Consider for instance the flow graph shown in Fig. 2-62. Here the input variable is the current I_i of the network shown in Fig. 2-63. The output variable is the voltage V_i. Z_i is the impedance seen looking into the terminals i–i', with each source of the network replaced by its internal impedance. $Z_i(x = 0)$ is the same impedance when a specified element x of the network is made equal to zero. With I_i and V_i as the input and output variables, respectively, of the flow graph, (2-98) becomes

$$\frac{Z_i}{Z_i(x = 0)} = \frac{F_x^0}{F_x}$$ [2-99]

In this particular case the return differences corresponding to the input and output variables take on a special significance. Recall that the null return difference F_x^0 is defined as the return difference with zero at the output node. Here this means that V_i must equal zero or, in other words, that terminals i–i' in Fig. 2-63 are shorted. Similarly the return difference F_x is defined with zero at the input node, which in this case means that I_i equals zero, or the terminals i–i' are open. Hence for this case we can rewrite (2-99) as

$$\frac{Z_i}{Z_i(x=0)} = \frac{F_x \text{ (with terminals shorted)}}{F_x \text{ (with terminals open)}} \qquad [2\text{-}100]$$

This expression for the impedance at any two terminals of an active network with feedback is frequently referred to as the Blackman impedance relation. Although it is thereby singled out as a separate relation it is basically no more than a restatement of (2-98) in terms of the signal-flow graph characterizing the driving point impedance of a general network. That it merits this special treatment is evident from the fact that it is very useful for the calculation of impedances in active networks. This is because impedances, like so many other basic parameters of an active network, are fundamentally affected by the measure of feedback applied to the network or, in other words, by the return difference quantities obtained with respect to the various active devices present in the network.

An illustrative example will demonstrate this last point. We consider the single-loop feedback amplifier shown in Fig. 2-64 and are interested in the impedance seen from terminal A to ground, namely Z_{AG}, with the gain μ of the forward amplifier used as the reference element for the evaluation of the return differences.

In operation, terminals A and B are tied together. From inspection of Fig. 2-64 we therefore obtain

$$Z_{AG}(\mu = 0) = \frac{Z_1 Z_2}{Z_1 + Z_2} = Z_p \qquad (2\text{-}101)$$

i.e., Z_p is the parallel combination of Z_1 and Z_2. Furthermore

$$F_\mu(A\text{--}G \text{ shorted}) = 1 \qquad (2\text{-}102a)$$

since the feedback is removed by the short circuit. Also

$$F_\mu(A\text{--}G \text{ open}) = 1 + \mu\beta \qquad (2\text{-}102b)$$

where $\mu\beta$ is the loop gain of the network. Therefore with (2-100) the impedance looking into A and G is given by

$$Z_{AG} = \frac{Z_p}{1 + \mu\beta} \qquad (2\text{-}103)$$

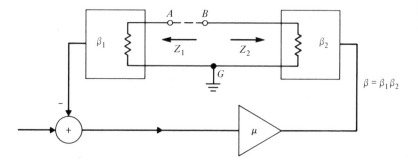

FIG. 2-64. Single-loop feedback amplifier.

Equation (2-103) states that the impedance measured across the feedback path of a single-loop feedback network is the impedance in the loop, measured with no active device in the loop, divided by the return difference.

Let us now break the feedback path at terminals A and B (see Fig. 2-64) and measure the impedance between these two terminals. By inspection we obtain

$$Z_{AB}(\mu = 0) = Z_1 + Z_2 = Z_s \qquad (2\text{-}104)$$

(where Z_s denotes Z_1 and Z_2 in series), and

$$F_\mu(A\text{--}B \text{ shorted}) = 1 + \mu\beta \qquad (2\text{-}105a)$$

Also

$$F_\mu(A\text{--}B \text{ open}) = 1 \qquad (2\text{-}105b)$$

since, when A and B are open, the feedback is removed. Hence, with (2-100) we have

$$Z_{AB} = Z_s(1 + \mu\beta) \qquad (2\text{-}106)$$

This states that the impedance in series with a feedback path of a single-loop network is the impedance in the loop, measured with no active device in the loop, multiplied by the return difference. Incidentally, notice that (2-103) and (2-106) point out one method of obtaining a negative impedance, namely, in the feedback path of a single-loop positive feedback network, in which case $\mu\beta$ is negative.

In his classical work on feedback analysis and design,[12] H. W. Bode derived and presented the above as well as many other basic theorems on feedback systems. One additional theorem that is of interest to active network design

12. H. W. Bode, *Network Analysis and Feedback Amplifier Design*, (New York: D. Van Nostrand Co. 1945).

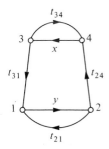

FIG. 2-65. Signal-flow graph for the derivation of (2-107).

and relates to the utilization of the return difference concept is most readily stated in terms of the following equation:

$$\frac{F_x}{F_y} = \frac{(F_x)_{y=0}}{(F_y)_{x=0}}$$ [2-107]

This expression is useful in the evaluation of the return difference with respect to one element when the return difference with respect to another is already known. Consider, for example, the signal-flow graph shown in Fig. 2-65. By inspection we have

$$F_x = 1 - x\left(t_{34} + y\,\frac{t_{31}t_{24}}{1 - yt_{21}}\right)$$ (2-108a)

$$F_y = 1 - y\left(t_{21} + x\,\frac{t_{24}t_{31}}{1 - xt_{34}}\right)$$ (2-108b)

From these two expression we obtain

$$(F_x)_{y=0} = 1 - xt_{34}$$ (2-109a)

and

$$(F_y)_{x=0} = 1 - yt_{21}$$ (2-109b)

Substitution of (2-108) and (2-109) into (2-107) demonstrates the applicability of the theorem to our graph.

2.2.7 The Bilinear Form of a Network Function

The transfer function of a generalized single-loop feedback system was derived in the discussion on root locus theory. Assuming a positive feedback

loop with forward gain μ and feedback transmission $t(s)$, one useful form
of this function was given by (2-8) namely,

$$T(s) = \frac{N(s)}{D(s)} = \frac{\mu}{1 - \mu t(s)} \tag{2-110}$$

From this expression the characteristic equation (2-9b) was shown to follow
directly, i.e.,

$$1 - \mu \frac{n(s)}{d(s)} = 0 \tag{2-111}$$

This equation is used to derive the root locus of a given system. If we compare
(2-110) with the transmission of a generalized flow graph (Fig. 2-58) given
by (2-93):

$$T(s, x) = t_{14}(s) + x \frac{t_{12}(s)t_{34}(s)}{1 - xt_{32}(s)} \tag{2-112}$$

the two functions are seen to possess a certain similarity. Equation (2-112)
was shown to express the transfer function of a general network in a fundamental form, namely, in terms of the essential transmittances of the network.
These were found to be basic to the network, since, among other things, they
permitted a direct derivation of the return differences of the network. Thus we
find that the form of transfer function represented by (2.110) and (2.112) is
very much more basic and useful to the designer than the equivalent conventional ratio of two polynomials with real coefficients

$$T(s) = \frac{\sum_{i=1}^{m} b_i s^i}{\sum_{j=1}^{n} a_j s^j} \tag{2-113}$$

with which we began our discussion of active networks in Chapter 1 (see
(1-2) and (1-3)). This basic form can be generalized if we expand (2-112) as
follows:

$$T(s, x) = \frac{t_{14} - x(t_{14}t_{32} - t_{12}t_{34})}{1 - xt_{32}} \tag{2-114}$$

The transmittances t_{ij} will in turn be rational functions in s, i.e., ratios of two
polynomials with real coefficients:

$$t_{ij}(s) = \frac{n_{ij}(s)}{d_{ij}(s)} \tag{2-115}$$

Since the functions t_{ij} represent the transmittances of the basic flow graph of
Fig. 2-58, which, it will be recalled, is drawn in such a way that x appears by

itself and in one branch only, it follows that the transmittances t_{ij} are independent of x. Thus, substituting (2-115) into (2-114) we obtain a general transfer function of the form

$$T(s) = \frac{N(s)}{D(s)} = \frac{A(s) + xB(s)}{U(s) + xV(s)} \qquad [2\text{-}116]$$

where x does not appear in any of the polynomials $A(s)$, $B(s)$, $U(s)$, and $V(s)$. Equation (2-116) is called the bilinear form of a network function. It was first shown by Bode that any network function can be expressed in this form.[13] The bilinear form of a network function is particularly convenient when the effects of a network parameter x on the numerator or denominator polynomial roots (zeros and poles, respectively) of the network function are to be found. The reason for this is easy to see. Consider a polynomial $P(s, x)$ with real coefficients p_i and assume that it can be written in bilinear form with respect to a parameter x:

$$P(s, x) = \sum_{i=1}^{n} p_i(x)s^i = P_1(s) + xP_2(s) \qquad (2\text{-}117)$$

In order to find the roots of $P(s, x)$ we set (2-117) equal to zero and write

$$1 + x\frac{P_1(s)}{P_2(s)} = 0 \qquad (2\text{-}118)$$

where $P_1(s)$ and $P_2(s)$ are polynomials in s, and x is a network parameter independent of s. This expression is in the form required to find the root locus of $P(s, x)$ as a function of x. Thus a network function in bilinear form with respect to x is directly amenable to root locus design with respect to x.

It may be useful to recall briefly the significance of whether the polynomial of interest $P(s, x)$ is actually the denominator $D(s, x)$ or the numerator $N(s, x)$ of a network function $T(s, x)$. The roots $D(s, x)$ are the poles or the *natural frequencies* of the network. Since the stability and selectivity of a network depends directly on its natural frequencies, the effects of any parameter x on these will generally be of most importance. Thus the root locus of a network will most often be the locus of its natural frequencies in the s plane with respect to x.

The roots of $N(s, x)$ are the transmission zeros of the network. Although they will have little effect on the stability of the network they, together with the poles, will determine the nature of its frequency response. The dependence of the network zeros on x may therefore also be of interest on occasion, in which case the root locus of the polynomial $N(s, x)$ will be required.

13. H. W. Bode, loc. cit., p. 10.

The parameter that is varied to obtain the root locus of $N(s, x)$ or $D(s, x)$ is most commonly the transmittance (e.g., gain) or other characteristic of one of the *active* devices of the network. This is particularly true when the locus of the natural frequencies is being investigated. The root locus method may however be extended to any other network parameter as well.

It is frequently useful to relate the polynomials of the bilinear form of a network function to the essential transmittances of Fig. 2-58. We can rewrite (2-116) as

$$T(s) = \frac{A(s)}{U(s)} + x \frac{[B(s)U(s) - A(s)V(s)]/U^2(s)}{1 + xV(s)/U(s)} \qquad (2\text{-}119)$$

Comparing (2-119) with (2-112) we obtain

$$t_{14} = \frac{A(s)}{U(s)} \qquad [2\text{-}120a]$$

$$t_{12}\,t_{34} = \frac{B(s)U(s) - A(s)V(s)}{U^2(s)} \qquad [2\text{-}120b]$$

$$t_{32} = -\frac{V(s)}{U(s)} \qquad [2\text{-}120c]$$

The corresponding flow graph is shown in Fig. 2-66. The return difference and null return difference can now also be directly related to the components of the bilinear form. By comparing (2-119) with (2-94) we see that

$$F_x(s) = 1 + x\frac{V(s)}{U(s)} = \frac{U(s) + xV(s)}{U(s)} = \frac{D(s)}{U(s)} \qquad [2\text{-}121]$$

Since

$$U(s) = D(s) - xV(s) \qquad (2\text{-}122)$$

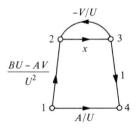

FIG. 2-66. Transmittances of first-order signal-flow graph in terms of the polynomials of the bilinear form.

(2-121) can also be written as

$$F_x(s) = \frac{1}{1 - x \dfrac{V(s)}{D(s)}} \qquad [2\text{-}123]$$

The null return difference follows directly by substituting the expressions of (2-120) into (2-95). We find that

$$F_x^0(s) = 1 + x \frac{B(s)}{A(s)} = \frac{A(s) + xB(s)}{A(s)} = \frac{N(s)}{A(s)} \qquad [2\text{-}124]$$

Since

$$A(s) = N(s) - xB(s) \qquad (2\text{-}125)$$

(2-124) can also be written as

$$F_x^0(s) = \frac{1}{1 - x \dfrac{B(s)}{N(s)}} \qquad [2\text{-}126]$$

The significance of these various expressions for return difference will become evident when network sensitivity is discussed in Chapter 4. On the other hand, certain observations on network behavior can already be made now based on the essential flow graph of Fig. 2-66 and inspection of the corresponding equations.

1. If $A(s) = 0$ in the bilinear form of the network function, the corresponding flow graph has no leakage path, i.e., $t_{14} = 0$.
2. When the essential flow graph has no leakage path, the zero return difference becomes infinitely large, i.e., it loses any physical meaning.

In conclusion it would seem beneficial to illustrate the derivation of a network transfer function in its bilinear form. Consider, for example, the active circuit shown in Fig. 2-67. Let us derive the voltage transfer function V_o/V_i in bilinear form with respect to the amplifier gain μ. The transfer function is given by

$$T(s) = \frac{V_o}{V_i}$$

$$= \frac{\mu Y_1 Y_2}{[Y_2 + Y_6 + Y_5(1 - \mu)][Y_1 + Y_4] +}$$

$$[Y_6 + Y_5(1 - \mu)][Y_2 + Y_3] + Y_2 Y_3(1 - \mu)$$

$$(2\text{-}127)$$

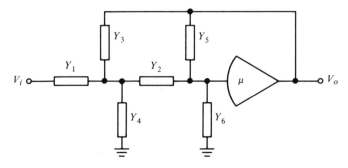

FIG. 2-67. Active network used to illustrate the bilinear form.

In bilinear form with respect to μ we obtain

$$T(s) = \frac{\mu B(s)}{U(s) + \mu V(s)} \tag{2-128}$$

where

$$A(s) \equiv 0 \tag{2-129a}$$

$$B(s) = Y_1 Y_2 \tag{2-129b}$$

$$U(s) = [Y_1 + Y_2 + Y_3 + Y_4][Y_5 + Y_6] + Y_2[Y_1 + Y_3 + Y_4] \tag{2-129c}$$

$$V(s) = - Y_5[Y_1 + Y_2 + Y_3 + Y_4] - Y_2 Y_3 \tag{2-129d}$$

If the bilinear form with respect to Y_1 is of interest we obtain

$$T(s) = \frac{Y_1 B(s)}{U(s) + Y_1 V(s)} \tag{2-130}$$

where

$$A(s) \equiv 0 \tag{2-131a}$$

$$B(s) = \mu Y_2 \tag{2-131b}$$

$$U(s) = [Y_2 + Y_3 + Y_4][(1 - \mu)Y_5 + Y_6] + Y_2[(1 - \mu)Y_3 + Y_4] \tag{2-131c}$$

$$V(s) = [Y_2 + Y_6 + Y_5(1 - \mu)] \tag{2-131d}$$

Notice that the polynomials of the bilinear form, namely $A(s)$, $B(s)$, $U(s)$, and $V(s)$ change as the parameter of interest (i.e., x in the general form of (2-116)) is changed.

Apart from demonstrating the derivation of the bilinear form, this example will very likely have stirred up a rather disturbing question in the reader's mind. In discussing all the elegant methods of root locus design and signal-flow graph reduction one tends to forget that the prerequisite for the use of these methods is the derivation of the network transfer function itself. The derivation of (2-127) in this last example can be most tedious if approached by the standard " node-and-mesh-equation " methods, and the reader may justifiably have wondered if there are no simple methods of bypassing this initial hurdle. Deriving the transfer function by the signal-flow graph method as described in Sections 2.2.2 and 2.2.3 is one possibility, but it will be remembered that this still entails going through the basic network equations and therefore does not necessarily simplify the procedure. Fortunately, and presumably to the reader's relief, there are some very direct and simple methods of analyzing circuits of the kind shown in Fig. 2-67. These will be discussed in the next chapter.

2.2.8 Summary

1. Because most active networks with complex poles are basically feedback configurations, numerous analytical tools originally developed in conjunction with control systems, e.g., root locus analysis and signal-flow graph theory, have been found very useful in active network design.

2. The *root locus* represents graphically the dependence of the pole location of a network on one of the network parameters (generally a parameter characterizing the active device in an active network). More specifically, it provides a graphical trial-and-error technique for the determination of the roots of the characteristic equation $1 \pm \mu t(s) = 0$ of a network for all values of μ.

3. Root locus theory is based on the *angle* and *magnitude* criteria. For negative feedback the angle criterion states that the root locus consists of all those points in the s plane for which the loop phase shift is $(2n + 1)$ times $180°$ ($180°$ root locus); for positive feedback the root locus consists of all those points in the s plane for which the loop phase shift is $2n$ times $180°$ ($0°$ root locus). The magnitude criterion states that for both feedback types the magnitude of the loop transmittance, evaluated at values of s that correspond to points on the root locus, must equal unity.

4. Eight basic rules for the design of root loci can be derived from the angle and magnitude criteria. The most important of these are:

a. The root locus with respect to a variable parameter μ starts (i.e., $\mu = 0$) at the open-loop poles and terminates (i.e., $\mu = \infty$) at the open-loop zeros.

b. The root locus exists at all points on the real axis which lie to the left of an *odd* number of singularities for *negative* feedback; to the left of an *even* number of singularities and to the right of the rightmost singularity for *positive* feedback.

c. The number of asymptotes of a root locus is determined by the number v of infinite zeros of loop transmittance, where v is the difference between the number of finite loop transmittance poles and zeroes. The angles of these asymptotes with the real axis are $\pm(2q-1)$ $180°/v$ for negative feedback and $\pm 2q\cdot180°/v$ for positive feedback, where $q = 1, 2, 3, \ldots$.

d. The breakaway points of the root locus from the negative real axis are determined by solving the equation

$$\frac{d\mu}{ds} = \frac{d}{ds}\left(-\frac{1}{t(s)}\right) = 0.$$

5. *The signal-flow graph* of a network is a graphical representation of the set of linear equations characterizing that network in terms of nodes and branches.

6. Of all the rules permitting the reduction of a complex signal-flow graph to a simpler equivalent, the general multiloop multipath rule (Mason's rule) is the most powerful and hence the most important. It states that

if T_{ij} is the overall transmittance from the source i to the sink j and P_k the transmittance of the kth forward path from source to sink, then

$$T_{ij} = \frac{1}{\Delta}\sum_{k} P_k \Delta_k$$

where

$\Delta = 1 - \sum L_1 + \sum L_2 - \sum L_3 + \cdots + (-1)^j\sum L_j$ ($=$ graph determinant),

L_1 are the loop gains (first-order loops),

L_2 are the loop gain products of all possible combinations of two non-touching loops (second-order loops),

L_3 are the loop gain products of all possible combinations of three non-touching loops (third-order loops),

L_j are the loop gain products of all possible combinations of j nontouching loops (jth-order loops), and

Δ_k is the value of Δ not touching the kth forward path.

(Two loops are considered nontouching when they have no common nodes between them.)

7. The complexity of a signal-flow graph is determined by the order of its corresponding essential signal-flow graph, i.e., by the number of its essential nodes.

8. The set of essential nodes of a signal-flow graph comprises the minimal set of nodes which must be removed to open or remove all its loops.

9. An essential signal-flow graph of order one (i.e., with a single (essential) node beside the input and output nodes) consists of a three-node graph with four transmittance paths.

10. The effect of feedback on the parameters of an active network can be measured by the return difference F_x with respect to a network element x and the null return difference F_x^0.

11. Blackman's impedance relation

$$\frac{Z_i}{Z_i(x=0)} = \frac{F_x \text{ (with terminals shorted)}}{F_x \text{ (with terminals open)}}$$

is a special case of the relation

$$\frac{T}{t_{14}} = \frac{F_x^0}{F_x}$$

where T is the overall transmittance of a first-order signal-flow graph and t_{14} its corresponding leakage path (i.e., the transmittance of the network when $x = 0$).

12. Any network transfer function can always be written in the bilinear form with respect to any component x, i.e.,

$$T(s) = \frac{A(s) + xB(s)}{U(s) + xV(s)}$$

where x does not appear in any of the polynomials $A(s)$, $B(s)$, $U(s)$, or $V(s)$.

CHAPTER

3

ANALYSIS OF LINEAR ACTIVE NETWORKS

INTRODUCTION

In the discussion of most of the topics covered so far it has been assumed that a transfer function is given in one form or another (ratio of two polynomials, product of poles and zeros, etc.). The characterization of those transfer functions or some particular aspects of them were then dealt with. Thus we have discussed the relevance of pole and zero locations per se, the boundaries on pole and zero locations for various types of passive RC networks, the significance of negative real vs. conjugate complex poles, the loci of poles and zeros as functions of an arbitrary network parameter, and finally the derivation and manipulation of signal-flow graphs describing the flow of signals in a linear system. It remains for us now to bridge the gap between the network itself and the above-mentioned techniques used to characterize network functions by showing how to obtain the network functions themselves. Of course we have had the classical methods of mesh and node analysis (Kirchhoff's voltage and current laws) at our disposal to obtain them; we are also now in a position to represent these equations graphically in a signal-flow graph and, using the reduction rules, to simplify the graph sufficiently to obtain the network functions almost by inspection. However, as was pointed out, there is a certain amount of arbitrariness in obtaining a signal-flow graph via the network equations, since the complexity of the resulting graph will depend very much on the set of equations chosen. Consequently, unless a certain degree of intuition and insight is brought to bear on the problem, the emerging graph may be quite unwieldy and its reduction cumbersome. This is why the main application of signal-flow graphs is not so much in the derivation of network transfer functions as in the evaluation of network feedback characteristics. This usually entails investigations into acceptable stability margins and degrees of sensitivity to element variations compatible with a given measure of feedback (i.e., return difference).

Fortunately there are some very direct ways in which to derive network functions from the network diagram which do not require the intermediate step of writing network equations. One of the most useful of these analysis methods uses the indefinite admittance matrix. This method provides a matrix of admittance parameters characterizing a general multiterminal or multipole network *by inspection of the network*. Herein lies the strength of the method. Expanding the matrix by conventional methods provides the desired network function. At that point the process of network function characterization as described in previous chapters can proceed.

In this chapter we shall discuss the indefinite admittance matrix in some detail, as well as its application to active networks and, in particular, to *RC* networks combined with general controlled sources and operational amplifiers. We shall also describe some other network theorems and methods of analysis that come in handy for various types of networks.

3.1 THE INDEFINITE ADMITTANCE MATRIX

A floating multipole network (shown in Fig. 3-1) is a network with n terminals whose potentials are referenced to a point (ground) not connected to the network.

Such a network has the following properties:

1. $\sum_{k=1}^{n} I_k = 0$ (Kirchhoff's current law).

2. Each current I_k depends linearly on the potential difference between the kth terminal and the reference point.

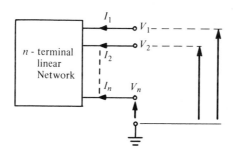

FIG. 3-1. Floating multipole network.

The network can be characterized in matrix form by the following set of equations:

$$
\begin{bmatrix} I_1 \\ I_2 \\ \vdots \\ I_n \end{bmatrix} = \begin{bmatrix} y_{11} & y_{12} & \cdots & y_{1n} \\ y_{21} & y_{22} & \cdots & y_{2n} \\ \vdots & \vdots & & \vdots \\ y_{n1} & y_{n2} & \cdots & y_{nn} \end{bmatrix} \begin{bmatrix} V_1 \\ V_2 \\ \vdots \\ V_n \end{bmatrix} + \begin{bmatrix} I_1^0 \\ I_2^0 \\ \vdots \\ I_n^0 \end{bmatrix}
\qquad [3\text{-}1]
$$

or, in matrix notation,

$$
[I] = [Y][V] + [I^0] \qquad [3\text{-}2]
$$

I_k^0 is the current flowing into the kth terminal when all terminals are connected to the reference node. If all I_k^0 are zero, the multipole is called *inactive*. If any one is not zero, the multipole is called *self-exciting*. We shall be concerned here with inactive networks only; hence all I_k^0 terms will be considered equal to zero. The matrix $[Y]$ is called the indefinite admittance matrix because no network terminal is taken as reference.

It may seem at this point that two conflicting statements regarding the indefinite admittance matrix have been made. On the one hand (3-1) represents the complete set of equations representing a given network. On the other hand it was stated that the indefinite admittance matrix of a network can be derived by inspection of the network diagram "without going through the network equations." Actually both statements are true, since we can indeed obtain the matrix by inspection and thereby automatically derive the coefficients of the complete set of equations describing the network.

The rules for the derivation of the matrix of a passive[1] network are very simple. They are based on an important property of the admittance matrix of a network, namely, that if the nodal equations (where voltages are independent variables) for a network are written, then the resulting admittance matrix bears a one-to-one relationship to the topology of that network. This is best illustrated by an example. Let us consider the network shown in Fig. 3-2. The corresponding admittance matrix is given by

$$
[Y] = \begin{array}{c} \\ \end{array}\begin{bmatrix} Y_1 + Y_4 + Y_6 & -Y_6 & -Y_4 & -Y_1 \\ -Y_6 & Y_3 + Y_5 + Y_6 & -Y_5 & -Y_3 \\ -Y_4 & -Y_5 & Y_2 + Y_4 + Y_5 & -Y_2 \\ -Y_1 & -Y_3 & -Y_2 & Y_1 + Y_2 + Y_3 \end{bmatrix} \begin{array}{c} 1 \\ 2 \\ 3 \\ 4 \end{array}
$$

$$
(3\text{-}3)
$$

It was obtained in the following steps:

1. The nodes of the network are labeled $1, 2, 3, \ldots, n$ (e.g., the four terminals in Fig. 3-2).

----

1. We shall consider the inclusion of active devices later.

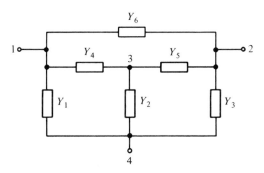

FIG. 3-2. Passive network.

2. The matrix rows and columns are labeled corresponding to the voltage nodes of the network (i.e., there are as many rows and columns as there are voltage nodes).

3. The sum of the admittances entering node i is the same as the term appearing at the intersection of the ith row and the ith column. (For example, Y_2, Y_4, and Y_5 enter node 3 in Fig. 3-2 and their sum appears at the intersection of row 3 and column 3.)

4. The admittances going from node i to node j appear in the matrix as the negative of the term at the intersection of the ith row and jth column. (For example, admittance Y_5 going from node 3 to node 2 appears as $-Y_5$ at the intersection of row 3 and column 2.)

It should be clear that the matrix derivation for a reciprocal (e.g., passive) network is particularly simple because it results in a symmetrical matrix. In the case of a nonreciprocal (e.g., active) network, where matrix symmetry no longer holds, we can derive the admittance matrix from the basic definition of its terms:

$$y_{ij} = \frac{I_i}{V_j}\bigg|_{\substack{V_\mu = 0 \\ j \neq \mu}} \qquad\qquad [3\text{-}4]$$

Hence the coefficient y_{ij} can be interpreted as the current I_i that flows through terminal i when all the terminals $1 \leq \mu \leq n$ except the jth terminal are grounded and a potential of 1 V is applied to j with respect to ground. This definition must of course also hold for the passive matrix, as indeed the above-mentioned four steps imply. From the definition given in (3-4) it follows that y_{ij} has the dimension of an admittance. It also follows that the coefficients y_{ij} completely specify the external behavior of the network.

When the admittance matrix of a network has been obtained directly by inspection, its determinant Δ can be expanded by conventional means and any desired network function derived. This last step may, however, involve tedious multiplications and numerous cross products, many of which may drop out in the final answers. Much of this can be avoided and the necessary analyses simplified appreciably if we look at the properties of the indefinite admittance matrix a little more closely.

3.1.1 Properties of the Indefinite Admittance Matrix

The indefinite admittance matrix has the following important properties:

1. The sum of the elements in each column equals zero:

$$y_{1j} + y_{2j} + \cdots + y_{nj} = 0, \quad j = 1, 2, \ldots, n$$

2. The sum of the elements in each row equals zero:

$$y_{i1} + y_{i2} + \cdots + y_{in} = 0, \quad i = 1, 2, \ldots, n$$

3. The determinant of the matrix equals zero.

4. All first cofactors are equal (equicofactor matrix).

The first property is a result of Kirchhoff's current law, the second a result of the fact that the node currents are dependent on potential differences rather than on the absolute potentials of the nodes. A matrix with these first two properties, i.e., that the sum of the elements of every row and of every column equals zero, is called a zero-sum matrix. As a consequence of this zero-sum property the admittance matrix is singular, i.e., its determinant is equal to zero (property 3). Another consequence of the zero-sum property is the fact that all first-order (i.e., first) cofactors Y_j^i of the indefinite admittance matrix are equal (property 4).

The first cofactor Y_j^i of a matrix is obtained by deleting the ith row and the jth column from the determinant of $[Y]$ to first produce the minor Y_j^i. Affixing the sign $(-1)^{i+j}$ to the minor results in the desired cofactor Y_j^i. (The cofactor is thus a "signed" minor.) Thus the fourth property above states that

$$Y_j^i = Y_n^n, \quad i = 1, 2, \ldots, n \qquad (3\text{-}5)$$
$$j = 1, 2, \ldots, n$$

This property does not hold for the second- or higher-order cofactors. It results from the fact that any first cofactor is obtainable from any other by elementary transformations involving addition of rows and columns while taking account of the zero-sum property of the matrix.

3.1.2 Manipulation of the Indefinite Admittance Matrix

One of the reasons the indefinite admittance matrix is such a useful tool for network analysis is the flexibility resulting from a floating or nonoriented network representation. This initial independence of orientation permits the characteristics of any particular network orientation to be obtained with ease as a special case. This will become evident when we consider the steps required to manipulate the indefinite admittance matrix into a definite matrix or vice versa, or when we consider the suppression, contraction, or addition of nodes or the method of obtaining the matrix of networks in parallel. The proofs of these manipulations are a direct result of the zero-sum property of the indefinite admittance matrix and will not be discussed here. They are well documented in the literature.[2]

Derivation of the Definite Matrix   If one of the network terminals is to be " grounded," i.e., the reference point is to be assigned to it, then the indefinite matrix is made into a definite matrix by deleting the row and column corresponding to the grounded terminal. Similarly, a grounded terminal can be floated and the indefinite matrix obtained by adding a row and column such that the zero-sum properties are satisfied.

Suppose, for example, that we have a three-terminal network, as shown in Fig. 3-3a. The indefinite admittance matrix is given by

$$[Y] = \begin{bmatrix} y_{11} & y_{12} & y_{13} \\ y_{21} & y_{22} & y_{23} \\ y_{31} & y_{32} & y_{33} \end{bmatrix} \tag{3-6}$$

If terminal 3 is to be grounded, the new matrix is obtained by deleting row 3 and column 3,[3] i.e.,

$$[y]_3 = \begin{bmatrix} y_{11} & y_{12} \\ y_{21} & y_{22} \end{bmatrix} \tag{3-7}$$

Equation (3-7) will be recognized as the short-circuit admittance matrix of a three-pole or two-port network, where terminals 1 and 3 constitute one port and terminals 2 and 3 the other. In precisely the same way terminals 1 or 2 can be grounded and the short-circuit admittance matrices for the corresponding two-port configurations obtained.

Reversing the process, suppose that we have a grounded three-terminal network as shown in Fig. 3-3b. Its matrix is given by (3-7). If terminal 3 is now to be floating (see Fig. 3-3a) the corresponding indefinite matrix will be

$$[Y] = \begin{bmatrix} y_{11} & y_{12} & -(y_{11} + y_{12}) \\ y_{21} & y_{22} & -(y_{21} + y_{22}) \\ -(y_{11} + y_{21}) & -(y_{12} + y_{22}) & y_{11} + y_{21} + y_{12} + y_{22} \end{bmatrix} \tag{3-8}$$

Reduction of Multipoles   A higher-order multipole may be reduced to a lower order by *contraction* or *suppression*.

---

2. A well written and detailed discussion on the indefinite admittance matrix, which has influenced parts of this chapter, can be found in: S. K. Mitra, *Analysis and Synthesis of Linear Active Networks* (New York: John Wiley & Sons, 1969).
3. The indefinite admittance matrix of an n-pole has so far been designated $[Y]$. The corresponding definite matrix resulting when terminal k is grounded will be designated $[y]_k$, the first minor as Y_m^i, the corresponding first cofactor as $\underline{Y}_m^i$, the second as $\underline{Y}_{nm}^{ij}$, and so on.

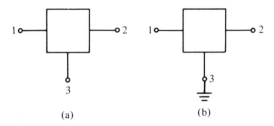

FIG. 3-3. Conversion from an indefinite (a) to a definite (b) two-port.

Contraction is the joining together of two or more terminals to form a single terminal. The new matrix is obtained by adding the columns and rows of the terminals being joined to form a new column and row for the new terminal. Referring to Fig. 3-4, in which terminals 3 and 4 are joined to form the new terminal 3′, we obtain

$$
\begin{bmatrix}
y_{11} & y_{12} & y_{13} & y_{14} \\
y_{21} & y_{22} & y_{23} & y_{24} \\
y_{31} & y_{32} & y_{33} & y_{34} \\
y_{41} & y_{42} & y_{43} & y_{44}
\end{bmatrix}
\rightarrow
\begin{bmatrix}
y_{11} & y_{12} & y_{13}+y_{14} \\
y_{21} & y_{22} & y_{23}+y_{24} \\
y_{31}+y_{41} & y_{32}+y_{42} & y_{33}+y_{34}+y_{43}+y_{44}
\end{bmatrix}
$$

(3-9)

Notice that the resulting matrix is of a lower order by one than the original.

Suppression is the operation of making some terminals inaccessible. The current associated with the suppressed terminals will be zero. The admittance equations for an *n*-pole with $n - i$ suppressed terminals can be written in matrix notation as

$$[I_1] = ([Y_{11}] - [Y_{12}][Y_{22}]^{-1}[Y_{21}])[V_1]$$ [3-10]

$[I_1]$ is the column vector of order *i* representing the currents associated with the *i* accessible terminals, $[V_1]$ is the column vector of the voltages at those *i* terminals, and $[Y_{11}] - [Y_{12}][Y_{22}]^{-1}[Y_{21}]$ is the new indefinite admittance matrix of the multipole with $n - i$ suppressed terminals.

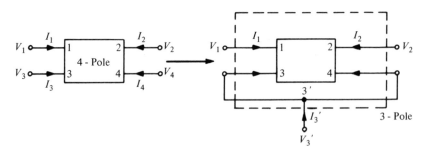

FIG. 3-4. Multipole reduction by contraction.

The submatrices $[Y_{ij}]$ are derived from a partitioning of the original matrix as follows: In (3-11) we suppose that terminals $j = i+1$ through n of the original n terminals are to be suppressed:

$$
\text{accessible}
\left\{
\begin{pmatrix}
\begin{bmatrix} I_1 \\ I_2 \\ \vdots \\ I_i \end{bmatrix} [I_1]
\end{pmatrix}
\right.
\\
\text{suppressed}
\left\{
\begin{pmatrix}
\begin{bmatrix} I_j \\ \vdots \\ I_n \end{bmatrix}
\end{pmatrix}
\right.
=
\begin{bmatrix}
\begin{array}{cccc|cccc}
y_{11} & y_{12} & \cdots & y_{1i} & y_{1j} & \cdots & y_{1n} \\
y_{21} & y_{22} & \cdots & y_{2i} & y_{2j} & \cdots & y_{2n} \\
\vdots & \vdots & [Y_{11}] \;\; \vdots & \vdots & [Y_{12}] \;\; \vdots & & \vdots \\
y_{i1} & y_{i2} & \cdots & y_{ii} & y_{ij} & \cdots & y_{in} \\
\hline
y_{j1} & y_{j2} & \cdots & y_{ji} & y_{jj} & \cdots & y_{jn} \\
\vdots & \vdots & [Y_{21}] \;\; \vdots & \vdots & [Y_{22}] \;\; \vdots & & \vdots \\
y_{n1} & y_{n2} & \cdots & y_{ni} & y_{nj} & \cdots & y_{nn}
\end{array}
\end{bmatrix}
\begin{bmatrix}
\begin{array}{c} V_1 \\ V_2 \\ \vdots \\ V_i \\ \hline V_j \\ \vdots \\ V_n \end{array}
\end{bmatrix}
[V_1]
$$

$$
(3\text{-}11)
$$

Partitioning (3-11) according to the number of accessible and suppressed terminals we obtain the submatrices of (3-10) as follows:

$$
[Y_{11}] = \begin{bmatrix} y_{11} & \cdots & y_{1i} \\ \vdots & & \vdots \\ y_{i1} & \cdots & y_{ii} \end{bmatrix}
\qquad
[Y_{12}] = \begin{bmatrix} y_{1j} & \cdots & y_{1n} \\ \vdots & & \vdots \\ y_{ij} & \cdots & y_{in} \end{bmatrix}
$$

$$
(3\text{-}12)
$$

$$
[Y_{21}] = \begin{bmatrix} y_{j1} & \cdots & y_{ji} \\ \vdots & & \vdots \\ y_{n1} & \cdots & y_{ni} \end{bmatrix}
\qquad
[Y_{22}] = \begin{bmatrix} y_{jj} & \cdots & y_{jn} \\ \vdots & & \vdots \\ y_{nj} & \cdots & y_{nn} \end{bmatrix}
$$

The plausibility of (3-10) becomes clearer if we consider the simplest possible case of terminal suppression, namely, the three-terminal network of which one terminal is to be suppressed. A typical example might be the grounded two-port shown in Fig. 3-3b, of which the output terminal is to be suppressed, i.e., I_2 is to be set equal to zero. The admittance matrix of the grounded two-port was already given by (3-7). Setting $I_2 = 0$, we have $V_2 = -(y_{21}/y_{22})V_1$. Substituting into the equation for I_1 we obtain the "admittance matrix" of the remaining one-port:

$$
I_1 = \frac{\Delta y}{y_{22}} = \left(y_{11} - \frac{y_{12} y_{21}}{y_{22}} \right) V_1
\qquad (3\text{-}13)
$$

The expression in parentheses will be recognized (see Table 3-4) as the input driving point admittance of an unloaded, grounded two-port. Analogously, (3-10) can be considered as an extension, in matrix form, of a general driving point admittance with respect to the i terminals of an n-pole, when the remaining $n - i$ terminals are open circuited.

As an application of (3-10) let us suppress terminal 3 of a three-pole net-work (see Fig. 3-5):

$$
[Y_{\text{original}}] = \begin{bmatrix} y_{11} & y_{12} & y_{13} \\ y_{21} & y_{22} & y_{23} \\ y_{31} & y_{32} & y_{33} \end{bmatrix}
\qquad (3\text{-}14a)
$$

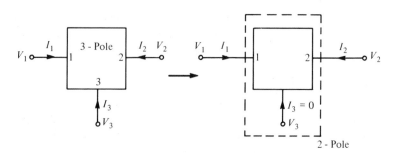

FIG. 3-5. Multipole reduction by suppression.

Partitioning (3-14a) we obtain the submatrices

$$[Y_{11}] = \begin{bmatrix} y_{11} & y_{12} \\ y_{21} & y_{22} \end{bmatrix}$$

$$[Y_{12}] = \begin{bmatrix} y_{13} \\ y_{23} \end{bmatrix} \tag{3-14b}$$

$$[Y_{21}] = \begin{bmatrix} y_{31} & y_{32} \end{bmatrix}$$

$$[Y_{22}] = \begin{bmatrix} y_{33} \end{bmatrix}$$

Therefore

$$[Y_{22}]^{-1} = \frac{1}{y_{33}}$$

and

$$[Y_{12}][Y_{21}] = \begin{bmatrix} y_{13} \\ y_{23} \end{bmatrix} \begin{bmatrix} y_{31} & y_{32} \end{bmatrix} = \begin{bmatrix} y_{13}y_{31} & y_{13}y_{32} \\ y_{23}y_{31} & y_{23}y_{32} \end{bmatrix} \tag{3-14c}$$

We obtain the new matrix for the three-pole with one suppressed terminal as:

$$[Y_{new}] = \begin{bmatrix} y_{11} - \dfrac{y_{13}y_{31}}{y_{33}} & y_{12} - \dfrac{y_{13}y_{32}}{y_{33}} \\ y_{21} - \dfrac{y_{23}y_{31}}{y_{33}} & y_{22} - \dfrac{y_{23}y_{32}}{y_{33}} \end{bmatrix} \tag{3-15}$$

Parallel Connection   The indefinite admittance matrix of multipoles connected in parallel is obtained by adding the corresponding elements of the individual indefinite admittance matrices. For a parallel connection of m multipoles we obtain

$$[I] = \left[\sum_{k=1}^{m} [Y]^{(k)} \right][V] \tag{3-16}$$

All m of the matrices must be of the same order. The order of a matrix may be increased by adding rows and columns of zeros.

The effect of connecting an admittance Y between any two terminals of a multipole can now also be summarized in simple terms; it is to add or subtract Y to the elements at the intersections of the corresponding rows and columns of the indefinite admittance matrix. Y is added to the corresponding elements on the main diagonal and subtracted from the corresponding other elements.

AN EXAMPLE: It may be helpful at this point to go through an illustrative example that will clarify some of the properties of the indefinite admittance matrix described so far. To do so, let us compute the Y matrix for the bridged-T shown in Fig. 3-6a. This circuit can be broken up into parallel parts, as shown in Fig. 3-7. We obtain the following submatrices by inspection:

$$[Y]^{(1)} = \begin{array}{c} \\ \\ \end{array} \begin{array}{cccc} 1 & 2 & 3 & 4 \\ \left[\begin{array}{cccc} G_1 & 0 & -G_1 & 0 \\ 0 & 0 & 0 & 0 \\ -G_1 & 0 & G_1 & 0 \\ 0 & 0 & 0 & 0 \end{array}\right] & \begin{array}{c} 1 \\ 2 \\ 3 \\ 4 \end{array} \end{array}$$

$$[Y]^{(2)} = \left[\begin{array}{cccc} 0 & 0 & 0 & 0 \\ 0 & G_2 & -G_2 & 0 \\ 0 & -G_2 & G_2 & 0 \\ 0 & 0 & 0 & 0 \end{array}\right] \begin{array}{c} 1 \\ 2 \\ 3 \\ 4 \end{array}$$

$$[Y]^{(3)} = \left[\begin{array}{cccc} sC_1 & -sC_1 & 0 & 0 \\ -sC_1 & sC_1 & 0 & 0 \\ 0 & 0 & 0 & 0 \\ 0 & 0 & 0 & 0 \end{array}\right] \begin{array}{c} 1 \\ 2 \\ 3 \\ 4 \end{array} \qquad (3\text{-}17)$$

$$[Y]^{(4)} = \left[\begin{array}{cccc} 0 & 0 & 0 & 0 \\ 0 & 0 & 0 & 0 \\ 0 & 0 & sC_2 & -sC_2 \\ 0 & 0 & -sC_2 & sC_2 \end{array}\right] \begin{array}{c} 1 \\ 2 \\ 3 \\ 4 \end{array}$$

where $[Y]^{(k)}$ is the indefinite admittance matrix of the corresponding four-pole N_k. Adding the matrices of (3-17) we obtain the indefinite admittance matrix of the complete bridged-T of Fig. 3-6a:

$$[Y] = \left[\begin{array}{cccc} G_1 + sC_1 & -sC_1 & -G_1 & 0 \\ -sC_1 & G_2 + sC_1 & -G_2 & 0 \\ -G_1 & -G_2 & G_1 + G_2 + sC_2 & -sC_2 \\ 0 & 0 & -sC_2 & sC_2 \end{array}\right] \qquad (3\text{-}18a)$$

To demonstrate how simple it is to account for added admittances connected between any two terminals of the original network in the corresponding indefinite admittance matrix, let us now consider the modified bridged-T shown in Fig. 3-6b. The admittance G has been added in parallel with C_1.

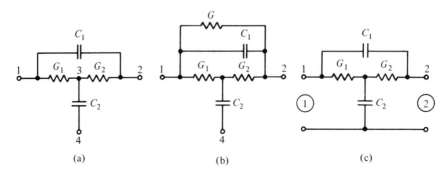

FIG. 3-6. (a) Bridged-T network, (b) Modified bridged-T network, (c) Network of part a, with terminal 4 as reference terminal.

The modified indefinite admittance matrix $[Y]'$ then has the form

$$[Y]' = \begin{bmatrix} G_1 + G + sC_1 & -(G + sC_1) & -G_1 & 0 \\ -(G + sC_1) & G_2 + G + sC_1 & -G_2 & 0 \\ -G_1 & -G_2 & G_1 + G_2 + sC_2 & -sC_2 \\ 0 & 0 & -sC_2 & sC_2 \end{bmatrix} \quad (3\text{-}18b)$$

Returning to the original bridged-T of Fig. 3-6a, let us now take terminal 4 as the reference terminal (Fig. 3-6c). Deleting row 4 and column 4, we obtain

$$[y]_4 = \begin{bmatrix} G_1 + sC_1 & -sC_1 & -G_1 \\ -sC_1 & G_2 + sC_1 & -G_2 \\ -G_1 & -G_2 & G_1 + G_2 + sC_2 \end{bmatrix} \quad (3\text{-}19)$$

In order to obtain the two-port short-circuit admittance matrix of the bridged-T we must suppress terminal 3. After partitioning (3-19) appropriately we obtain the submatrices

$$[Y_{11}] = \begin{bmatrix} G_1 + sC_1 & -sC_1 \\ -sC_1 & G_2 + sC_2 \end{bmatrix} \qquad [Y_{12}] = \begin{bmatrix} -G_1 \\ -G_2 \end{bmatrix}$$

$$[Y_{21}] = [-G_1 \quad -G_2] \qquad\qquad [Y_{22}] = [G_1 + G_2 + sC_2] \quad (3\text{-}20)$$

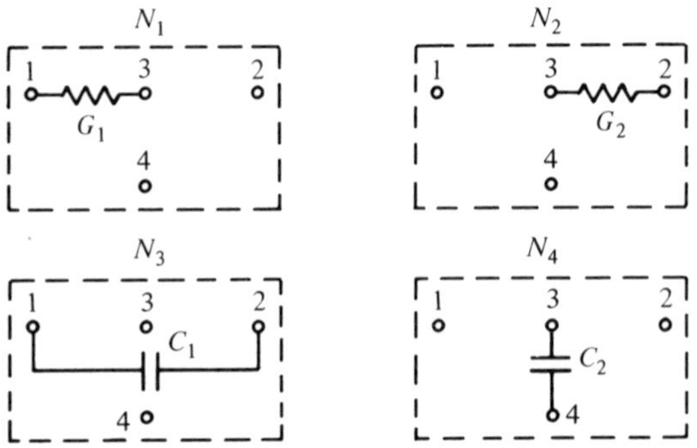

FIG. 3-7. Bridged-T decomposed into four parallel networks N_1, N_2, N_3, N_4.

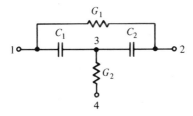

FIG. 3-8. Bridged-T Network (dual of Fig. 3-6).

Thus

$$[Y_{11}] - [Y_{12}][Y_{22}]^{-1}[Y_{21}]$$

$$= \begin{bmatrix} G_1 + sC_1 & -sC_1 \\ -sC_1 & G_2 + sC_1 \end{bmatrix} - \begin{bmatrix} \dfrac{G_1^2}{G_1 + G_2 + sC_2} & \dfrac{G_1 G_2}{G_1 + G_2 + sC_2} \\ \dfrac{G_1 G_2}{G_1 + G_2 + sC_2} & \dfrac{G_2^2}{G_1 + G_2 + sC_2} \end{bmatrix} \quad (3\text{-}21)$$

The short-circuit admittance matrix for the bridged-T of Fig. 3-6c follows:

$$[y] = \begin{bmatrix} \dfrac{s^2 C_1 C_2 + s(C_1 G_1 + C_1 G_2 + C_2 G_1) + G_1 G_2}{G_1 + G_2 + sC_2} \\[2ex] -\dfrac{s^2 C_1 C_2 + sC_1(G_1 + G_2) + G_1 G_2}{G_1 + G_2 + sC_2} \end{bmatrix}$$

$$\begin{bmatrix} -\dfrac{s^2 C_1 C_2 + sC_1(G_1 + G_2) + G_1 G_2}{G_1 + G_2 + sC_2} \\[2ex] \dfrac{s^2 C_1 C_2 + s(C_1 G_1 + C_1 G_2 + C_2 G_2) + G_1 G_2}{G_1 + G_2 + sC_2} \end{bmatrix} \quad (3\text{-}22)$$

It should be clear that the indefinite admittance matrix of the bridged-T given by (3-18a) could have been obtained directly by inspection of the network in Fig. 3-6a. Since it is a passive and nonreciprocal network the simple set of rules for the construction of the admittance matrix by inspection could have been used rather than the fairly complicated procedure of breaking up the bridged-T into four parallel subnetworks. Thus, if we consider the dual[4] of the bridged-T circuit in Fig. 3-6, which is shown in Fig. 3-8, and apply the aforementioned inspection rules, we obtain the indefinite admittance matrix directly:

$$[Y] = \begin{bmatrix} G_1 + sC_1 & -G_1 & -sC_1 & 0 \\ -G_1 & G_1 + sC_2 & -sC_2 & 0 \\ -sC_1 & -sC_2 & G_2 + s(C_1 + C_2) & -G_2 \\ 0 & 0 & -G_2 & G_2 \end{bmatrix} \quad (3\text{-}23)$$

4. $RC:CR$ duality will be discussed in Section 3.4. Although identical component designations are used in Figs. 3-6 and 3-8, G_1 in Fig. 3-6 need not equal G_1 in Fig. 3-8, and so on. On "duality" also see footnote 10. in Chapter 1.

In some instances, however, the complexity of a given network may justify the roundabout procedure used to derive (3-18), particularly if the network contains active elements (e.g., voltage or current sources) in it. Having obtained the two-port short-circuit admittance matrix by suppression, contraction, or whatever other procedures are required to reduce the indefinite admittance matrix of the initial multipole to the definite matrix of the corresponding three-pole or two-port, one can proceed to obtain any desired two-port network function in terms of y_{11}, y_{12}, y_{21}, and y_{22}. This is done by using the well known two-port matrix relationships summarized in Tables 3-1 to 3-4. In Table 3-1 the six most common two-port matrices are defined. The $[z_{ij}]$, $[y_{ij}]$ and $[ABCD]$ matrices for some typical two-port network configurations are given in Table 3-2. The interrelationships between the six basic two-port matrices are given in Table 3-3. Finally in Table 3-4 the impedance and gain relationships for terminated two-ports are given. Clearly the indefinite admittance matrices can be easily derived from the short-circuit admittance matrices given in Table 3-2 by using the zero-sum property. Conversely, having obtained the two-port short-circuit admittance matrix, as we did for the bridged-T, Table 3-4 provides the relationships necessary to calculate the most important network functions.

The preceding bridged-T example will not have concealed the fact that the amount of analysis necessary to go from the indefinite admittance matrix of a multipole to a two-port network function such as the voltage ratio or input impedance is still considerable—even if the indefinite admittance matrix is derived directly from a given network diagram. In our example this was mainly due to the rather cumbersome manipulations required to suppress terminal 3 of the bridged-T after having grounded terminal 4, in order to obtain the short-circuit admittance matrix. Fortunately it can be

TABLE 3-1. TWO-PORT NETWORK PARAMETERS

		For passive and reciprocal circuits
Impedance matrix $[z_{ij}]$ $\Delta z = z_{11}z_{22} - z_{12}z_{21}$	$\begin{bmatrix} V_1 \\ V_2 \end{bmatrix} = \begin{bmatrix} z_{11} & z_{12} \\ z_{21} & z_{22} \end{bmatrix} \begin{bmatrix} I_1 \\ I_2 \end{bmatrix}$	$z_{12} = z_{21}$
Admittance matrix $[y_{ij}]$ $\Delta y = y_{11}y_{22} - y_{12}y_{21}$	$\begin{bmatrix} I_1 \\ I_2 \end{bmatrix} = \begin{bmatrix} y_{11} & y_{12} \\ y_{21} & y_{22} \end{bmatrix} \begin{bmatrix} V_1 \\ V_2 \end{bmatrix}$	$y_{12} = y_{21}$
Hybrid matrix $[h_{ij}]$ $\Delta h = h_{11}h_{22} - h_{12}h_{21}$	$\begin{bmatrix} V_1 \\ I_2 \end{bmatrix} = \begin{bmatrix} h_{11} & h_{12} \\ h_{21} & h_{22} \end{bmatrix} \begin{bmatrix} I_1 \\ V_2 \end{bmatrix}$	$h_{12} = -h_{21}$
Hybrid matrix $[g_{ij}]$ $\Delta g = g_{11}g_{22} - g_{12}g_{21}$	$\begin{bmatrix} I_1 \\ V_2 \end{bmatrix} = \begin{bmatrix} g_{11} & g_{12} \\ g_{21} & g_{22} \end{bmatrix} \begin{bmatrix} V_1 \\ I_2 \end{bmatrix}$	$g_{12} = -g_{21}$
Transmission matrix $[ABCD]$ $\Delta A = AD - BC$	$\begin{bmatrix} V_1 \\ I_1 \end{bmatrix} = \begin{bmatrix} A & B \\ C & D \end{bmatrix} \begin{bmatrix} V_2 \\ -I_2 \end{bmatrix}$	$\Delta A = 1$
Reverse transmission matrix $[\mathscr{ABCD}]$ $\Delta \mathscr{A} = \mathscr{AD} - \mathscr{BC}$	$\begin{bmatrix} V_2 \\ I_2 \end{bmatrix} = \begin{bmatrix} \mathscr{A} & \mathscr{B} \\ \mathscr{C} & \mathscr{D} \end{bmatrix} \begin{bmatrix} V_1 \\ -I_1 \end{bmatrix}$	$\Delta \mathscr{A} = 1$

TABLE 3-2. COMMON TWO-PORT CONFIGURATIONS

Circuit type	$[z_{ij}]$	$[y_{ij}]$	$[ABCD]$
	∞	$\begin{bmatrix} \dfrac{1}{Z_1} & -\dfrac{1}{Z_1} \\[2mm] -\dfrac{1}{Z_1} & \dfrac{1}{Z_1} \end{bmatrix}$	$\begin{bmatrix} 1 & Z_1 \\ 0 & 1 \end{bmatrix}$
	∞	∞	$\begin{bmatrix} 1 & 0 \\[2mm] \dfrac{1}{Z_2} & 1 \end{bmatrix}$
	$\begin{bmatrix} Z_2 & Z_2 \\ Z_2 & Z_2 \end{bmatrix}$		
	$\begin{bmatrix} \dfrac{Z_1}{2}+2Z_2 & 2Z_2 \\[2mm] 2Z_2 & 2Z_2 \end{bmatrix}$	$\begin{bmatrix} \dfrac{2}{Z_1} & -\dfrac{2}{Z_1} \\[2mm] -\dfrac{2}{Z_1} & \dfrac{2}{Z_1}+\dfrac{1}{2Z_2} \end{bmatrix}$	$\begin{bmatrix} 1+\dfrac{Z_1}{4Z_2} & \dfrac{Z_1}{2} \\[2mm] \dfrac{1}{2Z_2} & 1 \end{bmatrix}$
	$\begin{bmatrix} 2Z_2 & 2Z_2 \\[2mm] 2Z_2 & \dfrac{Z_1}{2}+2Z_2 \end{bmatrix}$	$\begin{bmatrix} \dfrac{2}{Z_1}+\dfrac{1}{2Z_2} & -\dfrac{2}{Z_1} \\[2mm] -\dfrac{2}{Z_1} & \dfrac{2}{Z_1} \end{bmatrix}$	$\begin{bmatrix} 1 & \dfrac{Z_1}{2} \\[2mm] \dfrac{1}{2Z_2} & 1+\dfrac{Z_1}{4Z_2} \end{bmatrix}$
	$\begin{bmatrix} Z_1+Z_2 & Z_2 \\ Z_2 & Z_2+Z_3 \end{bmatrix}$	$\dfrac{1}{\dfrac{1}{Z_1}+\dfrac{1}{Z_2}+\dfrac{1}{Z_3}}\begin{bmatrix} \dfrac{1}{Z_1}\left(\dfrac{1}{Z_2}+\dfrac{1}{Z_3}\right) & -\dfrac{1}{Z_1 Z_3} \\[2mm] -\dfrac{1}{Z_1 Z_3} & \dfrac{1}{Z_3}\left(\dfrac{1}{Z_1}+\dfrac{1}{Z_2}\right) \end{bmatrix}$	$\begin{bmatrix} \dfrac{Z_1}{Z_2}+1 & Z_1+Z_3+\dfrac{Z_1 Z_3}{Z_2} \\[2mm] \dfrac{1}{Z_2} & 1+\dfrac{Z_3}{Z_2} \end{bmatrix}$

Continued

TABLE 3-2.—(Continued)

Circuit type	$[z_{ij}]$	$[y_{ij}]$	$[ABCD]$
Symmetric T: $\frac{Z_1}{2}$, Z_2, $\frac{Z_1}{2}$	$\begin{bmatrix} \frac{Z_1}{2}+Z_2 & Z_2 \\ Z_2 & \frac{Z_1}{2}+Z_2 \end{bmatrix}$	$\dfrac{2}{Z_1\left(\frac{Z_1}{2}+2Z_2\right)}\begin{bmatrix} \frac{Z_1}{2}+Z_2 & -Z_2 \\ -Z_2 & \frac{Z_1}{2}+Z_2 \end{bmatrix}$	$\dfrac{1}{Z_2}\begin{bmatrix} \frac{Z_1}{2}+Z_2 & \frac{Z_1}{2}\left(\frac{Z_1}{2}+2Z_2\right) \\ 1 & \frac{Z_1}{2}+Z_2 \end{bmatrix}$
Z_3, Z_2, Z_1	$\dfrac{1}{Z_1+Z_2+Z_3}\begin{bmatrix} Z_1(Z_2+Z_3) & Z_1Z_3 \\ Z_1Z_3 & Z_3(Z_1+Z_2) \end{bmatrix}$	$\begin{bmatrix} \frac{1}{Z_1}+\frac{1}{Z_2} & -\frac{1}{Z_2} \\ -\frac{1}{Z_2} & \frac{1}{Z_2}+\frac{1}{Z_3} \end{bmatrix}$	$\begin{bmatrix} \frac{Z_2}{Z_3}+1 & Z_2 \\ \frac{1}{Z_1}+\frac{1}{Z_3}+\frac{Z_2}{Z_1Z_3} & 1+\frac{Z_2}{Z_1} \end{bmatrix}$
$2Z_2$, Z_1, $2Z_2$	$\dfrac{2Z_2}{\frac{1}{2Z_2}+\frac{2}{Z_1}}\begin{bmatrix} \frac{1}{Z_1}+\frac{1}{2Z_2} & \frac{1}{Z_1} \\ \frac{1}{Z_1} & \frac{1}{Z_1}+\frac{1}{2Z_2} \end{bmatrix}$	$\begin{bmatrix} \frac{1}{Z_1}+\frac{1}{2Z_2} & -\frac{1}{Z_1} \\ -\frac{1}{Z_1} & \frac{1}{Z_1}+\frac{1}{2Z_2} \end{bmatrix}$	$Z_1\begin{bmatrix} \frac{1}{Z_1}+\frac{1}{2Z_2} & 1 \\ \frac{1}{2Z_2}\left(\frac{1}{2Z_2}+\frac{2}{Z_1}\right) & \frac{1}{Z_1}+\frac{1}{2Z_2} \end{bmatrix}$
Lattice: Z_1, Z_1 / Z_2, Z_2	$\begin{bmatrix} \frac{Z_1+Z_2}{2} & \frac{Z_2-Z_1}{2} \\ \frac{Z_2-Z_1}{2} & \frac{Z_1+Z_2}{2} \end{bmatrix}$	$\begin{bmatrix} \frac{1}{2Z_1}+\frac{1}{2Z_2} & -\left(\frac{1}{2Z_1}-\frac{1}{2Z_2}\right) \\ -\left(\frac{1}{2Z_1}-\frac{1}{2Z_2}\right) & \frac{1}{2Z_1}+\frac{1}{2Z_2} \end{bmatrix}$	$\dfrac{1}{Z_2-Z_1}\begin{bmatrix} Z_1+Z_2 & 2Z_1Z_2 \\ 2 & Z_1+Z_2 \end{bmatrix}$
Z_1, R / R, Z_2	$\dfrac{1}{Z_1(Z_1+2R)}\begin{bmatrix} RZ_1^2+2R^2Z_1+2R^3 & 2R^2(Z_1+R) \\ 2R^2(Z_1+R) & RZ_1^2+2R^2Z_1+2R^3 \end{bmatrix}$	$\begin{bmatrix} \frac{Z_1^2+2Z_1R+2R^2}{Z_1R(Z_1+2R)} & -\frac{2(Z_1+R)}{Z_1(Z_1+2R)} \\ -\frac{2(Z_1+R)}{Z_1(Z_1+2R)} & \frac{Z_1^2+2Z_1R+2R^2}{Z_1R(Z_1+2R)} \end{bmatrix}$	$\begin{bmatrix} \frac{Z_1^2+2Z_1R+2R^2}{2R(Z_1+R)} & \frac{Z_1(Z_1+2R)}{2(Z_1+R)} \\ \frac{Z_1(Z_1+2R)}{2R^2(Z_1+R)} & \frac{Z_1^2+2Z_1R+2R^2}{2R(Z_1+R)} \end{bmatrix}$

If $Z_1Z_2 = R^2$ then $Z_{\text{image}} = R$

TABLE 3-3. MATRIX INTERRELATIONSHIPS FOR TWO-PORT NETWORKS

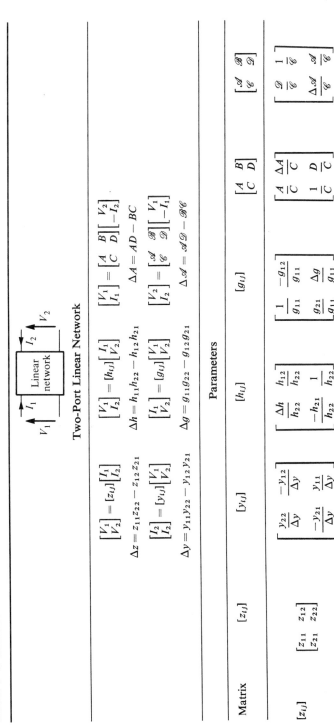

Two-Port Linear Network

$$\begin{bmatrix} V_1 \\ V_2 \end{bmatrix} = [z_{ij}]\begin{bmatrix} I_1 \\ I_2 \end{bmatrix} \qquad \begin{bmatrix} V_1 \\ I_2 \end{bmatrix} = [h_{ij}]\begin{bmatrix} I_1 \\ V_2 \end{bmatrix} \qquad \begin{bmatrix} V_1 \\ I_1 \end{bmatrix} = \begin{bmatrix} A & B \\ C & D \end{bmatrix}\begin{bmatrix} V_2 \\ -I_2 \end{bmatrix}$$

$$\Delta z = z_{11}z_{22} - z_{12}z_{21} \qquad \Delta h = h_{11}h_{22} - h_{12}h_{21} \qquad \Delta A = AD - BC$$

$$\begin{bmatrix} I_2 \\ I_2 \end{bmatrix} = [y_{ij}]\begin{bmatrix} V_1 \\ V_2 \end{bmatrix} \qquad \begin{bmatrix} I_1 \\ V_2 \end{bmatrix} = [g_{ij}]\begin{bmatrix} V_1 \\ I_1 \end{bmatrix} \qquad \begin{bmatrix} V_2 \\ I_2 \end{bmatrix} = \begin{bmatrix} \mathscr{A} & \mathscr{B} \\ \mathscr{C} & \mathscr{D} \end{bmatrix}\begin{bmatrix} V_1 \\ -I_1 \end{bmatrix}$$

$$\Delta y = y_{11}y_{22} - y_{12}y_{21} \qquad \Delta g = g_{11}g_{22} - g_{12}g_{21} \qquad \Delta\mathscr{A} = \mathscr{A}\mathscr{D} - \mathscr{B}\mathscr{C}$$

Parameters

Matrix	$[z_{ij}]$	$[y_{ij}]$	$[h_{ij}]$	$[g_{ij}]$	$\begin{bmatrix} A & B \\ C & D \end{bmatrix}$	$\begin{bmatrix} \mathscr{A} & \mathscr{B} \\ \mathscr{C} & \mathscr{D} \end{bmatrix}$
$[z_{ij}]$	$\begin{bmatrix} z_{11} & z_{12} \\ z_{21} & z_{22} \end{bmatrix}$	$\begin{bmatrix} \dfrac{y_{22}}{\Delta y} & \dfrac{-y_{12}}{\Delta y} \\ \dfrac{-y_{21}}{\Delta y} & \dfrac{y_{11}}{\Delta y} \end{bmatrix}$	$\begin{bmatrix} \dfrac{\Delta h}{h_{22}} & \dfrac{h_{12}}{h_{22}} \\ \dfrac{-h_{21}}{h_{22}} & \dfrac{1}{h_{22}} \end{bmatrix}$	$\begin{bmatrix} \dfrac{1}{g_{11}} & \dfrac{-g_{12}}{g_{11}} \\ \dfrac{g_{21}}{g_{11}} & \dfrac{\Delta g}{g_{11}} \end{bmatrix}$	$\begin{bmatrix} \dfrac{A}{C} & \dfrac{\Delta A}{C} \\ \dfrac{1}{C} & \dfrac{D}{C} \end{bmatrix}$	$\begin{bmatrix} \dfrac{\mathscr{D}}{\mathscr{C}} & \dfrac{1}{\mathscr{C}} \\ \dfrac{\Delta\mathscr{A}}{\mathscr{C}} & \dfrac{\mathscr{A}}{\mathscr{C}} \end{bmatrix}$

Continued

TABLE 3-3.—(Continued)

Matrix	$[z_{ij}]$	$[y_{ij}]$	$[h_{ij}]$	$[g_{ij}]$	$\begin{bmatrix} A & B \\ C & D \end{bmatrix}$	$\begin{bmatrix} \mathscr{A} & \mathscr{B} \\ \mathscr{C} & \mathscr{D} \end{bmatrix}$
$[y_{ij}]$	$\begin{bmatrix} \dfrac{z_{22}}{\Delta z} & -\dfrac{z_{12}}{\Delta z} \\[2mm] -\dfrac{z_{21}}{\Delta z} & \dfrac{z_{11}}{\Delta z} \end{bmatrix}$	$\begin{bmatrix} y_{11} & y_{12} \\ y_{21} & y_{22} \end{bmatrix}$	$\begin{bmatrix} \dfrac{1}{h_{11}} & -\dfrac{h_{12}}{h_{11}} \\[2mm] \dfrac{h_{21}}{h_{11}} & \dfrac{\Delta h}{h_{11}} \end{bmatrix}$	$\begin{bmatrix} \dfrac{\Delta g}{g_{22}} & \dfrac{g_{12}}{g_{22}} \\[2mm] -\dfrac{g_{21}}{g_{22}} & \dfrac{1}{g_{22}} \end{bmatrix}$	$\begin{bmatrix} \dfrac{D}{B} & -\dfrac{\Delta A}{B} \\[2mm] -\dfrac{1}{B} & \dfrac{A}{B} \end{bmatrix}$	$\begin{bmatrix} \dfrac{\mathscr{A}}{\mathscr{B}} & -\dfrac{1}{\mathscr{B}} \\[2mm] -\dfrac{\Delta\mathscr{A}}{\mathscr{B}} & \dfrac{\mathscr{D}}{\mathscr{B}} \end{bmatrix}$
$[h_{ij}]$	$\begin{bmatrix} \dfrac{\Delta z}{z_{22}} & \dfrac{z_{12}}{z_{22}} \\[2mm] -\dfrac{z_{21}}{z_{22}} & \dfrac{1}{z_{22}} \end{bmatrix}$	$\begin{bmatrix} \dfrac{1}{y_{11}} & -\dfrac{y_{12}}{y_{11}} \\[2mm] \dfrac{y_{21}}{y_{11}} & \dfrac{\Delta y}{y_{11}} \end{bmatrix}$	$\begin{bmatrix} h_{11} & h_{12} \\ h_{21} & h_{22} \end{bmatrix}$	$\begin{bmatrix} \dfrac{g_{22}}{\Delta g} & -\dfrac{g_{12}}{\Delta g} \\[2mm] -\dfrac{g_{21}}{\Delta g} & \dfrac{g_{11}}{\Delta g} \end{bmatrix}$	$\begin{bmatrix} \dfrac{B}{D} & \dfrac{\Delta A}{D} \\[2mm] -\dfrac{1}{D} & \dfrac{C}{D} \end{bmatrix}$	$\begin{bmatrix} \dfrac{\mathscr{B}}{\mathscr{A}} & \dfrac{1}{\mathscr{A}} \\[2mm] -\dfrac{\Delta\mathscr{A}}{\mathscr{A}} & \dfrac{\mathscr{C}}{\mathscr{A}} \end{bmatrix}$
$[g_{ij}]$	$\begin{bmatrix} \dfrac{1}{z_{11}} & -\dfrac{z_{12}}{z_{11}} \\[2mm] \dfrac{z_{21}}{z_{11}} & \dfrac{\Delta z}{z_{11}} \end{bmatrix}$	$\begin{bmatrix} \dfrac{\Delta y}{y_{22}} & \dfrac{y_{12}}{y_{22}} \\[2mm] -\dfrac{y_{21}}{y_{22}} & \dfrac{1}{y_{22}} \end{bmatrix}$	$\begin{bmatrix} \dfrac{h_{22}}{\Delta h} & -\dfrac{h_{12}}{\Delta h} \\[2mm] -\dfrac{h_{21}}{\Delta h} & \dfrac{h_{11}}{\Delta h} \end{bmatrix}$	$\begin{bmatrix} g_{11} & g_{12} \\ g_{21} & g_{22} \end{bmatrix}$	$\begin{bmatrix} \dfrac{C}{A} & -\dfrac{\Delta A}{A} \\[2mm] \dfrac{1}{A} & \dfrac{B}{A} \end{bmatrix}$	$\begin{bmatrix} \dfrac{\mathscr{C}}{\mathscr{D}} & -\dfrac{1}{\mathscr{D}} \\[2mm] \dfrac{\Delta\mathscr{A}}{\mathscr{D}} & \dfrac{\mathscr{B}}{\mathscr{D}} \end{bmatrix}$
$\begin{bmatrix} A & B \\ C & D \end{bmatrix}$	$\begin{bmatrix} \dfrac{z_{11}}{z_{21}} & \dfrac{\Delta z}{z_{21}} \\[2mm] \dfrac{1}{z_{21}} & \dfrac{z_{22}}{z_{21}} \end{bmatrix}$	$\begin{bmatrix} -\dfrac{y_{22}}{y_{21}} & -\dfrac{1}{y_{21}} \\[2mm] -\dfrac{\Delta y}{y_{21}} & -\dfrac{y_{11}}{y_{21}} \end{bmatrix}$	$\begin{bmatrix} -\dfrac{\Delta h}{h_{21}} & -\dfrac{h_{11}}{h_{21}} \\[2mm] -\dfrac{h_{22}}{h_{21}} & -\dfrac{1}{h_{21}} \end{bmatrix}$	$\begin{bmatrix} \dfrac{1}{g_{21}} & \dfrac{g_{22}}{g_{21}} \\[2mm] \dfrac{g_{11}}{g_{21}} & \dfrac{\Delta g}{g_{21}} \end{bmatrix}$	$\begin{bmatrix} A & B \\ C & D \end{bmatrix}$	$\begin{bmatrix} \dfrac{\mathscr{D}}{\Delta\mathscr{A}} & \dfrac{\mathscr{B}}{\Delta\mathscr{A}} \\[2mm] \dfrac{\mathscr{C}}{\Delta\mathscr{A}} & \dfrac{\mathscr{A}}{\Delta\mathscr{A}} \end{bmatrix}$
$\begin{bmatrix} \mathscr{A} & \mathscr{B} \\ \mathscr{C} & \mathscr{D} \end{bmatrix}$	$\begin{bmatrix} \dfrac{z_{22}}{z_{12}} & \dfrac{\Delta z}{z_{12}} \\[2mm] \dfrac{1}{z_{12}} & \dfrac{z_{11}}{z_{12}} \end{bmatrix}$	$\begin{bmatrix} -\dfrac{y_{11}}{y_{12}} & -\dfrac{1}{y_{12}} \\[2mm] -\dfrac{\Delta y}{y_{12}} & -\dfrac{y_{22}}{y_{12}} \end{bmatrix}$	$\begin{bmatrix} \dfrac{1}{h_{12}} & \dfrac{h_{11}}{h_{12}} \\[2mm] \dfrac{h_{22}}{h_{12}} & \dfrac{\Delta h}{h_{12}} \end{bmatrix}$	$\begin{bmatrix} -\dfrac{g_{22}}{g_{12}} & -\dfrac{1}{g_{12}} \\[2mm] -\dfrac{\Delta g}{g_{12}} & -\dfrac{g_{11}}{g_{12}} \end{bmatrix}$	$\begin{bmatrix} \dfrac{D}{\Delta A} & \dfrac{B}{\Delta A} \\[2mm] \dfrac{C}{\Delta A} & \dfrac{A}{\Delta A} \end{bmatrix}$	$\begin{bmatrix} \mathscr{A} & \mathscr{B} \\ \mathscr{C} & \mathscr{D} \end{bmatrix}$

TABLE 3-4. IMPEDANCE AND GAIN RELATIONSHIPS
FOR A TERMINATED TWO-PORT NETWORK

$$Z_s = \frac{1}{Y_s} \qquad \qquad Z_L = \frac{1}{Y_L}$$

Linear network, with source E_s, V_1, I_1 at input (Z_{in}) and V_2, I_2 at output (Z_{out}).

Terminated Linear Two-Port Network

Parameters

	$[z_{ij}]$	$[y_{ij}]$	$[h_{ij}]$	$[g_{ij}]$	$\begin{bmatrix} A & B \\ C & D \end{bmatrix}$	$\begin{bmatrix} \mathscr{A} & \mathscr{B} \\ \mathscr{C} & \mathscr{D} \end{bmatrix}$
Z_{in}	$\dfrac{\Delta z + z_{11}Z_L}{z_{22} + Z_L}$	$\dfrac{y_{22} + Y_L}{\Delta y + y_{11}Y_L}$	$\dfrac{\Delta h + h_{11}Y_L}{h_{22} + Y_L}$	$\dfrac{g_{22} + Z_L}{\Delta g + g_{11}Z_L}$	$\dfrac{AZ_L + B}{CZ_L + D}$	$\dfrac{\mathscr{B} + \mathscr{D}Z_L}{\mathscr{A} + \mathscr{C}Z_L}$
Z_{out}	$\dfrac{\Delta z + z_{22}Z_s}{z_{11} + Z_s}$	$\dfrac{y_{11} + Y_s}{\Delta y + y_{22}Y_s}$	$\dfrac{h_{11} + Z_s}{\Delta h + h_{22}Z_s}$	$\dfrac{\Delta g + g_{22}Y_s}{g_{11} + Y_s}$	$\dfrac{B + DZ_s}{A + CZ_s}$	$\dfrac{\mathscr{A}Z_s + \mathscr{B}}{\mathscr{C}Z_s + \mathscr{D}}$
* $A_i = \dfrac{-I_2}{I_1}$	$\dfrac{z_{21}}{z_{22} + Z_L}$	$\dfrac{-y_{21}Y_L}{\Delta y + y_{11}Y_L}$	$\dfrac{-h_{21}Y_L}{h_{22} + Y_L}$	$\dfrac{g_{21}}{\Delta g + g_{11}Z_L}$	$\dfrac{1}{CZ_L + D}$	$\dfrac{\Delta\mathscr{A}}{\mathscr{A} + \mathscr{C}Z_L}$
$A_v = \dfrac{V_2}{V_1}$	$\dfrac{z_{21}Z_L}{\Delta z + z_{11}Z_L}$	$\dfrac{-y_{21}}{y_{22} + Y_L}$	$\dfrac{-h_{21}}{\Delta h + h_{11}Y_L}$	$\dfrac{g_{21}Z_L}{g_{22} + Z_L}$	$\dfrac{1}{A + BY_L}$	$\dfrac{\Delta\mathscr{A}}{\mathscr{B}Y_L + \mathscr{D}}$

* The reason for the negative sign in the expression for A_i is that A_i is the ratio of the load current $(-I_2)$ to the input current (I_1).

shown that the most useful network functions can be obtained directly from the indefinite admittance matrix by reduction procedures that require nothing more than the deletion of rows and columns of $[Y]$ and matrix expansions of what remains. These procedures will now be described. As in this and the preceding section we refer to the well documented literature for the proofs of their legitimacy.

3.1.3 Network Functions of a Multipole

Formulas for the most important network functions of a multipole can be derived directly from the zero-sum and the equicofactor properties of the corresponding indefinite admittance matrix.

We first recall that the first cofactor $\underline{Y}^i_j$ of a matrix $[Y]$ is obtained by deleting the ith row and the jth column of $[Y]$ and prefixing the resulting minor with the sign $(-1)^{i+j}$. Furthermore, if $[Y]$ is a zero-sum matrix it is also an equicofactor matrix, at least with respect to the first cofactors (see (3-5)). Therefore in selecting a first cofactor one is at liberty to select the cofactor

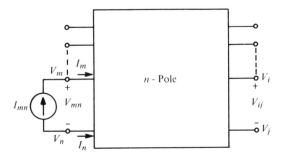

FIG. 3-9. Multipole with n terminals.

that results in the most simple expression. We now denote the second co-factor of $[Y]$ by $\underline{Y}_{ij}^{mn}$. It is obtained by deleting the mth and nth rows and the ith and jth columns of $[Y]$ and affixing the sign of $(-1)^{m+n+i+j}$ to the resulting minor. Thus

$$\underline{Y}_{ij}^{mn} = (-1)^{m+n+i+j} Y_{ij}^{mn} = (-1)^{m+n+i+j} \cdot \begin{pmatrix} \text{Minor obtained by deleting} \\ \text{the } m\text{th and } n\text{th rows and the} \\ i\text{th and } j\text{th columns of } [Y] \end{pmatrix}$$

[3-24]

With these definitions the most important network functions of a multipole can be expressed in relatively simple terms.

Transfer Impedance   Referring to the multipole in Fig. 3-9 we assume that the current I_{mn} flows into terminal m and out of terminal n and that all other currents are zero:

$$I_{mn} = I_m = -I_n \qquad (3\text{-}25)$$

Likewise the voltage V_{ij} is the voltage between terminals i and j, i.e.,

$$V_{ij} = V_i - V_j \qquad (3\text{-}26)$$

The transfer impedance V_{ij}/I_{mn}, which is the voltage drop across terminals i and j caused by the current source I_{mn} connected across terminals m and n, can then be derived utilizing the zero-sum property of the indefinite admittance matrix of the n-pole. One obtains

$$Z_{mn}^{ij} = \frac{V_{ij}}{I_{mn}} = \text{sgn}\,(m-n) \cdot \text{sgn}\,(i-j)\,\frac{\underline{Y}_{ij}^{mn}}{\underline{Y}_n^n} \qquad [3\text{-}27]$$

where the second cofactor $\underline{Y}_{ij}^{mn}$ is defined as in (3-24) (the first was defined at the end of Section 3.1.1) and[5] sgn $y = +1$ if $y > 0$, and -1 if $y < 0$. It should be emphasized once more that because of the equicofactor property of the

5. The abbreviation sgn y stands for "signum y"; it characterizes a step function that jumps from -1 to $+1$ at $y=0$. It is used here to represent the fact that the polarity of the network functions depends on the sequence of the corresponding terminals (e.g. $V_{ij} = -V_{ji}$).

indefinite matrix any other cofactor $\underline{Y}_j^i$ can be used instead of $\underline{Y}_n^n$ in (3-27) if it is simpler to expand.

Driving Point Impedance   The driving point impedance, defined as the ratio of the voltage across terminals m and n (see Fig. 3-9) resulting from the current source I_{mn} across those same terminals, follows directly from (3-27) if we let $i = m$ and $j = n$:

$$Z_{mn} = \frac{V_{mn}}{I_{mn}} = \frac{\underline{Y}_{mn}^{mn}}{\underline{Y}_n^n} \qquad [3\text{-}28]$$

Transfer Voltage Ratio   The transfer voltage ratio between terminals i, j and m, n can be obtained directly by dividing the transfer impedance in (3-27) by the driving point impedance in (3-28). We obtain

$$T_{mn}^{ij} = \frac{V_{ij}}{V_{mn}} = \operatorname{sgn}(m-n)\operatorname{sgn}(i-j)\frac{\underline{Y}_{ij}^{mn}}{\underline{Y}_{mn}^{mn}} \qquad [3\text{-}29]$$

Notice that the subscripts on Z_{mn}^{ij} and T_{mn}^{ij} indicate that the current I_{mn} is injected into the network and that $I_m = -I_n$; the superscripts correspond to the voltage measured between terminals i and j as a result of the current I_{mn}. This correspondence of input variable (current impressed on terminals m, n) and output variable (voltage measured between terminals i and j) to subscript and superscript is reversed in the second cofactors of the corresponding equations, (3-27) and (3-29).

The physical meaning of the network equations given above can be interpreted as follows. Using the indefinite admittance matrix we are permitted to choose any four terminals for the calculation of a transfer function; in general the reference voltage and current are at different terminals. By deleting column j we have set $V_j = 0$ and chosen terminal j as the voltage reference. By deleting row n we indicate that I_n is no longer determined by the network admittances and voltages but by an additional constraint; by using terminal n as the second terminal of the port through which current is supplied to the network we automatically invoke the constraint $I_n = -I_m$. Thus deleting column j and row n has resulted in the definite matrix of the $(n - 1)$-pole whose jth pole is now the reference terminal. The desired transfer functions can then be obtained directly using Cramer's rule, in terms of the determinant of the remaining equations (assuming that this determinant exists which for the nondegenerate networks considered here will be the case) and the appropriate cofactor.

AN EXAMPLE   As an example of formulas (3-27), (3-28), and (3-29), we shall apply them to the bridged-T in Fig. 3-8 whose indefinite admittance matrix (3-23) we obtained by inspection, namely,

$$[Y] = \begin{bmatrix} G_1 + sC_1 & -G_1 & -sC_1 & 0 \\ -G_1 & G_1 + sC_2 & -sC_2 & 0 \\ -sC_1 & -sC_2 & G_2 + s(C_1 + C_2) & -G_2 \\ 0 & 0 & -G_2 & G_2 \end{bmatrix}$$

We wish to compute the voltage ratio

$$T_{14}^{24} = \frac{V_{24}}{V_{14}}$$

Since $m = 1$, $n = 4$ and $i = 2$, $j = 4$, we obtain

$$\underline{Y}_{24}^{14} = - \begin{vmatrix} -G_1 & -sC_2 \\ -sC_1 & G_2 + s(C_1 + C_2) \end{vmatrix}$$

$$= s^2 C_1 C_2 + sG_1(C_1 + C_2) + G_1 G_2 \qquad (3\text{-}30)$$

and

$$\underline{Y}_{14}^{14} = \begin{vmatrix} G_1 + sC_2 & -sC_2 \\ -sC_2 & G_2 + s(C_1 + C_2) \end{vmatrix}$$

$$= s^2 C_1 C_2 + s[G_1(C_1 + C_2) + G_2 C_2] + G_1 G_2 \qquad (3\text{-}31)$$

Thus from (3-29) we have

$$T_{14}^{24} = \frac{s^2 C_1 C_2 + sG_1(C_1 + C_2) + G_1 G_2}{s^2 C_1 C_2 + s[G_1(C_1 + C_2) + G_2 C_2] + G_1 G_2} \qquad (3\text{-}32)$$

Notice how much simpler the procedure was here, where we obtained the indefinite admittance matrix (3-23) by inspection of the network in Fig. 3-8 and derived the transfer function by application of formula (3-29), than it was in the derivation of (3-22) for the bridged-T of Fig. 3-6a. (Having the short-circuit admittance matrix (3-22) the voltage transfer function follows directly, since $T(s) = -y_{12}/y_{22}$. See Table 3-4 for $Y_L = 0$).

3.1.4 The Three–Pole

Since most active devices are three-pole elements (e.g., transistors, vacuum tubes, FETs,) it is worth our while to spend a little time examining the special features of the three pole in somewhat more depth.

Consider the indefinite admittance matrix of the three-pole shown in Fig. 3-10. It is given by

$$[Y] = \begin{matrix} & A & B & C \\ & \begin{bmatrix} y_{AA} & y_{AB} & y_{AC} \\ y_{BA} & y_{BB} & y_{BC} \\ y_{CA} & y_{CB} & y_{CC} \end{bmatrix} & \begin{matrix} A \\ B \\ C \end{matrix} \end{matrix} \qquad (3\text{-}33a)$$

We recall that the short-circuit admittance matrix for the two-port resulting when terminal A, B, or C is common to the input and output can be readily obtained; the row and column corresponding to the common terminal are deleted from (3-33a). It can be shown that, for example y_{AA} is the short-circuit driving point admittance of the port at A, when terminal B or C is common to the input and output ports. Similar relations hold for y_{BB} and y_{CC}. Thus we may write

$$y_{AA} = y_{11B} = y_{11C}$$

$$y_{BB} = y_{11A} = y_{22C} \qquad (3\text{-}33b)$$

$$y_{CC} = y_{22A} = y_{22B}$$

where, for example, y_{11B} and y_{11C} are the short-circuit driving point admittance with terminals B or C, respectively, common to the input and output ports. With (3-33b) the definite matrix $[y]_k$, which is obtained by deletion of the kth row and column, will always appear in the following standard form, where k is identified with the terminal common to both input and output:

$$[y]_k = \begin{bmatrix} y_{11k} & y_{12k} \\ y_{21k} & y_{22k} \end{bmatrix} \qquad (3\text{-}34)$$

Notice that by using the notation of (3-33b) the input and output variables of the definite matrix resulting from (3-33a) automatically result in the form of (3-34).

It is of interest to point out here that the same basic scheme for identifying input and output terminals can be applied to the admittance matrix of a transistor.[6] This is based on the fact that the following relations hold for transistor parameters:

$$y_{\text{base}} = y_{11e} = y_{11c}$$

$$y_{\text{emitter}} = y_{11b} = y_{22c}$$

$$y_{\text{collector}} = y_{22b} = y_{22e} \qquad (3\text{-}35)$$

where, for example, y_{11e} and y_{11c} are the short-circuit driving point admittances when the emitter and collector, respectively, are used as the common terminal.

Comparison of (3-35) and (3-33b) reveals that the base b, emitter e, and collector c correspond to terminals A, B, and C, respectively. Consequently, if terminals b, e, and c are arranged to correspond with A, B, and C of Fig. 3-10, as shown in Fig. 3-11, and the indefinite admittance matrix is derived from this configuration, then the corresponding common emitter, common base, or common collector short-circuit admittance matrix can be obtained directly by deleting the corresponding row and column. Thus the transistor indefinite matrix should be written in the following form:

$$
\begin{array}{ccc}
b & e & c
\end{array}
$$
$$
[Y] = \begin{bmatrix} \text{Indefinite } y \text{ matrix} \\ \text{formed by either the} \\ CE,\ CB,\ \text{or } CC\ y \text{ parameters} \end{bmatrix} \begin{matrix} b \\ e \\ c \end{matrix} \qquad (3\text{-}36)
$$

An example illustrating the use of (3-36) will follow in the next section.

---

6. W. F. Kiss and R. A. Gilson, On the formulation of the indefinite matrix, *IEEE J. Solid-State Circuits*, **SC-3**, 307–308 (1968).

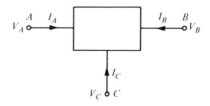

FIG. 3-10. General three-pole.

Returning to the general three-pole, we know that it is because of the zero-sum property that the knowledge of any 2 × 2 matrix gives all of the admittances necessary to form all of the possible 2 × 2 matrices included in (3-33a), simply by algebraic addition. Moreover, it is not necessary that the four known admittances form a submatrix. Thus, for instance, one can find all of the driving point and transfer admittances of a three-pole for all orientations by knowing the three driving point admittances y_{AA}, y_{BB}, and y_{CC}, and any one of its transfer admittances.

Let us assume that we know the definite matrix $[y]_C$ (terminal C grounded) of a *passive* reciprocal three-pole. Due to reciprocity, $y_{AB} = y_{BA}$. Thus

$$[y]_C = \begin{bmatrix} y_{AA} & y_{AB} \\ y_{AB} & y_{BB} \end{bmatrix} \tag{3-37}$$

We then obtain the indefinite admittance matrix

$$[Y] = \begin{matrix} & A & B & C & \\ \begin{bmatrix} y_{AA} & y_{AB} & -(y_{AA} + y_{AB}) \\ y_{AB} & y_{BB} & -(y_{AB} + y_{BB}) \\ -(y_{AA} + y_{AB}) & -(y_{AB} + y_{BB}) & y_{AA} + 2y_{AB} + y_{BB} \end{bmatrix} & \begin{matrix} A \\ B \\ C \end{matrix} \end{matrix} \tag{3-38}$$

We can now obtain the short-circuit admittance matrix for any orientation of our three-pole and, using the matrix interrelations of Table 3-3, any other matrix required. This has been carried out for three different orientations of a passive reciprocal network (where all three matrices are referred to the same initial configuration) and listed in Table 3-5. Not listed but obvious is the fact that due to reciprocity and the identities in (3-33b), input and output terminals can be interchanged merely by interchanging the diagonal elements in the y, z,

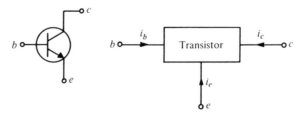

FIG. 3-11. Transistor as a three-pole.

TABLE 3-5. MATRIX PARAMETERS FOR THREE ORIENTATIONS OF A PASSIVE RECIPROCAL THREE-POLE*

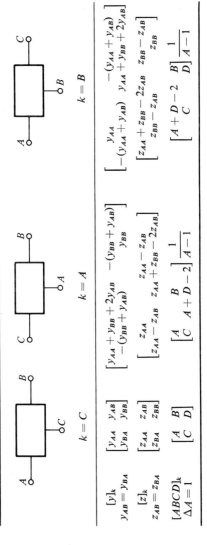

	$k = C$	$k = A$	$k = B$
$[y]_k$ $y_{AB} = y_{BA}$	$\begin{bmatrix} y_{AA} & y_{AB} \\ y_{BA} & y_{BB} \end{bmatrix}$	$\begin{bmatrix} y_{AA}+y_{BB}+2y_{AB} & -(y_{BB}+y_{AB}) \\ -(y_{BB}+y_{AB}) & y_{BB} \end{bmatrix}$	$\begin{bmatrix} y_{AA} & -(y_{AA}+y_{AB}) \\ -(y_{AA}+y_{AB}) & y_{AA}+y_{BB}+2y_{AB} \end{bmatrix}$
$[z]_k$ $z_{AB} = z_{BA}$	$\begin{bmatrix} z_{AA} & z_{AB} \\ z_{BA} & z_{BB} \end{bmatrix}$	$\begin{bmatrix} z_{AA} & z_{AA}-z_{AB} \\ z_{AA}-z_{AB} & z_{AA}+z_{BB}-2z_{AB} \end{bmatrix}$	$\begin{bmatrix} z_{AA}+z_{BB}-2z_{AB} & z_{BB}-z_{AB} \\ z_{BB}-z_{AB} & z_{BB} \end{bmatrix}$
$[ABCD]_k$ $\Delta A = 1$	$\begin{bmatrix} A & B \\ C & D \end{bmatrix}$	$\begin{bmatrix} A & B \\ C & A+D-2 \end{bmatrix}\dfrac{1}{A-1}$	$\begin{bmatrix} A+D-2 & B \\ C & D \end{bmatrix}\dfrac{1}{A-1}$

* M. N. S. Swamy, On the matrix parameters of a three terminal network, *Proc. IEEE*, 1081–1082 (1966).

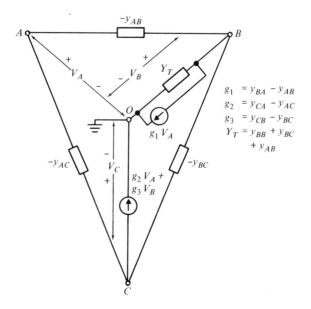

$$g_1 = y_{BA} - y_{AB}$$
$$g_2 = y_{CA} - y_{AC}$$
$$g_3 = y_{CB} - y_{BC}$$
$$Y_T = y_{BB} + y_{BC} + y_{AB}$$

FIG. 3-12. Equivalent circuit of the general three-pole. After: J. L. Stewart, An equivalent circuit for linear N-port active networks, *IRE Trans. Circuit Theory*, **CT-6**, (1959), 234–235.

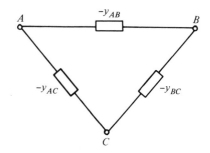

FIG. 3-13. Equivalent circuit of reciprocal three-pole.

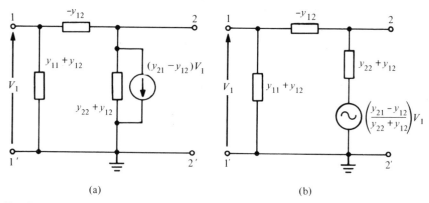

(a)   (b)

FIG. 3-14. The π-equivalent circuits of a grounded two-port. (a) With dependent current source. (b) With dependent voltage source.

140

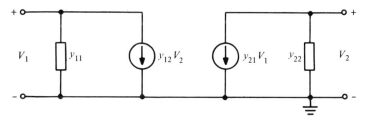

FIG. 3-15. Two-current-generator equivalent circuit of a grounded two port.

and $ABCD$ matrices while leaving the off-diagonal elements unchanged. Furthermore, we see that if t_{BA} is the voltage transfer function of a reciprocal three-pole when terminal C is grounded, interchanging the input (A) and the ground (C) terminals we obtain the complementary transfer function[7] $t_{BC} = 1 - t_{BA}$.

Finally, with the help of the indefinite admittance matrix let us consider the equivalent circuit of the three-pole. It is shown in Fig. 3-12. The circuit consists of four admittances and three voltage-controlled current sources. If the three-pole is reciprocal then the g_i defining the current sources become zero, and the equivalent diagram simplifies to that shown in Fig. 3-13. If we ground terminal C of Fig. 3-12 and generate a two-port, we obtain the two π-equivalent circuits of a two-port defined by their short-circuit admittance parameters, as shown in Fig. 3-14. This in turn can be represented by the more familiar two-current generator equivalent diagram shown in Fig. 3-15. This diagram will be recognized as the one frequently used to characterize transistors and other active elements. It will be used for this purpose in the following section.

3.1.5 The Indefinite Admittance Matrix of Networks Containing Active Elements

Actually there is no real need to consider the indefinite admittance matrix of a network separately when it contains active elements. Except for the symmetry property, which is only valid for reciprocal passive networks, all other properties, and thus all manipulation rules, remain the same. There are, however, certain aspects of the indefinite matrix that have been found useful when using the matrix in conjunction with commonly used active devices such as transistors or operational amplifiers. In the following, we shall therefore go through certain illustrative examples of applications of the indefinite admittance matrix to active networks, pointing out any useful or interesting considerations as we go along.

Transistors   The short-circuit admittance matrix of the common emitter transistor configuration is defined by the transistor equivalent diagram shown

---

7. This fact has been used to obtain the complementary network function from a given active network configuration; see. D. Hilberman. Input and ground as complements in active filters, *IEEE Trans. Circuit Theory*, CT-20, 540-547 (1973).

in Fig. 3-16. Notice that this equivalent circuit corresponds to the two-current-generator circuit described in the preceding section (see Fig. 3-15). The subscripts on the y parameters are commonly used with transistors; the first subscript designates the matrix element as follows:

i input
o output
f forward transfer
r reverse transfer

The second subscript identifies the common terminal used, in this case the emitter (b and c identify the base and collector, respectively, as the common terminals). The y or any other network parameters used to characterize the transistor (e.g., h and z parameters) are, of course, directly related to physical device parameters of the transistor.[8]

By inspection of Fig. 3-16 we obtain the transistor equations of the common emitter configuration in terms of its short-circuit admittance parameters as follows:

$$\begin{matrix} b & c \end{matrix}$$
$$\begin{bmatrix} I_b \\ I_c \end{bmatrix} = \begin{bmatrix} y_{ie} & y_{re} \\ y_{fe} & y_{oe} \end{bmatrix} \begin{bmatrix} V_b \\ V_c \end{bmatrix} \tag{3-39}$$

The indefinite admittance matrix may now be formed by adding the row and column, corresponding to the floating emitter terminal of the equivalent diagram shown in Fig. 3-17. We obtain

$$\begin{matrix} b & c & e \end{matrix}$$
$$\begin{bmatrix} I_b \\ I_c \\ I_e \end{bmatrix} = \begin{bmatrix} y_{ie} & y_{re} & -(y_{ie} + y_{re}) \\ y_{fe} & y_{oe} & -(y_{fe} + y_{oe}) \\ -(y_{ie} + y_{fe}) & -(y_{re} + y_{oe}) & y_{ie} + y_{re} + y_{fe} + y_{oe} \end{bmatrix} \begin{bmatrix} V_b \\ V_c \\ V_e \end{bmatrix} \tag{3-40}$$

Rearranging the indefinite matrix so that the definite matrix resulting from deletion of any one of the rows and columns always appears in the following form (see (3-34)):

$$[y]_k = \begin{bmatrix} y_{11} & y_{12} \\ y_{21} & y_{22} \end{bmatrix} = \begin{bmatrix} y_i & y_r \\ y_f & y_o \end{bmatrix} \tag{3-41}$$

we obtain from (3-36) and (3-40)

$$\begin{matrix} b & e & c \end{matrix}$$
$$[Y] = \begin{bmatrix} y_{ie} & -(y_{ie} + y_{re}) & y_{re} \\ -(y_{ie} + y_{fe}) & y_{ie} + y_{re} + y_{fe} + y_{oe} & -(y_{re} + y_{oe}) \\ y_{fe} & -(y_{fe} + y_{oe}) & y_{oe} \end{bmatrix} \tag{3-42}$$

8. E.g., see, C. L. Searle et al., *Elementary Circuit Properties of Transistors,* (New York: John Wiley & Sons, 1964), or M. S. Ghausi, *Principles and Design of Linear Active Circuits* (New York: McGraw-Hill, 1965), Chapter 9.

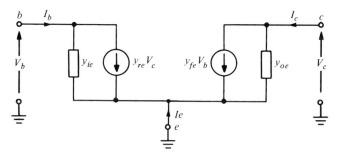

FIG. 3-16. Equivalent circuit for the common-emitter transistor configuration in terms of its y parameters.

Equation (3-42) now provides us with the indefinite admittance matrix of the transistor in terms of the short-circuit admittance parameters of the common emitter configuration. By deleting any one of the rows and columns in order to obtain the definite admittance matrix of the common emitter, common base, or common collector configuration the resulting short-circuit admittance matrix will appear in the form of (3-41).

Very often transistors are specified in terms of their h parameters (corresponding to the equivalent diagram of Fig. 3-18) because these parameters are particularly easy to measure. From Table 3-3 we can find the y parameters of (3-39) in terms of the common emitter h parameters:

$$[y]_e = \begin{matrix} b & c \\ \begin{bmatrix} \dfrac{1}{h_{ie}} & -\dfrac{h_{re}}{h_{ie}} \\[2ex] \dfrac{h_{fe}}{h_{ie}} & \dfrac{\Delta h_e}{h_{ie}} \end{bmatrix} & \begin{matrix} b \\[2ex] c \end{matrix} \end{matrix} \qquad (3\text{-}43)$$

where

$$\Delta h_e = h_{ie} h_{oe} - h_{re} h_{fe} \qquad (3\text{-}44)$$

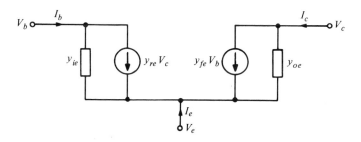

FIG. 3-17. Three-pole equivalent circuit of the common-emitter transistor.

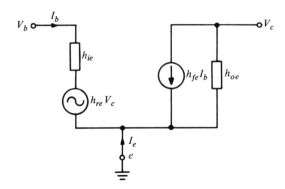

FIG. 3-18. Equivalent circuit for the common-emitter transistor in terms of its h parameters.

Substituting (3-43) into (3-42) we obtain the indefinite admittance matrix of a transistor in terms of its common emitter h parameters:

$$[Y] = \frac{1}{h_{ie}} \begin{bmatrix} 1 & h_{re} - 1 & -h_{re} \\ (-h_{fe} + 1) & h_{fe} + \Delta h_e + 1 - h_{re} & h_{re} - \Delta h_e \\ h_{fe} & -(h_{fe} + \Delta h_e) & \Delta h_e \end{bmatrix} \begin{matrix} b \\ e \\ c \end{matrix} \qquad [3\text{-}45]$$

Deleting any row and column from (3-45), the definite matrix will again appear in the form of (3-41).

Another common way of characterizing a transistor is in terms of its equivalent-T circuit, as shown in Fig. 3-19. As was already shown in Chapter 2, Section 2.2.1 (see Fig. 2-23) this equivalent circuit is often very useful in the analysis of transistor circuits. Its most noteworthy feature is that its parameters are physical elements of a transistor biased in the normal mode.

Proceeding as before we can obtain the short-circuit admittance matrix of Fig. 3-19 in terms of the T parameters:

$$[y]_e = \frac{1}{r_b r_e + r_e r_c + (1 - \alpha) r_b r_c} \begin{bmatrix} r_e - (\alpha - 1) r_c & -r_e \\ \alpha r_c - r_e & r_b + r_e \end{bmatrix} \begin{matrix} b \\ c \end{matrix} \qquad (3\text{-}46)$$

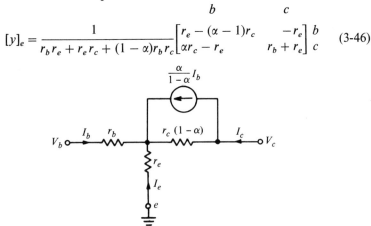

FIG. 3-19. Equivalent-T circuit for the common-emitter transistor.

Expanding and rearranging we obtain the corresponding indefinite admittance matrix:

$$[Y] = \frac{1}{r_b r_e + r_e r_c + (1 - \alpha)r_b r_c} \begin{bmatrix} r_e - (\alpha - 1)r_c & (\alpha - 1)r_c & -r_e \\ -r_c & r_b + r_c & -r_b \\ \alpha r_c - r_e & -\alpha r_c - r_b & r_b + r_e \end{bmatrix} \begin{matrix} b \\ e \\ c \end{matrix}$$

$$\begin{matrix} b & e & c \end{matrix}$$

[3-47]

Here again we obtain the short-circuit admittance matrix for any transistor configuration in the form of (3-41) by deleting the corresponding row and column in (3-47).

EXAMPLES: Various applications of the transistor matrices derived above will illustrate their usefulness. Consider first the current-feedback pair (also referred to as shunt-series feedback pair) shown in Fig. 3-20. This amplifier can be decomposed into the three parallel multiports shown in Fig. 3-21. By inspection of the figures and assuming that the two transistors are identical we obtain, with (3-42), the following y matrices for the partial networks in Fig. 3-21.

$$[y]^a = \begin{bmatrix} y_{ie} & y_{re} & 0 & 0 \\ y_{fe} & y_{oe} + G_c & 0 & 0 \\ 0 & 0 & 0 & 0 \\ 0 & 0 & 0 & 0 \end{bmatrix} \begin{matrix} 1 \\ 3 \\ 4 \\ 2 \end{matrix} \qquad (3\text{-}48)$$

$$\begin{matrix} 1 & 3 & 4 & 2 \end{matrix}$$

$$[y]^b = \begin{bmatrix} G_1 & 0 & -G_1 & 0 \\ 0 & 0 & 0 & 0 \\ -G_1 & 0 & G_1 + G_2 & 0 \\ 0 & 0 & 0 & 0 \end{bmatrix} \begin{matrix} 1 \\ 3 \\ 4 \\ 2 \end{matrix} \qquad (3\text{-}49)$$

$$\begin{matrix} 1 & 3 & 4 & 2 \end{matrix}$$

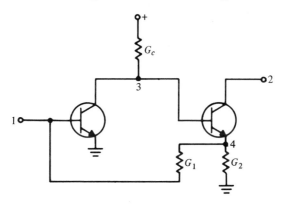

FIG. 3-20. Transistor current-feedback pair. After: E. M. Cherry, An engineering approach to the design of transistor feedback amplifiers, *J. Brit. Inst. Radio Engrs.* **25**, 127–144 (1963); **27**, 349–352 (1964).

$$
[y]^c = \begin{array}{cccc}
1 & 3 & 4 & 2 \\
\begin{bmatrix}
0 & 0 & 0 & 0 \\
0 & y_{ie} & -(y_{ie} + y_{re}) & y_{re} \\
0 & -(y_{ie} + y_{fe}) & y_{ie} + y_{re} + y_{fe} + y_{oe} & -(y_{re} + y_{oe}) \\
0 & y_{fe} & -(y_{fe} + y_{oe}) & y_{oe}
\end{bmatrix}
& \begin{array}{c} 1 \\ 3 \\ 4 \\ 2 \end{array}
\end{array}
\quad (3\text{-}50)
$$

The superscript k on $[y]^k$ in the expressions above refer to the circuits a, b, and c, shown in Fig. 3-21. Generally the following approximations hol y_rd: $_e \approx 0$,

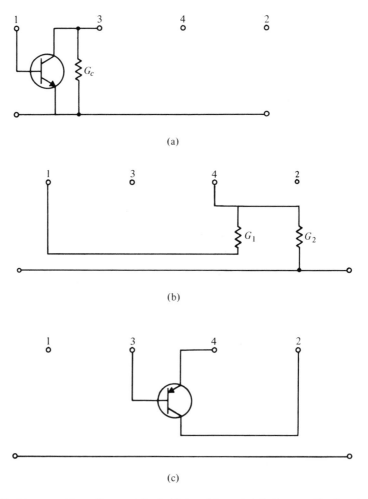

(a)

(b)

(c)

FIG. 3-21. Decomposition of current-feedback transistor pair into three multipole networks in parallel.

$y_{oe} \ll y_{ie}$, $y_{ie} \ll y_{fe}$, $y_{oe} \ll G_1 + G_2$, $y_{oe} \ll G_c$. Consequently, the overall Y matrix, which is found by simply adding (3-48), (3-49), and (3-50), becomes

$$[y]^{abc} = \begin{array}{cccc} 1 & 3 & 4 & 2 \end{array}$$

$$[y]^{abc} = \begin{bmatrix} y_{ie} + G_1 & 0 & -G_1 & 0 \\ y_{fe} & y_{ie} + G_c & -y_{ie} & 0 \\ -G_1 & -y_{fe} & G_1 + G_2 + y_{fe} & -y_{oe} \\ 0 & y_{fe} & -y_{fe} & y_{oe} \end{bmatrix} \begin{array}{c} 1 \\ 3 \\ 4 \\ 2 \end{array} \qquad (3\text{-}51)$$

With (3-51) we have the definite Y matrix for the feedback pair of Fig. 3-20. The indefinite Y matrix can be obtained by adding a row and column that fulfills the zero-sum condition. Then, forming the appropriate first and second cofactors of the resulting matrix, network functions such as those given by (3-27), (3-28), and (3-29) can be obtained.

It may sometimes be advantageous to proceed in a less systematic manner, whereby some insight into the nature of the admittance matrix is required initially and maintained throughout the calculations. As we shall see shortly, this is particularly helpful when working with active devices possessing uncommon characteristics, such as ideal active devices or operational amplifiers. In our present example, terminals 3 and 4 are internal, all network functions of interest being with respect to terminals 1 (input), 2 (output), and ground. Proceeding systematically, we could suppress terminals 3 and 4 simultaneously using the suppression formula (3-10). Notice that to do so, the rows and columns of the expanded (indefinite) version of (3-51) must be rearranged in order to permit matrix partitioning into accessible and suppressed submatrices (see (3-11)). However, proceeding as we set out to do, directly, or "with insight," we observe that I_3 and I_4 must both be zero, since the corresponding terminals are internal and no external current is being injected into them. Thus V_3 and V_4 can be found in terms of V_1 and V_2, and the resulting values substituted into the equations for I_1 and I_2 in (3-51). In this way V_3 and V_4 can be eliminated and I_1 and I_2 obtained as functions of V_1 and V_2. Assuming that $G_c \approx 0$, the resulting short-circuit admittance parameters of the feedback pair with respect to terminals 1, 2, and ground are thereby readily found:

$$y_{11} = y_i + G_1 + \frac{y_f^2}{y_i}\frac{G_1}{G_1 + G_2} \approx \frac{y_f^2}{y_i}\frac{G_1}{G_1 + G_2} \qquad (3\text{-}52a)$$

$$y_{12} = -\frac{y_0 G_1}{G_1 + G_2} \qquad (3\text{-}52b)$$

$$y_{21} = -\frac{y_f^2}{y_i} \qquad (3\text{-}52c)$$

$$y_{22} = y_0 \qquad (3\text{-}52d)$$

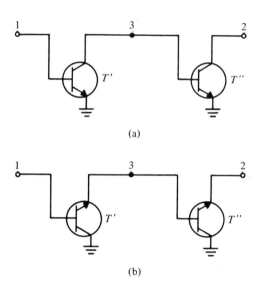

(a)

(b)

Fig. 3-22. Cascaded transistor pairs. (a) Common-emitter pair. (b) Common-collector (Darlington) pair.

To obtain, for example, the overall current gain for the current-feedback pair (which is essentially a current amplifier with low-input and high-output impedance) we have

$$G_I = -\frac{I_2}{I_1} = -\frac{y_{21}}{y_{11}} = 1 + \frac{G_2}{G_1} = 1 + \frac{R_1}{R_2} \qquad (3\text{-}53)$$

where $G_1 = 1/R_1$ and $G_2 = 1/R_2$.

As another example, we shall find the current gain of the cascaded transistor pair shown in Fig. 3-22a. Generally the preferable way to find the response of two-port networks in cascade is to use the $[ABCD]$ or transmission matrices, since the overall matrix is the product of the individual matrices. It is more likely for the y parameters of transistors to be available, however, than their transmission parameters, in which case it is simpler to proceed as follows.

The cascade connection of Fig. 3-22a may be decomposed as shown in Fig. 3-23. Referring to (3-41), we obtain the following matrices for the transistors of Fig. 3-23 where, for convenience, subscript e, denoting common emitter y parameters, has been omitted:

$$[y]'_e = \begin{bmatrix} y'_i & y'_r & 0 \\ y'_f & y'_o & 0 \\ 0 & 0 & 0 \end{bmatrix} \begin{matrix} 1 \\ 3 \\ 2 \end{matrix} \qquad (3\text{-}54)$$

$$\begin{matrix} 1 & 3 & 2 \end{matrix}$$

and

$$[y]''_e = \begin{array}{c} \\ \\ \\ \end{array} \begin{array}{ccc} 1 & 3 & 2 \\ \begin{bmatrix} 0 & 0 & 0 \\ 0 & y''_i & y''_r \\ 0 & y''_f & y''_o \end{bmatrix} & \begin{array}{c} 1 \\ 3 \\ 2 \end{array} \end{array} \tag{3-55}$$

Simple matrix addition gives the overall y matrix of Fig. 3-22a as

$$[y]_e = \begin{array}{ccc} 1 & 3 & 2 \\ \begin{bmatrix} y'_i & y'_r & 0 \\ y'_f & y'_o + y''_i & y''_r \\ 0 & y''_f & y''_o \end{bmatrix} & \begin{array}{c} 1 \\ 3 \\ 2 \end{array} \end{array} \tag{3-56}$$

Since terminal 3 is internal we can set $I_3 = 0$ and obtain

$$y'_f V_1 + (y'_o + y''_i)V_3 + y''_r V_2 = 0 \tag{3-57}$$

Solving for V_3, we have

$$V_3 = -\frac{y'_f}{y'_o + y''_i} V_1 - \frac{y''_r}{y'_o + y''_i} V_2 \tag{3-58}$$

Substituting (3-58) into the expressions for I_1 and I_2 in (3-56) (thereby eliminating V_3), we obtain the short-circuit admittance matrix for the cascaded transistor pair of Fig. 3-22a:

$$[y]_e = \begin{bmatrix} y'_i - \dfrac{y'_r y'_f}{y'_o + y''_i} & -\dfrac{y'_r y''_r}{y'_o + y''_i} \\[3mm] -\dfrac{y'_f y''_f}{y'_o + y''_i} & y''_o - \dfrac{y''_r y''_f}{y'_o + y''_i} \end{bmatrix} \tag{3-59}$$

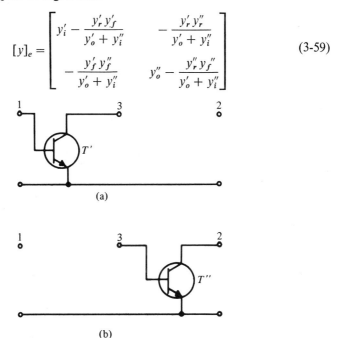

(a)

(b)

FIG. 3-23. Decomposition of cascaded transistor pair into two networks in parallel.

Often the h parameters are more readily available than the y parameters. Thus we can substitute (3-43) into (3-59) and obtain

$$[y]_e = \begin{bmatrix} \dfrac{1}{h_i'}\left(1 + \dfrac{h_r' h_f' h_i''}{h_i' + h_i'' \Delta h'}\right) & -\dfrac{h_r' h_r''}{h_i' + h_i'' \Delta h'} \\[4mm] -\dfrac{h_f' h_f''}{h_i' + h_i'' \Delta h'} & \dfrac{1}{h_i''}\left(\Delta h'' + \dfrac{h_i'' h_f'' h_i'}{h_i' + h_i'' \Delta h'}\right) \end{bmatrix} \tag{3-60}$$

where

$$\Delta h' = h_i' h_o' - h_f' h_r' \tag{3-61a}$$

and

$$\Delta h'' = h_i'' h_o'' - h_f'' h_r'' \tag{3-61b}$$

The current gain results directly; from (3-59) we have

$$G_I = -\frac{I_2}{I_1} = -\frac{y_{21}}{y_{11}} = \frac{y_f' y_f''}{y_i'(y_o' + y_i'') - y_r' y_f'} \tag{3-62}$$

and, from (3-60),

$$G_I = -\frac{y_{21}}{y_{11}} = \frac{h_f' h_f'' h_i'}{h_i' + h_i'' \Delta h' + h_i'' h_r' h_f'} \tag{3-63}$$

This example demonstrates once more that it is useful to keep the physical meaning of the admittance matrix in mind in order to take certain shortcuts in reducing it to the desired transfer function. Typically, internal nodes, and the fact that no external currents are entering them, permit simplifications in the analysis.

It should be noted that the results of the preceding cascaded transistor example are directly applicable to the Darlington pair (i.e., cascaded common-collector transistor pair) shown in Fig. 3-22b; we need only replace the common emitter parameters (subscript e) used in (3-54) and (3-55) by the corresponding common collector parameters (subscript c). The latter can be obtained in terms of the common emitter parameters by deleting row, and column c in (3-42), (3-45), or (3-47).

As a final example of the application of the admittance matrix to transistor circuits we shall consider the transistor amplifier shown in Fig. 3-24. As in the previous examples we can decompose this circuit into two parallel networks N_a and N_b, as shown in Fig. 3-25. The short-circuit admittance matrix of N_a is given by

$$[y]^a = \begin{bmatrix} \dfrac{1}{R_A} + \dfrac{1}{R_B} + \dfrac{1}{R_F} & -\dfrac{1}{R_F} \\[4mm] -\dfrac{1}{R_F} & \dfrac{1}{R_F} + \dfrac{1}{R_c} \end{bmatrix} \tag{3-64}$$

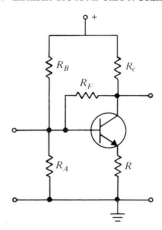

FIG. 3-24. Transistor amplifier.

The short-circuit admittance matrix of N_b is obtained from (3-47) by replacing r_e by $r_e + R$ and deleting the row and column e:

$$[y]^b = \frac{1}{\Delta} \begin{bmatrix} r_e + R - (\alpha - 1)r_c & -(r_e + R) \\ \alpha r_c - (r_e + R) & r_b + r_e + R \end{bmatrix} \begin{matrix} b \\ c \end{matrix} \qquad (3\text{-}65)$$

where

$$\Delta = (r_b + r_c)(r_e + R) + (1 - \alpha)r_b r_c \qquad (3\text{-}66)$$

Adding the two matrices $[y]^a$ and $[y]^b$ we obtain the short-circuit admittance matrix of the amplifier in Fig. 3-24:

$$[y]^{ab} = \begin{bmatrix} \dfrac{1}{R_A} + \dfrac{1}{R_B} + \dfrac{1}{R_F} + \dfrac{r_e + R - (\alpha - 1)r_c}{\Delta} & -\dfrac{1}{R_F} - \dfrac{(r_e + R)}{\Delta} \\[4mm] -\dfrac{1}{R_F} + \dfrac{\alpha r_c - (r_e + R)}{\Delta} & \dfrac{1}{R_F} + \dfrac{1}{R_c} + \dfrac{r_b + r_e + R}{\Delta} \end{bmatrix}$$

$$(3\text{-}67)$$

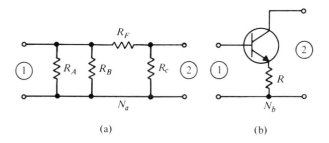

(a)   (b)

FIG. 3-25. Decomposition of transistor amplifier into two parallel networks N_a and N_b.

As $r_c \rightarrow \infty$, and letting $\alpha \approx 1$ this matrix simplifies to

$$[y] = \begin{bmatrix} \dfrac{1}{R_A} + \dfrac{1}{R_B} + \dfrac{1}{R_F} & -\dfrac{1}{R_F} \\[3mm] -\dfrac{1}{R_F} + \dfrac{1}{r_e + R} & \dfrac{1}{R_F} + \dfrac{1}{R_c} \end{bmatrix} \tag{3-68}$$

It is interesting to note that if we let $r_e + R \approx R = R_F/2$, then $y_{21} = 1/R_F = -y_{12}$, and as we shall see in Chapter 5, (3-68) then corresponds to the matrix of a nonideal gyrator.

Controlled Sources   In using the admittance matrix for the analysis of active networks one of the problems that arises is that the admittance matrix of most *ideal* active elements does not exist. Consider, for example, one category of ideal active elements that is very useful in the modeling of most electronic circuits and devices, namely, controlled sources.

The most common *independent* source is a two-terminal current or voltage source. It generally constitutes the input to a network (e.g., the signal genera-tor). In contrast the controlled source is a *dependent* four-terminal source consisting of a current or voltage source whose value at the two controlled terminals depends on the current or voltage at the two controlling terminals. Thus there are four possible controlled-source combinations; the voltage-controlled voltage source (VVS), the voltage-controlled current source (VCS), the current-controlled voltage source (CVS), and the current-controlled current source (CCS). The dependence in each case is unilateral, i.e., there is no transmission in the reverse direction. The corresponding circuit models for the ideal controlled sources and their characterizing equations are shown in Fig. 3-26. The models are ideal, since all input, output, and feedback impedances are considered either zero or infinite. The two-port parameters

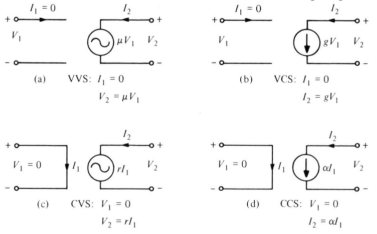

FIG. 3-26. The four possible controlled sources and their defining equations.

for the four ideal controlled sources are listed in Table 3-6. Note that the admittance matrix only exists for the VCS. It is noteworthy, too, that only the transmission matrix exists for all the ideal controlled sources, and incidentally for most other ideal active elements as well. This situation can be remedied if nonideal active elements are considered. This means either assuming finite impedances or considering the element to consist of an ideal element with parasitic impedances. This is no real disadvantage, since the nonideal active element corresponds more closely to practical reality. Thus the nonideal versions of the ideal controlled sources may be represented as in Fig. 3-27, where finite input and output impedances have been added to the ideal elements of Fig. 3-26. With these impedances the corresponding y matrices now exist (Fig. 3-28) and multipole analysis of active networks containing controlled sources can proceed as above, using the indefinite admittance matrix.

In some cases the analysis of a network is greatly simplified by assuming ideal active elements. In such cases a common procedure is to start out with the nonideal element in order to obtain a y matrix for the element and to compensate for the parasitic impedances of the nonideal element with negative impedances attached to the element terminals. For example, Fig. 3-29a shows a nonideal CVS and Fig. 3-29b shows negative resistances attached to the nonideal two-port such that the overall equivalent network represents an ideal CVS. The same procedure can be followed with any other ideal active element.

TABLE 3-6. TWO-PORT PARAMETERS OF IDEAL
CONTROLLED SOURCES

	Parameters				
Source	$[z_{ij}]$	$[y_{ij}]$	$[h_{ij}]$	$[g_{ij}]$	$[ABCD]$
VVS	—	—	—	$\begin{bmatrix} 0 & 0 \\ \mu & 0 \end{bmatrix}$	$\begin{bmatrix} \dfrac{1}{\mu} & 0 \\ 0 & 0 \end{bmatrix}$
VCS	—	$\begin{bmatrix} 0 & 0 \\ g & 0 \end{bmatrix}$	—	—	$\begin{bmatrix} 0 & \dfrac{1}{g} \\ 0 & 0 \end{bmatrix}$
CVS	$\begin{bmatrix} 0 & 0 \\ r & 0 \end{bmatrix}$	—	—	—	$\begin{bmatrix} 0 & 0 \\ \dfrac{1}{r} & 0 \end{bmatrix}$
CCS	—	—	$\begin{bmatrix} 0 & 0 \\ \alpha & 0 \end{bmatrix}$	—	$\begin{bmatrix} 0 & 0 \\ 0 & \dfrac{1}{\alpha} \end{bmatrix}$

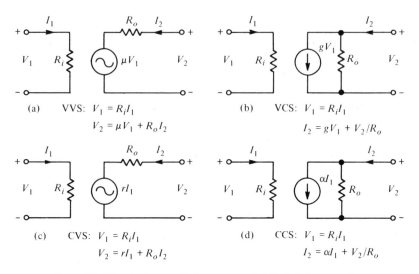

FIG. 3-27. Nonideal controlled sources and their defining equations.

Operational Amplifiers The operational amplifier (opamp) is basically a special kind of controlled source, specifically a very-high-gain VVS (the ideal opamp has infinite gain). Ideally it can be represented by the controlled source shown in Fig. 3-30 (where the conventional symbol for an opamp is also shown). The opamp can be used in the differential mode (signal source between terminals 1 and 2) or in the single-ended mode (signal connected between either terminal 1 or 2 and ground, the other terminal being grounded). The output voltage V_o is of polarity opposite to that of one of the input voltages (e.g., V_1) and of the same polarity as the other (V_2). The ideal opamp has an infinite input impedance (no loading on the input signal source), zero output impedance, and a zero output signal when the input voltage is zero (zero offset).

$$[y]_{VVS} = \begin{bmatrix} \dfrac{1}{R_i} & 0 \\[2ex] -\dfrac{\mu}{R_0} & \dfrac{1}{R_0} \end{bmatrix} \qquad [y]_{VCS} = \begin{bmatrix} \dfrac{1}{R_i} & 0 \\[2ex] g & \dfrac{1}{R_0} \end{bmatrix}$$

$$[y]_{CVS} = \begin{bmatrix} \dfrac{1}{R_i} & 0 \\[2ex] \dfrac{-r}{R_i R_0} & \dfrac{1}{R_0} \end{bmatrix} \qquad [y]_{CCS} = \begin{bmatrix} \dfrac{1}{R_i} & 0 \\[2ex] \dfrac{\alpha}{R_i} & \dfrac{1}{R_0} \end{bmatrix}$$

FIG. 3-28. Admittance matrices of nonideal controlled sources.

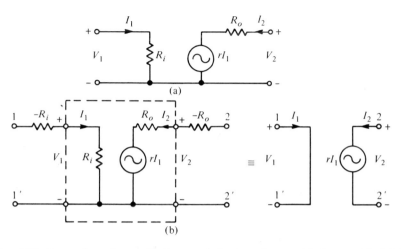

FIG. 3-29. Conversion of a nonideal controlled source into an ideal one. (a) Original nonideal CVS. (b) Addition of negative terminal resistances results in an ideal CVS.

The defining equations for the opamp are

$$I_1 = I_2 = 0 \tag{3-69}$$

$$V_o = A_o(V_2 - V_1) = -A_o V_d \tag{3-70}$$

(where $A_o \to \infty$), and $(V_1 - V_2) = V_d$. The zero-offset condition states that

$$V_o \big|_{V_d=0} = 0 \tag{3-71}$$

From (3-69) it follows that the indefinite admittance matrix does not exist for the ideal operational amplifier. As with other ideal controlled sources, finite (parasitic) input and/or output impedances must be introduced in order for the indefinite matrix to exist. In addition, in the opamp case, the gain A_o must be assumed to be finite. The resulting more realistic nonideal-controlled-source representation of the opamp is shown in Fig. 3-31. It is defined by the

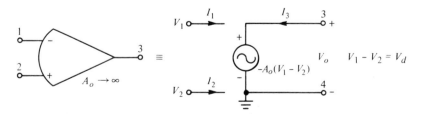

FIG. 3-30. The ideal operational amplifier and its equivalent circuit.

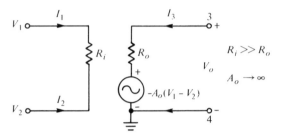

FIG. 3-31. Equivalent circuit of nonideal operational amplifier.

following equations in terms of its definite admittance parameters ("definite" because terminal 4 is tied to ground):

$$
\begin{bmatrix} I_1 \\ I_2 \\ I_3 \end{bmatrix} = \begin{bmatrix} \dfrac{1}{R_i} & -\dfrac{1}{R_i} & 0 \\ -\dfrac{1}{R_i} & \dfrac{1}{R_i} & 0 \\ \dfrac{A_o}{R_o} & -\dfrac{A_o}{R_o} & \dfrac{1}{R_o} \end{bmatrix} \begin{bmatrix} V_1 \\ V_2 \\ V_o \end{bmatrix} \tag{3-72}
$$

where for the ideal opamp $R_i \to \infty$, $R_o \to 0$, $A_o \to \infty$.

With (3-72), calculations involving the definite or indefinite admittance matrix can proceed as before. If ideal opamp conditions are of interest we can let $A_o \to \infty$, $R_i \to \infty$ and $R_o \to 0$ *after* the calculations involving the admittance matrix have been completed. Ideal opamp conditions may, in fact, very often be assumed, since even a practical opamp generally has gain and impedance characteristics that approximate the ideal closely enough to allow for network functions that are essentially independent of these opamp characteristics.

AN EXAMPLE: To illustrate the procedure outlined above we shall find the voltage transfer function V_o/V_i for the active network incorporating an opamp shown in Fig. 3-32. Using the nonideal-controlled-source opamp representation of Fig. 3-31, Fig. 3-32 may be decomposed into the two parallel networks of Fig. 3-33. The indefinite admittance matrix of the passive five-pole of Fig. 3-33a is

$$
[Y]^a = \begin{array}{c} \\ \end{array}
\begin{bmatrix}
Y_a & -Y_a & 0 & 0 & 0 \\
-Y_a & Y_a + Y_b + Y_c + Y_d & -Y_c & -Y_d & -Y_b \\
0 & -Y_c & Y_c + Y_e & -Y_e & 0 \\
0 & -Y_d & -Y_e & Y_d + Y_e & 0 \\
0 & -Y_b & 0 & 0 & Y_b
\end{bmatrix}
\begin{array}{c} 1 \\ 2 \\ 3 \\ 4 \\ 5 \end{array} \tag{3-73}
$$

with column headings $1 \quad 2 \quad 3 \quad 4 \quad 5$

The indefinite admittance matrix of the inverting opamp configuration shown in Fig. 3-33b results directly from (3-72) by first deleting the row and

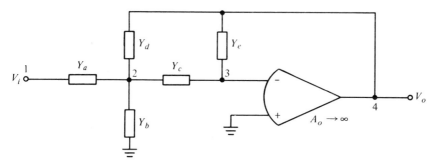

FIG. 3-32. Network incorporating an operational amplifier. After: A Bridgman and R. Brennan, Simulation of transfer functions using only one operational amplifier, *1957 IRE WESCON Conventional Record*, Vol. 1, Part 4, pp. 273–278.

column pertaining to I_2 and V_2, respectively, (node 2 of the opamp is grounded) and then expanding as follows:

$$[Y]^b = \begin{bmatrix}
 & 1 & 2 & 3 & 4 & 5 & \\
0 & 0 & 0 & 0 & 0 & 1 \\
0 & 0 & 0 & 0 & 0 & 2 \\
0 & 0 & \dfrac{1}{R_i} & 0 & -\dfrac{1}{R_i} & 3 \\
0 & 0 & \dfrac{A_o}{R_o} & \dfrac{1}{R_o} & -\dfrac{(A_o + 1)}{R_o} & 4 \\
0 & 0 & -\dfrac{1}{R_i} - \dfrac{A_o}{R_o} & -\dfrac{1}{R_o} & \dfrac{1}{R_i} + \dfrac{1}{R_o} + \dfrac{A_o}{R_o} & 5
\end{bmatrix}$$

(3-74)

Adding the matrices of (3-73) and (3-74) we obtain the indefinite matrix of the total circuit shown in Fig. 3-32:

$$Y = \begin{bmatrix}
1 & 2 & 3 & 4 & 5 & \\
Y_a & -Y_a & 0 & 0 & 0 & 1 \\
-Y_a & Y_T & -Y_c & -Y_d & -Y_b & 2 \\
0 & -Y_c & Y_c + Y_e + \dfrac{1}{R_i} & -Y_e & -\dfrac{1}{R_i} & 3 \\
0 & -Y_d & -Y_e + \dfrac{A_o}{R_o} & Y_d + Y_e + \dfrac{1}{R_o} & -\dfrac{A_o + 1}{R_o} & 4 \\
0 & -Y_b & -\dfrac{1}{R_i} - \dfrac{A_o}{R_o} & -\dfrac{1}{R_o} & Y_b + \dfrac{1}{R_i} + \dfrac{1}{R_o} + \dfrac{A_o}{R_o} & 5
\end{bmatrix}$$

(3-75)

where
$$Y_T = Y_a + Y_b + Y_c + Y_d$$
From (3-29) we have:

$$-\frac{V_o}{V_i} = \frac{V_{45}}{V_{15}} = \text{sgn}\,(4-5)\,\text{sgn}\,(1-5)\,\frac{\underline{Y}_{45}^{15}}{\underline{Y}_{15}^{15}} \tag{3-76}$$

and from (3-75) we obtain

$$\underline{Y}_{45}^{15} = -\begin{array}{ccc} 1 & 2 & 3 \\ \left|\begin{array}{ccc} -Y_a & Y_T & -Y_c \\ 0 & -Y_c & Y_c + Y_e + \dfrac{1}{R_i} \\ 0 & -Y_d & -Y_e + \dfrac{A_o}{R_o} \end{array}\right| & \begin{array}{c} 2 \\ 3 \\ 4 \end{array} \end{array} \tag{3-77}$$

and

$$\underline{Y}_{15}^{15} = \begin{array}{ccc} 2 & 3 & 4 \\ \left|\begin{array}{ccc} Y_T & -Y_c & -Y_d \\ -Y_c & Y_c + Y_e + \dfrac{1}{R_i} & -Y_e \\ -Y_d & -Y_e + \dfrac{A_o}{R_o} & Y_d + Y_e + \dfrac{1}{R_o} \end{array}\right| & \begin{array}{c} 2 \\ 3 \\ 4 \end{array} \end{array} \tag{3-78}$$

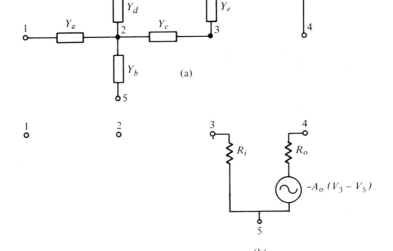

FIG. 3-33. Decomposition of the active network shown in Fig. 3-32 into two parallel multipole networks.

Instead of going through the complete expansion of the determinants given by (3-77) and (3-78), we can utilize certain features peculiar to this and other active networks containing operational amplifiers in order to simplify the neccessary calculations considerably. The circuit of Fig. 3-32 requires an operational amplifier to provide a network function that is practically *independent* of the actual amplifier gain. It other words, if the amplifier gain A_o is sufficiently high, the resulting network function depends only on the passive admittances Y_i, and since an operational amplifier is used in the circuit, this indicates that A_o is intended to be very large. If this is so, we can let A_o, and even more so A_o/R_o, approach infinity. In terms of (3-77) and (3-78) this means that any term not multiplied by A_o (or A_o/R_o) can be considered negligibly small. Thus only those terms in (3-77) and (3-78) that are multiplied by A_o or A_o/R_o need be expanded; all other terms may be neglected. We therefore obtain directly from (3-77)

$$\underline{Y}_{45}^{15}\Bigg|_{\frac{A_o}{R_o}\to\infty} = -Y_a\left(Y_c\cdot\frac{A_o}{R_o}\right)$$

(3-79)

and from (3-78)

$$\underline{Y}_{15}^{15}\Bigg|_{\frac{A_o}{R_o}\to\infty} = Y_T\left(Y_e\frac{A_o}{R_o}\right) + Y_d\left(Y_c\frac{A_o}{R_o}\right)$$

(3-80)

Substituting these expressions into (3-76) and cancelling out A_o/R_o we obtain, for the network of Fig. 3-32,

$$\frac{V_o}{V_i} = -\frac{Y_a Y_c}{Y_e(Y_a + Y_b + Y_c + Y_d) + Y_c Y_d}$$

(3-81)

3.2 Constrained–Network Analysis Applied to Networks Containing Operational Amplifiers

The calculations above using the indefinite admittance matrix were simplified somewhat by making use of the inherent operational amplifier property of near infinite gain. On the other hand the method is still somewhat cumbersome in that, as with most ideal active elements, at least one parasitic impedance must be added to the ideal amplifier model and finite gain must initially be assumed in order for the indefinite admittance matrix to exist. Utilizing some results from the analysis of constrained networks,[9] Nathan has shown how these limitations can be eliminated and the analysis of active networks incorporating operational amplifiers simplified significantly.

---

9. A. Nathan, Matrix analysis of constrained networks, *Proc. Inst. Elect. Engrs. Part. C*, **108**, 98–106 (1961).

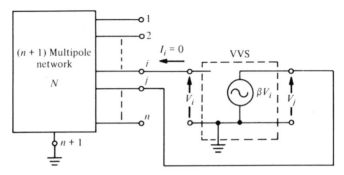

FIG. 3-34. Grounded $(n + 1)$-terminal network connected to a grounded ideal VVS.

3.2.1 Operational Amplifiers with Finite Gain

Consider the grounded, ideal, voltage-controlled voltage source (VVS) with gain β, tied to nodes i and j of the grounded $(n + 1)$-terminal network[10] shown in Fig. 3-34. The VVS introduces a constraint on the network by forcing the voltage V_j of node j with respect to the reference node (e.g., ground) to follow that of node V_i so as to maintain the relation

$$V_j = \beta V_i \qquad (3\text{-}82)$$

Since an ideal VVS is assumed, its input impedance is infinite (no loading of the driving terminal i) and its output impedance zero (isolation of the current in node j from effects of any terminal sources other than that at i, which is the VVS itself). The constraint imposed by this ideal VVS on the network N is precisely the type dealt with by Nathan. He shows that in this case a simple rule can be formulated by which the required definite matrix[11] $[y]$ of the constrained network can be derived directly from the definite admittance matrix $[y']$ of the network without the constraint:

β *finite*: In the unconstrained matrix $[y']$ add column j multiplied by β to column i and subsequently delete row and column j. The result is the constrained matrix $[y]$.

The justification for the first step of this rule involving columns i and j should be evident, since constraint (3-82) implies a form of node contraction between nodes i and j of the multipole N. (Consider, for example, the case of $\beta = 1$.) The second step (deletion of row j) follows from the zero output impedance of the VVS. With it, the current in node j is isolated from the effects of any

---

10. As we shall see in the following discussion, Nathan's method deals with grounded networks and therefore involves definite rather than indefinite matrices.
11. The symbol $[y]$ is being used here for the definite admittance matrix of a constrained network, the symbol $[y']$ for the definite admittance matrix of the same network without constraints. As before, $[Y]$ denotes the indefinite admittance matrix of a network.

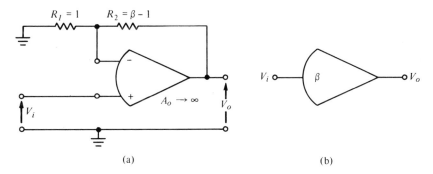

(a)   (b)

FIG. 3-35. Opamp in the noninverting mode to provide VVS characteristics. (a) Circuit configuration. (b) Symbol.

terminal sources other than the VVS itself. Thus I_j depends on the constraint and not on the network equations, and the jth equation must be deleted.[12]

The operational amplifier is especially well suited to providing close-to-ideal VVS properties, and therefore very frequently used for this purpose, in the so-called noninverting mode as shown in Fig. 3-35a. The gain of the amplifier in this configuration, in which it is driven from the positive or "noninverting" terminal, and the gain-determining feedback network is tied to the negative or "inverting" terminal, is given by

$$\frac{V_o}{V_i} = \frac{R_1 + R_2}{R_1} \tag{3-83}$$

where high open-loop gain A_o is assumed. Letting $R_2 = (\beta - 1)R$ and $R_1 = R$ it follows from (3-83) that the voltage gain $V_o/V_i = \beta$. More will be said about the operational amplifier itself, used both in this and other modes, in Chapters 5 and 7. Let it suffice to say here, that with the assumption of high open-loop gain A_o, the input impedance of this amplifier configuration is extremely high, the input current therefore negligible, and the output impedance also negligibly small. Consequently this configuration, symbolized by Fig. 3-35b, provides an excellent practical approximation to an ideal VVS, and it is small wonder that it is so frequently used as such.

EXAMPLES: In the following examples we shall analyze several active networks in which an opamp is used as a VVS. One such network is shown in Fig. 3-36. The unconstrained matrix $[y']$ can be obtained by inspection:

$$[y'] = \begin{array}{cccc} \quad 1 \qquad\quad 3 \qquad\qquad 4 \qquad\quad 2 \end{array}$$
$$\begin{bmatrix} Y_1 & -Y_1 & 0 & 0 \\ -Y_1 & Y_1 + Y_2 + Y_3 + Y_4 & -Y_2 & -Y_3 \\ 0 & -Y_2 & Y_2 + Y_5 + Y_6 & -Y_5 \\ 0 & -Y_3 & -Y_5 & Y_3 + Y_5 \end{bmatrix} \begin{array}{c} 1 \\ 3 \\ 4 \\ 2 \end{array} \tag{3-84}$$

12. Nathan explains this step as follows: I_j represents an external current source that injects exactly the current I_j into the network necessary to satisfy (3-82) and the other network equations. Consequently the jth equation is dependent on the other network equations and can be deleted.

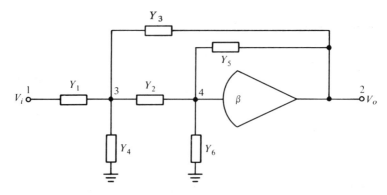

FIG. 3-36. Active network in which an opamp is used as a VVS.

The constrained node 2 can now be eliminated, after adding column 2, multiplied by β, to column 4, since

$$V_2 = \beta \cdot V_4 \qquad (3\text{-}85)$$

Thus

$$[y] = \begin{bmatrix} & 1 & 3 & 4 \\ & Y_1 & -Y_1 & 0 \\ & -Y_1 & Y_1 + Y_2 + Y_3 + Y_4 & -(Y_2 + \beta Y_3) \\ & 0 & -Y_2 & Y_2 + Y_6 + Y_5(1 - \beta) \end{bmatrix} \qquad (3\text{-}86)$$

The voltage gain V_4/V_1 can now be obtained from (3-86) by application of Cramer's rule. Thus, in general, if V_q and V_p represent the voltages at the qth and pth terminals, respectively, with respect to the reference node of a network, as shown in Fig. 3-37, then we have

$$V_q = \frac{\bar{y}_q^p}{\Delta} I_p \qquad (3\text{-}87)$$

and

$$V_p = \frac{\bar{y}_p^p}{\Delta} I_p \qquad (3\text{-}88)$$

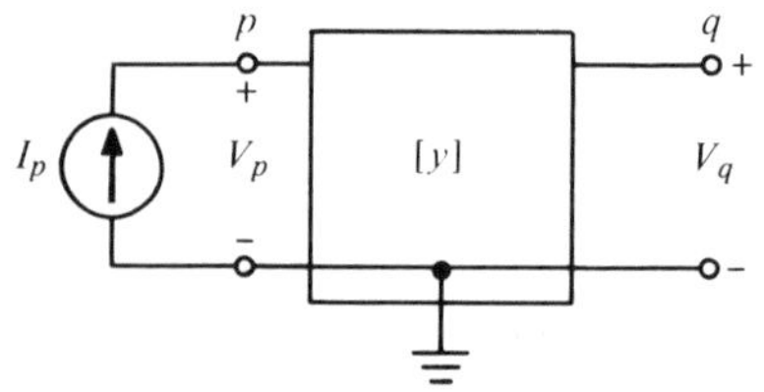

FIG. 3-37. Network with driving node p and driven node q.

provided p is not a driven node and q is not a driving node. The terms $\bar{y}^i_j$ denote the first cofactors of the definite, constrained matrix $[y]$ and Δ its determinant. Thus from (3-87) and (3-88) we obtain[13]

$$\frac{V_q}{V_p} = \frac{\bar{y}^p_q}{\bar{y}^p_p} \tag{3-89}$$

Returning now to our example of Fig. 3-36, we obtain the cofactors $\bar{y}^i_j$ of the definite matrix (3-86):[14]

$$\bar{y}^1_4 = Y_1 Y_2 \tag{3-90}$$

and

$$\bar{y}^1_1 = \begin{vmatrix} Y_1 + Y_2 + Y_3 + Y_4 & -(Y_2 + \beta Y_3) \\ -Y_2 & Y_2 + Y_6 + Y_5(1 - \beta) \end{vmatrix}$$
$$= (Y_1 + Y_4)[Y_2 + Y_6 + Y_5(1 - \beta)] + (Y_2 + Y_3)[Y_6 + Y_5(1 - \beta)]$$
$$+ Y_2 Y_3(1 - \beta) \tag{3-91}$$

To obtain the voltage-transfer ratio V_2/V_1 we first substitute (3-90) and (3-91) into (3-89) and combine the resulting ratio V_4/V_1 with the constraint (3-85):

$$T(s) = \frac{V_o}{V_i} = \frac{V_2}{V_1} = \beta \cdot \frac{V_4}{V_1} = \beta \frac{Y_1 Y_2}{D(s)} \tag{3-92}$$

where

$$D(s) = (Y_1 + Y_4)[Y_2 + Y_6 + Y_5(1 - \beta)] + (Y_2 + Y_3)[Y_6 + Y_5(1 - \beta)]$$
$$+ Y_2 Y_3(1 - \beta) \tag{3-93}$$

Incidentally, (3-92) is the general network function for the "controlled-source" active realization of various second-order filter networks. By replacing the Y_j by resistors or capacitors appropriately, any one of the basic second-order filter functions (low-pass, high-pass, bandpass, etc.), can be obtained. This is discussed in detail elsewhere.[15]

As a second example of an active network in which the opamp is used as a VVS, consider the network shown in Fig. 3-38. By inspection we obtain the unconstrained admittance matrix

$$[y'] = \begin{array}{c} \\ \\ \\ \\ \\ \end{array} \begin{array}{ccccc} 1 & 2 & 3 & 4 & 5 \\ \end{array}$$
$$[y'] = \begin{bmatrix} G_1 & 0 & -G_1 & 0 & 0 \\ 0 & sC_1 & -sC_1 & 0 & 0 \\ -G_1 & -sC_1 & G_1 + sC_1 & 0 & 0 \\ 0 & 0 & 0 & G_2 & -G_2 \\ 0 & 0 & 0 & -G_2 & G_2 + sC_2 \end{bmatrix} \begin{array}{c} 1 \\ 2 \\ 3 \\ 4 \\ 5 \end{array} \tag{3-94}$$

---

13. This expression for a definite y matrix can also be derived from the expression for the indefinite y matrix given by (3-29) if nodes j and n are both assumed grounded, i.e., the corresponding columns and nodes deleted from $[Y]$.
14. The minor $y_q{}^p$ of the cofactor $\bar{y}_q{}^p$ must be multiplied by $(-1)^{p+q}$. However, here $\bar{y}_4{}^1$ applies to the first row and *third* column of $[y]$, the subscript 4 referring only to the number of the corresponding node.
15. See *Linear Integrated Networks: Design.* Chapter 2.

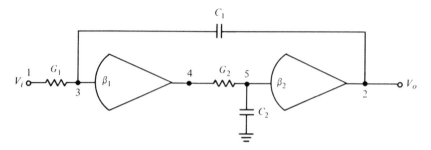

FIG. 3-38. Active low-pass filter using opamps as VVSs.

In this example we have *two* constrained nodes defined by

$$V_4 = \beta_1 V_3 \tag{3-95a}$$

and

$$V_2 = \beta_2 V_5 \tag{3-95b}$$

The constrained node 4 can be eliminated, after adding column 4 multiplied by β_1 to column 3; the constrained node 2 can be eliminated in the same way, involving β_2 and column 5. Similarly currents I_2 and I_4 depend on the network constraints, so that rows 2 and 4 may be deleted. Thus we obtain the constrained matrix

$$[y] = \begin{array}{ccc} 1 & 3 & 5 \\ \begin{bmatrix} G_1 & -G_1 & 0 \\ -G_1 & G_1 + sC_1 & -\beta_2 sC_1 \\ 0 & -\beta_1 G_2 & G_2 + sC_2 \end{bmatrix} & \begin{array}{c} 1 \\ 3 \\ 5 \end{array} \end{array} \tag{3-96}$$

Since nodes 3 and 5 are internal we can again in this example utilize the fact that they will be subjected to no excitation by external current sources or in other words that $I_3 = I_5 = 0$. Thus we have

$$\begin{bmatrix} I_1 \\ 0 \\ 0 \end{bmatrix} = \begin{bmatrix} G_1 & -G_1 & 0 \\ -G_1 & G_1 + sC_1 & -\beta_2 sC_1 \\ 0 & -\beta_1 G_2 & G_2 + sC_2 \end{bmatrix} \begin{bmatrix} V_1 \\ V_3 \\ V_5 \end{bmatrix} \tag{3-97}$$

Solving for V_1 and V_5 by using Cramer's rule we obtain

$$V_1 = \frac{\begin{vmatrix} I_1 & -G_1 & 0 \\ 0 & G_1 + sC_1 & -\beta_2 sC_1 \\ 0 & -\beta_1 G_2 & G_2 + sC_2 \end{vmatrix}}{\Delta}$$

$$= \frac{I_1 [(G_1 + sC_1)(G_2 + sC_2) - \beta_1 \beta_2 sG_2 C_1]}{\Delta} \tag{3-98}$$

and

$$V_5 = \cfrac{\begin{vmatrix} G_1 & -G_1 & I_1 \\ -G_1 & G_1 + sC_1 & 0 \\ 0 & -\beta_1 G_2 & 0 \end{vmatrix}}{\Delta}$$

$$= I_1 \frac{\beta_1 G_1 G_2}{\Delta} \tag{3-99}$$

Thus with (3-95b) we have

$$\frac{V_o}{V_i} = \frac{V_2}{V_1} = \frac{\beta_1 \beta_2 G_1 G_2}{s^2 C_1 C_2 + s[G_1 C_2 + G_2 C_1 - \beta_1 \beta_2 G_2 C_1] + G_1 G_2} \tag{3-100}$$

Notice that for the special case that $\beta_1 = \beta_2 = 1$, (3-100) becomes a function of passive components only. Note also how the calculations were shortened appreciably by taking into account that nodes 3 and 5 are internal and therefore not subject to external current excitation.

Incidentally the transfer function (3-100) derived for the network of Fig. 3-38 corresponds to a second-order low-pass filter. The characteristics of representative second-order functions are discussed in more detail in Chapter 1 of *Linear Integrated Network: Design.*

The steps taken to obtain the transfer function (3-100) represent the detailed procedure summarized by equations (3-87) to (3-89). At the same time they may have helped to explain the rationale behind those equations; it is based on an application of Cramer's rule and on the fact that a current is injected only into the input terminal so that all the other terms of the column matrix $[I]$ are zero. After some practice the procedure can be shortened appreciably, beyond even the explicit use of equations (3-87) to (3-89). To demonstrate this, consider for example, the network in Fig. 3-39. After having numbered the nodes, the nonconstrained matrix $[y']$ is obtained by inspection:

$$[y'] = \begin{array}{c} \\ \\ \\ \\ \\ \end{array} \begin{matrix} 1 & 2 & 3 & 4 \\ \begin{bmatrix} G_1 & -G_1 & 0 & 0 \\ -G_1 & G_1 + G_2 + s(C_1 + C_2) & -sC_2 & -G_2 \\ 0 & -sC_2 & G_3 + sC_2 & 0 \\ 0 & -G_2 & 0 & G_2 \end{bmatrix} \begin{matrix} 1 \\ 2 \\ 3 \\ 4 \end{matrix} \end{matrix} \tag{3-101}$$

The voltage constraint $V_4 = \beta V_3$ permits us to delete row and column 4 by multiplying column 4 by β and adding it to column 3 thereby providing the constrained matrix $[y]$:

$$[y] = \begin{matrix} 1 & 2 & 3 \\ \begin{bmatrix} G_1 & -G_1 & 0 \\ -G_1 & G_1 + G_2 + s(C_1 + C_2) & -(\beta G_2 + sC_2) \\ 0 & -sC_2 & G_3 + sC_2 \end{bmatrix} \begin{matrix} 1 \\ 2 \\ 3 \end{matrix} \end{matrix} \tag{3-102}$$

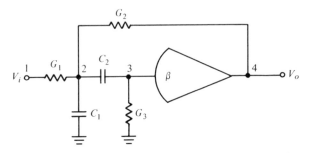

FIG. 3-39. Active bandpass filter using an opamp as VVS.

From here the transfer function can be obtained by inspection. Combining Cramer's rule and the fact that nodes 2 and 3 are internal (i.e., $I_2 = I_3 = 0$) we know that, according to (3-89),

$$\frac{V_3}{V_1} = \frac{\bar{y}_3^1}{\bar{y}_1^1} \tag{3-103}$$

and therefore, by inspection of (3-102), we obtain directly

$$\frac{V_o}{V_i} = \frac{V_4}{V_1} = \frac{\beta s G_1 C_2}{(G_3 + sC_2)[G_1 + G_2 + s(C_1 + C_2)] - sC_2(\beta G_2 + sC_2)} \tag{3-104}$$

Observe that we have already taken the node constraint $V_4 = \beta V_3$ into account. For the purpose of identifying second-order network functions, it is a good idea to write (3-104) in the form

$$T(s) = \frac{N(s)}{s^2 + (A - \beta B)s + C} \tag{3-105}$$

We obtain

$$\frac{V_o}{V_i} = \frac{\beta(G_1/C_1)s}{s^2 + \left(\dfrac{G_1 + G_2 + G_3}{C_1} + \dfrac{G_3}{C_2} - \beta\dfrac{G_2}{C_1}\right)s + \dfrac{G_3(G_1 + G_2)}{C_1 C_2}} \tag{3-106}$$

As discussed in Chapter 1 of *Linear Integrated Networks: Design*, this is the transfer function of a bandpass network.

We mentioned earlier that the methods of circuit analysis discussed in this chapter permit us to derive a network transmission function (e.g., (3-106) above) from which we can then proceed to various investigations involving perhaps, the necessity for flow-graph representations or root locus designs. These may be required, for example, to examine the sensitivity and stability of the network with respect to variations of some or all of its components. Although we have not yet discussed sensitivity, let us at least briefly apply

some of the feedback techniques studied in the previous chapter to our transmission function (3-106). To do so it is useful to rewrite (3-106) in feedback
form, i.e., with the characteristic equation in the denominator. For this
purpose, let

$$\hat{d}(s) = s^2 + \left[\frac{G_1 + G_2 + G_3}{C_1} + \frac{G_3}{C_2} \right] s + \frac{G_3(G_1 + G_2)}{C_1 C_2} \tag{3-107}$$

where $\hat{d}(s)$ is the denominator of the passive RC network associated with
(3-106). It is readily obtained by setting the active element (e.g., β) equal to
zero. The term $\hat{d}(s)$ must, by definition, have two negative real poles $\hat{p}_1$ and $\hat{p}_2$:

$$\hat{d}(s) = (s - \hat{p}_1)(s - \hat{p}_2) \tag{3-108}$$

Dividing (3-106) by $\hat{d}(s)$ we obtain

$$\frac{V_o}{V_i} = T(s) = \frac{(G_1/C_1)s}{\hat{d}(s)} \cdot \frac{\beta}{1 - \beta \dfrac{G_2/C_1}{\hat{d}(s)} s} \tag{3-109}$$

Letting

$$\hat{t}_1(s) = \frac{(G_1/C_1)s}{\hat{d}(s)} \tag{3-110a}$$

and

$$\hat{t}_2(s) = \frac{(G_2/C_1)s}{\hat{d}(s)} \tag{3-110b}$$

we obtain

$$T(s) = \hat{t}_1(s) \frac{\beta}{1 - \beta \hat{t}_2(s)} \tag{3-111}$$

The root locus of the poles of $T(s)$ with respect to β result directly from (3-111).
The open-loop poles are the poles of $\hat{t}_2(s)$, namely, $\hat{p}_1$ and $\hat{p}_2$; the open-loop
zeros are the zeros of $\hat{t}_2(s)$, one of which is at the origin of the s plane, the
other at infinity. Applying the design rules of the $0°$ root locus (because of the
negative sign in the denominator of $T(s)$, which implies a positive feedback
network) we readily obtain the root locus shown in Fig. 3-40.

The constrained-network analysis method requires the use of an ideal VVS
(i.e., operational amplifier). Thus, if the effect of the nonideal characteristics
of an operational amplifier on the performance of a circuit are to be examined,
then this method cannot be used. In such cases the method outlined earlier
(Section 3.1.5) is useful in that it requires the operational amplifier to be
introduced into the circuit in its nonideal form.

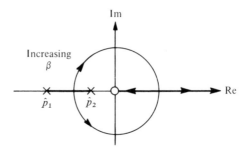

FIG. 3-40. Root locus with respect to β of the poles generated by the positive feedback network in Fig. 3-39.

3.2.2 Operational Amplifiers with Infinite Gain

We consider here applications in which the operational amplifier is used in the inverting mode e.g., in the previously discussed network shown in Fig. 3-32. Here the amplifier is used in the open-loop mode (i.e., no feedback network such as R_1 and R_2 in Fig. 3-35) and it is required that A_o, or β in (3-82), go to infinity. To avoid introducing an infinite quantity into the constrained y matrix, Nathan suggests the following simple modification of his analysis procedure:[16] The source currents operating into node j (see Fig. 3-34) are eliminated as before, but the voltage V_j is retained and V_i, the voltage of the *driving* node, is eliminated. The rule is

$\beta \to \infty$: To obtain the constrained matrix $[y]$, delete the row corresponding to the driven node (e.g., node j) and the column corresponding to the driving node (e.g., node i) in the nonconstrained matrix $[y']$.

The rationale for this modification is simple to explain. With β approaching infinity and V_j remaining finite, and assuming a *negative* feedback loop, we obtain from our constraint in (3-82)

$$V_i \Big|_{\beta \to \infty} = 0 \qquad (3\text{-}112)$$

Thus terminal i is a "virtual ground" and column i in the nonconstrained definite matrix $[y']$ can be deleted. As in the preceding case the current into the jth node depends directly on the constraint (3-82), i.e., in this case it must take on a value such that (3-112) holds. The current I_j is therefore predetermined and the jth row must be deleted from the $[y']$ matrix.

16. A. Nathan, Matrix analysis of networks having infinite-gain operational amplifiers, *Proc. IEEE*, **49**, 1577–1578 (1961).

EXAMPLES: To show how much this procedure simplifies the analysis of circuits incorporating operational amplifiers with infinite gain, we return first to the network given in Fig. 3-32 and calculate once more the transfer function V_o/V_i using Nathan's method. The nonconstrained y matrix is obtained by inspection:

$$[y'] = \begin{array}{c} \\ \end{array}\overset{\quad 1 \qquad\qquad 2 \qquad\qquad\qquad 3 \qquad\qquad 4}{\begin{bmatrix} Y_a & -Y_a & 0 & 0 \\ -Y_a & Y_a + Y_b + Y_c + Y_d & -Y_c & -Y_d \\ 0 & -Y_c & Y_c + Y_e & -Y_e \\ 0 & -Y_d & -Y_e & Y_d + Y_e \end{bmatrix}} \begin{matrix} 1 \\ 2 \\ 3 \\ 4 \end{matrix} \tag{3-113}$$

Next the driven row (4) and the driving column (3) are deleted, leaving the constrained matrix

$$[y] = \overset{\quad 1 \qquad\qquad 2 \qquad\qquad\qquad 4}{\begin{bmatrix} Y_a & -Y_a & 0 \\ -Y_a & Y_a + Y_b + Y_c + Y_d & -Y_d \\ 0 & -Y_c & -Y_e \end{bmatrix}} \begin{matrix} 1 \\ 2 \\ 3 \end{matrix} \tag{3-114}$$

Using Cramer's rule as above we obtain from (3-87)

$$V_4 = \frac{Y_a Y_c}{\Delta} I_1 \tag{3-115}$$

and

$$V_1 = \frac{-Y_e(Y_a + Y_b + Y_c + Y_d) - Y_c Y_d}{\Delta} I_1 \tag{3-116}$$

Dividing (3-116) into (3-115) we obtain

$$\frac{V_o}{V_i} = \frac{V_4}{V_1} = -\frac{Y_a Y_c}{Y_e(Y_a + Y_b + Y_c + Y_d) + Y_c Y_d} \tag{3-117}$$

This, of course, is the same result as we obtained in (3-81), but the analysis required here is considerably simpler. Notice futhermore that negative feedback is automatically accounted for by (3-112).

The simplicity of Nathan's method hardly suffers as the network analyzed becomes more complicated. Consider, for example, the network shown in Fig. 3-41 and observe that, although it is somewhat more complex than the one in Fig. 3-32, its analysis requires no more effort. The nonconstrained matrix $[y']$ can be obtained by inspection:

$$[y'] = \overset{\quad 1 \qquad\qquad 2 \qquad\qquad\qquad 3 \qquad\qquad\qquad 4}{\begin{bmatrix} Y_1 + Y_6 & -Y_1 & -Y_6 & 0 \\ -Y_1 & Y_1 + Y_2 + Y_3 + Y_4 & -Y_3 & -Y_4 \\ -Y_6 & -Y_3 & Y_3 + Y_5 + Y_6 & -Y_5 \\ 0 & -Y_4 & -Y_5 & Y_4 + Y_5 \end{bmatrix}} \begin{matrix} 1 \\ 2 \\ 3 \\ 4 \end{matrix} \tag{3-118}$$

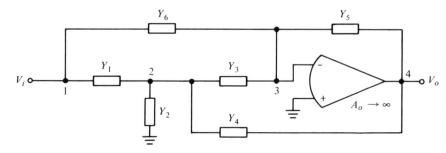

FIG. 3-41. General network incorporating "infinite-gain" operational amplifier.

The constrained matrix $[y]$ results by deleting row 4 and column 3:

$$[y] = \begin{array}{c} \quad\; 1 \qquad\qquad\quad 2 \qquad\qquad 4 \\ \begin{bmatrix} Y_1 + Y_6 & -Y_1 & 0 \\ -Y_1 & Y_1 + Y_2 + Y_3 + Y_4 & -Y_4 \\ -Y_6 & -Y_3 & -Y_5 \end{bmatrix} \begin{array}{c} 1 \\ 2 \\ 3 \end{array} \end{array} \qquad (3\text{-}119)$$

Using Cramer's rule we obtain

$$\frac{V_o}{V_i} = \frac{V_4}{V_1} = \frac{\bar{y}_4^1}{\bar{y}_1^1} = \frac{\begin{vmatrix} -Y_1 & Y_1 + Y_2 + Y_3 + Y_4 \\ -Y_6 & -Y_3 \end{vmatrix}}{\begin{vmatrix} Y_1 + Y_2 + Y_3 + Y_4 & -Y_4 \\ -Y_3 & -Y_5 \end{vmatrix}} \qquad (3\text{-}120)$$

Expanding, we obtain the desired voltage transfer function

$$\frac{V_o}{V_i} = -\frac{Y_1 Y_3 + Y_6 (Y_1 + Y_2 + Y_3 + Y_4)}{Y_3 Y_4 + Y_5 (Y_1 + Y_2 + Y_3 + Y_4)} \qquad (3\text{-}121)$$

As in one of the previous examples, let us pursue this example a little further in order to demonstrate how we might now continue our investigations of this network in terms of some of the feedback concepts discussed in the previous chapter. Consider, for example, that we wish to know the return difference and the null return difference of the network with respect to Y_3. For this purpose we may write (3-121) in its bilinear form with respect to Y_3:

$$\frac{V_o}{V_i} = -\frac{Y_6 (Y_1 + Y_2 + Y_4) + Y_3 (Y_1 + Y_6)}{Y_5 (Y_1 + Y_2 + Y_4) + Y_3 (Y_4 + Y_5)} = \frac{A(s) + xB(s)}{U(s) + xV(s)} = \frac{N(s)}{D(s)} \qquad (3\text{-}122\text{a})$$

$$A(s) = Y_6 (Y_1 + Y_2 + Y_4)$$
$$B(s) = Y_1 + Y_6$$
$$U(s) = Y_5 (Y_1 + Y_2 + Y_4)$$
$$V(s) = Y_4 + Y_5$$

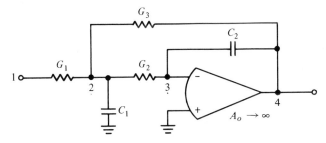

FIG. 3-42. Active low-pass filter incorporating "infinite-gain" operational amplifier.

The return difference with respect to Y_3 then results directly:

$$F_{Y_3} = \frac{D(s)}{U(s)} = 1 + Y_3 \frac{Y_4 + Y_5}{Y_5(Y_1 + Y_2 + Y_4)} \qquad (3\text{-}122b)$$

as does the null return difference with respect to Y_3:

$$F_{Y_3}^{\circ} = \frac{N(s)}{A(s)} = 1 + Y_3 \frac{Y_1 + Y_6}{Y_6(Y_1 + Y_2 + Y_4)} \qquad (3\text{-}122c)$$

As a final, more practical, application of Nathan's "infinite gain" rule, consider the active network shown in Fig. 3-42. Writing the nonconstrained matrix $[y']$ we obtain by inspection

$$[y'] = \begin{array}{c} \\ \\ \\ \\ \end{array} \left[\begin{array}{cccc} \overset{1}{G_1} & \overset{2}{-G_1} & \overset{3}{0} & \overset{4}{0} \\ -G_1 & G_1 + G_2 + G_3 + sC_1 & -G_2 & -G_3 \\ 0 & -G_2 & G_2 + sC_2 & -sC_2 \\ 0 & -G_3 & -sC_2 & G_3 + sC_2 \end{array} \right] \begin{array}{c} 1 \\ 2 \\ 3 \\ 4 \end{array} \qquad (3\text{-}123)$$

Deleting the driven row 4 and the driving column 3 we obtain the constrained matrix

$$[y] = \left[\begin{array}{ccc} \overset{1}{G_1} & \overset{2}{-G_1} & \overset{4}{0} \\ -G_1 & G_1 + G_2 + G_3 + sC_1 & -G_3 \\ 0 & -G_2 & -sC_2 \end{array} \right] \begin{array}{c} 1 \\ 2 \\ 3 \end{array} \qquad (3\text{-}124)$$

and with Cramer's rule the transmittance from 1 to 4 follows:

$$\frac{V_4}{V_1} = \frac{\bar{y}_4^1}{\bar{y}_1^1} = \frac{G_1 C_2}{\begin{vmatrix} G_1 + G_2 + G_3 + sC_1 & -G_3 \\ -G_2 & -sC_2 \end{vmatrix}}$$

$$= \frac{-G_1 G_2}{s^2 C_1 C_2 + sC_2(G_1 + G_2 + G_3) + G_2 G_3} \qquad (3\text{-}125)$$

This is the voltage transfer function of a low-pass filter. Notice that the topology of the network is identical to that of the general network in Fig. 3-32.

Thus (3-125) could have been obtained by substituting the admittances given in Fig. 3-42 into the general transfer function given by (3-81). The procedure followed here is simple enough, however, to warrant a direct analysis, in order to remain independent of general network results that may not be readily at hand.

3.2.3 Operational Amplifiers in the Differential–Input Mode

Let us now consider another common constraint imposed on networks using operational amplifiers. Here we consider the case of a network using the operational amplifier in the differential-input infinite-gain mode, as shown in Fig. 3-43. Condition (3-112), that the input voltage be zero, becomes in this case

$$V_i = V_k \qquad (3\text{-}126)$$

This suggests that the sum of columns i and k should replace column i and that column k and row j should be deleted in the nonconstrained admittance matrix. These two steps can be summarized by the following simple rule:

Differential mode, $\beta \to \infty$: To obtain the constrained matrix $[y]$, add the two columns corresponding to the two driving nodes (e.g., nodes i and k), delete a column corresponding to one of the two driving nodes (e.g., node k), and delete the row corresponding to the driven node (e.g., node j) in the nonconstrained matrix $[y']$.

AN EXAMPLE: This rule will be illustrated by one example. We consider the active network shown in Fig. 3-44, for which we wish to determine the voltage transfer ratio V_o/V_i. The nonconstrained $[y']$ matrix is obtained by inspection:

$$[y'] = \begin{matrix} & 1 & 2 & 3 & 4 & \\ \begin{bmatrix} Y_a + Y_c & -Y_a & -Y_c & 0 \\ -Y_a & Y_a + Y_b & 0 & 0 \\ -Y_c & 0 & Y_c + Y_d & -Y_d \\ 0 & 0 & -Y_d & Y_d \end{bmatrix} & \begin{matrix} 1 \\ 2 \\ 3 \\ 4 \end{matrix} \end{matrix} \qquad (3\text{-}127)$$

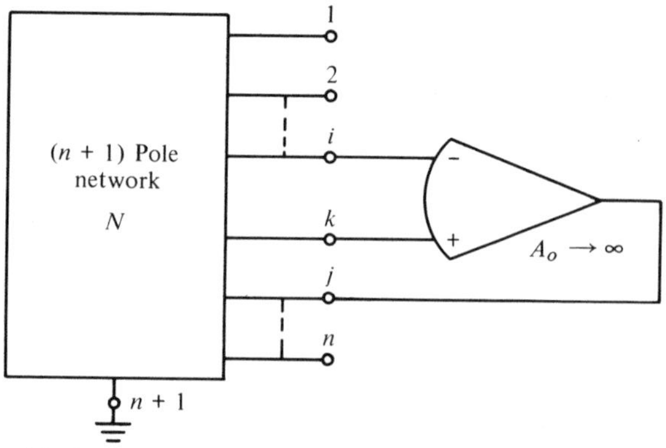

FIG. 3-43. Active multipole incorporating an operational amplifier in the differential mode.

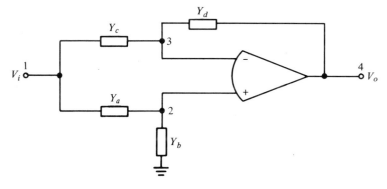

FIG. 3-44. Active network with an operational amplifier in the differential mode.

Adding columns 2 and 3 and deleting column 3 and row 4, we obtain the constrained matrix

$$
[y] = \begin{bmatrix} Y_a + Y_c & -(Y_a + Y_c) & 0 \\ -Y_a & Y_a + Y_b & 0 \\ -Y_c & Y_c + Y_d & -Y_d \end{bmatrix} \begin{matrix} 1 \\ 2 \\ 3 \end{matrix}
\qquad \begin{matrix} 1 & \quad 2 & \quad 4 \end{matrix}
$$

(3-128)

From (3-128) we have

$$
\frac{V_o}{V_i} = \frac{V_4}{V_1} = \frac{\bar{y}_4^1}{\bar{y}_1^1} = \frac{\begin{vmatrix} -Y_a & Y_a + Y_b \\ -Y_c & Y_c + Y_d \end{vmatrix}}{\begin{vmatrix} Y_a + Y_b & 0 \\ Y_c + Y_d & -Y_d \end{vmatrix}} = \frac{Y_a Y_d - Y_b Y_c}{Y_d(Y_a + Y_b)}
$$

(3-129)

Equation (3-129) represents the desired voltage transfer ratio. Because the amplifier gain A was assumed to be infinitely large, the transfer function is independent of the operational amplifier characteristics.

3.3 THE ANALYSIS OF LADDER NETWORKS

Since ladder networks occur frequently in practice we shall briefly discuss some useful methods of coping with them analytically. Let us first consider the two-terminal ladder structure shown in Fig. 3-45a. Notice that the series branches are characterized by impedances, the shunt branches by admittances. We define Z_i as the input impedance of the ladder looking into series branch i, and Y_j as the input admittance looking into shunt branch j. Thus Z_1 is the input impedance of the ladder.

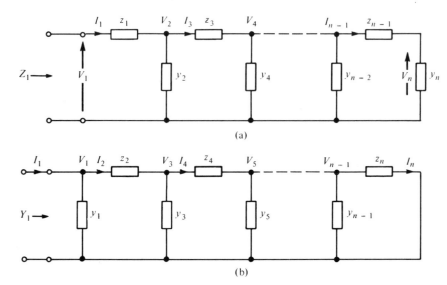

FIG. 3-45. Two-terminal ladder network (a) and its dual (b).

3.3.1 Driving Point Functions

To calculate the input or driving point impedance Z_1 of the ladder network in Fig. 3-45a, we start at the far end and can say immediately that $Y_n = y_n$. Z_{n-1} is the impedance of the series connection of branches n and $n-1$ and therefore equals $z_{n-1} + (1/y_n)$. Similarly $Y_{n-2} = y_{n-2} + 1/(z_{n-1} + 1/y_n)$. Thus, by iteration, the Z_i and Y_i are given, respectively,

$$Z_i = z_i + \frac{1}{Y_{i+1}} \tag{3-130a}$$

$$Y_j = y_j + \frac{1}{Z_{j+1}} \tag{3-130b}$$

Applying this iterative procedure to the entire ladder network, we obtain the input impedance as

$$Z_1 = z_1 + \cfrac{1}{y_2 + \cfrac{1}{z_3 + \cfrac{1}{y_4 + \cfrac{}{\ddots + \cfrac{1}{z_{n-1} + \cfrac{1}{y_n}}}}}} \tag{3-131}$$

The right side of (3-131) is called a *continued fraction*, since the fractional notation is continued from term to term. Since the form of (3-131) is cumbersome, it is generally written more compactly in the form

$$Z_1 = z_1 + \cfrac{1}{y_2} + \cfrac{1}{z_3} + \cfrac{1}{y_4} + \cdots + \cfrac{1}{z_{n-1}} + \cfrac{1}{y_n} \qquad [3\text{-}132]$$

Note that the signs of addition (and subtraction) are in the denominator.

A very useful feature of (3-132) is that there is a one-to-one relationship between the branch impedances and admittances of the ladder network and the terms in the continued fraction expansion. Thus it is a very simple matter to find a ladder corresponding to a given continued fraction, or, conversely to find a fraction corresponding to a given ladder network.

An important characteristic of ladder networks is that the dual of any given ladder is another ladder whose series impedances and shunt admittances are, respectively, the shunt admittances and series impedances of the original. Thus, the dual of the ladder shown in Fig. 3-45a is the ladder in Fig. 3-45b. The input admittance Y_1 of the dual ladder follows directly:

$$Y_1 = y_1 + \cfrac{1}{z_2} + \cfrac{1}{y_3} + \cfrac{1}{z_4} + \cdots + \cfrac{1}{y_{n-1}} + \cfrac{1}{z_n} \qquad (3\text{-}133)$$

3.3.2 Transfer Functions

Having briefly discussed the driving point functions of ladder networks we shall now consider their transfer characteristics.

For this we return to the ladder networks shown in Fig. 3-45, but now consider them as two-ports. Thus in Fig. 3-45a we consider the voltage transfer function V_n/V_1, where V_1 is the voltage at the input terminals and V_n the voltage across y_n. Likewise, in Fig. 3-45b we consider the current transfer function I_n/I_1. Notice that only the currents through the impedances and the voltages across the admittances are given. They are numbered consecutively in the same way as the corresponding series impedances and shunt admittances. Using Kirchhoff's voltage and current laws alternately, and starting at the input, the following set of simultaneous equations, represented in matrix form, is obtained:

$$
\begin{bmatrix} V_1 \\ I_1 \\ V_2 \\ I_3 \\ \vdots \\ V_{n-2} \\ I_{n-1} \end{bmatrix}
=
\begin{bmatrix} I_1 \\ V_2 \\ I_3 \\ V_4 \\ \vdots \\ I_{n-1} \\ V_n \end{bmatrix}
\begin{bmatrix}
z_1 & 1 & 0 & 0 & 0 & \cdots & 0 \\
0 & y_2 & 1 & 0 & 0 & \cdots & 0 \\
0 & 0 & z_3 & 1 & 0 & \cdots & 0 \\
0 & 0 & 0 & y_4 & 1 & \cdots & 0 \\
\vdots & \vdots & & & & & \vdots \\
& & & & & & 0 \\
& & & & 0 & z_{n-1} & 1 \\
0 & & \cdots & & 0 & 0 & y_n
\end{bmatrix}
\qquad [3\text{-}134a]
$$

This can be rewritten in the form

$$
\begin{bmatrix} V_1 \\ 0 \\ 0 \\ 0 \\ \vdots \\ 0 \\ 0 \end{bmatrix} = \begin{bmatrix} I_1 \\ V_2 \\ I_3 \\ V_4 \\ \vdots \\ I_{n-1} \\ V_n \end{bmatrix} \begin{bmatrix} z_1 & 1 & 0 & 0 & 0 & \cdots & 0 \\ -1 & y_2 & 1 & 0 & 0 & \cdots & 0 \\ 0 & -1 & z_3 & 1 & 0 & \cdots & 0 \\ 0 & 0 & -1 & y_4 & 1 & \cdots & \vdots \\ & & & & & & 0 \\ 0 & & & & -1 & z_{n-1} & 1 \\ 0 & & \cdots & & 0 & -1 & y_n \end{bmatrix}
\qquad \text{[3-134b]}
$$

Solving this set of equations for V_n using Cramer's rule, we obtain

$$
V_n = \frac{\begin{vmatrix} z_1 & 1 & 0 & 0 & 0 & \cdots & & V_1 \\ -1 & y_2 & 1 & 0 & 0 & \cdots & & 0 \\ 0 & -1 & z_3 & 1 & 0 & \cdots & & 0 \\ 0 & 0 & -1 & y_4 & 1 & \cdots & & 0 \\ \vdots & & & & & & & \vdots \\ & & & & & -1 & z_{n-1} & 0 \\ 0 & & & \cdots & & 0 & -1 & 0 \end{vmatrix}}{\begin{vmatrix} z_1 & 1 & 0 & 0 & 0 & & \cdots & 0 \\ -1 & y_2 & 1 & 0 & 0 & & & 0 \\ 0 & -1 & z_3 & 1 & 0 & & & 0 \\ 0 & 0 & -1 & y_4 & 1 & & & \\ \vdots & & & & & & & \vdots \\ & & & & -1 & z_{n-1} & & 1 \\ 0 & & & \cdots & & 0 & -1 & y_n \end{vmatrix}}
\qquad (3\text{-}135)
$$

Since V_1 is the only term in the last column in the numerator, it is easy to expand the numerator by minors along the last column. The single cofactor of V_1 is always -1, and therefore the entire numerator is always equal to V_1 when the original equations are in the form of (3-134b). As a result it is more convenient, in calculating the transfer function of a ladder network, to consider V_i/V_o (or in the present case V_1/V_n) then V_o/V_i (or V_n/V_1). From (3-135) we therefore obtain, for the general ladder network of Fig. 3-45a,

$$
\frac{V_1}{V_n} = \begin{vmatrix} z_1 & 1 & 0 & 0 & \cdots & & 0 \\ -1 & y_2 & 1 & 0 & \cdots & & 0 \\ 0 & -1 & z_3 & 1 & \cdots & & 0 \\ 0 & 0 & -1 & y_4 & \cdots & & 0 \\ \vdots & & & & & & \vdots \\ & & & & 0 & z_{n-1} & 1 \\ 0 & & \cdots & & -1 & -1 & y_n \end{vmatrix}
\qquad (3\text{-}136)
$$

Clearly, with this general expression, the determinant defining the reciprocal transfer function of an arbitrary ladder network can be obtained by inspection. The network impedances and admittances appear only in the main diagonal of the determinant and progress alternately from z_1 to y_n.

3.3.3 Continuants

The determinant in (3-136) has a special form, in that the variables occur only in the main diagonal. The lower adjacent diagonal consists only of -1 elements, while the upper adjacent diagonal is made up only of $+1$ elements. All other elements of the determinant are zeros. Determinants which have this peculiar form are called continuants.

Passive Ladder Networks   As an example, to show the ease of obtaining the input–output voltage transfer function by use of the continuant, we consider the RC ladder network shown in Fig. 3-46. Writing the corresponding continuant by inspection we obtain

$$\frac{V_1}{V_6} = \begin{vmatrix} R_1 & 1 & 0 & 0 & 0 & 0 \\ -1 & sC_2 & 1 & 0 & 0 & 0 \\ 0 & -1 & R_3 & 1 & 0 & 0 \\ 0 & 0 & -1 & sC_4 & 1 & 0 \\ 0 & 0 & 0 & -1 & R_5 & 1 \\ 0 & 0 & 0 & 0 & -1 & sC_6 \end{vmatrix} \tag{3-137}$$

In the reverse situation, if the continuant (3-137) were given, it would be easy enough to derive the ladder network from it; the component values for the network elements correspond directly to the elements of the main diagonal of the continuant. Thus here again, as with the continued-fraction representation of the driving point function of a two-terminal ladder network, a one-to-one relationship exists between the continuant and the physical ladder network. The continuant notation is perfectly general and applies to any linear passive ladder network.

Thanks to the duality property of ladder networks mentioned earlier, it is a simple matter to derive the continuant for the dual of a ladder network. The dual of the ladder network in Fig. 3-47a is shown in Fig. 3-47b. The continuant of this dual network has exactly the same form as that of Fig. 3-47a except that all quantities are duals. Thus the first element is a shunt admittance, and

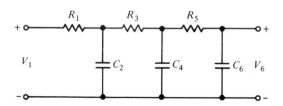

Fig. 3-46. Two-port RC ladder network.

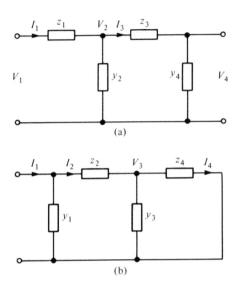

FIG. 3-47. Two-port ladder network (a) and its dual (b).

we calculate the reciprocal current, instead of the reciprocal voltage, transfer function. By inspection we obtain

$$\frac{I_1}{I_4} = \begin{vmatrix} y_1 & 1 & 0 & 0 \\ -1 & z_2 & 1 & 0 \\ 0 & -1 & y_3 & 1 \\ 0 & 0 & -1 & z_4 \end{vmatrix} \qquad (3\text{-}138)$$

Having identified the (reciprocal) transfer function of a ladder network as a continuant, it is not unreasonable to expect that this provides us with some analytical advantages. This is indeed the case, as we shall see if we consider the theory of continuants in a little detail.[17] For this purpose let us define the general ladder network shown in Fig. 3-48. In terms of its elements, $a_1, a_3, \ldots, a_{n-1}$ are the impedances of the series branches and $a_2, a_4, \ldots a_n$ are the admittances of the shunt branches. We now define the continuant, which describes the input–output voltage ratio of the ladder, by $K(1, n)$, so that

$$\frac{V_1}{V_2} = \begin{vmatrix} a_1 & 1 & 0 & \cdots & \cdots & & 0 \\ -1 & a_2 & 1 & & \cdots & & 0 \\ \vdots & & & & & & \vdots \\ & & & & -1 & a_{n-1} & 1 \\ 0 & & & \cdots & & -1 & a_n \end{vmatrix} = K(1, n) \qquad [3\text{-}139]$$

17. T. Muir and W. H. Metzler, *A Treatise on the Theory of Determinants*, (New York: Dover, 1960).

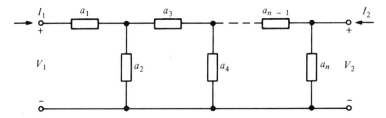

FIG. 3-48. General ladder network.

Similarly $K(i, n - j)$ is the continuant resulting from (3-139) when the first $i - 1$ and the last $n - j$ rows and columns are removed. Its determinant is of $(n - j - i + 1)$th order.

$K(1, n)$ is called a continuant of order n. It can be expanded by using Euler's rule, which states:

A continuant of order n is expanded by writing down the product of the main diagonal $a_1 a_2 a_3 \cdots a_n$. This is the first term of the expansion. To find the remaining terms, omit one or more pairs of consecutive factors from this product in all possible ways, always replacing each consecutive pair of factors omitted by unity.

An example will clarify this rule. Consider a continuant of order five:

$$K(1, 5) = \begin{vmatrix} a_1 & 1 & 0 & 0 & 0 \\ -1 & a_2 & 1 & 0 & 0 \\ 0 & -1 & a_3 & 1 & 0 \\ 0 & 0 & -1 & a_4 & 1 \\ 0 & 0 & 0 & -1 & a_5 \end{vmatrix} \qquad (3\text{-}140)$$

The first term of the expansion, the product of the main diagonal, is $a_1 a_2 a_3 a_4 a_5$. The following consecutive pairs, and groups of pairs can be omitted from this product: $a_1 a_2$, $a_2 a_3$, $a_3 a_4$, $a_4 a_5$, $a_1 a_2 a_3 a_4$, $a_1 a_2 a_4 a_5$, $a_2 a_3 a_4 a_5$. As a result we obtain the following expansion of (3-140):

$$K(1, 5) = a_1 a_2 a_3 a_4 a_5 + a_3 a_4 a_5 + a_1 a_4 a_5$$
$$+ a_1 a_2 a_5 + a_1 a_2 a_3 + a_5 + a_3 + a_1 \qquad (3\text{-}141)$$

If the order of the continuant is even, the last term of its expansion will be unity.

A continuant can be decomposed into the sum of lower-order continuants according to the rule:

$$K(1, n) = a_1 K(2, n) + K(3, n) \qquad [3\text{-}142]$$

where $K(k, n)$ is the continuant remaining after the first $k - 1$ rows and columns have been removed from $K(1, n)$. This follows directly by expanding (3-139) along the first row or column. Since the following symmetry property holds:

$$K(a_1, a_2, \ldots a_n) = K(a_n, a_{n-1}, \ldots a_1) \qquad (3\text{-}143)$$

the decomposition rule of (3-142) can also be written

$$K(1, n) = a_n K(1, n - 1) + K(1, n - 2) \qquad [3\text{-}144]$$

Employing the rules of (3-142) and (3-144) in the opposite direction, i.e., to ascend from lower- to higher-order continuants, they become recursion formulas. Note that a zero-order continuant equals unity.

The continuant of a ladder network is closely related to its $[ABCD]$ or transmission matrix. From Table 3-1 and Fig. 3-48 we find, if the first branch is a series branch, that

$$\frac{V_1}{V_2} = K(1, n) = A \qquad [3\text{-}145]$$

The current gain with the output short-circuited (i.e., $V_2 = 0$), given by $D = -I_1/I_2$ of the transmission matrix, is obtained for the ladder network of Fig. 3-48 by removing the first and last branch and calculating the corresponding continuant $K(2, n - 1)$. (Notice that the ladder then has the form of the ladder in Fig. 3-47b, which was characterized by the current ratio (3-138). Obviously this current ratio is not changed by adding a series branch in front, and a shunt branch at the end of the ladder, since I_1 flows from a current source and V_2 is equal to zero.) Thus

$$-\frac{I_1}{I_2}\bigg| = K(2, n - 1) = D \qquad [3\text{-}146]$$

Similar reasoning provides the other two transmission matrix parameters B and C in terms of the continuant of the ladder network in Fig. 3-48. This is summarized in the first row of Table 3-7. The remaining rows give the relationship between transmission matrix parameters and continuants for the other basic types of ladder configurations.

One other feature of continuants is noteworthy.[18] From a consideration of the recursion formulas (3-142) and (3-144) we can readily derive three of the transmission parameters of a ladder network from the fourth. Thus, having calculated $A = K(1, n)$ for the even-order ladder network whose first branch is in series, we obtain the parameters B, C, and D directly from A:

$$A = K(1, n); \qquad B = \frac{\partial A}{\partial a_n}$$

$$C = \frac{\partial A}{\partial a_1}; \qquad D = \frac{\partial^2 A}{\partial a_1 \partial a_n} \qquad [3\text{-}147]$$

Similarly, having derived the parameter $C = K(1, n)$ for the odd-order ladder whose first branch is in shunt, the parameters A, B, and D can be derived from it, using the relations (3-142) and (3-144). The resulting expressions for the four basic ladder networks have been included in Table 3-7. Incidentally,

18. M. N. S. Swamy, Continuants and ladder networks, *Proc. IEEE*, **54**, 1110–1111 (1966).

TABLE 3-7. TRANSMISSION MATRIX PARAMETERS OF LADDER NETWORKS IN TERMS OF THEIR CONTINUANTS*

Ladder type	n	Parameter			
		A	B	C	D
First branch is a *series* branch	even	$K(1, n)$	$K(1, n-1) = \dfrac{\partial A}{\partial a_n}$	$K(2, n) = \dfrac{\partial A}{\partial a_1}$	$K(2, n-1) = \dfrac{\partial^2 A}{\partial a_1 \partial a_n}$
First branch is a *series* branch	odd	$K(1, n-1) = \dfrac{\partial B}{\partial a_n}$	$K(1, n)$	$K(2, n-1) = \dfrac{\partial^2 B}{\partial a_1 \partial a_n}$	$K(2, n) = \dfrac{\partial B}{\partial a_1}$
First branch is a *shunt* branch	odd	$K(2, n) = \dfrac{\partial C}{\partial a_1}$	$K(2, n-1) = \dfrac{\partial^2 C}{\partial a_1 \partial a_n}$	$K(1, n)$	$K(1, n-1) = \dfrac{\partial C}{\partial a_n}$
First branch is a *shunt* branch	even	$K(2, n-1) = \dfrac{\partial^2 D}{\partial a_1 \partial a_n}$	$K(2, n) = \dfrac{\partial D}{\partial a_1}$	$K(1, n-1) = \dfrac{\partial D}{\partial a_n}$	$K(1, n)$

$$\Delta A = AD - BC = 1$$

* L. Storch, Continuants—a superior tool in the analysis of ladder networks and ladders of recurrent networks, *IEEE Trans. Circuit Theory*, CT-**12**, 444–446 (1965).

TABLE 3-8. THE z AND y PARAMETERS OF TWO LADDER CONFIGURATIONS IN TERMS OF THEIR CONTINUANTS

		Parameter	
Ladder Type	n	$[z_{ij}]$	$[y_{ij}]$
First branch is a *series* branch	even	$\dfrac{1}{K(2,n)}\begin{bmatrix} K(1,n) & 1 \\ 1 & K(2,n-1) \end{bmatrix}$ $\Delta z = z_{11}z_{22} - z_{12}z_{21} = \dfrac{K(1,n-1)}{K(2,n)}$	$\dfrac{1}{K(1,n-1)}\begin{bmatrix} K(2,n-1) & -1 \\ -1 & K(1,n) \end{bmatrix}$ $\Delta y = y_{11}y_{22} - y_{12}y_{21} = \dfrac{K(2,n)}{K(1,n-)}$
First branch is a *shunt* branch	odd	$\dfrac{1}{K(1,n)}\begin{bmatrix} K(2,n) & 1 \\ 1 & K(1,n-1) \end{bmatrix}$ $\Delta z = z_{11}z_{22} - z_{12}z_{21} = \dfrac{K(2,n-1)}{K(1,n)}$	$\dfrac{1}{K(2,n-1)}\begin{bmatrix} K(1,n-1) & -1 \\ -1 & K(2,n) \end{bmatrix}$ $\Delta y = y_{11}y_{22} - y_{12}y_{21} = \dfrac{K(1,n)}{K(2,n-)}$

once the continuant of a ladder network is known, it is a simple matter to derive the sensitivities of its network functions to coefficient variations.[19]

Having obtained the transmission parameters it is easy enough, via the matrix interrelationships given in Table 3-3, to obtain other matrix parameters in terms of continuants. The z and y parameters for two basic ladder configurations are given in Table 3-8. The continuants for the remaining two ladder configurations can be obtained in the same manner.

Active Ladder Networks   So far we have considered passive ladder networks only. As we shall see in later chapters many active networks consist of nothing more than passive ladder networks with voltage feedback from the output to one or more of the internal (shunt) admittances. This feedback can be accounted for in the form of a controlled voltage source $\beta_i V_0$ in series with the corresponding admittance y_i. An example of an active ladder network of this kind is shown in Fig. 3-49. Ideal voltage sources with zero output impedance and individually adjustable gains β_2, β_4, etc., are assumed. As we know, operational amplifiers in the noninverting mode comply excellently with these assumptions. Thus a practical realization of the active ladder in Fig. 3-49 may have the form shown in Fig. 3-50.

Extending the use of continuants to active ladder networks and thereby making them as easily accessible to analysis by inspection as passive ladder networks, we proceed as follows [20] Writing Kirchhoff's voltage and current laws alternately, we obtain a set of equations of the same form as those in (3-134) except that each equation containing the admittance y_i will also have

19. Y. Ceyhun: On the properties of continuants and sensitivity computations, *IEEE Trans. Circuit Theory*, **CT-20**, 167–169 (1973).
20. J. G. Holbrook, The recurrent–continuant method of transfer function synthesis, *Radio Electron. Engr*, **38**, 73–79 (1969).

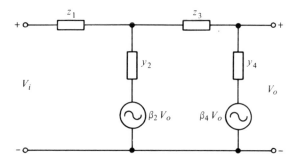

FIG. 3-49. General active ladder network.

the term $-\beta_i y_i V_o$ added to it. This fact can be used as the basis for the following two simple rules for writing the continuant of an active ladder network by inspection:

1. Assume that the gain of each amplifier is zero and write the normal continuant by inspection.
2. In each row in which y_i appears, add the term $-\beta_i y_i$ at the intersection of that row and the *last* column.

Using these rules, the modified continuant for the ladder of Fig. 3-49 can be easily written by inspection. We obtain

$$\frac{V_i}{V_o} = \begin{vmatrix} z_1 & 1 & 0 & 0 \\ -1 & y_2 & 1 & -\beta_2 y_2 \\ 0 & -1 & z_3 & 1 \\ 0 & 0 & -1 & y_4(1-\beta_4) \end{vmatrix} \quad (3\text{-}148)$$

If some of the voltage sources β_i are absent, then the intersection of the corresponding rows with the last column remains zero as in the passive case.

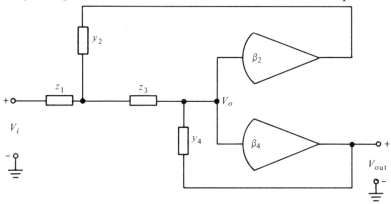

FIG. 3-50. Active ladder network using noninverting operational amplifiers. After: G. G. Bloodworth and N. R. S. Nesbitt, Active filters operating below 1 Hz, *IEEE Trans. Instrum. Meas.*, **IM-16**, 115-120, (1967).

Clearly the form of (3-148) is not that of a continuant as defined by (2-139), and therefore cannot be expanded directly using the theory of continuants, unless it can be rearranged into normal continuant form. If this is not the case the determinant in (3-148) is expanded by conventional means.

3.3.4 Recurrents and Continuants

One procedure that has been found useful in connection with the continuants of passive ladder networks is the conversion from a recurrent into a continuant.[21] A recurrent is a determinant characterizing a general polynomial of the form

$$P_n(s) = \sum_{i=0}^{n} C_i s^i = C_n s^n + C_{n-1} s^{n-1} + \cdots + C_1 s + C_0 \qquad [3\text{-}149]$$

Consider, for example, a fifth-order polynomial

$$P_5(s) = C_5 s^5 + C_4 s^4 + C_3 s^3 + C_2 s^2 + C_1 s + C_0 \qquad (3\text{-}150)$$

This can be written in the following determinant form:

$$P_5(s) = \begin{vmatrix} C_5 & C_4 & C_3 & C_2 & C_1 & C_0 \\ -1 & s & 0 & 0 & 0 & 0 \\ 0 & -1 & s & 0 & 0 & 0 \\ 0 & 0 & -1 & s & 0 & 0 \\ 0 & 0 & 0 & -1 & s & 0 \\ 0 & 0 & 0 & 0 & -1 & s \end{vmatrix} \qquad (3\text{-}151)$$

This determinant is called a recurrent. Notice how similar it is in form to that of a continuant. It is characterized by the fact that the coefficients of the polynomial make up its first row, the main diagonal (except for the first element) consists of the polynomial variable s, the lower adjacent diagonal is made up entirely of -1 elements, and all other elements are zero. Generally the term C_0 equals unity.

It can be shown that a recurrent can be converted into a continuant by a process involving nothing more than elementary arithmetic operations. The implications of this procedure are obvious. Assuming that a desired voltage transfer function is given in polynomial form (e.g., the input–output ratio of a low-pass function will have the form $V_i/V_o = D(s)/N_0$ where $D(s)$ is a polynominal of nth order and N_0 a constant), it can be expressed in the form of a recurrent. Converting the recurrent into a continuant, the elements of the corresponding ladder network result in terms of the polynominal coefficients, by inspection.

21. J. G. Holbrook, op. cit.

Assume, for example, that the following third-order low-pass function is desired:

$$\frac{V_i}{V_o} = s^3 + a_2 s^2 + a_1 s + 1 \qquad (3\text{-}152)$$

By inspection the corresponding recurrent is:

$$\frac{V_i}{V_o} = \begin{vmatrix} 1 & a_2 & a_1 & 1 \\ -1 & s & 0 & 0 \\ 0 & -1 & s & 0 \\ 0 & 0 & -1 & s \end{vmatrix} \qquad (3\text{-}153)$$

Dividing column 2 and multiplying row 3 by a_2 (which does not change the value of the determinant) we obtain:

$$\frac{V_i}{V_o} = \begin{vmatrix} 1 & 1 & a_1 & 1 \\ -1 & s/a_2 & 0 & 0 \\ 0 & -1 & a_2 s & 0 \\ 0 & 0 & -1 & s \end{vmatrix} \qquad (3\text{-}154)$$

Now, multiplying column 1 by $-a_1$ and adding it to column 3 and subtracting column 2 from column 4 we obtain

$$\frac{V_i}{V_o} = \begin{vmatrix} 1 & 1 & 0 & 0 \\ -1 & s/a_2 & a_1 & -s/a_2 \\ 0 & -1 & a_2 s & 1 \\ 0 & 0 & -1 & s \end{vmatrix} \qquad (3\text{-}155)$$

Multiplying row 4 by $1/a_2$ and adding to row 2, we get

$$\frac{V_i}{V_o} = \begin{vmatrix} 1 & 1 & 0 & 0 \\ -1 & \dfrac{s}{a_2} & a_1 - \dfrac{1}{a_2} & 0 \\ 0 & -1 & a_2 & 1 \\ 0 & 0 & -1 & s \end{vmatrix} \qquad (3\text{-}156)$$

The upper adjacent diagonal can now be converted to unity values by dividing row 2 by $(a_1 - 1/a_2)$ and multiplying column 1 by the same quantity:

$$\frac{V_i}{V_o} = \begin{vmatrix} a_1 - \dfrac{1}{a_2} & 1 & 0 & 0 \\ -1 & \dfrac{s}{a_1 a_2 - 1} & 1 & 0 \\ 0 & -1 & a_2 s & 1 \\ 0 & 0 & -1 & s \end{vmatrix} \qquad (3\text{-}157)$$

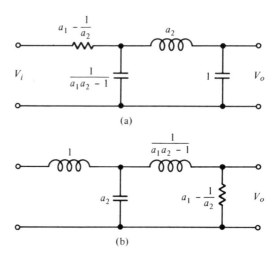

FIG. 3-51. Two ladder networks corresponding to the continuant of (3-157).

Thus we have obtained the continuant corresponding to the desired transfer function (3-152). The elements of the ladder network realizing this function are given in terms of the polynominal coefficients and can be obtained by inspection. The corresponding ladder network is shown in Fig. 3-51a. Notice that numerous other ladder networks can be obtained from a continuant by carrying out simple operations such as row and column exchanges that do not change its value. Furthermore, the symmetry property (3-143) permits us to reverse the main diagonal of a continuant end-for-end without changing its value. The equivalent ladder network resulting from such a reversal is shown in Fig. 3-51b.

The preceding example showed that the recurrent–continuant method of synthesizing ladder networks is particularly useful for low-pass functions, since these involve only one polynomial, rather than the ratio of two polynomials. In the case of high-pass, bandpass, or band-rejection networks the corresponding modifications of the low-pass prototype network must be made.

3.4 SCALING AND DUALITY OF RC NETWORKS

3.4.1 Scaling

We consider here a passive RC network N. The loop equations in matrix form are given by[22]

22. Part of this discussion is based on unpublished notes by A. L. Pappas at Bell Telephone Laboratories.

$$\begin{bmatrix} V_1 \\ V_2 \\ \vdots \\ V_n \end{bmatrix} = \begin{bmatrix} \varkappa_{11} & \varkappa_{12} & \cdot & \cdot & \cdot & \varkappa_{1n} \\ \varkappa_{21} & \varkappa_{22} & & & & \varkappa_{2n} \\ \vdots & \vdots & & & & \vdots \\ \varkappa_{n1} & \varkappa_{n2} & \cdot & \cdot & \cdot & \varkappa_{nn} \end{bmatrix} \begin{bmatrix} I_1 \\ I_2 \\ \vdots \\ I_n \end{bmatrix} \tag{3-158}$$

where $\varkappa_{ij}$ is of the form

$$\varkappa_{ij} = \varkappa_{ij}(R_{ij}, C_{ij}, s) = R_{ij} + \frac{1}{sC_{ij}} \tag{3-159}$$

Equation (3-158) can be written more compactly in the form

$$[V] = [\varkappa][I] \tag{3-160}$$

Assuming that N is a two-port, i.e., that only V_1 and V_2 are nonzero and only I_1 and I_2 flow through the corresponding external ports, we can write

$$I_1 = \frac{\Delta_{11}}{\Delta} V_1 + \frac{\Delta_{21}}{\Delta} V_2 = y_{11} V_1 + y_{12} V_2$$

$$I_2 = \frac{\Delta_{12}}{\Delta} V_1 + \frac{\Delta_{22}}{\Delta} V_2 = y_{21} V_1 + y_{22} V_2 \tag{3-161}$$

Δ is the determinant of the matrix $[\varkappa]$ and Δ_{ij} is the cofactor of the element in the ith row and jth column. The y parameters in (3-161) are, of course, the short-circuit admittance parameters of the resulting two-port. The open-circuit impedance or z parameters, which would have resulted from the *nodal* equations characterizing the network N, are related to the y parameters as follows (see Table 3-3):

$$\begin{bmatrix} z_{11} & z_{12} \\ z_{21} & z_{22} \end{bmatrix} = \frac{1}{\Delta y} \begin{bmatrix} y_{22} & -y_{12} \\ -y_{21} & y_{11} \end{bmatrix} \tag{3-162}$$

where $\Delta y = y_{11} y_{22} - y_{12} y_{21}$. In the case of a reciprocal network we have $y_{12} = y_{21}$ and $z_{12} = z_{21}$ (see Table 3-1).

Let us now scale each resistor in N by the factor r and each capacitor by the factor c. Then a typical entry $\varkappa$ in the loop matrix $[\varkappa]$ transforms as follows:

$$\varkappa(R, C, s) \rightarrow \varkappa(rR, cC, s) \tag{3-163}$$

With (3-159)

$$\varkappa(rR, cC, s) = r\left(R + \frac{1}{srcC}\right) = r\varkappa(R, C, rcs) \tag{3-164}$$

Thus

$$\varkappa(rR, cC, s) = r\varkappa(R, C, rcs) \tag{3-165}$$

Notice that if $r = 1/c$ the impedance level is changed but the frequency behavior is not.

Letting $[\varkappa']$ be the loop matrix of the scaled network N', we can now express the corresponding cofactors $\Delta'_{ij}(s)$ and the determinant $\Delta'(s)$ by the cofactors $\Delta_{ij}(s)$ and by the determinant $\Delta(s)$ of the original $n \times n$ matrix $[\varkappa]$, as follows:

$$\Delta'(s) = r^n \Delta(rcs) \tag{3-166}$$

and

$$\Delta'_{ij}(s) = r^{n-1} \Delta_{ij}(rcs) \tag{3-167}$$

Observe that the cofactors correspond to $(n-1) \times (n-1)$ submatrices of $[\varkappa]$.

The y parameters of the scaled network N' follow immediately. Using primes to indicate the transformed quantities, we have

$$y'_{ij}(s) = \frac{\Delta'_{ji}(s)}{\Delta(s)} = \frac{r^{n-1} \Delta_{ji}(rcs)}{r^n \Delta(rcs)} = \frac{1}{r} y_{ij}(rcs) \tag{3-168}$$

The transformation of the z parameters follows from (3-162) and (3-168):

$$z'_{ij}(s) = r z_{ij}(rcs) \tag{3-169}$$

Letting $T(s)$ denote any of the dimensionless (e.g., voltage or current) transfer functions given in Table 3-4, we conclude that

$$T'(s) = T(rcs) \tag{3-170}$$

These results can be summarized as follows:

Let N be a passive RC network and N' the network obtained from N by scaling each resistor by the factor r and each capacitor by the factor c. Let $F(R_i, C_j, s)$ denote a network function of N. Then the corresponding network function $F'(R_i, C_j, s)$ of N' results:

$$F'(R_i, C_j, s) = F(rR_i, cC_j, s) = k'F(R_i, C_j, rcs) \tag{3-171}$$

where

$$k' = \begin{cases} \dfrac{1}{r} & \text{when } F \text{ is an admittance function} \\ 1 & \text{when } F \text{ is a transfer function} \\ r & \text{when } F \text{ is an impedance function} \end{cases}$$

When $r = 1/c$, N' is an *impedance scaled* version of N with the same frequency behavior as N; for all other cases, N' is both *impedance* and *frequency scaled*.

3.4.2 RC Duality and the RC:CR Transformation

It is noteworthy that in the discussion above, the scaling factors r and c were not required to be real. Let us then consider the special case for which r and c are complex, namely $r = c = 1/s$, where $s = \sigma + j\omega$. Then

$$R_i \to \frac{R_i}{s}; \quad sC_j \to C_j \tag{3-172}$$

Thus from (3-159)

$$R_i + \frac{1}{sC_j} \to \frac{R_i}{s} + \frac{1}{C_j} = \frac{1}{C_j} + \frac{1}{s(1/R_i)} \tag{3-173a}$$

or

$$\tilde{R}_i = \frac{1}{C_j}; \quad \tilde{C}_j = \frac{1}{R_i} \tag{3-173b}$$

where we use "$\sim$" instead of a prime to denote scaling by $1/s$ instead of by a constant. Consequently a network function $F(R_i, C_j, s)$ becomes

$$F(R_i, C_j, s) \to F\left(\frac{1}{C_j}, \frac{1}{R_i}, s\right) \tag{3-174}$$

Thus we find that for the special case in which we scale each resistor and capacitor of a network N by $1/s$, we obtain a new network $\tilde{N}$ in which each resistor R_i is replaced by a capacitor of value $1/R_i$ and each capacitor C_j by a resistor of value $1/C_j$. The network $\tilde{N}$ is known as the *RC-dual* of the original network N, and the process of scaling by $1/s$ as the *RC:CR trans-formation*. The network function $\tilde{F}(R_i, C_j, s)$ pertaining to $\tilde{N}$ results directly from (3-171) and (3-172):

$$\tilde{F}(R_i, C_j, s) = F\left(\frac{1}{C_j}, \frac{1}{R_i}, s\right) = k(s)F\left(R_i, C_j, \frac{1}{s}\right) \qquad [3\text{-}175]$$

where

$$k(s) = \begin{cases} s & \text{when } F \text{ is an admittance function} \\ 1 & \text{when } F \text{ is a transfer function} \\ \dfrac{1}{s} & \text{when } F \text{ is an impedance function} \end{cases}$$

If we scale the resistors and capacitors by a constant *and* the $1/s$ term, i.e., $R \to (r/s)R$ and $C \to (c/s)C$ we conclude from (3-171) and (3-175) that

$$\tilde{F}'(R_i, C_j, s) = F\left(\frac{1}{cC_j}, \frac{1}{rR_i}, s\right) = k'(s)F\left(R_i, C_j, \frac{rc}{s}\right) \qquad [3\text{-}176]$$

where

$$k'(s) = \begin{cases} \dfrac{s}{r} & \text{when } F \text{ is an admittance function} \\[2mm] 1 & \text{when } F \text{ is a transfer function} \\[2mm] \dfrac{r}{s} & \text{when } F \text{ is an impedance function} \end{cases}$$

The network resulting from this composite transformation is the *scaled RC-dual* of the original network.

As an example of the scaling procedures described above, consider the *RC* network shown in Fig 3-52a, where the resistor values are given in ohms (Ω), the capacitor values in farads (F). Scaling the resistors by $r = \frac{1}{5}$, the capacitors by $c = \frac{1}{3}$ we obtain the scaled network shown in Fig. 3-52b. Carrying out the $RC{:}CR$ transformation on this network, we obtain the scaled *RC*-dual network in Fig. 3-52c.

From (3-175) we can summarize the $RC{:}CR$ transformation of a network N as follows:

Replacing each resistance R_i in N by a capacitor of value $1/R_i$ F and each capacitor C_j by a resistor of value $1/C_j$ Ω, the driving point (transfer) admittance $\tilde{Y}(s)$ of the resulting *RC*-dual network $\tilde{N}$ in terms of the driving point (transfer) admittance of the original network N is given by

$$\tilde{Y}(s) = sY\left(\frac{1}{s}\right) \tag{3-177}$$

Similarly the driving point (transfer) impedances are related by

$$\tilde{Z}(s) = \frac{1}{s}Z\left(\frac{1}{s}\right) \tag{3-178}$$

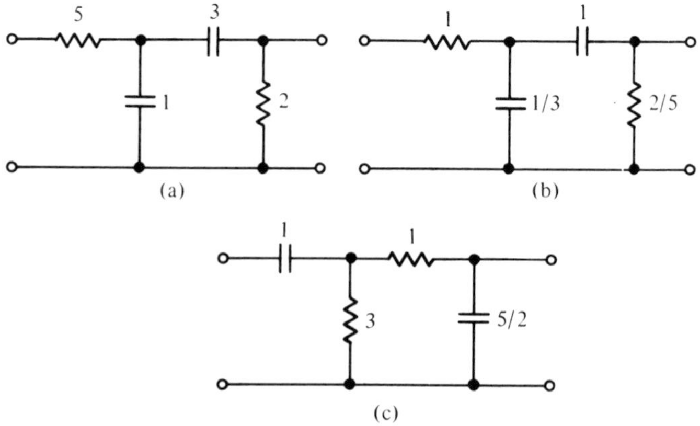

FIG. 3-52. The scaled *RC*-dual of a network. (a) Original. (b) Scaling by $r = \frac{1}{5}$, $c = \frac{1}{3}$. (c) *RC:CR* transformation.

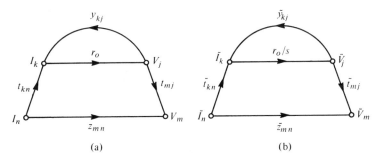

FIG. 3-53. Signal-flow graph of an active RC network incorporating an ideal current-controlled voltage source (a) and its RC-dual (b).

and the transfer voltage (current) ratios by

$$\tilde{T}(s) = T\left(\frac{1}{s}\right) \qquad [3\text{-}179]$$

The $RC:CR$ transformation can now readily be extended to active RC networks.[23] Only networks incorporating controlled sources need be considered, since any active element can be represented in terms of controlled sources. Consider, for example, an active RC network N containing an ideal current-controlled voltage source (CVS) described by a transimpedance of r_0 Ω. Assuming that N is characterized by the signal-flow graph shown in Fig. 3-53a, then we obtain the transfer impedance function $Z_{mn}(s) = V_m/I_n$ by inspection as follows:

$$Z_{mn}(s) = \frac{V_m}{I_n} = z_{mn}(s) + \frac{r_0 t_{kn}(s) t_{mj}(s)}{1 - r_0 y_{kj}(s)} \qquad (3\text{-}180)$$

We now transform the active RC network N into its dual $\tilde{N}$ according to the following rule: Each resistor R_i of N is replaced by a capacitor $\tilde{C}_i = 1/R_i$, and each capacitor C_j by a resistor $\tilde{R}_j = 1/C_j$. Furthermore the CVS of N is replaced by a CVS having a transimpedance r_0/s. Fig. 3-53b shows the signal-flow graph corresponding to $\tilde{N}$, from which we obtain

$$\tilde{Z}_{mn}(s) = \frac{\tilde{V}_m}{\tilde{I}_n} = \tilde{z}_{mn}(s) + \frac{\dfrac{r_0}{s} \tilde{t}_{kn}(s) \tilde{t}_{mj}(s)}{1 - \dfrac{r_0}{s} \tilde{y}_{kj}(s)} \qquad (3\text{-}181)$$

Since $\tilde{z}_{mn}(s)$, $\tilde{y}_{kj}(s)$, $\tilde{t}_{kn}(s)$, and $\tilde{t}_{mj}(s)$ are transfer functions of passive RC networks within $\tilde{N}$ they are related to the corresponding passive RC networks

23. S. K. Mitra, A network transformation for active RC networks, *Proc. IEEE*, 55, 2021–2022 (1967).

of the original network N by (3-177) and (3-179). Thus (3-181) can be re-written as

$$\tilde{Z}_{mn}(s) = \frac{1}{s} z_{mn}\left(\frac{1}{s}\right) + \frac{\dfrac{r_0}{s} t_{kn}\left(\dfrac{1}{s}\right) t_{mj}\left(\dfrac{1}{s}\right)}{1 - \dfrac{r_0}{s} s y_{kj}\left(\dfrac{1}{s}\right)}$$

$$= \frac{1}{s} Z_{mn}\left(\frac{1}{s}\right) \tag{3-182}$$

This relation is also valid if the pertinent network function is a driving point impedance. Furthermore, if other types of controlled sources are substituted for the CVS, similar relations are obtained. The transformation of the controlled sources is included in Table 3-9, in which the generalized $RC:CR$ trans-

TABLE 3-9. GENERALIZED RC:CR
TRANSFORMATION
FOR ACTIVE RC NET-
WORKS

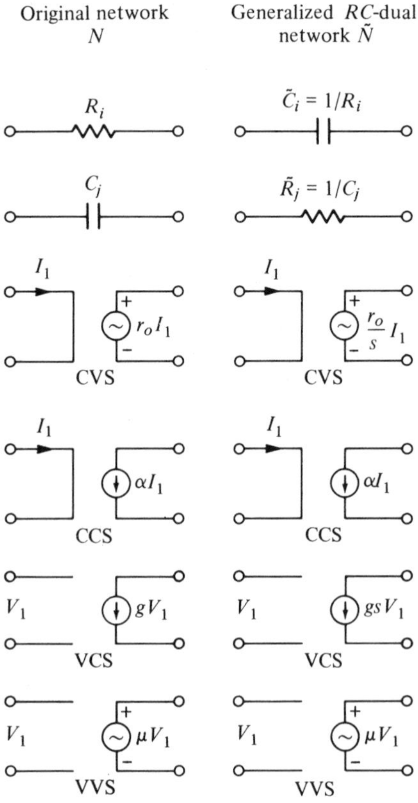

Original network N	Generalized RC-dual network $\tilde{N}$

formation is summarized. Together with the transformations indicated in this table, the network functions of the transformed network $\tilde{N}$ are obtained from the original network N by using the relationships (3-177), (3-178), and (3-179).

3.5 SYMMETRICAL AND POTENTIALLY SYMMETRICAL NETWORKS

One network characteristic that is very useful in connection with RC networks is that of symmetry and, closely related to it, that of potential symmetry. We shall discuss these concepts in this section. Consider the cascade connection of the two networks N_a and N_b shown in Fig. 3-54. By multiplying out the transmission parameters of the two networks and converting these to open-circuit impedance parameters the overall z matrix results as follows:

$$[z]_{ab} = \frac{1}{z_{22a} + z_{11b}} \cdot \begin{bmatrix} z_{11a}z_{11b} + \Delta z_a & z_{12a}z_{12b} \\ z_{21a}z_{21b} & z_{22a}z_{22b} + \Delta z_b \end{bmatrix} \quad (3\text{-}183)$$

If the two networks N_a and N_b are identical and connected in mirror-image symmetry with respect to a vertical center line, as shown in Fig. 3-55, then the total network is said to be structurally and electrically symmetrical. The open-circuit impedance matrix for a physically symmetrical network in terms of the z parameters of one of the symmetrical network halves results:

$$[z]_s = \begin{bmatrix} z_{11} - \dfrac{z_{21}^2}{2z_{22}} & \dfrac{z_{12}^2}{2z_{22}} \\ \dfrac{z_{21}^2}{2z_{22}} & z_{11} - \dfrac{z_{21}^2}{2z_{22}} \end{bmatrix} \quad (3\text{-}184)$$

Therefore

$$z_{11s} = z_{22s} \quad (3\text{-}185)$$

for a symmetrical network. If the symmetrical network is also reciprocal, then

$$z_{12s} = z_{21s} \quad (3\text{-}186)$$

The open-circuit impedance parameters z_{ijs} of the overall network can be obtained in terms of the open- and short-circuit impedances of half the total

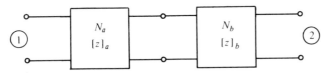

FIG. 3-54. Two networks N_a and N_b in cascade.

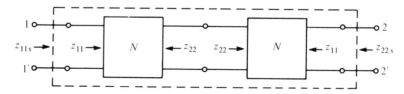

FIG. 3-55. Structurally and electrically symmetrical network generated by two identical networks connected in mirror-image symmetry.

symmetrical network by applying Bartlett's bisection theorem.[24] This theorem says that

$$z_{11s} + z_{21s} = z_{oc} \tag{3-187}$$

and

$$z_{11s} - z_{21s} = z_{sc} \tag{3-188}$$

where z_{oc} and z_{sc} are, respectively, the open- and short-circuit driving point impedances of half of the original symmetrical network as defined in Fig. 3-56. Notice that the two halves of the symmetrical network may be interconnected at an arbitrary number of points. The z matrix of a symmetrical reciprocal network can now be expressed as

$$[z]_s = \begin{bmatrix} \dfrac{z_{oc} + z_{sc}}{2} & \dfrac{z_{oc} - z_{sc}}{2} \\[3mm] \dfrac{z_{oc} - z_{sc}}{2} & \dfrac{z_{oc} + z_{sc}}{2} \end{bmatrix} \tag{3-189}$$

Substituting the expressions given by (3-184) into (3-187) and (3-188), we can also write

$$z_{11s} + z_{21s} = z_{11} \tag{3-190}$$

$$z_{11s} - z_{21s} = \frac{\Delta z}{z_{22}} = \frac{1}{y_{11}} \tag{3-191}$$

A potentially symmetrical network can be defined in terms of a structurally symmetrical network whose right half has been impedance-scaled by a constant factor ρ. This is shown in Fig. 3-57. With (3-183) the corresponding open-circuit impedance matrix then becomes

$$[z]_{ps} = \begin{bmatrix} z_{11} - \dfrac{1}{1+\rho} \cdot \dfrac{z_{21}^2}{z_{22}} & \dfrac{\rho}{1+\rho} \cdot \dfrac{z_{21}^2}{z_{22}} \\[3mm] \dfrac{\rho}{1+\rho} \cdot \dfrac{z_{21}^2}{z_{22}} & \rho z_{11} - \dfrac{\rho^2}{1+\rho} \cdot \dfrac{z_{21}^2}{z_{22}} \end{bmatrix} \tag{3-192}$$

24. E. A. Guillemin, *Synthesis of Passive Networks* (New York: John Wiley & Sons, 1957), pp. 196–200.

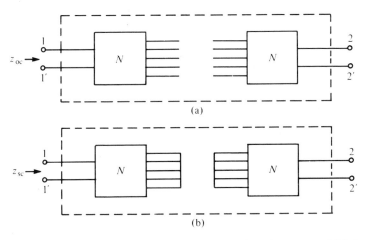

(a)

(b)

FIG. 3-56. Definition of (a) the open-circuit driving point impedance z_{oc} and (b) the short-circuit driving point impedance z_{sc}, pertaining to Bartlett's bisection theorem.

Utilizing the expressions given by (3-184), the open-circuit impedance matrix of the potentially symmetrical network can be given in terms of the impedance parameters of the symmetrical network:

$$[z]_{ps} = \begin{bmatrix} z_{11s} + \dfrac{\rho - 1}{\rho + 1} z_{21s} & \dfrac{2\rho}{1 + \rho} z_{21s} \\ \dfrac{2\rho}{1 + \rho} z_{21s} & \rho\left(z_{11s} - \dfrac{\rho - 1}{\rho + 1} z_{21s}\right) \end{bmatrix} \tag{3-193}$$

Finally, using Bartlett's theorem, the impedance matrix in terms of the open- and short-circuit impedances of one-half the symmetrical network becomes

$$[z]_{ps} = \begin{bmatrix} \dfrac{\rho z_{oc} + z_{sc}}{1 + \rho} & \dfrac{\rho}{1 + \rho}(z_{oc} - z_{sc}) \\ \dfrac{\rho}{1 + \rho}(z_{oc} - z_{sc}) & \dfrac{\rho}{1 + \rho}(z_{oc} + \rho z_{sc}) \end{bmatrix} \tag{3-194}$$

As we would expect, for $\rho = 1$ the terms of this matrix are identical with those of a symmetrical network as given by (3-189).

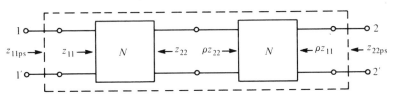

FIG. 3-57. Potentially symmetrical network.

3.6 THE CHARACTERISTIC EQUATION OF LINEAR ACTIVE RC NETWORKS

3.6.1 The Coefficients of the Characteristic Equation

Frequently, rather than requiring the complete transfer function of an active RC network, we need only the characteristic equation, from which some, or all, of the poles can then readily be derived. As we shall see in what follows, the characteristic equation of an RC network can be obtained relatively easily by inspection of the network.[25] To demonstrate this, consider the active RC network shown in Fig. 3-58. It contains only the three capacitors C_1, C_2, and C_3, which are connected to the ports 1, 2, and 3, respectively. No other energy storage elements are assumed within the network.

It can be shown that the coefficients of the characteristic equation of an (active) RC network consist of combinations of the capacitors in the network and the low-frequency short-circuit conductances and open-circuit resistances seen at the terminal pairs associated with the capacitors. To find these conductances and resistances, we remove the capacitors from the network and set any independent sources remaining within the network to zero. This network can then be represented by the node equations (in matrix form)[26]:

$$
\begin{bmatrix} I_1 \\ I_2 \\ I_3 \end{bmatrix} = \begin{bmatrix} g_{11} & g_{12} & g_{13} \\ g_{21} & g_{22} & g_{23} \\ g_{31} & g_{32} & g_{33} \end{bmatrix} \begin{bmatrix} V_1 \\ V_2 \\ V_3 \end{bmatrix}
\qquad [3\text{-}195]
$$

By definition, the conductance seen at the terminal pair j with all other terminal pairs *shorted* is g_{jj}. We also require the resistance at the terminal pair j with all other terminal pairs *open*, which we shall designate R_j^0. For this purpose we can solve (3-195) to obtain the voltage V_j in terms of the current I_j with all other currents set equal to zero. Using Cramer's rule we then obtain (as in (3-88) in Section 3-2)

$$
V_j = I_j \frac{\bar{g}_j^j}{\Delta g}
\qquad (3\text{-}196)
$$

where $\bar{g}_j^j$ is the first cofactor of the jth element in the conductance matrix $[g]$ given by (3-195) and Δg is its determinant. Thus the open-circuit resistance at the terminal pair j is

$$
R_j^0 = \frac{\bar{g}_j^j}{\Delta g}
\qquad [3\text{-}197]
$$

25. R. D. Thornton et al., *Multistage Transistor Circuits*, (New York: John Wiley & Sons, 1965), Chapter 1.
26. We use g_{ij} instead of y_{ij} as the elements of the admittance matrix here to denote the fact that the elements are frequency independent, since the network contains no storage elements.

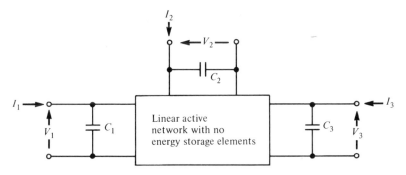

FIG. 3-58. Linear active RC network containing three capacitors.

If now we return the capacitors to the network, we obtain the y matrix of the network:

$$[y] = \begin{bmatrix} g_{11} + sC_1 & g_{12} & g_{13} \\ g_{21} & g_{22} + sC_2 & g_{23} \\ g_{31} & g_{32} & g_{33} + sC_3 \end{bmatrix} \qquad (3\text{-}198)$$

Since the capacitors are connected only across the terminals of the network their susceptances appear only in the principal diagonal of the matrix.

With the determinant Δy of our matrix $[y]$ we obtain the characteristic equation $\Delta y = 0$ of the network. This is because the characteristic equation determines the natural frequencies of a network, i.e., those frequencies at which voltages can exist at the network terminals when the terminal currents are zero (e.g., $I_1 = I_2 = I_3 = 0$). This condition can exist only if $\Delta y = 0$.

Inspection of (3-198) shows that Δy will generally contain all powers of s from zero to three. In a general network containing n capacitors the highest power of s in the characteristic equation will be n, *provided there are no capacitor loops in the network*. If there are capacitor loops the degree of $\Delta y = 0$ will decrease accordingly. Furthermore, assuming that the network is stable, all the natural frequencies (i.e., poles) will have negative real parts, and there will be no missing terms in the middle of the characteristic equation. Thus, in our three-capacitor network, the degree of $\Delta y = 0$ will be 3 (assuming there are no capacitor loops) and the characteristic equation will have the form

$$\Delta y = 0 = a_3 s^3 + a_2 s^2 + a_1 s + a_0 \qquad (3\text{-}199)$$

If there is a capacitor loop, a_3 will be zero. However, if $a_3 \neq 0$ and $a_0 \neq 0$ then, for stability, a_1 and a_2 must also be nonzero.

As pointed out above, we can now identify the coefficients a_i of the characteristic equation of a network in terms of its short-circuit conductances and open-circuit impedances. The procedure to be followed will be illustrated using the example of our three-capacitor network. From (3-199) it follows that a_0 results from Δy when $s = 0$, i.e., when all the capacitors are removed from the network. Thus

$$a_0 = \Delta g \qquad (3\text{-}200)$$

By inspection of (3-198) we see that the coefficient of the first power of s, i.e., a_1, can be seen to have the following form:

$$a_1 = C_1 \bar{g}_1^1 + C_2 \bar{g}_2^2 + C_3 \bar{g}_3^3 \qquad (3\text{-}201)$$

Similarly, by expanding (3-198) and comparing coefficients with (3-199), we find the coefficient of the second power of s to be

$$a_2 = C_2 C_3 g_{11} + C_1 C_3 g_{22} + C_1 C_2 g_{33} \qquad (3\text{-}202)$$

and the coefficient of the third power results from multiplying the susceptive parts of the elements on the principal diagonal of $[y]$; thus for our three-capacitor network

$$a_3 = C_1 C_2 C_3 \qquad (3\text{-}203)$$

Taking the ratio a_1/a_0 for our example, we obtain

$$\frac{a_1}{a_0} = C_1 \frac{\bar{g}_1^1}{\Delta g} + C_2 \frac{\bar{g}_2^2}{\Delta g} + C_3 \frac{\bar{g}_3^3}{\Delta g} \qquad (3\text{-}204)$$

Notice that the coefficient of C_j in (3-204) is the open-circuit resistance seen by C_j as given by (3-197):

$$\frac{a_1}{a_0} = R_1^0 C_1 + R_2^0 C_2 + R_3^0 C_3 = \sum_{j=1}^{3} \tau_{jo} \qquad (3\text{-}205)$$

where τ_{jo} is the time constant of the jth capacitor resulting from the network when all other capacitors are removed. By removing all but one capacitor, a complicated network is generally divided up into relatively simple parts, so that the individual open-circuit resistors R_j^0 can be found by inspection.

Turning our attention to an nth-order active RC network without capacitor loops, the characteristic equation is given by

$$\Delta y = 0 = a_n s^n + a_{n-1} s^{n-1} + \cdots + a_k s^k + \cdots + a_1 s + a_0 \qquad (3\text{-}206)$$

and, as in (3-205),

$$\frac{a_1}{a_0} = R_1^0 C_1 + R_2^0 C_2 + \cdots + R_n^0 C_n = \sum_{j=1}^{n} R_j^0 C_j = \sum_{j=1}^{n} \tau_{jo} \qquad (3\text{-}207)$$

In order to find the general coefficient a_k for the nth-order network we must introduce an additional resistance term, namely, R_j^i. This is the driving point resistance seen by C_j when C_i is *shorted* and all other capacitors removed. Similarly, the presence of more than one superscript to R_j indicates that the corresponding capacitors are shorted (infinite-frequency reactance) while the remaining capacitors are opened (zero-frequency reactance) and, in the limit,

$$R_j^{12\cdots\mu\cdots n}\bigg|_{\mu \neq j} = R_j^s = \frac{1}{g_{jj}} \qquad [3\text{-}208]$$

where R_j^s is the driving point resistance seen by C_j when *all* other capacitors are short circuited. This, by definition, is the reciprocal of the short-circuit admittance g_{jj}. With these resistance terms we obtain the general coefficient of (3-206) divided by the constant term a_0:[27]

$$\frac{a_k}{a_0} = \sum_{i=1}^{n-(k-1)} \sum_{j=i+1}^{n-(k-2)} \cdots \sum^n C_i C_j \cdots C_n R_i^0 R_j^i \cdots R_n^{ij\cdots(n-1)} \qquad [3\text{-}209]$$

Note that there are k summation signs in (3-209) and k C-factors and R-factors in each term of a_k/a_0. For the highest, i.e., nth, term, we have (see, e.g., (3-203)),

$$a_n = C_1 C_2 \cdots C_n \qquad (3\text{-}210)$$

To illustrate the use of (3-209), consider the term a_2/a_0 of an nth-degree characteristic equation. It has the form

$$\frac{a_2}{a_0} = \sum_{i=1}^{n-1} \sum_{j=i+1}^{n} C_i C_j R_i^0 R_j^i \qquad (3\text{-}211)$$

which, for $n = 4$, results in

$$\frac{a_2}{a_0} = C_1 C_2 R_1^0 R_2^1 + C_1 C_3 R_1^0 R_3^1 + C_1 C_4 R_1^0 R_4^1$$

$$+ C_2 C_3 R_2^0 R_3^2 + C_2 C_4 R_2^0 R_4^2 + C_3 C_4 R_3^0 R_4^3 \qquad (3\text{-}212)$$

Similarly, we obtain a_3/a_0 for the nth-degree characteristic equation as

$$\frac{a_3}{a_0} = \sum_{i=1}^{n-2} \sum_{j=i+1}^{n-1} \sum_{k=j+1}^{n} C_i C_j C_k R_i^0 R_j^i R_k^{ij} \qquad (3\text{-}213)$$

and, for $n = 4$,

$$\frac{a_3}{a_0} = C_1 C_2 C_3 R_1^0 R_2^1 R_3^{12} + C_1 C_2 C_4 R_1^0 R_2^1 R_4^{12}$$

$$+ C_1 C_3 C_4 R_1^0 R_3^1 R_4^{13} + C_2 C_3 C_4 R_2^0 R_3^2 R_4^{23} \qquad (3\text{-}214)$$

Finally, the fourth coefficient a_4/a_0 for an nth-degree characteristic equation is

$$\frac{a_4}{a_0} = \sum_{i=1}^{n-3} \sum_{j=i+1}^{n-2} \sum_{k=j+1}^{n-1} \sum_{l=k+1}^{n} C_i C_j C_k C_l R_i^0 R_j^i R_k^{ij} R_l^{ijk} \qquad (3\text{-}215)$$

27. B. L. Cochrun and A. Grabel, A method for the determination of the transfer function of electronic circuits, *IEEE Trans. Circuit Theory*, **CT-20**, 16–20 (1973).

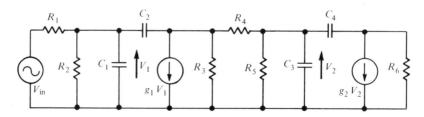

FIG. 3-59. High-frequency equivalent circuit of two-stage amplifier.

which, for $n = 4$, becomes

$$\frac{a_4}{a_0} = C_1 C_2 C_3 C_4 R_1^0 R_2^1 R_3^{12} R_4^{123} \tag{3-216}$$

EXAMPLE: To provide a practical example of the foregoing, consider the active RC network shown in Fig. 3-59. Typically this may represent the equivalent circuit of a two-stage amplifier at high frequencies. The network has four capacitors, none of which are in a loop; thus the characteristic equation will be of fourth degree, and the expressions for a_i, $i = 1, 2, 3, 4$, given above can be used. Starting out with a_4 we obtain the driving point resistances required for (3-216) by inspection of Fig. 3-59:

$$R_1^0 = R_1 \left\| R_2 = \frac{R_1 R_2}{R_1 + R_2} \right. \tag{3-217a}$$

$$R_2^1 = R_3 \left\| (R_4 + R_5) = \frac{R_3 (R_4 + R_5)}{R_3 + R_4 + R_5} \right. \tag{3-217b}$$

$$R_3^{12} = R_4 \left\| R_5 = \frac{R_4 R_5}{R_4 + R_5} \right. \tag{3-217c}$$

$$R_4^{123} = R_6 \tag{3-217d}$$

With (3-215) we therefore obtain

$$\frac{a_4}{a_0} = \frac{R_1 R_2 R_3 R_4 R_5 R_6 C_1 C_2 C_3 C_4}{(R_1 + R_2)(R_3 + R_4 + R_5)} \tag{3-218}$$

Furthermore, from (3-210) we have

$$a_4 = C_1 C_2 C_3 C_4 \tag{3-219a}$$

and therefore

$$a_0 = \frac{(R_1 + R_2)(R_3 + R_4 + R_5)}{R_1 R_2 R_3 R_4 R_5 R_6} \tag{3-219b}$$

To calculate a_3/a_0 from (3-214) we require some driving point resistances in addition to those in (3-217). Some of these involve the current sources g_1 and

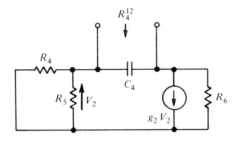

FIG. 3-60. Network in Fig. 3-59 modified to calculate R_4^{12}.

g_2 of the network (remember that only the *independent* sources are set to zero). These sources are taken into account as follows. Consider the calculation of R_4^{12}. Shorting capacitors C_1 and C_2, and removing C_3 and C_4, we are left with the circuit shown in Fig. 3-60. To calculate the driving point resistance seen across the terminals of C_4, we inject a current I at those terminals and calculate the resulting voltage V, thereby obtaining

$$R_4^{12} = \frac{V}{I} = \frac{R_4 R_5}{R_4 + R_5}(1 + g_2 R_6) + R_6$$

$$= R_3^{12}(1 + g_2 R_6) + R_6 \qquad (3\text{-}220a)$$

Similarly,

$$R_2^0 = R_1^0(1 + g_1 R_2^1) + R_2^1 \qquad (3\text{-}220b)$$

R_3^2 requires a little more analysis, but is obtained in precisely the same way. Following this procedure, the coefficients a_2 and a_1 of the characteristic equation are also readily obtained. Solving the resulting equation for its roots results in the natural frequencies, or poles, of the network.

Let us now briefly examine how the procedure outlined above must be modified if capacitor loops are present in the network. If, for example, three capacitors C_1, C_2, and C_3 are in a loop, then shorting any two of them automatically shorts out the third. As a consequence the resistances R_1^{23}, R_2^{13}, and R_3^{12} will be zero and the degree of the characteristic equation will be reduced by one, that is, $a_n = 0$. The next term, a_{n-1}, will exist but may be modified on account of the loop; in the case of a second loop, it too will be zero, and so on.

As an example of a network with a capacitor loop (i.e., C_1, C_2, and C_3), consider the network in Fig. 3-61. This could represent the high-frequency equivalent circuit of a two-stage amplifier comprising FET-type active devices. The network has four capacitors, but if we try to calculate a_4/a_0 from (3-216) we find

$$R_1^0 = R_1, \qquad R_2^1 = R_2, \qquad R_4^{123} = R_3, \qquad R_3^{12} = 0 \qquad (3\text{-}221)$$

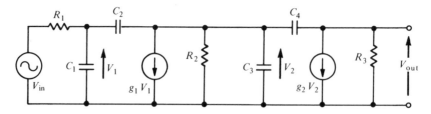

FIG. 3-61. High-frequency equivalent circuit of two-stage amplifier containing a capacitor loop.

Consequently $a_4 = 0$ because the driving point resistance across the terminals of C_3 is zero when C_1 and C_2 are shorted. The term a_3/a_0 is therefore also modified, i.e., from (3-214) we have

$$\frac{a_3}{a_0} = C_1 C_2 C_4 R_1^0 R_2^1 R_4^{12} + C_1 C_3 C_4 R_1^0 R_3^1 R_4^{13} + C_2 C_3 C_4 R_2^0 R_3^2 R_4^{23} \quad (3\text{-}222)$$

The coefficient of $C_1 C_2 C_3$ is missing because $R_3^{12} = 0$. By contrast the terms a_2/a_0 and a_1/a_0 will remain unchanged because they do not involve the shorting of two capacitors.

3.6.2 Approximate Pole Locations of the Characteristic Equation

When the characteristic equation of a network has been found, the poles can be determined from its solution when it is set equal to zero (see (3-199)). In many cases, however, only the lowest pole, i.e., the one closest to the origin, and in others only the highest pole is required, and we shall show here that these can frequently be related in an approximate manner to the ratios of adjacent coefficients in the characteristic equation.[28] To see this we consider the third-order polynomial $P(s)$, whose roots are p_1, p_2, and p_3. We can write

$$P(s) = a_0 + a_1 s + a_2 s^2 + a_3 s^3 = \left(1 - \frac{s}{p_1}\right)\left(1 - \frac{s}{p_2}\right)\left(1 - \frac{s}{p_3}\right) \quad (3\text{-}223)$$

where

$$a_0 = 1 \qquad (3\text{-}224a)$$

$$a_1 = -\left(\frac{1}{p_1} + \frac{1}{p_2} + \frac{1}{p_3}\right) \qquad (3\text{-}224b)$$

$$a_2 = \frac{1}{p_1 p_2} + \frac{1}{p_1 p_3} + \frac{1}{p_2 p_3} \qquad (3\text{-}224c)$$

28. R. D. Thornton et al., loc. cit.

$$a_3 = -\frac{1}{p_1 p_2 p_3} \tag{3-224d}$$

$$\frac{a_2}{a_3} = -(p_1 + p_2 + p_3) \tag{3-224e}$$

To obtain the desired relationships we must make the assumption that the roots of $P(s)$ are widely separated, i.e.,

$$p_1 \ll p_2 \ll p_3 \tag{3-225}$$

Then we have from (3-224a) and (3-224b):

$$p_1 \approx -\frac{a_0}{a_1} \tag{3-226a}$$

from (3-224e):

$$p_3 \approx -\frac{a_2}{a_3} \tag{3-226b}$$

and from (3-224d):

$$p_2 \approx -\frac{a_1}{a_2} \tag{3-226c}$$

In general, assuming that the poles are widely separated, we can extend these results to a network with n poles and obtain the approximate position of the jth pole as

$$p_j \approx -\frac{a_{j-1}}{a_j} \tag{3-227}$$

To gain some insight into the accuracy of this result, consider the extreme case of three identical poles at $s = -1$. Applying (3-227) to the coefficients of the corresponding equation $(s + 1)^3 = 0$ we obtain $p_1 = -\frac{1}{3}$, $p_2 = -1$ and $p_3 = -3$. We conclude that the poles should be separated by at least a factor of 3 from one another in order to obtain a reasonably good approximation from (3-227).

Comparing (3-224a) and (3-224b) with (3-205) we find

$$\frac{a_1}{a_0} = R_1^0 C_1 + R_2^0 C_2 + R_3^0 C_3 = -\left(\frac{1}{p_1} + \frac{1}{p_2} + \frac{1}{p_3}\right) \tag{3-228}$$

and for the general case

$$\frac{a_1}{a_0} = -\sum_{j=1}^{n} \frac{1}{p_j} = \sum_{j=1}^{n} \tau_{jo} \tag{3-229}$$

Thus, the sum of the open-circuit time constants is equal to the negative sum of the reciprocals of the poles of an active RC network. This relationship is exact. If one pole p_l is much *smaller* in magnitude than any of the others we obtain

$$p_l \bigg|_{\substack{p_l \ll p_j \\ j \neq l}} \approx -\frac{a_0}{a_1} = -\frac{1}{\sum\limits_{j=1}^{n} \tau_{jo}} \qquad j = 1, 2, \ldots, n \qquad [3\text{-}230]$$

That is, the lowest pole of the network can be approximated by taking the reciprocal of the negative sum of the open-circuit time constants of the network as defined by (3-207).

Just as the two lowest coefficients a_0 and a_1 have a significant meaning with respect to the characteristic equation, the same is true of the two highest coefficients a_{n-1} and a_n. These can be written in the following form (see, e.g., (3-224e)):

$$\frac{a_{n-1}}{a_n} = -\sum_{j=1}^{n} p_j \qquad (3\text{-}231)$$

The ratio a_{n-1}/a_n can also be written in terms of network time constants. To see this we return to the coefficients found for the characteristic equation of the three-capacitor network shown in Fig. 3-58. Dividing (3-202) by (3-203) we obtain

$$\frac{a_2}{a_3} = \frac{g_{11}}{C_1} + \frac{g_{22}}{C_2} + \frac{g_{33}}{C_3} \qquad (3\text{-}232)$$

and for an nth-order network

$$\frac{a_{n-1}}{a_n} = \frac{g_{11}}{C_1} + \frac{g_{22}}{C_2} + \cdots + \frac{g_{nn}}{C_n} \qquad (3\text{-}233)$$

Remember that each term g_{jj} corresponds to the conductance seen across the jth terminal pair (that is across the capacitor C_j) when all other terminal pairs (that is, all other capacitors) are short circuited. Thus with (3-208) we have

$$\frac{C_j}{g_{jj}} = R_j^s C_j = \tau_{js} \qquad (3\text{-}234)$$

where τ_{js} is the time constant of the jth capacitor calculated with all other capacitors shorted. As in the calculation of the open-circuit time constant τ_{jo}, when shorting all but the jth capacitor of a network, the network is broken up into simple subnetworks whose time constants can generally be obtained by inspection. It is clear that where τ_{jo} may be interpreted as the *zero-frequency time constant* associated with C_j (where all other C's are zero), τ_{js} may be interpreted as the *infinite-frequency time constant* obtained when all capacitors, with the exception of C_j, take on infinite values.

Combining (3-231), (3-233), and (3-234) we obtain

$$\frac{a_{n-1}}{a_n} = -\sum_{j=1}^{n} p_j = \sum_{j=1}^{n} \frac{1}{\tau_{js}} \qquad [3\text{-}235]$$

which is analogous to the expression for a_1/a_0 given by (3-229). It implies that for an active RC network the sum of the reciprocal short-circuit time constants is equal to the negative sum of the poles.

Assuming now that one pole is far *larger* than any of the others, we can approximate the highest pole p_h from (3-235) as follows:

$$p_h \bigg|_{\substack{p_h \gg p_j \\ j \neq h}} \approx -\frac{a_{n-1}}{a_n} = -\sum_{j=1}^{n} \frac{1}{\tau_{js}} \qquad [3\text{-}236]$$

i.e., the highest pole of a network can be approximated by taking the negative sum of the reciprocals of the short-circuit time constants of the network as defined by (3-234).

The results given by (3-230) and (3-236) are intuitively meaningful if we recall the zero-frequency and infinite-frequency interpretation of the τ_{jo} and τ_{js} time-constants, respectively. Thus the lowest pole frequency, i.e., the lowest natural frequency, corresponds to the inverse sum of the zero-frequency time constants, where each C_j is combined with the zero-frequency driving point resistance associated with it. Likewise the highest pole, or the highest natural frequency, corresponds to the sum of the inverse infinite-frequency time constants whereby each C_j is combined with the infinite-frequency driving point resistance associated with it.

3.7 SUMMARY

1. Each of the elements of the indefinite admittance matrix is a (positive or negative) short-circuit admittance.

2. To find the self admittances y_{jj} of terminal j and the mutual admittances y_{ij} between terminals i and j of a network, connect all terminals except terminal j to the chosen reference (also called datum) point. Apply a unit voltage to terminal j (i.e., $V_j(s) = 1$). Then y_{jj} is the current $I_j(s)$ entering terminal j and y_{ij} is the current $I_i(s)$ entering terminal i.

3. The indefinite admittance matrix is symmetrical for a passive reciprocal network and in general is unsymmetrical for an active network.

4. The indefinite admittance matrix of a passive reciprocal network can be obtained as follows: The sum of the admittances entering node i is entered in position ii of the matrix; the negative value of the admittance of a branch between terminals i and j is entered in positions ij and ji. When no branch connects any two terminals (zero admittance), a zero is entered at the appropriate positions in the matrix.

5. The sum of the elements in each column is equal to zero. This is a direct consequence of Kirchhoff's current law.

6. The sum of the elements in each row is equal to zero. This follows from the fact that the node currents are dependent on potential differences between the various nodes rather than on the potential of these nodes with respect to the reference node. It is not necessary to make use of this fact in deriving this property for a passive network that obeys reciprocity (symmetrical matrix). It must be used, however, to establish the property for an active network.

7. The determinant of $[Y]$ is zero.

8. The first cofactors are all equal. This property is true for any zero-sum matrix. Because of it the indefinite admittance matrix is also called *the equicofactor matrix*.

9. The indefinite matrix $[Y]$ may be simply transformed into a definite (i.e., node-to-datum) admittance matrix by deleting any row k and column k; this has the effect of making terminal k the reference terminal and making the network an $(n - 1)$-port with all ports possessing terminal k as a common terminal. Conversely, any definite admittance matrix can be converted into an indefinite admittance matrix by adding to it one row and one column such that the sum of the elements in each row and column of the new matrix is zero.

10. The indefinite (and definite) matrix of a higher-order multipole may be reduced to a lower order by contraction of terminal pairs or suppression of individual terminals.

11. The indefinite matrix of networks in parallel can be obtained by adding the corresponding terms of the individual matrices.

12. Network functions of a multipole can be obtained as ratios of first- and second-order cofactors of the indefinite admittance matrix of that multipole.

13. The most commonly used multipole in active network design is the three-pole.

14. The indefinite admittance matrices of active networks are treated in the same way as those of passive networks. The active devices are described in terms of their y parameters.

15. Most active devices can be readily modeled using combinations of the four basic controlled sources: VVS, VCS, CVS, CCS.

16. Operational amplifiers are commonly used to provide a very high-gain inverting VVS when used in the inverting mode, and a conventional (low-gain) VVS when used in the noninverting mode.

17. A very useful method of analyzing active networks incorporating operational amplifiers used in any mode (inverting, noninverting, or differential) is to apply the matrix analysis methods of constrained networks.

18. Ladder networks represent a special and frequently used category of networks. As such, special methods of analyzing them have been developed.

19. The driving point immittances of ladder networks may be simply calculated using continued fractions.

20. A one-to-one relationship exists between the branch impedances and admittances of a ladder network and the terms of its continued-fraction expansion.

21. The transmission functions of ladder networks can readily be calculated using the theory of continuants.

22. A one-to-one relationship exists between the branch impedances and admittances of a ladder network and the diagonal elements of its continuant.

23. To obtain the RC dual of an active RC network, its resistors, capacitors, and transmittances are scaled by $1/s$.

24. Symmetrical and potentially symmetrical networks can be readily analyzed using Bartlett's bisection theorem.

25. The degree of the characteristic equation of a linear active RC network is equal to the number of capacitors in the network, provided the capacitors do not form any loops. The degree is reduced by the number of capacitor loops in the network.

26. The coefficients of the characteristic equation of an active RC network are combinations of the capacitors and the driving point resistances seen at the terminals of those capacitors, where various combinations of capacitors are shorted and opened. The coefficients can be obtained by inspection of the network.

27. Assuming that the poles of a network are widely separated, they can be approximated by the coefficients of the characteristic equation.

CHAPTER

4

THE SENSITIVITY OF LINEAR ACTIVE NETWORKS

INTRODUCTION

In the preceding chapters we have discussed the characterization and analysis of linear networks. We must now turn our attention to the fact that in practice the elements of a network are subjected to a number of influences that will vary their nominal values. Manufacturing tolerances will cause the component values to deviate from nominal initially, while aging and ambient conditions (e.g., temperature and humidity) will cause them to vary further during operation. The network designer must therefore address himself to the question of how these variations, many of which set in after his design is realized, will affect the properties of his network, and, more specifically, if his design will still meet the specified requirements after these variations have taken place. In other words, the designer must know how sensitive his network is to variations in the values of its components. Clearly there are many possible ways of defining this sensitivity, some of which are more suitable for one type of component or network characteristic than for another.

In Chapter 6 we shall examine various component types currently used for hybrid-integrated circuit design and the variations that can be expected from them with age, temperature, and other environmental changes. First, however, in this chapter, we must learn how to cope with component variations analytically and in a general manner. In so doing we shall discuss the various sensitivity concepts that have recently been introduced to characterize the stability[1] of linear active networks, and consider how they can be utilized in practice.

1. Stability is here to be understood in the sense of invariability or insensitivity rather than in the feedback-sense of proneness to instability or to oscillations.

208

4.1 TRANSMISSION SENSITIVITY

The potential instability of active RC networks, as well as the fact that mere drift of the transmission zeros and poles within the LHP influences the network transmission characteristics directly and possibly drastically, has revived a concept first introduced by Bode, namely that of transmission sensitivity. It is defined as follows[2]

$$S_x^{T(s)} = \frac{d \ln T(s)}{d \ln x} = \frac{dT(s)/T(s)}{dx/x} \tag{4-1}$$

This expression is the ratio of the relative change of the network transmission function $T(s)$ to the relative change of the network element x that caused the change in $T(s)$.

4.1.1 Three Basic Expressions for Transmission Sensitivity

In Terms of Poles and Zeros   To obtain the transmission sensitivity of an nth-order transmission function given in terms of its poles and zeros we proceed as follows. Starting out with the transmission function $T(s)$, with both numerator and denominator polynomials factored in terms of their zeros and poles, respectively, we have

$$T(s) = K \frac{\displaystyle\prod_{i=1}^{m} (s - z_i)}{\displaystyle\prod_{j=1}^{n} (s - p_j)} \tag{4-2}$$

Assuming only simple poles and zeros, and taking the natural logarithm of (4-2), we have

$$\ln T(s) = \ln K + \sum_{i=1}^{m} \ln (s - z_i) - \sum_{j=1}^{n} \ln (s - p_j) \tag{4-3}$$

Taking the total differential of (4-3) with respect to K, z_i, and p_j we obtain

$$d \ln T(s) = \frac{dT(s)}{T(s)} = \frac{dK}{K} - \sum_{i=1}^{m} \frac{dz_i}{s - z_i} + \sum_{j=1}^{n} \frac{dp_j}{s - p_j} \tag{4-4}$$

Dividing both sides of (4-4) by the relative change of the network element x, we obtain the transmission sensitivity

$$\frac{d \ln T(s)}{d \ln x} = \frac{dT(s)/T(s)}{dx/x} = \frac{dK/K}{dx/x} - \sum_{i=1}^{m} \frac{\dfrac{dz_i}{dx/x}}{s - z_i} + \sum_{j=1}^{n} \frac{\dfrac{dp_j}{dx/x}}{s - p_j} \tag{4-5}$$

2. H. W. Bode, *Network Analysis and Feedback Amplifier Design*, (New York: Van Nostrand, 1945), p. 52. The definition given by Bode is the reciprocal of that given in (4-1).

With (4-1) this can be rewritten as:

$$S_x^{T(s)} = S_x^K - \sum_{i=1}^{m} \frac{\mathscr{S}_x^{z_i}}{s - z_i} + \sum_{j=1}^{n} \frac{\mathscr{S}_x^{p_j}}{s - p_j} \qquad [4\text{-}6]$$

where:

$$\mathscr{S}_x^{z_i} = \frac{dz_i}{d \ln x} = \frac{dz_i}{dx/x} \qquad [4\text{-}7]$$

$$\mathscr{S}_x^{p_j} = \frac{dp_j}{d \ln x} = \frac{dp_j}{dx/x} \qquad [4\text{-}8]$$

and

$$S_x^K = \frac{1}{K} \cdot \mathscr{S}_x^K = \frac{d \ln K}{d \ln x} = \frac{dK/K}{dx/x} \qquad (4\text{-}9)$$

The expressions given by (4-7) and (4-8) give a measure of the change in the s plane location of a zero or pole caused by the relative change in a network element x.[3] They may therefore be interpreted as the defining functions for zero and pole sensitivity, respectively.[4] Similarly (4-9) represents the sensitivity of the scale factor K.

With the above definitions, and bearing in mind that we have assumed simple poles and zeros, the overall sensitivity of the transmission function of a network to differentially small variations of an element x as given by (4-6) can be interpreted as a partial fraction expansion, in which all critical network frequencies become poles and the residues are the pole and zero sensitivities. This demonstrates the not surprising fact that since the poles and zeros themselves determine the transmission characteristics, the pole and zero sensitivities determine the transmission sensitivity of a network (up to an additive constant determined by the sensitivity of the multiplier K).

Another important observation can be made from inspection of the transmission sensitivity given by (4-6). To begin with, the term S_x^K is generally of little importance, since changes in K involve only the gain level of the network $T(s)$, which is of little consequence in most active networks; it can usually be controlled quite easily. This leaves the terms of the partial fraction expansion of $S_x^{T(s)}$. Since the critical frequencies of this expansion are the poles and zeros of $T(s)$, they will be determined by the $T(s)$ function initially required to satisfy the specifications of a given network. Rarely will it be possible to select these poles and zeros to fulfill any other criterion, such as one that would affect (i.e., minimize) the corresponding transmission sensitivity. This leaves only the residues, i.e., the pole and zero sensitivities as defined by (4-7) and

3. Note that while the transmission sensitivity (4-1) is a function of s, the root sensitivities (4-7) and (4-8) are real or complex numbers.
4. J. G. Truxal and I. M. Horowitz, Sensitivity considerations in active network synthesis, *Proc. 2nd Midwest Symp. Circuit Theory*, Michigan State Univ., Dec. 1965

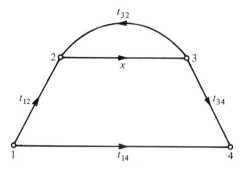

Fig. 4-1. Basic signal-flow graph of a feedback network in which a particular network element x appears only in one branch.

(4-8), with which to exercise any control over $S_x^{T(s)}$. Indeed, we shall see later that much of sensitivity theory revolves around the problem of finding network configurations that (1) realize the poles and zeros of a specified network function $T(s)$ and (2) simultaneously possess minimum pole and zero sensitivities with respect to the most variable or unstable network elements x_j.

In Terms of the Return and Null-Return Differences Because of the feedback character of most active networks, it is to be expected that the transmission sensitivity can be related to the amount of feedback present in a network. By generalizing the basic return difference concept of feedback networks, discussed in Chapter 2, a very simple expression can indeed be derived.

It was pointed out in Chapter 2 that the signal-flow graph of a network can always be reduced to the basic form shown in Fig. 4-1, in which a particular network element x appears as the transmittance of only one branch. The transmittance from the input to the origin of the x branch is given by $t_{12}(s)$, and from the terminal of the x branch to the output terminal by $t_{34}(s)$. The transmittance of the feedback branch around the x branch is given by $t_{32}(s)$. In general there is also a leakage transmittance directly from the input terminal to the output terminal; it is given by $t_{14}(s)$ and is equal to the overall system transmittance when $x = 0$. For an arbitrary value of x the overall transmission function was shown to be

$$T(s) = t_{14}(s) + \frac{x t_{12}(s) t_{34}(s)}{1 - x t_{32}(s)} \tag{4-10}$$

The return difference referred to the network element x was obtained as

$$F_x(s) = 1 - x t_{32}(s). \tag{4-11}$$

and the null return difference with respect to x as

$$F_x^0(s) = 1 - x\left(t_{32} - \frac{t_{34} t_{12}}{t_{14}}\right) \tag{4-12}$$

The transmission sensitivity as defined by (4-1) can then be expressed in terms of the return difference as follows:

$$S_x^{T(s)} = \frac{1}{F_x(s)}\left[1 - \frac{t_{14}(s)}{T(s)}\right] \tag{4-13}$$

With (4-12) we obtain

$$S_x^{T(s)} = \frac{1}{F_x(s)} - \frac{1}{F_x^0(s)} \tag{4-14}$$

For the frequent case that the leakage transmittance $t_{14}(s)$ is equal to zero[5] (4-13) and (4-14) simplify to

$$S_x^{T(s)} = \frac{1}{F_x(s)} \qquad \text{for} \quad T(s, x = 0) = 0 \tag{4-15}$$

Thus, in a network with no leakage transmittance, i.e., a network in which the removal of the reference element x leads to zero transmission from source to sink, the transmission sensitivity is the reciprocal of the return difference with respect to x.

For a given value $x = x_0$, (4-15) is a function of s. Assuming that $t_{32}(s)$ in $F_x(s)$ is the transfer function of an LLF network (see Chapter 1, Section 1.1), it approaches a constant value or zero at high frequencies. As a consequence of (4-11), the sensitivity function given by (4-15) must therefore approach a constant value (including unity). In other words, *for the important case where there is no leakage transmittance (simple feedback loop) the sensitivity function must have an equal number of zeros and poles.*

In Terms of the Bilinear Form   Because of its fundamental importance, a more convenient expression than either (4-6) or (4-14) is desirable for the actual calculation of the transmission sensitivity of a network. This can be obtained very simply from the bilinear form of the transmission function:

$$T(s) = \frac{N(s)}{D(s)} = \frac{A(s) + xB(s)}{U(s) + xV(s)} \tag{4-16}$$

The transmission sensitivity $S_x^{T(s)}$ can now be expressed in terms of the polynomials $A(s)$, $B(s)$, $U(s)$, and $V(s)$. We found previously that

$$F_x(s) = \frac{D(s)}{U(S)} \tag{4-17a}$$

$$= \frac{1}{1 - x\dfrac{V(s)}{D(s)}} \tag{4-17b}$$

---

5. The fact that there is no leakage transmittance implies that the network represented by the signal-flow graph in Fig. 4-1 consists of a single feedback loop. Therefore the closed-loop zeros are independent of the network element x.

and

$$F_x^0(s) = \frac{N(s)}{A(s)}$$ (4-18a)

$$= \frac{1}{1 - x\,\dfrac{B(s)}{N(s)}}$$ (4-18b)

With (4-14), (4-17a), and (4-18a) we therefore have

$$S_x^{T(s)} = \frac{U(s)}{D(s)} - \frac{A(s)}{N(s)}$$ [4-19a]

Similarly, with (4-14), (4-17b), and (4-18b) we obtain

$$S_x^{T(s)} = -x\left[\frac{V(s)}{D(s)} - \frac{B(s)}{N(s)}\right]$$ [4-19b]

Notice that if either $A(s)$ or $B(s)$ equals zero in the bilinear form of $T(s)$, the transmission sensitivity $S_x^{T(s)}$ depends only on the denominator $D(s)$ of $T(s)$, (i.e., on its poles), apart from the fact that the poles of $S_x^{T(s)}$ *are* the poles of the corresponding transmittance $T(s)$. Networks for which this is true (i.e., either $A(s)$ or $B(s)$ is equal to zero) have been termed type I networks[6] since they have some basically different characteristics with respect to sensitivity than type II networks for which both $A(s)$ and $B(s)$ exist. Type I networks either possess no leakage path (i.e., $A(s) = 0$) or only a leakage path (i.e., $B(s) = 0$). Either way, as shown in Fig. 4-2, there is only one forward path between source and sink in the signal-flow graph corresponding to the basic form of Fig. 4-1. Incidentally, it is useful to note that single feedback loops, i.e., networks whose transmission sensitivity is the reciprocal of the return difference with respect to x as defined by (4-15), are a special case of type I networks. More is said about type I and II networks in Chapter 1 of *Linear Integrated Networks: Design*, in the discussion of the sensitivity minimization of second-order networks.

Comparisons and Applications   The object of sensitivity theory is first to analyze the sensitivity of any particular network parameter or function with respect to variations of a network element x and then, if necessary, to find methods of modifying the network in order to minimize this sensitivity or at least to reduce it to some acceptable or specified value. Each of the three basically different expressions for $S_x^{T(s)}$ given by (4-6), (4-14), and (4-19), respectively, is useful in its own way in providing insight into some aspect of $S_x^{T(s)}$ and thereby in assisting to identify and subsequently to minimize the transmission sensitivity of a network characterized by $T(s)$.

6. G. S. Moschytz, Second-order pole–zero pair selection for *n*th-order minimum sensitivity networks, *IEEE Trans. Circuit Theory*, **CT-17**, 527–534 (1970).

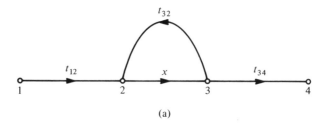

(a)

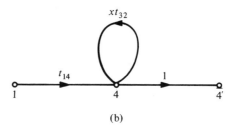

(b)

FIG. 4-2. Signal-flow graphs of type I networks. (a) $A(s) = 0$, so that $t_{14} = 0$, (b) $B(s) = 0$ so that $t_{12} t_{34} = t_{14} t_{32}$.

Equation (4-6) expresses $S_x^{T(s)}$ as a partial fraction expansion with the poles equal to the poles and zeros of $T(s)$ and the residues equal to the corresponding root sensitivities. Since the poles are determined by $T(s)$, they cannot be modified. Thus (4-6) emphasizes the contribution of the root sensitivities to the transmission sensitivity and indicates that one way of minimizing transmission sensitivity is to minimize the individual root sensitivities. As we shall see later this can be achieved, for example, by synthesizing networks specifically for minimum pole and zero sensitivities, or else by realizing appropriate network configurations with components selected in such a way as to cancel out individual root-sensitivity contributions.

Equation (4-14) relates transmission sensitivity directly to the amount of feedback present in a network, i.e., to the return difference with respect to a network element x. Feedback is seen to affect the influence of parameter variations on the transfer function of a network directly, and is, in fact, often used for the sole purpose of controlling such variations. Expression (4-14) is therefore very useful in identifying feedback loops in a network and in ascertaining their effects on the overall transmission sensitivity. It is particularly useful when an active network is actually represented in the form of a feedback system with recognizable feedback loops, or in signal-flow graph form, in which the loops are readily obtainable. To clarify these remarks consider the

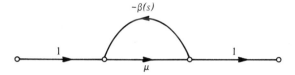

FIG. 4-3. Signal-flow graph of an active feedback network.

single-loop feedback network given in signal-flow graph form in Fig. 4-3. In order to make any sensitivity studies on amplifiers given in this form, the sensitivity expression (4-14) is obviously the most convenient one. Let us, for example, derive the sensitivity of the amplifier transmittance $T(s)$ with respect to the forward gain μ and with respect to the feedback factor $\beta(s)$. We have, by inspection,

$$T(s) = \frac{\mu}{1 + \mu\beta(s)} \tag{4-20}$$

Since $T(s)$ is zero when $\mu = 0$ (type I network) we have, from (4-15),

$$S_\mu^{T(s)} = \frac{1}{F_\mu(s)} = \frac{1}{1 + \mu\beta(s)} \tag{4-21}$$

If, on the other hand, the sensitivity with respect to β is of interest, we see from (4-20) that $T(s)|_{\beta=0} = t_{14} = \mu$. Thus with (4-14)[7] (or more directly with (4-13)) we have

$$S_\beta^{T(s)} = \frac{1}{1 + \mu\beta} \left[1 - \frac{\mu}{\mu/(1 + \mu\beta)} \right] = \frac{-\mu\beta}{1 + \mu\beta} \tag{4-22}$$

Clearly, (4-14) provides an excellent indicator for the interaction between feedback and sensitivity in a network that is given in some convenient topological form, e.g., by signal-flow graph or block diagram. In the present example of a feedback amplifier it points out the following important characteristic of single-loop feedback systems: from (4-21) and (4-22) we have

$$\frac{\Delta T}{T} = S_\mu^T \frac{\Delta\mu}{\mu} + S_\beta^T \frac{\Delta\beta}{\beta} = \frac{\Delta\mu/\mu}{1 + \mu\beta} - \frac{\mu\beta}{1 + \mu\beta} \frac{\Delta\beta}{\beta} \tag{4-23a}$$

and for $\mu\beta \gg 1$ we have

$$\frac{\Delta T}{T} \approx -\frac{\Delta\beta}{\beta} \tag{4-23b}$$

Thus, as the loop gain $\mu\beta$ is increased, the sensitivity of $T(s)$ to variations in μ approaches zero, while that to variations in β approaches minus one. Feedback is seen to exert a considerable influence on the sensitivity of $T(s)$ with respect to the forward path but almost none with respect to the feedback path of a single-loop feedback system.

7. It is easy to see that $F_x(s) = 1 + \mu\beta$ and $F_x^0(s) = 1$.

While on the subject of single-loop feedback networks, we can become still a little more specific. The feedback around the reference element of such a network is measured quantitatively by the magnitude of the return difference and is commonly expressed in decibels. When this magnitude is greater than unity, the feedback is negative; when it is less than unity, the feedback is positive. This should be clear from stability considerations. Consequently it is with negative feedback ($|F_\mu| > 1$) that the sensitivity in (4-21) is reduced. Positive feedback, for which $|F_\mu| < 1$, generally increases the sensitivity, although there are some exceptions; e.g., when the sensitivity of $|T(s)|$, rather than that of the complex quantity, is considered.

Another interesting example in this context is illustrated in Fig. 4-4. Figure 4-4a represents the signal-flow graph of a multiloop amplifier consisting of n feedback loops L_j. Each loop contains m amplifier stages with gains $\mu_i (i = 1, 2, \ldots, m)$ in the forward path and a feedback path β_j. Thus the gain of a single feedback stage T_j is

$$T_j = \frac{\prod_{i=1}^{m} \mu_i}{1 - \beta_j \prod_{i=1}^{m} \mu_i} \tag{4-24}$$

The overall gain T_{io} then results:

$$T_{io} = \prod_{j=1}^{n} T_j = \prod_{j=1}^{n} \left(\frac{\prod_{i=1}^{m} \mu_i}{1 - \beta_j \prod_{i=1}^{m} \mu_i} \right)_j \tag{4-25}$$

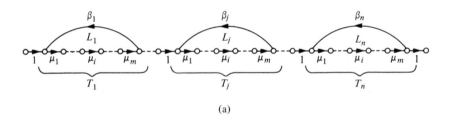

(a)

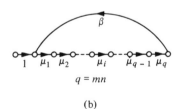

$q = mn$

(b)

FIG. 4-4. Multiloop amplifier. (a) n feedback loops with m gain stages μ_i in each. (b) A single feedback loop with $q = mn$ stages in it.

From (4-21) the sensitivity of T_j to variations of an amplifier stage μ_i is given by

$$S_{\mu_i}^{T_j} = \frac{1}{1 - \beta_j \prod\limits_{i=1}^{m} \mu_i} \tag{4-26}$$

If we assume that all stages μ_i are identical, then the gain of a stage T_j results from (4-24) as

$$T_j = \frac{\mu^m}{1 - \beta_j \mu^m} \tag{4-27}$$

and the sensitivity of T_j to variations of all m amplifier stages μ will be

$$S_{\mu}^{T_j} = \sum_{i=1}^{m} S_{\mu_i}^{T_j} \bigg|_{\mu_i = \mu} = \frac{m}{1 - \beta_j \mu^m} \tag{4-28}$$

Furthermore if all loops are identical (i.e., $\beta_j = \beta$ and $T_j = T$) then the overall gain results from (4-25) as

$$T_{io} = T^n = \left(\frac{\mu^m}{1 - \beta \mu^m} \right)^n \tag{4-29}$$

and the sensitivity of T_{io} to variations of all mn stages μ is

$$S_{\mu}^{T_{io}} = \sum_{j=1}^{n} S_{\mu}^{T_j} \bigg|_{T_j = T} = \frac{mn}{1 - \beta \mu^m} = mn \frac{\sqrt[n]{T_{io}}}{\mu^m} \tag{4-30}$$

From (4-30) we conclude that for a given number of amplifier stages α, where $\alpha = mn$, the sensitivity of T_{io} to variations of the α amplifier stages decreases as m increases.[8] Thus the amplifier with α stages of gain μ has minimum sensitivity to variations in μ when $\alpha = m$, i.e., when there is only one feedback loop over all the amplifiers, as illustrated in Fig. 4-4b.

A numerical example will demonstrate this point. Consider a four-stage amplifier ($\alpha = mn = 4$) in which each stage has a gain of $\mu = 100$ and the overall gain is $T_{io} = 10,000$. The following three configurations can then be obtained:

1. $n = 4$, $m = 1$, i.e., four feedback stages with one amplifier per stage (see Fig. 4-5a). Then each stage $T = 10$, $(1 - \mu\beta) = (1 - 100\beta) = 10$, $\beta = -0.09$, and $S_{\mu}^{T_{io}} = 0.4$.
2. $n = 2$, $m = 2$, i.e., two feedback stages with two amplifiers per stage (see Fig. 4-5b). Here each stage $T = 100$, $(1 - \mu^2\beta) = (1 - 10^4\beta) = 100$, $\beta = -0.0099$ and $S_{\mu}^{T_{io}} = 0.04$.
3. $n = 1$, $m = 4$, i.e., one feedback stage with four amplifiers in its forward path (see Fig. 4-5c). Then we have $T = T_{io} = 10^4$, $(1 - \mu^4\beta) = (1 - 10^8\beta) = 10^4$, $\beta = -9.999 \times 10^{-5}$ and $S_{\mu}^{T_{io}} = 0.0004$.

8. This is only true so long as $T_{io} < \mu^\alpha$, i.e., as long as the closed-loop gain is smaller than the open-loop forward gain. This will always be the case whenever *negative* feedback is used to increase the stability of a network because the closed loop gain is thereby automatically decreased. (Some of the available gain is used to stabilize the network).

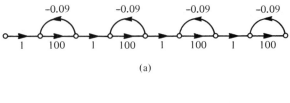

(a)

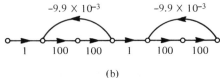

(b)

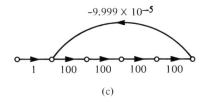

(c)

FIG. 4-5. The three possible feedback configurations containing four gain stages in the forward path.

Thus, assuming that all four amplifier stages μ vary by the same amount, the overall gain of the configuration consisting of a single loop around all four stages will vary one thousand times less than the one consisting of four individual feedback loops, and one hundred times less than the one in which there are two loops with two gain stages in each.

Of the three basic types of sensitivity expressions discussed earlier, the third, given by (4-19), is particularly useful when the sensitivity function of a given network has to be calculated in functional form. In such cases, the transfer function of the network is generally available and can be written in bilinear form by inspection. It is then almost trivial to go from that step to the one of obtaining (4.19).

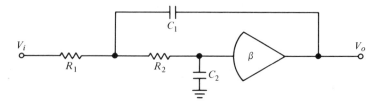

FIG. 4-6. Active low-pass filter incorporating a voltage-controlled voltage source.

Let us, for example, consider the active low-pass filter shown in Fig. 4-6. This is a special case of the general controlled-source (i.e., VVS) network discussed in Chapter 3 (see Fig. 3-36). Substituting $Y_1 = 1/R_1$, $Y_2 = 1/R_2$, $Y_3 = sC_1$, $Y_6 = sC_2$, and $Y_4 = Y_5 = 0$ into the general transfer function (3-92), we obtain the voltage transfer function in bilinear form with respect to the gain β:

$$T(s) = \frac{\beta}{s^2 R_1 R_2 C_1 C_2 + s[(R_1 + R_2)C_2 + R_1 C_1] + 1 - \beta s C_1 R_1} \quad (4\text{-}31)$$

Comparing with (4-16) we have

$$A(s) = 0 \quad (4\text{-}32a)$$

$$B(s) = 1 \quad (4\text{-}32b)$$

$$U(s) = s^2 R_1 R_2 C_1 C_2 + s[R_1 C_1 + (R_1 + R_2)C_2] + 1 \quad (4\text{-}32c)$$

$$V(s) = -sC_1 R_1 \quad (4\text{-}32d)$$

With (4-32a) this is a type I network and since $T(s, \beta = 0) = 0$, we obtain the transmission sensitivity with respect to β directly, using (4-15) and (4-19a):

$$S_\beta^{T(s)} = \frac{U(s)}{D(s)} = \frac{s^2 R_1 R_2 C_1 C_2 + s[R_1 C_1 + (R_1 + R_2)C_2] + 1}{s^2 R_1 R_2 C_1 C_2 + s[R_1 C_1(1 - \beta) + (R_1 + R_2)C_2] + 1} \quad (4\text{-}33)$$

Notice that, as was pointed out earlier, the sensitivity function has an equal number of finite poles and zeros because (4-15) is valid here.

Once we have the transfer function $T(s)$ in terms of all the network elements, as given by (4-31), it is a simple matter to obtain the remaining sensitivities, e.g., those with respect to the resistors and capacitors of the network. Observe that the network of Fig. 4-6 is type I with respect to all its components, i.e., either $A(s)$ or $B(s)$ equals zero for any particular case. Consequently $S_x^{T(s)}$ has the same denominator as $T(s)$ for all values of x, namely,

$$D(s) = s^2 R_1 R_2 C_1 C_2 + s[(R_1 + R_2)C_2 + R_1 C_1(1 - \beta)] + 1 \quad (4\text{-}34)$$

It is convenient to use (4-19a) when $A(s) = 0$ and (4-19b) when $B(s) = 0$. By inspection of (4-31) we obtain

$$S_{R_1}^{T(s)} = -R_1 \frac{s^2 R_2 C_1 C_2 + s[C_1(1 - \beta) + C_2]}{D(s)} \quad (4\text{-}35a)$$

$$S_{R_2}^{T(s)} = -R_2 \frac{s^2 R_1 C_1 C_2 + sC_2}{D(s)} \quad (4\text{-}35b)$$

$$S_{C_1}^{T(s)} = -C_1 \frac{s^2 R_1 R_2 C_2 + sR_1(1 - \beta)}{D(s)} \quad (4\text{-}35c)$$

$$S_{C_2}^{T(s)} = -C_2 \frac{s^2 R_1 R_2 C_1 + s(R_1 + R_2)}{D(s)} \quad (4\text{-}35d)$$

Notice that

$$S_{R_1}^{T(s)} + S_{R_2}^{T(s)} = S_{C_1}^{T(s)} + S_{C_2}^{T(s)} \tag{4-36}$$

This is no coincidence, and will be shown to be generally true in what follows.

4.1.2 Some Useful Transmission Sensitivity Relations

RLC Networks[9]   Consider an *RLC* network containing l inductors L_i, c capacitors C_i, r resistors R_i, and a network function that depends on these elements $F(L_i, C_i, R_i, s)$. From the impedance scaling property of linear networks (see Chapter 3, Section 3.4) it then follows that

$$F\left(aL_i, \frac{C_i}{a}, aR_i, s\right) = \lambda \cdot F(L_i, C_i, R_i, s) \tag{4-37}$$

where

$$\lambda = \begin{cases} a \text{ if } F(s) \text{ is an impedance function} \\ 1 \text{ if } F(s) \text{ is a voltage or current transfer function} \\ \dfrac{1}{a} \text{ if } F(s) \text{ is an admittance function} \end{cases}$$

and a is an arbitrary parameter. Differentiating both sides of (4-37) with respect to a we have

$$\sum_{i=1}^{l} \frac{\partial F(aL_i, C_i/a, aR_i, s)}{\partial aL_i} L_i + \sum_{i=1}^{r} \frac{\partial F(aL_i, C_i/a, aR_i, s)}{\partial aR_i} R_i$$

$$+ \sum_{i=1}^{c} \frac{\partial F(aL_i, C_i/a, aR_i, s)}{\partial(a/C_i)} \cdot \frac{1}{C_i} = F(L_i, C_i, R_i, s) \cdot \frac{d\lambda}{da} \tag{4-38}$$

Setting $a = 1$ and dividing both sides of (4-38) by $F(s)$ we obtain

$$\sum_{i=1}^{l} S_{L_i}^{F(s)} + \sum_{i=1}^{r} S_{R_i}^{F(s)} - \sum_{i=1}^{c} S_{C_i}^{F(s)} = \kappa \tag{[4-39]}$$

where

$$\kappa = \begin{cases} 1 \text{ if } F(s) \text{ is an impedance function} \\ 0 \text{ if } F(s) \text{ is a voltage or current transfer function} \\ -1 \text{ if } F(s) \text{ is an admittance function} \end{cases}$$

9. Many of these relations were first derived by M. L. Blostein, Some bounds on sensitivity in *RLC* Networks, *Proc. 1st Allerton Conf. Circuit and System Theory*, Univ. of Illinois, pp. 488–501 (1963): also Dissertation: *Sensitivity Considerations in RLC Networks*, Univ. of Illinois, 1963.

Note that the negative sign on the capacitance sensitivity term results from the fact that

$$S_{1/x_i}^{F(s)} = -S_{x_i}^{F(s)} \qquad (4\text{-}40)$$

From (4-39) we can write, for an RC network function $T(s)$,

$$\sum_{i=1}^{r} S_{R_i}^{F(s)} = \sum_{i=1}^{c} S_{C_i}^{F(s)} + \kappa \qquad [4\text{-}41]$$

Thus the identity obtained in (4-36) for the circuit of Fig. 4-6 reflects the general rule stated in (4-39).

Another useful expression can be obtained by frequency scaling the general network function $F(L_i, C_i, R_i, s)$. Since only the inductors and capacitors are frequency dependent, we have (see Chapter 3, Section 3-4)

$$F(aL_i, aC_i, R_i, s) = F(L_i, C_i, R_i, as) \qquad (4\text{-}42)$$

where, again, a is an arbitrary parameter. Proceeding as before, i.e., differentiating both sides of (4-42) with respect to a, and setting $a = 1$, we obtain

$$\sum_{i=1}^{l} S_{L_i}^{F(s)} + \sum_{i=1}^{c} S_{C_i}^{F(s)} = \frac{d \ln F(s)}{d \ln s} \qquad [4\text{-}43]$$

Thus with (4-39) we have, for an RC network,

$$\sum_{i=1}^{c} S_{C_i}^{F(s)} = \frac{d \ln F(s)}{d \ln s} \qquad [4\text{-}44a]$$

and

$$\sum_{i=1}^{r} S_{R_i}^{F(s)} = \frac{d \ln F(s)}{d \ln s} + \kappa \qquad [4\text{-}44b]$$

where, again,

$$\kappa = \begin{cases} 1 \text{ if } F(s) \text{ is an impedance function} \\ 0 \text{ if } F(s) \text{ is a voltage or current transfer function} \\ -1 \text{ if } F(s) \text{ is an admittance function} \end{cases}$$

With (4-44a) and (4-44b) we obtain

$$\sum_{i=1}^{r} S_{R_i}^{F(s)} + \sum_{i=1}^{c} S_{C_i}^{F(s)} = 2 \frac{d \ln F(s)}{d \ln s} + \kappa \qquad [4\text{-}45]$$

Equations (4-39) and (4-43) represent fundamental relations between sensitivity functions and are valid for any RLC network. The restriction to RLC components is, however, not essential to the proof of these relations. In fact it can be shown[10] that (4-39) results because $F(L_i, C_i, R_i, s)$ is a homogeneous

10. C. Belove, Sensitivity functions for homogeneous functions, *IEEE Trans. Circuit Theory*, **CT-11**, 171 (1964); also: The sensitivity function in variability analysis, *IEEE Trans. Reliability*, **R-15**, 70–76, (1966).

function of order one, zero, or minus one in the elements L_i, C_i, and R_i, and that the sum of sensitivities of a homogeneous function is a constant, equal to the order of homogeneity. This fact may sometimes simplify sensitivity calculations and will therefore be elaborated here briefly.

The Sensitivity of Homogeneous Functions   To find out whether a function $A(x_1, x_2, \ldots, x_n)$ is homogeneous, each of its elements x_i is multiplied by an arbitrary parameter α. If a power of α can be factored out of the resulting expression such that

$$A(\alpha x_1, \alpha x_2, \ldots, \alpha x_n) = \alpha^m A(x_1, x_2, \ldots, x_n) \tag{4-46}$$

then $A(x_1, x_2, \ldots, x_n)$ is called a *homogeneous function* and m is its *order of homogeneity*. Differentiating (4-46) with respect to α, setting $\alpha = 1$, and dividing by A, we obtain

$$\frac{x_1}{A}\frac{\partial A}{\partial x_1} + \frac{x_2}{A}\frac{\partial A}{\partial x_2} + \cdots + \frac{x_n}{A}\frac{\partial A}{\partial x_n} = m \tag{4-47}$$

This is known as Euler's formula for homogeneous functions. The left side of (4-47) will also be recognized as the sum of sensitivities of the function A to the elements x_i, so that we can write

$$\sum_{i=1}^{n} S_{x_i}^A = m \tag{4-48}$$

Since *dimensional homogeneity* is necessary for all physical equations it is clear that we can always apply this result to the dimensional factors of a network function. Note that if all the x_i vary by the same fractional amount dx/x, then the function A will vary as

$$\frac{dA}{A} = \left(\sum_{i=1}^{n} S_{x_i}^A\right)\frac{dx}{x} = m\frac{dx}{x} \tag{4-49}$$

As an example of (4-49), consider a resistor network. The resistance R_{in} seen at any pair of terminals, must have the dimension of ohms. It must therefore be homogeneous of order one with respect to the individual resistances R_i. We can therefore write

$$\sum_{i=1}^{r} S_{R_i}^{R_{\text{in}}} = 1 \tag{4-50}$$

Thus if all the resistors drift by the same percentage due to ambient changes (temperature, humidity, etc.), i.e., $dR_i/R_i = dR/R$, then we obtain from (4-49)

$$\frac{dR_{\text{in}}}{R_{\text{in}}} = \frac{dR}{R} \tag{4-51}$$

i.e., the input resistance will vary by the same percentage as the individual resistors R_i.

On the other hand, the voltage ratio T between any pair of terminals of the same resistive network will be dimensionless. T will therefore be homogeneous of order zero in the resistances, so that

$$\sum_{i=1}^{r} S_{R_i}^{T} = 0 \qquad (4\text{-}52)$$

With closely tracking resistors, drifting by the same amount dR/R, we therefore have

$$\frac{dT}{T} = 0 \qquad (4\text{-}53)$$

regardless of the complexity of the network.

It should be pointed out here that the voltage transfer function $T(s)$ of a frequency-dependent network is homogeneous of order zero with respect to *impedance* (or admittance) rather than with respect to, say, resistors or capacitors. If we consider a voltage divider consisting of a series impedance $Z_1(s)$ and a shunt impedance $Z_2(s)$ we have $T(s) = Z_2(s)/[Z_1(s) + Z_2(s)]$, which is dimensionless and homogeneous of order zero with respect to Z. However, if we let $Z_1(s) = R$ and $Z_2(s) = 1/sC$ we have $T(s) = 1/(1 + sCR)$ which is homogeneous with respect to neither R nor C. On the other hand, letting $Z_1' = \lambda Z_1 = \lambda R$ and $Z_2' = \lambda Z_2 = \lambda/sC$, which says that $R' = \lambda R$ and $C' = C/\lambda$, we have $T'(s) = T(s)$, which demonstrates the homogeneity with respect to impedance. As we shall see in practical examples later, the situation is different for frequency-independent functions such as $Q(R_i, C_i)$ or $\omega(R_i, C_i)$ which are homogeneous of order zero and minus one, respectively, with respect to resistors R_i or capacitors C_i.

If active elements are included in the network functions, these are still homogeneous in the elements, provided the active elements can be represented by an equivalent circuit containing dimensionless quantities or transfer immittances (e.g., VVS, CCS, VCS, CVS). Thus an equation analogous to (4-39) can be obtained for the active case.[11] Clearly (4-44) is valid for both active and passive networks.

Rules for Calculating with Sensitivity Functions   In carrying out sensitivity calculations it is very useful to be aware of some general rules which hold for the sensitivity of certain basic types of algebraic functions. A list of these rules or sensitivity relations is given in Table 4-1. To illustrate their use we shall demonstrate how much simpler it is now than it was with the expressions used earlier (see Section 4.1.1) to show that the single-loop amplifier containing m identical stages of gain μ (see Fig. 4-7a) is less sensitive to variations in μ than the multiloop feedback stage (see Fig. 4-7b), where both

---

11. S. K. Mitra, *Analysis and Synthesis of Linear Active Networks* (New York: John Wiley & Sons, 1969), p. 189.

TABLE 4-1. SENSITIVITY RELATIONS†

1. $S_x^x = 1$

2.‡ $S_x^{cx} = 1$

3. $S_x^{cx^n} = n$

4.§ $S^{y^n} = nS_x^y$

5. $S_{x^n}^y = \dfrac{1}{n} S_x^y$

6. $S_x^{cy} = S_x^y$

7. $S_x^y = S_{u_1}^y \cdot S_x^{u_1} + S_{u_2}^y \cdot S_x^{u_2} + \cdots$

   where

   $y = y(u_1, u_2, \ldots, u_n)$

8. $S_x^y = S_x^{|y|} + j\phi_y S_x^{\phi_y}$

9. $S_x^{|y|} = \text{Re } S_x^y$

10. $S_x^{y*} = (S_x^y)^*$

11. $S_x^{\phi_y} = \dfrac{1}{\phi_y} \text{ Im } S_x^y$

12. $S_x^{u \cdot v \cdots} = S_x^u + S_x^v + \cdots$

13. $S_x^{u/v} = S_x^u - S_x^v$

14. $S_x^{1/y} = - S_x^y$

15. $S_{1/x}^y = - S_x^y$

16. $S_x^{y+c} = \dfrac{y}{y+c} S_x^y$

17. $S_x^{u+v+\cdots} = \dfrac{1}{u+v+\cdots}(uS_x^u + vS_x^v + \cdots)$

18. $S_{cx}^y = S_x^y$

19. $S_{u/v}^y = \tfrac{1}{2}(S_u^y - S_v^y) = S_u^{\sqrt{y}} - S_v^{\sqrt{y}}$

20. $S_x^{e^y} = yS_x^y$

21. $S_x^{\ln y} = \dfrac{1}{\ln y} \cdot S_x^y$

22. $S_x^{\sin y} = y(\cot y) S_x^y$

23. $S_x^{\cos y} = y(\tan y) S_x^y$

24. $S_x^{\sinh y} = y(\coth y) S_x^y$

25. $S_x^{\coth y} = y(\tanh y) S_x^y$

---

† Many of these relations have been adapted from: J. Gorski-Popiel, Classical sensitivity—a collection of formulas, *IEEE Trans. Circuit Theory*, CT-10, 300–302 (1963); P. R. Geffe, Active filters, quart. rep., Westinghouse Res. & Dev. Rept. ECOM–0363–4.
‡ c and n are constants.
§ y, u, v are single-valued differentiable functions of x; also $y = |y| \cdot e^{j\phi_y}$

have the same number of gain stages μ and the same overall gain. Referring to Fig. 4-7a, let T_S and T_M be the overall gain of the single and multiloop amplifiers, respectively. Then

$$T_S = \frac{\mu^m}{1 - \beta_1 \mu^m} \tag{4-54}$$

and

$$T_M = T^m = \left(\frac{\mu}{1 - \beta_2 \mu}\right)^m \tag{4-55}$$

Since $T_S = T_M$, we have

$$1 - \beta_1 \mu^m = (1 - \beta_2 \mu)^m \tag{4-56}$$

From (4-54), (4-21), and Table 4-1 (formula 5) we have

$$S_{\mu^m}^{T_s} = \frac{1}{m} S_\mu^{T_s} = \frac{1}{1 - \beta_1 \mu^m} \tag{4-57}$$

Therefore, considering (4-56),

$$S_\mu^{T_s} = \frac{m}{1 - \beta_1 \mu^m} = \frac{m}{(1 - \beta_2 \mu)^m} \tag{4-58}$$

From (4-55), (4-21), and Table 4-1 (formula 4), we have

$$S_\mu^{T_M} = S_\mu^{T^m} = mS_\mu^T = \frac{m}{1 - \beta_2 \mu} \qquad (4\text{-}59)$$

Comparing (4-58) and (4-59) it follows that, for $|\beta_2 \mu| > 1$,

$$S_\mu^{T_S} < S_\mu^{T_M} \qquad (4\text{-}60)$$

which is what we found with considerably more effort in Section 4.1.1. At various opportunities in other sections of this book, the reader's attention will be drawn to the covenience of using the formulas listed in Table 4-1.

4.1.3 Transmission Sensitivity of RC Networks with Tracking Components

One of the important properties of hybrid-integrated circuits[12] is that components of a kind can be expected to track very closely indeed. As we shall see presently, networks with this tracking property may have certain very noteworthy characteristics. Furthermore, the tracking property permits some of the preceding general expressions to be simplified appreciably.

In what follows we shall consider only active RC networks, since only they are compatible with present-day integrated-circuit techniques; we shall also assume that the active devices used can be represented by transfer immittances or dimensionless quantities. The fractional change in the network function $T(s)$ can then be written in terms of the sensitivity summations given by (4-39), and in terms of the incremental changes in the resistors R_i and capacitors C_i. We obtain:

$$\frac{dT(s)}{T(s)} = \sum_{i=1}^{r} S_{R_i}^{T(s)} \frac{dR_i}{R_i} + \sum_{i=1}^{c} S_{C_i}^{T(s)} \frac{dC_i}{C_i} \qquad [4\text{-}61]$$

Assuming that all resistor variations and capacitor variations are equal, i.e.,

$$\frac{dR_i}{R_i} = \frac{dR}{R} \qquad i = 1, 2, \ldots, r \qquad (4\text{-}62a)$$

and

$$\frac{dC_i}{C_i} = \frac{dC}{C} \qquad i = 1, 2, \ldots, c \qquad (4\text{-}62b)$$

we obtain, from (4-61),

$$\frac{dT(s)}{T(s)} = \frac{dR}{R} \sum_{i=1}^{r} S_{R_i}^{T(s)} + \frac{dC}{C} \sum_{i=1}^{c} S_{C_i}^{T(s)} \qquad (4\text{-}63)$$

12. These will be discussed in detail in Chapters 6 and 7.

With (4-41) this becomes

$$\frac{dT(s)}{T(s)} = \left(\frac{dR}{R} + \frac{dC}{C}\right) \cdot \sum_{i=1}^{r} S_{R_i}^{T(s)} - \kappa \frac{dC}{C} \qquad \text{[4-64a]}$$

or

$$\frac{dT(s)}{T(s)} = \left(\frac{dR}{R} + \frac{dC}{C}\right) \cdot \sum_{i=1}^{c} S_{C_i}^{T(s)} + \kappa \frac{dR}{R} \qquad \text{[4-64b]}$$

It is clear from the last two expressions that the transmission sensitivity of networks with closely tracking components can be minimized if the drift of the resistors can be guaranteed to be equal and opposite to that of the capacitors. Since drift is generally due to ambient temperature variations this becomes a question of utilizing resistors and capacitors with equal and opposite temperature coefficients. This sensitivity minimization is most effective for voltage or current transfer ratios for which κ equals zero. When $T(s)$ is an immittance function the absolute drift or temperature coefficient values should also be as small as possible.

A useful expression can be obtained for networks with tracking components by combining expressions (4-44a) and (4-44b) with (4-63). We obtain

$$\frac{dT(s)}{T(s)} = \left(\frac{dR}{R} + \frac{dC}{C}\right) \frac{d \ln T(s)}{d \ln s} + \kappa \frac{dR}{R} \qquad \text{[4-65]}$$

For the case that

$$\frac{dC}{C} = \delta \cdot \frac{dR}{R} \qquad (4\text{-}66)$$

we obtain

$$\frac{dT(s)}{T(s)} = \frac{dR}{R} \left[(1 + \delta) \frac{d \ln T(s)}{d \ln s} + \kappa \right] \qquad (4\text{-}67)$$

The relevance of this interesting expression will be discussed in Section 4.1.6.

4.1.4 Sensitivity to Large Parameter Variations

The sensitivity function $S_x^{T(s)}$ defined above is based on incrementally small changes of the network parameter x. It will be shown in what follows[13] that this function can be generalized to include finite changes in the parameter x as well.

Let $T_0(s)$ designate the transmission function of a network when the variable parameter x takes on the reference value x_0. Then (4-10) becomes

$$T(s, x_0) = T_0(s) = t_{14} + x_0 \frac{t_{12} t_{34}}{1 - x_0 t_{32}} \qquad (4\text{-}68)$$

13. E. J. Angelo, Jr., Design of feedback systems, Research Report R–449–55, Polytechnic Institute of Brooklyn–379.

If x_0 is increased by the increment Δx then

$$T(s, x_0 + \Delta x) = T_0(s) + \Delta T(s) \qquad (4\text{-}69)$$

With (4-68) this becomes

$$T_0(s) + \Delta T(s) = t_{14} + (x_0 + \Delta x) \cdot \frac{t_{12} t_{34}}{1 - (x_0 + \Delta x)t_{32}} \qquad (4\text{-}70)$$

Subtracting (4-68) from (4-70) we obtain

$$T(s, x_0 + \Delta x) - T_0(s) = \Delta T(s) = \frac{\Delta x t_{12} t_{34}}{(1 - x_0 t_{32})[1 - (x_0 + \Delta x)t_{32}]} \qquad (4\text{-}71)$$

From (4-11) it follows that the return difference with respect to the perturbed parameter $x_0 + \Delta x$, i.e., the final value of the return difference, is given by

$$F_{x_0 + \Delta x} = 1 - (x_0 + \Delta x)t_{32} \qquad (4\text{-}72)$$

Thus, after some additional algebra, we obtain the incremental sensitivity

$$\frac{\Delta T(s)/T_0(s)}{\Delta x/x_0} = \frac{1}{F_{x_0 + \Delta x}} \left[1 - \frac{t_{14}(s)}{T_0(s)} \right] \qquad [4\text{-}73a]$$

and for $t_{14} = 0$

$$\frac{\Delta T(s)/T_0(s)}{\Delta x/x_0} = \frac{1}{F_{x_0 + \Delta x}} \qquad [4\text{-}73b]$$

This expression is the sensitivity of $T_0(s)$ to finite changes in x. When the increment of x is infinitesimal, (4-73) becomes identical with (4-13). Thus we find the important result that the sensitivity function $S_x^{T(s)}$ can be generalized to include finite as well as infinitesimal changes in the variable parameter. This adds a great deal of significance to sensitivity calculations and to the sensitivity function itself, since in practice the kind of parameter changes we are concerned with will generally be small but finite.[14]

4.1.5 Transmission and Parameter Variation

Having shown that the sensitivity function can be generalized to include finite parameter changes of a network, we are now in a position to introduce an additional concept to sensitivity theory by considering the relative variation V of a

---

14. If $T(s)$ is given as the ratio of the two polynomials $N(s)/D(s)$, then the relationship between incremental and differential sensitivity is: $(\Delta T/T)/(\Delta x/x) = S_x^T/(1 + \Delta x D'/D)$ where D' denotes the derivative of $D(s)$ with respect to x (see J. K. Fidler, C. Nightingale, "Differential-Incremental-Sensitivity Relationships" Electronics Letters, Vol. 8, No. 25 pp. 626-627 (1972), and T. Downs, "A note on the computation of large-change sensitivities" *IEEE Trans. Circuit Theory*, **CT-20** pp. 741-742, (1973)).

network function $F(s)$ resulting from the relative change $\Delta x/x$ of one of its components. It is defined by[15]

$$V_x^F = \frac{\Delta F}{F} = S_x^F \cdot \frac{\Delta x}{x} \qquad [4\text{-}74]$$

F may, of course, be any network function or characteristic (e.g., gain, frequency, pole Q, etc.) and x is any network element that is liable to change. Thus the variation of F with n of its elements x_i is given by

$$\frac{\Delta F}{F} = \sum_{i=1}^{n} V_{x_i}^F = \sum_{i=1}^{n} S_{x_i}^F \frac{\Delta x_i}{x_i} \qquad [4\text{-}75]$$

If the variation of F with one of its elements x_m exceeds the variations with respect to any other elements, we have

$$\frac{\Delta F}{F} \approx V_{x_m}^F \qquad (4\text{-}76)$$

where

$$V_{x_m}^F \gg V_{x_i}^F \qquad i \neq m$$

Although the emphasis in sensitivity considerations is usually on minimizing sensitivity, it is really the variation V that must be minimized. Thus, in some networks the sensitivity may be very small and the relative change $\Delta x/x$ of some particular element large, whereas in others the reverse may be true. Based on sensitivity considerations alone the former networks are superior to the latter; based on a comparison of functional variations, which are the true measures of network stability, they may be equivalent to the latter, or even inferior.

4.1.6 Gain and Phase Sensitivity

So far we have derived a number of expressions for the transmission sensitivity of networks and discussed their properties. The reader may well have been questioning the significance of the derivations, since they resulted in sensitivity functions in s, the relevance of which to an actual network may not be clear at all. Whereas a specified network function $T(s)$ will by now readily be accepted as a function defined by its poles and zeros in the s plane, poles and zeros of a sensitivity function will presumably require some interpretation. The fact is that the sensitivity of a network function will generally not be

---

15. The usage of difference and differential values throughout this text is a very loose one, and is dictated not by mathematical accuracy but by a desire to distinguish between practical measurable, and therefore finite variations (e.g., $\Delta x/x$) and differentially small variations (e.g., dx/x) which better justify the use of the conventional, first-order sensitivity function, which is defined in differential terms. Thus both $\Delta F/F$ and dF/F are obtained as first-order approximations using the same differential quantity $S_x{}^F$; however, $\Delta F/F$ is intended to indicate a quantity obtained in terms of measured variations $\Delta x_i/x_i$, whereas dF/F is assumed to have been obtained by calculating the total differential of the network function $F(s)$.

specified in terms of poles and zeros, but much rather in terms of acceptable tolerances of gain and phase over a given frequency band, such as over the pass band of a bandpass filter. This implies that the sensitivity of a network with respect to sinusoidal input signals is generally of interest and requires sensitivity to be defined in terms of gain and phase rather than as a function in s. As we shall see in the following, the properties of the transmission sensitivity function which were derived in the s plane can readily be rewritten in terms of gain and phase sensitivity.

The Sinusoidal Steady State   Since a linear network is generally expected to operate in its sinusoidal steady state, it will be specified over a limited frequency range and for operation with sinusoidal input signals. Thus we are more concerned with the variations of $T(j\omega)$ than with those of $T(s)$, with respect to variations of the network parameter x. Then

$$T(j\omega) = |T(j\omega)| \cdot e^{j\phi(\omega)} \tag{4-77a}$$

and

$$\ln T(j\omega) = \ln |T(j\omega)| + j \arg T(j\omega)$$
$$= \alpha(\omega) + j\phi(\omega) \tag{4-77b}$$

where $\alpha(\omega)$ and $\phi(\omega)$ are the *gain* and *phase functions*, respectively. The transmission sensitivity then follows:

$$S_x^{T(j\omega)} = \frac{d\alpha(\omega)}{dx/x} + j\frac{d\phi(\omega)}{dx/x} \tag{4-78}$$

We now define the *gain sensitivity*:[16]

$$\mathscr{S}_x^{\alpha(\omega)} = \frac{d\alpha(\omega)}{dx/x} \tag{4-79a}$$

and the *phase sensitivity*

$$\mathscr{S}_x^{\phi(\omega)} = \frac{d\phi(\omega)}{dx/x} \tag{4-79b}$$

Then we can rewrite the transmission sensitivity as

$$S_x^{T(j\omega)} = \mathscr{S}_x^{\alpha(\omega)} + j\mathscr{S}_x^{\phi(\omega)} \tag{4-80}$$

If, as is generally true, x is real, then the transmission sensitivity is related to the gain and phase sensitivity as follows:

$$\mathscr{S}_x^{\alpha(\omega)} = S_x^{|T(j\omega)|} = \operatorname{Re} S_x^{T(j\omega)} = \operatorname{Ev} S_x^{T(s)} \Big|_{s=j\omega} \tag{4-81a}$$

$$\mathscr{S}_x^{\phi(\omega)} = S_x^{\exp \phi(\omega)} = \operatorname{Im} S_x^{T(j\omega)} = \frac{1}{j} \operatorname{Od} S_x^{T(s)} \Big|_{s=j\omega} \tag{4-81b}$$

where Ev and Od stand for "even" and "odd," respectively.

---

16. This is the change of $|T(j\omega)|$ in nepers due to a relative change in x. To obtain the change of $|T(j\omega)|$ in dB, $\mathscr{S}_x{}^{\alpha(\omega)}$ must be multiplied by $20/\ln 10 = 8.68$.

Using the identities above we can now rewrite as functions of ω some of the expressions for transmission sensitivity that were derived earlier as functions of s. Letting $s = j\omega$ in (4-45) we have, with (4-81),

$$\text{Re}\left[\sum_{i=1}^{r} S_{R_i}^{T(j\omega)} + \sum_{i=1}^{c} S_{C_i}^{T(j\omega)}\right] = 2\frac{d\alpha(\omega)}{d\ln\omega} + \kappa \qquad [4\text{-}82]$$

and

$$\text{Im}\left[\sum_{i=1}^{r} S_{R_i}^{T(j\omega)} + \sum_{i=1}^{c} S_{C_i}^{T(j\omega)}\right] = 2\frac{d\phi(\omega)}{d\ln\omega} \qquad [4\text{-}83]$$

In terms of ω these two expressions are very interesting. Since $d\alpha(\omega)/d\ln\omega$ is the derivative of the logarithmic gain function with respect to the logarithm of the frequency, it is directly proportional to the slope of the frequency response plotted on semilogarithmic coordinates in dB/octave (or dB/decade). Thus (4-82) indicates that the sensitivity of the gain function $\alpha(\omega)$ to changes in the resistors R_i and capacitors C_i will be greatest in those ranges of frequency where $\alpha(\omega)$ has the steepest skirts. In general this is at the band edges of a filter function. Since the slope of a filter function is proportional to the pole Q's, this also indicates that the sensitivity of a network to component changes is proportional to Q. The same considerations apply to (4-83), where the phase sensitivities are shown to be proportional to the phase slope (e.g., envelope delay) when the phase curve is plotted on semilogarithmic coordinates.

Networks with Tracking Components   If we now consider a network whose resistors and capacitors track we have, from (4-65),

$$\frac{d|T(j\omega)|}{|T(j\omega)|} = d\alpha(\omega) = \left(\frac{dR}{R} + \frac{dC}{C}\right)\cdot\frac{d\alpha(\omega)}{d\ln\omega} + \kappa\frac{dR}{R} \qquad [4\text{-}84]$$

and

$$d\arg T(j\omega) = d\phi(\omega) = \left(\frac{dR}{R} + \frac{dC}{C}\right)\frac{d\phi(\omega)}{d\ln\omega} \qquad [4\text{-}85]$$

These expressions relate the variations in gain and phase functions of a network directly to its selectivity, i.e., to gain and phase slopes, as well as to the sum of the resistor and capacitor variations. As mentioned earlier, resistors and capacitors with equal but opposite drift coefficients (e.g., temperature, humidity, aging, etc.) can minimize any variations in the gain and phase responses. Letting $\Delta C/C = \delta(\Delta R/R)$ we then have

$$\Delta\alpha(\omega) = \frac{\Delta R}{R}\left[(1+\delta)\frac{d\alpha(\omega)}{d\ln\omega} + \kappa\right] \qquad (4\text{-}86)$$

and

$$\Delta\phi(\omega) = \frac{\Delta R}{R}(1+\delta)\cdot\frac{d\phi(\omega)}{d\ln\omega} \qquad (4\text{-}87)$$

Assuming that $\Delta C/C = \Delta R/R$, i.e., $\delta = 1$, we have from (4-86)

$$\Delta\alpha(\omega) = 2\frac{\Delta R}{R}\frac{d\alpha(\omega)}{d\ln\omega} + \kappa\frac{\Delta R}{R}$$ (4-88)

A similar expression can be found for $\Delta\phi(\omega)$ from (4-87).

Type I Networks Let us finally consider the gain sensitivity of a network to variations of a component x when the leakage path with respect to x is negligible (i.e., type I network for which $A(s)$ in (4-16) equals zero). It will be recalled that the transmission sensitivity of such a network is given by the reciprocal of the return difference with respect to x. Letting $s = j\omega$, the return difference can be written as

$$F_x(j\omega) = \xi(\omega, x) + j\zeta(\omega, x)$$ (4-89)

where

$$\xi(\omega, x) = \mathrm{Re}\, F_x(j\omega)$$

and

$$\zeta(\omega, x) = \mathrm{Im}\, F_x(j\omega)$$

With (4-15), and (4-81a) we then have

$$\mathscr{S}_x^{\alpha(\omega)} = \mathrm{Re}\,\frac{1}{F_x(j\omega)} = \frac{\xi(\omega, x)}{\xi^2(\omega, x) + \zeta^2(\omega, x)}$$ [4-90]

Thus, to minimize the gain sensitivity of a network whose transmission sensitivity is given by (4-15) it is only necessary for the return difference $F_x(j\omega)$ to have a small real part. In terms of (4-11) this means that the real part of the loop gain $xt_{32}(s)$ must be as close to unity as possible.[17] More will be said about this with respect to sensitivity minimization in Section 4-3.

4.2 ROOT SENSITIVITY

4.2.1 Expressions for the Root Sensitivity

Root Sensitivity and Transmission Sensitivity The root sensitivities of a network function $T(s)$ were defined in (4-7) and (4-8) in conjunction with the transmission sensitivity of $T(s)$ as given by (4-6). We recall that the root sensitivities are the residues of the transmission sensitivity given by a partial

17. J. G. Truxal and I. M. Horowitz, Sensitivity considerations in active network synthesis, *Proc. 2nd Midwest Symp. Circuit Theory*, Michigan State Univ., East Lansing, December 1956.

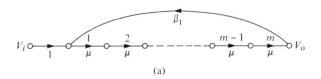

(a)

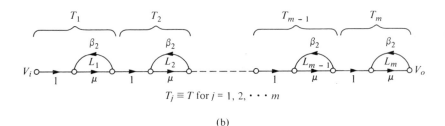

$T_j \equiv T$ for $j = 1, 2, \cdots m$

(b)

FIG. 4-7. Multistage feedback amplifiers. (a) Single loop. (b) Multiloop.

fraction expansion; the poles of the latter are the simple poles and zeros of $T(s)$.[18] The sensitivity of the jth pole p_j follows from (4-6):

$$\mathscr{S}_x^{p_j} = \operatorname{Res} S_x^{T(s)} \bigg|_{s=p_j} = (s - p_j) S_x^{T(s)} \bigg|_{s=p_j} \qquad [4\text{-}91]$$

Like the transmission sensitivity, the root sensitivity can be very simply expressed in terms of the return and null return difference. Noting that $\operatorname{Res} 1/F_x^0\big|_{s=p_j} = 0$ we have, from (4-14) and (4-19),

$$\mathscr{S}_x^{p_j} = \operatorname{Res} \frac{1}{F_x(s)} \bigg|_{s=p_j} = (s - p_j) \frac{U(s)}{D(s)} \bigg|_{s=p_j} = -x(s - p_j) \frac{V(s)}{D(s)} \bigg|_{s=p_j} \qquad [4\text{-}92a]$$

From this it follows that[19]

$$\mathscr{S}_x^{p_j} = \frac{U(s)}{D'(s)} \bigg|_{s=p_j} = -x \frac{V(s)}{D'(s)} \bigg|_{s=p_j} \qquad [4\text{-}92b]$$

Similarly, the zero sensitivity, derived from an expression corresponding to (4-91), can be expressed in terms of the null return difference: since $\operatorname{Res} 1/F_x\big|_{s=z_i} = 0$ we have

$$\mathscr{S}_x^{z_i} = \operatorname{Res} \frac{1}{F_x^0(s)} \bigg|_{s=z_i} = -(s - z_i) \frac{A(s)}{N(s)} \bigg|_{s=z_i} = x(s - z_i) \frac{B(s)}{N(s)} \bigg|_{s=z_i} \qquad [4\text{-}93a]$$

from which it follows that

$$\mathscr{S}_x^{p_j} = -\frac{A(s)}{N'(s)} \bigg|_{s=z_i} = x \frac{B(s)}{N'(s)} \bigg|_{s=z_i} \qquad [4\text{-}93b]$$

18. Only simple roots are considered here; multiple roots will be considered in Section 4.2.5. In practice we shall generally be concerned with the root sensitivity of second-order networks with pole Q's larger than 0.5, i.e., with single complex conjugate pole pairs.
19. If p_j is a simple root of the polynomial $D(s)$, then $\lim_{s \to p_j} [D(s) - D(p_j)]/(s - p_j) = D'(s)\big|_{s=p_j}$, where $D'(s)$ denotes the derivative of $D(s)$ with respect to s.

With (4-92) and (4-93) the root sensitivities can readily be calculated for a network function given in the bilinear form.

The Coefficients of a Polynomial and the Sensitivity of its Roots   A useful relationship between the coefficients a_j of a polynomial $P(s, x)$ and the sensitivities of its roots q_j can be derived.[20] Expressing $P(s, x)$ in the form

$$P(s, x) = \sum_{j=0}^{n} a_j s^j = \prod_{j=1}^{n} (s - q_j)$$

$$= (\alpha_n + x\beta_n)s^n + (\alpha_{n-1} + x\beta_{n-1})s^{n-1} + \cdots + (\alpha_0 + x\beta_0) \quad (4\text{-}94)$$

It follows that[21]

$$\sum_{j=1}^{n} q_j = -\frac{\alpha_{n-1} + x\beta_{n-1}}{\alpha_n + x\beta_n} \quad (4\text{-}95)$$

Differentiating both sides with respect to x we obtain

$$\sum_{j=1}^{n} \frac{dq_j}{dx} = -\frac{\beta_{n-1}(\alpha_n + x\beta_n) - \beta_n(\alpha_{n-1} + x\beta_{n-1})}{(\alpha_n + x\beta_n)^2}$$

Thus

$$\sum_{j=1}^{n} \mathscr{S}_x^{q_j} = x \frac{\beta_n \alpha_{n-1} - \beta_{n-1} \cdot \alpha_n}{(\alpha_n + x\beta_n)^2} \quad (4\text{-}96)$$

In particular

$$\sum_{j=1}^{n} \mathscr{S}_x^{q_j} \bigg|_{\beta_n = 0} = -x \frac{\beta_{n-1}}{\alpha_n} \quad [4\text{-}97a]$$

and

$$\sum_{j=1}^{n} \mathscr{S}_x^{q_j} \bigg|_{\alpha_n = 0} = \frac{\alpha_{n-1}}{x\beta_n} \quad [4\text{-}97b]$$

These expressions are often useful as a check in root sensitivity calculations. Another useful expression relates the sensitivities of a pair of conjugate complex roots q_j and q_j^* to each other:

$$\mathscr{S}_x^{q_j^*} = (\mathscr{S}_x^{q_j})^* \quad [4\text{-}98]$$

Coefficient and Root Sensitivity   A useful sensitivity measure is the sensitivity of a coefficient of a network polynomial to variations in a network element x. As we shall now show, certain relations between the coefficient sensitivity and the root sensitivity of the polynomial can readily be obtained.

20. F. F. Kuo, Pole–zero sensitivity in network functions, *IRE Trans. Circuit Theory*, **CT-5**, 372–373, (1958).
21. From the general relation concerning the sum of the roots of a polynomial; see (4-101) below.

Defining the sensitivity of a coefficient a_j in the polynomial given by (4-94) to variations of x by

$$S_x^{a_j} = \frac{d \ln a_j}{d \ln x} = \frac{da_j/a_j}{dx/x} \tag{4-99}$$

we obtain

$$S_x^{a_j} = x \frac{\beta_j}{a_j} \tag{4-100}$$

Now we require the following two well-known relations concerning the sum and product of the roots of a polynomial: referring to (4-94),

$$\sum_{j=1}^{n} q_j = -\frac{a_{n-1}}{a_n} \tag{4-101}$$

and

$$\prod_{j=1}^{n} q_j = (-1)^n \frac{a_0}{a_n}$$

From (4-101) we obtain, by simple derivation, $\qquad$ (4-102)

$$\sum_{j=1}^{n} \mathscr{S}_x^{q_j} = (S_x^{a_n} - S_x^{a_{n-1}}) \cdot \frac{a_{n-1}}{a_n} \tag{4-103}$$

and from (4-102)

$$\sum_{j=1}^{n} S_x^{q_j} = \sum_{j=1}^{n} \frac{\mathscr{S}_x^{q_j}}{q_j} = [S_x^{a_0} - S_x^{a_n}] \tag{4-104}$$

These two expressions relating the sum of the root sensitivities to certain coefficient sensitivities may readily be applied to second-order systems containing a pair of conjugate complex poles. Consider, for example, a pair of poles defined by the characteristic equation

$$(s - p)(s - p^*) = s^2 + 2\sigma s + \omega_p^2$$

$$= s^2 + \frac{\omega_p}{q} s + \omega_p^2 \tag{4-105}$$

From (4-103) and Table 4-1 we have

$$\mathscr{S}_x^p + \mathscr{S}_x^{p^*} = -2\sigma S_x^{2\sigma} = -2\sigma S_x^{\sigma} = -\frac{\omega_p}{q}(S_x^{\omega_p} - S_x^q)$$

or

$$S_x^{2\sigma} = S_x^{\sigma} = S_x^{\omega_p} - S_x^q \tag{4-106}$$

Similarly, from (4-104) and Table 4-1 we obtain

$$\frac{1}{p}\mathscr{S}_x^p + \frac{1}{p^*}\mathscr{S}_x^{p^*} = S_x^p + S_x^{p^*} = S_x^{\omega_p^2} = 2S_x^{\omega_p} \qquad (4\text{-}107)$$

Root Sensitivity and Root Locus Frequently it is of interest to relate the root sensitivity $\mathscr{S}_x^{q_j}$ to the root locus of q_j with respect to x. Since the root sensitivity $\mathscr{S}_x^{q_j}$ is an incremental vector in the complex plane, we may express it in terms of its real and imaginary parts:

$$\mathscr{S}_x^{q_j} = \operatorname{Re}\mathscr{S}_x^{q_j} + j\operatorname{Im}\mathscr{S}_x^{q_j} = |\mathscr{S}_x^{q_j}|e^{j\phi} \qquad (4\text{-}108)$$

The actual root displacement dq_j, which is also a differentially small vector in the s plane, can now be expressed in terms of the root sensitivity given by (4-108):

$$dq_j = |dq_j|\exp j\arg dq_j = |\mathscr{S}_x^{q_j}|\frac{dx}{x}e^{j\phi} \qquad (4\text{-}109)$$

Since the element x is assumed to be a physical network parameter, dx/x must be real. It therefore follows from (4-109) that

$$\arg dq_j = \arg \mathscr{S}_x^{q_j} = \phi \qquad [4\text{-}110]$$

Since dq_j is the (infinitesimal) displacement of the root q_j along the root locus of q_j with respect to the element x, (4-110) relates the root locus of a root q_j to its sensitivity function. In effect it says that *the angle of the tangent to the root locus at q_j is equal to the argument of the pole sensitivity of q_j*. This is illustrated in Fig. 4-8. If we consider a pole given by $p_j = -\sigma + j\omega_c$, it therefore follows that

$$\mathscr{S}_x^\sigma = \operatorname{Re}\mathscr{S}_x^{p_j} \qquad (4\text{-}111a)$$

and

$$\mathscr{S}_x^{\omega_c} = \operatorname{Im}\mathscr{S}_x^{p_j} \qquad (4\text{-}111b)$$

The pole displacements for maximum and minimum stability then follow in terms of the σ and ω_c sensitivity as shown in Fig. 4-9. Maximum stability is obtained when a pole drifts parallel to the $j\omega$ axis; minimum stability when a pole heads directly for the $j\omega$ axis. In the latter case the likelihood of intercepting the $j\omega$ axis and thereby causing self oscillation is largest. For a high-Q pole (close to the $j\omega$ axis) it may therefore be necessary not only to minimize the pole sensitivity but to ensure that its argument is close to 90° as well. In any event, if two networks realizing the same transfer function also have pole sensitivities with equal magnitudes, the network whose $\mathscr{S}_x^p$ argument is closer to 90° will be the preferable one to use from a stability point of view.

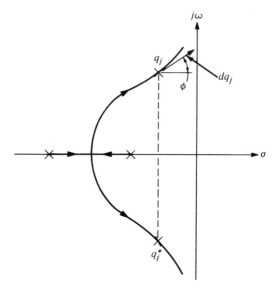

FIG. 4-8. Relationship between the root locus and the root sensitivity of a root q_j with respect to a network element x.

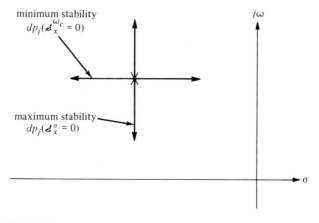

FIG. 4-9. Pole displacements corresponding to maximum and minimum stability.

Root Sensitivity of RLC Networks   Much as we did for the sum of transmission sensitivities we can now consider the sum of root sensitivities of an RLC network containing l inductors, c capacitors, and r resistors. Impedance-scaling the network by a factor a we obtain, for a root q_j of the numerator or denominator of the network function,

$$q_j\left(aL_i, \frac{C_i}{a}, aR_i\right) = q_j(L_i, C_i, R_i) \qquad (4\text{-}112)$$

Differentiating both sides of (4-112) with respect to a and setting a equal to unity we obtain

$$\sum_{i=1}^{l} \mathscr{S}_{L_i}^{q_j} - \sum_{i=1}^{c} \mathscr{S}_{C_i}^{q_j} + \sum_{i=1}^{r} \mathscr{S}_{R_i}^{q_j} = 0 \qquad [4\text{-}113]$$

For an RC network function we therefore have

$$\sum_{i=1}^{c} \mathscr{S}_{C_i}^{q_j} = \sum_{i=1}^{r} \mathscr{S}_{R_i}^{q_j} \qquad [4\text{-}114a]$$

and, for an LC network,

$$\sum_{i=1}^{l} \mathscr{S}_{L_i}^{q_j} = \sum_{i=1}^{c} \mathscr{S}_{C_i}^{q_j} \qquad [4\text{-}114b]$$

Similarly, frequency-scaling the network by a, we have, for the root q_j,

$$q_j\left(\frac{L_i}{a}, \frac{C_i}{a}, R_i\right) = aq_j(L_i, C_i, R_i) \qquad (4\text{-}115)$$

Going through the same process as above, we obtain

$$\sum_{i=1}^{l} \mathscr{S}_{L_i}^{q_j} + \sum_{i=1}^{c} \mathscr{S}_{C_i}^{q_j} = -q_j \qquad [4\text{-}116]$$

For an RC network this simplifies to

$$\sum_{i=1}^{c} \mathscr{S}_{C_i}^{q_j} = \sum_{i=1}^{r} \mathscr{S}_{R_i}^{q_j} = -q_j \qquad [4\text{-}117a]$$

and, for an LC network,

$$\sum_{i=1}^{l} \mathscr{S}_{L_i}^{q_j} = \sum_{i=1}^{c} \mathscr{S}_{C_i}^{q_j} = -q_j/2 \qquad [4\text{-}117b]$$

These expressions also follow directly from the order of homogeneity of an RC or an LC network.[22] As pointed out in the section on transmission

22. For an RC network in which $[\omega] \approx (RC)^{-1}$ (where $[\omega]$ denotes the physical dimension of ω), we have $\sum_{i=1}^{r} S_{R_i}{}^{q_j} = \sum_{i=1}^{c} S_{C_i}{}^{q_j} = -1$; in an LC network, $[\omega] \approx (LC)^{-1/2}$, and therefore $\sum_{i=1}^{l} S_{L_i}{}^{q_j} = \sum_{i=1}^{c} S_{C_i}{}^{q_j} = -\frac{1}{2}$.

sensitivity, these expressions are also valid for active networks, if the active elements can be represented by equivalent circuits containing dimensionless quantities or transfer immittances.

4.2.2 Root Sensitivity of Networks with Tracking Components

Here we consider the root sensitivity of active RC networks with tracking resistors and capacitors, where it is more convenient to work with the root sensitivity $S_x^{q_j}$ than $\mathscr{S}_x^{q_j}$, the two being related by

$$S_x^{q_j} = \frac{d \ln q_j}{d \ln x} = \frac{1}{q_j} \mathscr{S}_x^{q_j} \qquad (4\text{-}118)$$

As mentioned earlier, the case of tracking components has become increasingly important with the advent of hybrid-integrated circuits.

The relative variation of the jth root q_j due to variations in the c capacitors, r resistors, and g active elements G_i of an active RC network is given by

$$\frac{dq_j}{q_j} = \sum_{i=1}^{c} S_{C_i}^{q_j} \frac{dC_i}{C_i} + \sum_{i=1}^{r} S_{R_i}^{q_j} \frac{dR_i}{R_i} + \sum_{i=1}^{g} S_{G_i}^{q_j} \frac{dG_i}{G_i} \qquad (4\text{-}119)$$

Assuming uniform passive component variations and combining the effects of the g active elements into one element G we obtain, with (4-117),

$$\frac{dq_j}{q_j} = -\left(\frac{dR}{R} + \frac{dC}{C}\right) + S_G^{q_j} \frac{dG}{G} \qquad [4\text{-}120]$$

The relevance of this result with respect to sensitivity minimization is discussed elsewhere.[23] We shall now examine the practical significance of the root, or, more specifically, the pole sensitivity concept.

4.2.3 Pole, Frequency, and Q Variation

Consider an nth-order network possessing $n/2$ conjugate complex pole pairs.[24] It will be clear by now that the dominant pole pair, i.e., the pair closest to the ω axis, has the greatest effect on the overall transmission sensitivity as defined by (4-6). This also follows readily from the theory of the functions of a complex variable. Sensitivity studies will therefore frequently involve investigations into the sensitivity of a single (dominant) pole pair even though the overall network order is higher than two. Furthermore, when the $n/2$ complex pole pairs are realized by a noninteracting cascade of $n/2$ second-order filter sections, sensitivity studies will be altogether confined to single pole pairs.

23. See *Linear Integrated Networks: Design*, Chapter 4.
24. Clearly, n is assumed to be an even number here. If it is odd, this merely means the addition of a negative real pole, which has no bearing on the following discussion.

It is desirable in characterizing root, or in this case pole sensitivity to variations in x, to do so in conceptually more accessible and physically more easily measurable terms than, say, the sensitivity of the real part (e.g., σ) and of the imaginary part (e.g., ω_c) of the pole $p = -\sigma + j\omega_c$. Two quantities that lend themselves very well to laboratory measurements are the frequency and Q of a pole pair. Since any high-Q pole (and, by definition, the dominant pole pair has the highest Q of an nth-order network) imposes bandpass properties on the frequency response in its vicinity, measuring frequency and Q implies the measurement of peak or resonant frequency and 3 dB bandwidth as discussed in Section 1.6 of Chapter 1. Again, when second-order sections are cascaded, the characterization of individual pole pairs by these two measurements is all the more accurate. We shall therefore now show how the pole variation $\Delta p/p$ translates directly into variations of the undamped natural frequency ω_p and of Q.

Consider the (dominant) pole $p = -\sigma + j\omega_c$ and its displacement dp in Fig. 4-10.[25] The pole p and its conjugate p^* are defined by the characteristic equation

$$(s + p)(s + p^*) = s^2 + 2\sigma s + \omega_p^2 = 0 \qquad (4\text{-}121)$$

where

$$\omega_p^2 = \sigma^2 + \omega_c^2 \qquad (4\text{-}122)$$

The term p can also be defined in terms of ω_p and Q:[26]

$$s^2 + \frac{\omega_p}{Q} s + \omega_p^2 = 0 \qquad (4\text{-}123)$$

where

$$Q = \frac{\omega_p}{2\sigma} \qquad (4\text{-}124)$$

2σ is the 3 dB bandwidth of the bandpass response determined by p and p^*, ω_p is the center, or peak, frequency, and Q, defined by (4-124), is given in geometrical terms by the magnitude of p, which is ω_p, divided by twice its real part σ (see Fig. 4-10).

The pole variation can be written as

$$\frac{dp}{p} = \text{Re}\,\frac{dp}{p} + j\,\text{Im}\,\frac{dp}{p} \qquad (4\text{-}125)$$

Recall now from the theory of complex variables that when dp/p is real, the direction of dp and p coincide. As a result the pole p is shifted in the radial direction by an amount $d\omega_p$, while Q remains constant. Similarly, when dp/p

25. The following discussion is based on: G. S. Moschytz, A note on pole, frequency, and Q sensitivity, *IEEE J. Solid-State Circuits*, SC-6, 267–269 (1971).
26. Q is used here instead of q because bandpass characteristics are assumed in the vicinity of a dominant pole pair.

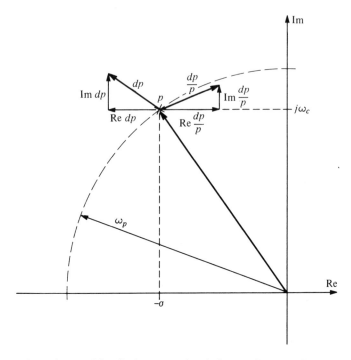

FIG. 4-10. The pole p and its displacement dp relative to the Re p, Im p axes and the Re dp/p, Im dp/p axes.

is imaginary, dp is perpendicular to p, i.e., p is shifted by dp tangentially along the circle centered at the origin; in other words ω_p remains constant but Q is varied by some amount dQ. Clearly the quantities ω_p and Q represent an alternative orthogonal coordinate system to the σ, $j\omega$ system defined by $dp = \mathrm{Re}\, dp + j\, \mathrm{Im}\, dp$. Not only are ω_p and Q readily measurable (in contrast to σ and $j\omega_c$) but, as we shall now show, $d\omega_p/\omega_p$ and dQ/Q are related in a very simple manner to the real and imaginary parts, respectively, of dp/p.

To calculate (4-125) we have

$$\frac{dp}{p} = \frac{-d\sigma + j\, d\omega_c}{-\sigma + j\omega_c} \tag{4-126}$$

By straightforward analysis we obtain

$$\mathrm{Re}\,\frac{dp}{p} = \frac{\sigma\, d\sigma + \omega_c\, d\omega_c}{\omega_p^2} = \frac{d\omega_p}{\omega_p} \tag{4-127}$$

and

$$\mathrm{Im}\,\frac{dp}{p} = \frac{\omega_c\, d\sigma - \sigma\, d\omega_c}{\omega_p^2} = -\frac{\sigma}{\omega_c}\frac{dQ}{Q} = \frac{dQ/Q}{\sqrt{4Q^2 - 1}} \tag{4-128}$$

Thus

$$\frac{dp}{p} = \frac{d\omega_p}{\omega_p} - j\frac{dQ/Q}{\sqrt{4Q^2 - 1}} \qquad \text{[4-129a]}$$

and, for $Q \gg 1$,

$$\frac{dp}{p} \approx \frac{d\omega_p}{\omega_p} - j\frac{dQ/Q}{2Q} \qquad \text{[4-129b]}$$

Thus, calculating the relative pole variation dp/p (rather than merely the pole shift dp) we obtain the relative frequency and Q variations, $d\omega_p/\omega_p$ and dQ/Q, directly from its real and imaginary parts, respectively. This is shown qualitatively in Fig. 4-11. Clearly the characterization of a pole p by the two measurable quantities

$$\omega_p = |p| \qquad (4\text{-}130)$$

and

$$Q = \frac{|p|}{2 \operatorname{Re} p} \qquad (4\text{-}131)$$

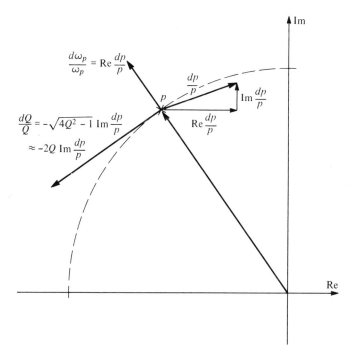

FIG. 4-11. Relations between $\operatorname{Re} dp/p$ and $d\omega_p/\omega_p$, and between $\operatorname{Im} dp/p$ and dQ/Q.

is much more useful than its characterization in terms of the two physically less accessible terms

$$-\sigma = \operatorname{Re} p \qquad (4\text{-}132)$$

and

$$\omega_c = \operatorname{Im} p \qquad (4\text{-}133)$$

Since frequency and Q sensitivity are also more meaningful to the circuit designer than pole (or root) sensitivity (because the former are directly measurable), we shall now show how these three quantities are also interrelated in a simple manner.

4.2.4 Pole, Frequency, and Q Sensitivity

Active RC Networks   Restricting ourselves to active RC networks, we can express the variation in the location of a pole with the summation of sensitivities given by (4-119). For convenience, using variations $V_{x_i}^p$ as defined by (4-74), we have, from (4-119),

$$\frac{dp}{p} = \operatorname{Re} \sum_{i=1}^{c} V_{C_i}^p + \operatorname{Re} \sum_{i=1}^{r} V_{R_i}^p + \operatorname{Re} V_G^p + j\left(\operatorname{Im} \sum_{i=1}^{c} V_{C_i}^p + \operatorname{Im} \sum_{i=1}^{r} V_{R_i}^p + \operatorname{Im} V_G^p \right)$$

$$(4\text{-}134)$$

Similarly[27]

$$\frac{d\omega_p}{\omega_p} = \sum_{i=1}^{c} V_{C_i}^{\omega_p} + \sum_{i=1}^{r} V_{R_i}^{\omega_p} + V_G^{\omega_p} \qquad (4\text{-}135)$$

and

$$\frac{dQ}{Q} = \sum_{i=1}^{c} V_{C_i}^{Q} + \sum_{i=1}^{r} V_{R_i}^{Q} + V_G^{Q} \qquad (4\text{-}136)$$

With (4-129) it follows that

$$\sum_{i=1}^{c} V_{C_i}^{\omega_p} + \sum_{i=1}^{r} V_{R_i}^{\omega_p} + V_G^{\omega_p} = \operatorname{Re} \sum_{i=1}^{c} V_{C_i}^p + \operatorname{Re} \sum_{i=1}^{r} V_{R_i}^p + \operatorname{Re} V_G^p \quad (4\text{-}137\text{a})$$

and

$$\sum_{i=1}^{c} V_{C_i}^{Q} + \sum_{i=1}^{r} V_{R_i}^{Q} + V_G^{Q} = -\sqrt{4Q^2 - 1}\left(\operatorname{Im} \sum_{i=1}^{c} V_{C_i}^p + \operatorname{Im} \sum_{i=1}^{r} V_{R_i}^p + \operatorname{Im} V_G^p \right)$$

$$(4\text{-}137\text{b})$$

---

27. As in (4-120), we have here combined the effects of the g active elements into one active element G.

Consequently, for any component x we find

$$V_x^{\omega_p} = \operatorname{Re} V_x^p \qquad [4\text{-}138\text{a}]$$

and

$$V_x^Q = -\sqrt{4Q^2 - 1}\, \operatorname{Im} V_x^p \qquad [4\text{-}138\text{b}]$$

The frequency and Q sensitivities in terms of the corresponding pole sensitivity therefore result as

$$S_x^{\omega_p} = \operatorname{Re} S_x^p \qquad [4\text{-}139]$$

and

$$S_x^Q = -\sqrt{4Q^2 - 1}\, \operatorname{Im} S_x^p \qquad [4\text{-}140\text{a}]$$

For $Q \gg 1$ we obtain

$$S_x^Q \approx -2Q\, \operatorname{Im} S_x^p \qquad [4\text{-}140\text{b}]$$

These expressions are useful in comparing theory with practice since it is often more convenient to calculate the pole sensitivities of a network than its frequency and Q sensitivities, but more convenient to measure the latter.

Networks with Tracking Components   Let us now consider a hybrid-integrated network for which we may assume tracking of all resistors and capacitors on a substrate. Equations (4-135) and (4-136) then simplify to

$$\frac{d\omega_p}{\omega_p} = \frac{dC}{C} \sum_{i=1}^{c} S_{C_i}^{\omega_p} + \frac{dR}{R} \sum_{i=1}^{r} S_{R_i}^{\omega_p} + S_G^{\omega_p} \frac{dG}{G} \qquad (4\text{-}141)$$

and

$$\frac{dQ}{Q} = \frac{dC}{C} \sum_{i=1}^{c} S_{C_i}^{Q} + \frac{dR}{R} \sum_{i=1}^{r} S_{R_i}^{Q} + S_G^{Q} \frac{dG}{G} \qquad (4\text{-}142)$$

Remember that ω_p is a homogeneous function of order -1 and Q a homogeneous function of order zero in the resistors R_i and the capacitors C_i. From (4-48) it therefore follows that

$$\sum_{i=1}^{r} S_{R_i}^{\omega_p} = \sum_{i=1}^{c} S_{C_i}^{\omega_p} = -1 \qquad [4\text{-}143]$$

and

$$\sum_{i=1}^{r} S_{R_i}^{Q} = \sum_{i=1}^{c} S_{C_i}^{Q} = 0 \qquad [4\text{-}144]$$

For hybrid integrated networks (4-141) and (4-142) therefore simplify to

$$\frac{d\omega_p}{\omega_p} = -\left(\frac{dR}{R} + \frac{dC}{C}\right) + S_G^{\omega_p}\frac{dG}{G} \qquad [4\text{-}145]$$

and

$$\frac{dQ}{Q} = S_G^Q\frac{dG}{G} \qquad [4\text{-}146]$$

These expressions are very useful for the calculation of frequency and Q drift in hybrid-integrated networks.

4.2.5 Multiple–Root Sensitivity

Up to this point we have considered the sensitivity of only simple roots of a network function. Indeed, the definitions of root sensitivity as given above (see, e.g. (4-92) and (4-93)) break down in the case of multiple roots, since the corresponding root displacements seemingly become infinitely large. We shall show here that this is in fact not so; a root of multiplicity m splits up into m distinct roots equispaced on a circle of finite radius that is centered at the original multiple root.

Consider, for example, a network function $T(s) = N(s)/D(s)$ having m multiple poles. The denominator can be written in the form

$$D(s, x) = U(s) + xV(s) = (s - p_j)^m D_0(s, x) \qquad (4\text{-}147)$$

Notice that the pole sensitivity according to (4-92a) gives $\mathscr{S}_x^{p_j} = (s - p_j)/F_x(s)$ $= U(s)/(s - p_j)^{m-1}D_0(s, x)$ evaluated at $s = p_j$, where, however, it goes to infinity. To circumvent this erroneous result we proceed as follows.

We assume that Δp_j is the displacement of the multiple pole p_j due to a small change Δx of the network element x from its nominal value x_0. Thus we obtain from (4-147)

$$D(p_j + \Delta p_j, x_0) = (\Delta p_j)^m D_0(p_j + \Delta p_j, x_0) \qquad (4\text{-}148)$$

Furthermore, letting

$$\Delta D(p_j + \Delta p_j, \Delta x) = D(p_j + \Delta p_j, x_0 + \Delta x) - D(p_j + \Delta p_j, x_0) \qquad (4\text{-}149)$$

we have

$$D(p_j + \Delta p_j, x_0) + \Delta D(p_j + \Delta p_j, \Delta x) = 0 \qquad (4\text{-}150)$$

Substituting (4-148) into (4-150) and solving for Δp_j we obtain[28]

$$\Delta p_j = \left[-\frac{\Delta D(p_j + \Delta p_j, \Delta x)}{D_0(p_j + \Delta p_j, x_0)}\right]^{1/m} \qquad (4\text{-}151)$$

---

28. See also: C. S. Phan and H. K. Kim, New definition of multiple-root sensitivity, *Electron. Lett.* **8**, 497 (1972).

or, to a good approximation,

$$\Delta p_j \approx \left[- \frac{\Delta D(p_j, \Delta x)}{D_0(p_j, x_0)} \right]^{1/m} \tag{4-152}$$

In this expression $D_0(p_j, x_0)$ results directly from the given polynomial (4-147), and $\Delta D(p_j, \Delta x)$ follows from (4-149). However, $\Delta D(p_j, \Delta x)$ can be simplified further if we recall that the poles of $T(s) = N(s)/D(s)$ are also the zeros of the corresponding return difference $F_x(s)$. Thus $F_x(s)$ can also be written in a form corresponding to the rightmost term in (4-147), and the zeros of $F_x(p_j)$ will also be displaced by Δp_j because of the element change Δx. Calculating Δp_j in terms of the return difference, we obtain an expression analogous to (4-152) in which the term $\Delta F_x(p_j, \Delta x)$ appears instead of $\Delta D(p_j, \Delta x)$. From the defining expression for $F_x(s)$ (see (4-17)) we obtain

$$\Delta F_x(s) = \Delta x \, \frac{V(s)}{U(s)} \tag{4-153}$$

and since

$$D(p_j) = U(p_j) + xV(p_j) = 0 \tag{4-154}$$

it follows that

$$\Delta F_x(p_j) = - \frac{\Delta x}{x} \tag{4-155}$$

Thus in terms of the return difference we obtain[29]

$$\Delta p_j = \left[L_m \frac{\Delta x}{x} \right]^{1/m} \tag{4-156a}$$

where

$$L_m = \left. \frac{(s - p_j)^m}{F_x(s)} \right|_{s=p_j} \tag{4-156b}$$

and L_m corresponds to the term $[D_0(p_j, x_0)]^{-1}$ in (4-152). Strictly speaking, L_m is the coefficient of the term $(s - p_j)^{-m}$ in the Laurent expansion of $1/F_x(s)$ about the multiple singularity p_j. Note that for $m = 1$ we have

$$L_1 = \left. \frac{s - p_j}{F_x(s)} \right|_{s=p_j} = \left. \text{Res} \, \frac{1}{F_x(s)} \right|_{s=p_j} = \mathscr{S}_x^{p_j} \tag{4-157}$$

i.e., (4-156) is general enough to include the expression for single-root sensitivity, whereas, as we pointed out above, this statement in reverse does not hold. Since $L_m(p_j)$ will in general be complex, (4-156) defines an m-valued quantity corresponding to the splitting up of the multiple pole p_j into m poles located equispaced on a circle of radius $|\Delta p_j|$ centered at p_j.

29. A. Papoulis, Displacement of the zeros of the impedance $Z(p)$ due to an incremental variation in the network elements, *Proc. IRE*, **43**, 70–82 (1955).

4.3 SENSITIVITY MINIMIZATION

We come now to an important part of our sensitivity discussion, namely, to the question of minimizing the sensitivity of a network function to variations of one or more of its components or, in short, to the subject of sensitivity minimization. Referring to the sensitivity of the transfer function given by (4-6), the partial fraction expansion given there can be interpreted as a summation of partial sensitivities. The minimization of S_x^T for all s is generally both difficult and often unnecessary; in many cases it may be quite sufficient to minimize *some* of the partial sensitivities rather than to minimize them all. This type of *partial desensitization* may be relatively simple because the effect of certain poles or pole pairs on the transmission characteristics may be more readily observable or predictable than that of the totality of network poles and zeros together. Furthermore some of the partial sensitivities given by (4-6) may affect the overall transmission sensitivity more than others, i.e., by their location, the sensitivity of some poles to element variations may be very much more critical than the sensitivity of others. We already pointed out in Section 4.2.3, for example, that the partial sensitivities related to the dominant poles of a network are the ones that should primarily be minimized or at least controlled the most stringently, since they influence the overall transmission characteristics the most strongly. As we know, the actual poles of the partial transmission sensitivities cannot be chosen freely since they are determined by the specified transfer function $T(s)$. This leaves the residues, i.e., the root sensitivities, to be minimized individually, depending on the position of the respective roots in the s plane. These questions are discussed in more detail elsewhere.[30]

4.3.1 Sensitivity Minimization of Type I Networks

For the present, let us look at networks whose leakage path with respect to a critical element x is zero (type I networks). For these networks we found (see (4-11) and (4-15)) that

$$S_x^{T(s)} = \frac{1}{F_x(s)} = \frac{1}{1 - xt_{32}(s)} \tag{4-158}$$

and also that $S_x^{T(s)}$ has an equal number of zeros and poles, so that it must have the form

$$S_x^{T(s)} = \frac{(s - z_{S1})(s - z_{S2}) \cdots (s - z_{Sn})}{(s - p_{S1})(s - p_{S2}) \cdots (s - p_{Sn})} \tag{4-159}$$

30. *Linear Integrated Networks: Design*, Chapters 1 and 4.

These two expressions show up some of the fundamental difficulties encountered in active network synthesis when one is attempting to minimize sensitivity.

If in the expression (4-158) x is assumed to be the gain of an active element, it would seem that the transmission sensitivity can be minimized to any desired amount by simply increasing x sufficiently. However, as we shall discuss in Chapter 7, to remain stable, i.e., to maintain a sufficient phase margin, the gain element must have a wide enough bandwidth to allow for a frequency roll-off smaller or equal to about 33 dB/decade.[31] The required bandwidth therefore increases rapidly, and far beyond the specified transmission frequency band, with increasing gain. The amount of desensitization obtainable by increasing gain is therefore generally limited either by the gain-bandwidth product of available active elements or by the fact that the type of application in question does not justify the cost of the required wideband device.

The expression for transmission sensitivity given by (4-159) may, at first sight, suggest a means of evading the inherent conflict between sensitivity minimization (by increasing gain), and system instability. It shows that, to keep the sensitivity low (at least within a specified frequency band, and assuming some freedom can be exercised on the choice of the poles and zeros of $T(s)$ with a view to minimizing $S_x^{T(s)}$), the poles of $S_x^{T(s)}$ should be placed as far away as possible from that segment of the $j\omega$ axis corresponding to the specified band, and the zeros should be placed as close as possible to this segment. This is indicated qualitatively in Fig. 4-12. However, since the poles of S_x^T coincide with those of the closed-loop transmission function, they are located on the root locus of the network poles with respect to x defined by $F_x(s) = 1 - xt_{32}(s) = 0$ for $0 \leq x \leq \infty$. On the other hand, the zeros of $S_x^{T(s)}$ coincide with the poles of the loop transmittance $t_{32}(s)$. Thus, whereas the closed-loop poles are conjugate complex and are situated close to the $j\omega$ axis (this is the only case of interest, i.e., when sensitivity becomes critical) the poles of $t_{32}(s)$ will very likely be on the negative real axis.[32] It is therefore impossible to follow this second procedure for the sensitivity minimization of highly selective (i.e., high-Q) active RC networks. The pole–zero diagram of a typical sensitivity function shown in Fig. 4-13 serves to illustrate this point. The illustration shows that in general the sensitivity zeros will be further away from the specified transmission frequency band on the $j\omega$ axis than the sensitivity poles, which is the reverse of the situation desired to minimize sensitivity.

Minimization of Partial Sensitivities   One method of minimizing the transmission sensitivity of type-I networks has been suggested[33] that may be

---

31. A roll off of 33 dB/decade, which is equivalent to 10 dB/octave, gives a phase margin of 30°. (Physically, only multiples of 20 dB/decade (6 dB/octave) can be attained.)
32. This assumes that x is the active network element which is generally the most variable and thus the most critical. In that case, $t_{32}(s)$ is the transmittance of a passive RC network.
33. J. G. Truxal and I. M. Horowitz, op. cit.

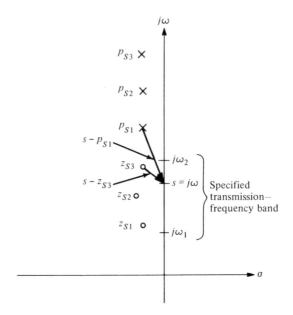

FIG. 4-12. Pole–zero distribution for minimum transmission sensitivity of type I networks in the frequency pass band.

practical in some special cases. As in the general class of networks, it consists of breaking the total transmission sensitivity into the sum of partial sensitivities so as to minimize the more critical sensitivities to the detriment of the less critical ones. The opportunity of taking such an approach often presents itself when a given application can be restricted to a particular class of input signals. It is particularly useful when only sinusoidal input signals need be

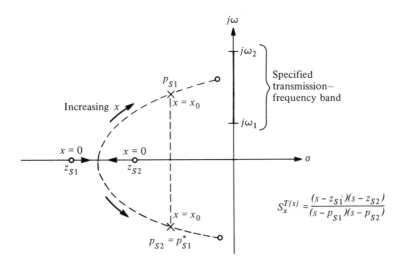

FIG. 4-13. Pole–zero diagram of a typical second-order sensitivity function.

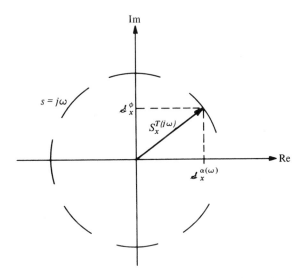

FIG. 4-14. Transmission sensitivity decomposed into gain and phase sensitivity components.

considered. In this case the transmission function evaluated along the $j\omega$ axis is of interest. It can then be expressed as the sum of gain and phase sensitivities as given by (4-80). Whichever of these two partial sensitivities is the more critical for a given application, its value can be minimized by an appropriate configuration. This is illustrated by Fig. 4-14. The sensitivity function corresponding to a given transmission function is shown for a specific input signal frequency ω. Whereas the magnitude of $S_x^{T(s)}$ is thereby determined, the real and imaginary parts depend on the chosen network configuration. As shown in Section 4.1.6 (see (4-89) and (4-90)), Truxal and Horowitz have utilized this fact by demonstrating that the gain sensitivity (i.e., the real part of $S_x^{T(s)}$) can be minimized for certain type I networks without requiring high loop gain. They thereby showed that in some circumstances the conflict between low sensitivity and high stability can be avoided.

4.3.2 Minimizing Denominator Sensitivity

Another practical method of minimizing the transmission sensitivity of active RC networks has been suggested by Herbst.[34] It is based on the fact that in general the sensitivities of the poles affect the overall sensitivity most. Herbst therefore considers the sensitivity of the denominator polynomial $D(s)$ defining these poles. With $D(s)$ given in the bilinear form, ie.,

$$D(s) = U(s) + xV(s) \qquad (4\text{-}160)$$

34. N. M. Herbst, Optimization of pole sensitivity in active RC networks, Res. Rep. EE 569, Cornell University, Ithaca, N. Y., 1963.

we obtain the sensitivity of $D(s)$ to variations of x as:[35]

$$S_x^{D(s)} = \frac{d \ln D(s)}{d \ln x} = x \frac{V(s)}{D(s)} \qquad [4\text{-}161]$$

For sinusoidal signals, for which $s = j\omega$, (4-161) can be minimized by introducing $j\omega$ zeros in $V(j\omega)$ in the frequency band of interest. Since the zeros of $V(s)$ are the transmission zeros of the feedback path of any general feedback network (see Chapter 2, Fig. 2-66), this requires a network with transmission zeros, such as a twin-T network, in the feedback path. Active networks of this kind are in use today, as described elsewhere.[36]

Type I Networks   If the network used to minimize $V(j\omega)$ is type I, then the transmission sensitivity and the pole-polynomial sensitivity are directly related. For the case that $A(s)$ in (4-16) equals zero, we have, from (4-19a),

$$S_x^{T(s)} = 1 - S_x^{D(s)} \qquad [4\text{-}162]$$

If $B(s)$ equals zero, we have, from (4-19b),

$$S_x^{T(s)} = -S_x^{D(s)} \qquad [4\text{-}163]$$

In the second case, minimizing the pole-polynomial sensitivity directly minimizes the transmission sensitivity as well. Realizing $j\omega$-axis zeros with $V(j\omega)$ in the frequency band of interest is then very effective in minimizing $S_x^{T(j\omega)}$.

Consider, for example, the network shown in Fig. 4-15. It consists of an inverting operational amplifier in the forward path and a twin-T null network in series with a noninverting operational amplifier β in the feedback path. The corresponding signal-flow graph is shown in Fig. 4-16. The voltage transfer ratio follows:

$$
\begin{aligned}
T(s) &= -\mu \frac{s^2 + (\omega_p/\hat{q})s + \omega_p^2}{s^2 + (\omega_p/\hat{q})s + \omega_p^2 + \mu\beta(s^2 + \omega_p^2)} \\
&= -\frac{\mu}{1 + \mu\beta} \cdot \frac{s^2 + (\omega_p/\hat{q})s + \omega_p^2}{s^2 + [\omega_p/\hat{q}(1 + \mu\beta)]s + \omega_p^2}
\end{aligned}
\qquad (4\text{-}164)
$$

The transmission sensitivity with respect to μ follows from (4-19a), where $A(s) = 0$. Thus, with (4-162),

$$
\begin{aligned}
S_\mu^{T(s)} &= \frac{1}{1 + \mu\beta} \cdot \frac{s^2 + (\omega_p/\hat{q})s + \omega_p^2}{s^2 + [\omega_p/\hat{q}(1 + \mu\beta)]s + \omega_p^2} \\
&= 1 - S_\mu^{D(s)}
\end{aligned}
\qquad (4\text{-}165)
$$

35. Similarly, the sensitivity of the numerator polynomial $N(s) = A(s) + xB(s)$ can of course be defined. One obtains $S_x^{N(s)} = xB(s)/N(s)$. This must be so, since $S_x^{T(s)} = S_x^{N(s)/D(s)} = S_x^{N(s)} - S_x^{D(s)}$ which must result in (4-19a) and (4-19b).
36. *Linear Integrated Networks: Design*, Chapters 2, 3, and 6.

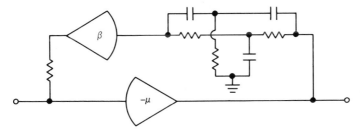

FIG. 4-15. Type I negative feedback network with inband transmission zeros in the feedback loop.

The pole-polynomial sensitivity follows directly from (4-164):

$$S_\mu^{D(s)} = \frac{\mu\beta}{1 + \mu\beta} \frac{(s^2 + \omega_p^2)}{s^2 + [\omega_p/\hat{q}(1 + \mu\beta)]s + \omega_p^2} \qquad (4\text{-}166)$$

$|T(j\omega)|$, $|S_\mu^{T(j\omega)}|$, and $|S_\mu^{D(j\omega)}|$ are plotted in Fig. 4-17. Notice that the transmission sensitivity is less than or equal to unity for all frequencies. On the other hand, it is not equal to the more desirable response $|S_\mu^{D(j\omega)}|$ because $S_\mu^{T(s)}$ corresponds to the case of (4-162), i.e., $A(s) = 0$, rather than to (4-163). The situation is reversed if we require the sensitivity with respect to the gain β, since then $B(s) = 0$. $S_\beta^{T(s)}$ now corresponds to (4-163), and we have, with (4-19b),

$$S_\beta^{T(s)} = -\frac{\mu\beta}{1 + \mu\beta} \cdot \frac{s^2 + \omega_p^2}{s^2 + [\omega_p/\hat{q}(1 + \mu\beta)]s + \omega_p^2} = -S_\beta^{D(s)} \qquad (4\text{-}167)$$

Comparing (4-167) with (4-166) we find that

$$S_\beta^{T(s)} = S_\mu^{D(s)} \qquad (4\text{-}168)$$

Referring to Fig. 4-17 we have here the very desirable situation in which the sensitivity to variations in β is zero in the vicinity of the pole frequency ω_p, and, as in the case of $S_\mu^{T(j\omega)}$, less than unity for all other frequencies.

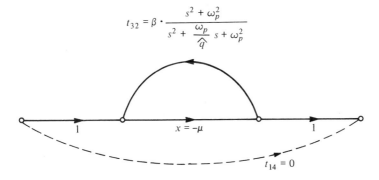

FIG. 4-16. Signal-flow graph of the negative feedback network shown in Fig. 4-15.

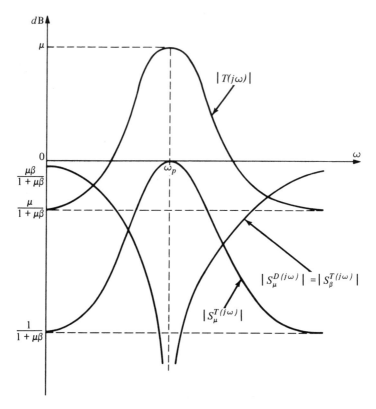

FIG. 4-17. Frequency response of amplitude, transmission sensitivity, and pole-polynomial sensitivity for the negative feedback network in Fig. 4-15.

4.4 SUMMARY

1. Sensitivity theory deals with the effects of network component variations on the specified network characteristics.

2. Transmission sensitivity is defined as the ratio of the relative change of the network transmission function $T(s)$ to the relative change of a network element x causing the change in $T(s)$.

3. Transmission sensitivity can be expressed: (1) as a partial fraction expansion, in which all critical network frequencies become poles and the residues are the pole and zero sensitivities; (2) as the difference between the inverse of the return difference and the inverse of the null return difference; and (3) in terms of the polynomials of the bilinear form of the transmission function. These expressions are useful in determining, respectively, (1) the effects of pole and zero perturbations, (2) the effects of feedback on the stability

(in the sense of insensitivity) of the network, and (3) the transmission sensitivity in functional form.

4. Networks whose equivalent essential signal-flow graphs have only one forward path between source and sink are called type I networks. Those with an additional (e.g., leakage) path are called type II networks.

5. In the bilinear form $T(s) = [A(s) + xB(s)]/[U(s) + xV(s)]$, type I networks have either $A(s)$ or $B(s)$ identically equal to zero, whereas both polynomials are nonzero for type II networks.

6. The transmission sensitivity of a network that has no leakage path (e.g. type I) is equal to the inverse of the return difference with respect to the changing element. The transmission function of such a network goes to zero as the reference element x is set equal to zero.

7. By impedance- and frequency-scaling a network function, useful transmission sensitivity relations can be obtained. The same expressions are obtained by considering the homogeneity of a network function and applying Euler's formula for homogeneous functions.

8. The sensitivity expressions of networks whose components track with ambient variations can be simplified considerably.

9. To minimize the variation of RC networks with closely tracking resistors and capacitors, the temperature coefficients of the resistors (TCR) should be equal to and opposite those of the capacitors (TCC).

10. The sensitivity of the gain function $\alpha(\omega)$ of a network to changes in its components is largest in those frequency ranges in which the slope of the gain function is steepest.

11. The root sensitivity of a network polynomial is closely related to the coefficient sensitivities of the polynomial.

12. The angle of the tangent to the root locus of a polynomial at the root q_j is equal to the argument of the root sensitivity of q_j.

13. The real part of the variation dp/p of a high-Q pole p is equal to the frequency variation $d\omega_p/\omega_p$; the imaginary part is approximately equal to $(1/2Q)(dQ/Q)$.

14. The transmission sensitivity of a network function can be minimized by minimizing its root sensitivities.

CHAPTER

5

SOME BASIC NETWORK ELEMENTS

INTRODUCTION

It was pointed out earlier (and is shown in *Linear Integrated Networks: Design*, chapter 2) that most active-filter synthesis techniques can be classified into a positive or negative feedback scheme of one kind or another. Thus, in principle, all we need to design active filters is a general purpose amplifier as the active device, combined with a passive *RC* network that will provide the necessary open-loop poles and zeros. As we shall see, the operational amplifier is extremely useful in this respect, particularly since we can simulate any controlled source with it. Basing all techniques on feedback, however, is only one possible interpretation of active network configurations and it may be convenient in some cases to interpret them differently. It is well known, for example, that oscillator theory can be based on the negative resistance principle on the one hand and on Nyquist's gain-and-phase principle on the other. A third possibility is to consider the characteristic equation of the corresponding feedback network. Each of these views or interpretations is valid, but one may be more convenient than another for a given application.

We are discussing more here than merely a question of viewpoint, since the interpretation of a synthesis technique may very well affect the realization of an actual device, or of the configuration used to realize a given network function. For example, if we take the feedback point of view, in which all active networks are based on some fundamental feedback principle, then a unit-gain voltage amplifier may be realized by a high-gain amplifier (such as an operational amplifier) with 100 percent feedback. If, however, we prefer to consider the network matrix, say the admittance or chain matrix, as the basic characterization of a device, then we will endeavor to design a circuit realizing the required matrix parameters without any regard for feedback principles. The outcome may then be quite different, e.g. an emitter follower or a Darlington pair (which admittedly can also be explained in terms of feedback). Other nonfeedback alternatives are also possible. The point is that whereas most active network techniques can be realized by feedback amplifiers or can at least be based on feedback principles, very much simpler or more useful devices and configurations may emerge in some cases if we look at the network realization from a different point of view.

A very useful point of view is that of considering the general matrix parameters of a two-port and examining what kind of circuits are capable of realizing certain specific cases of this matrix. We shall see in the following that certain sets of matrix parameters define very special types of two-port networks which we shall call voltage or current converters or inverters. Since only two of these, namely, the negative impedance converter and the gyrator, have been practically used in active filter design, these two will be discussed separately and in detail later.

5.1 NETWORK ELEMENTS DERIVED FROM THE TRANSMISSION MATRIX

5.1.1 Converters

All possible types of converters can be derived from the transmission or chain matrix (also often simply called the $ABCD$ matrix) of the general two-port shown in Fig. 5-1. The transmission matrix is defined by

$$V_1 = AV_2 - BI_2$$
$$I_1 = CV_2 - DI_2$$

or

$$\begin{bmatrix} V_1 \\ I_1 \end{bmatrix} = \begin{bmatrix} A & B \\ C & D \end{bmatrix} \begin{bmatrix} V_2 \\ -I_2 \end{bmatrix} \tag{5-1}$$

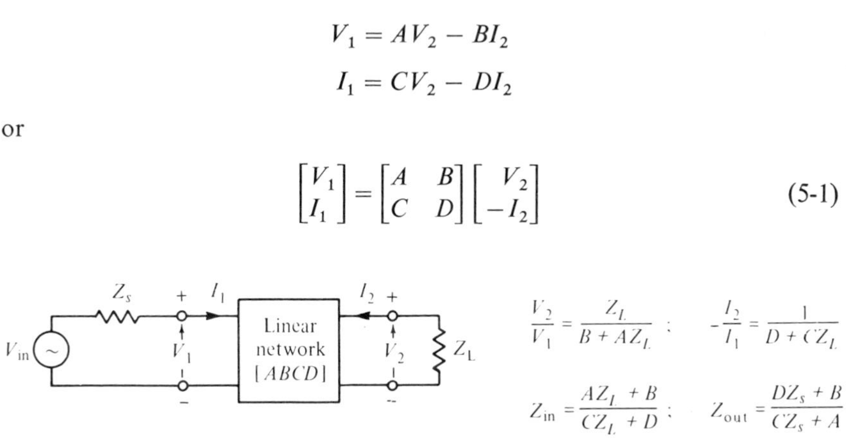

$$\frac{V_2}{V_1} = \frac{Z_L}{B + AZ_L} \quad ; \quad -\frac{I_2}{I_1} = \frac{1}{D + CZ_L}$$

$$Z_{in} = \frac{AZ_L + B}{CZ_L + D} \quad ; \quad Z_{out} = \frac{DZ_s + B}{CZ_s + A}$$

Fig. 5-1. General two-port characterized in terms of the transmission matrix.

The input impedance of the two-port in Fig. 5-1 with a load Z_L is given by

$$Z_{in} = \frac{V_1}{I_1} = \frac{AZ_L + B}{CZ_L + D} \tag{5-2}$$

If

$$B = C = 0$$
$$A, D \neq 0 \tag{5-3}$$

then

$$Z_{in} = \frac{A}{D} Z_L \tag{5-4}$$

The resulting network is called an *impedance converter* because the input impedance is directly proportional to the load impedance, i.e.,

$$Z_{in} = K(s)Z_L \tag{5-5a}$$

where

$$K(s) = \frac{A(s)}{D(s)} \tag{5-5b}$$

The equations for the impedance converter are

$$V_1 = AV_2$$
$$I_1 = -DI_2 \tag{5-6}$$

where A and D represent voltage and current gain, respectively. Their identification yields the various subclasses of impedance converters.

If A/D is the negative ratio of two driving point impedances, i.e.,

$$\frac{A}{D} = -\frac{Z_a(s)}{Z_b(s)} \tag{5-7}$$

then we have a generalized impedance converter (GIC). Suppose we identify A and D as follows

$$A = -\frac{Z_a}{Z_b}$$
$$D = 1 \tag{5-8a}$$

Then

$$V_1 = -\frac{Z_a}{Z_b} V_2 \tag{5-8b}$$
$$I_1 = -I_2$$

The current flows through the network undisturbed but the voltage is scaled and inverted. (Note that with the definition of current directions for the general two-port as shown in Fig. 5-1, the reference direction of the output current is reversed for the chain matrix.) Thus this device is termed a voltage-inversion type GIC, or VGIC.

If we let

$$A = 1$$

$$D = -\frac{Z_b}{Z_a}$$ (5-9a)

then we have

$$V_1 = V_2$$

$$I_1 = \frac{Z_b}{Z_a} I_2$$ (5-9b)

In this case the voltage is transmitted through the network unchanged and the current is inverted and scaled. This type of GIC is termed a current inversion type GIC, or CGIC. Note that for both CGIC and VGIC the conversion factor $K(s)$ is a function of s.

If the conversion factor is made to be a real constant, several other types of converters are obtained. The two most important ones follow. As before they depend on the identification of A and D.

A voltage-inversion negative-impedance converter (VNIC) is obtained if we let

$$A = -k_1$$

$$D = 1/k_2$$ (5-10a)

where $k_1, k_2 > 0$. Then we have

$$V_1 = -k_1 V_2$$

$$I_1 = -I_2/k_2$$ (5-10b)

Here the voltage is inverted from input to output and the current is not. Thus

$$Z_{in} = -k_1 k_2 Z_L$$ (5-11)

If $k_1 = k_2 = 1$ then the voltage is simply inverted and we have a unity-gain VNIC (UVNIC).

A current-inversion negative-impedance converter (CNIC) is obtained when

$$A = k_1$$

$$D = -1/k_2$$ (5-12a)

where $k_1, k_2 > 0$. We then have

$$V_1 = k_1 V_2$$

$$I_1 = I_2/k_2$$ (5-12b)

Here the current is inverted and the voltage is not. The input impedance, given by (5-11), is the same as for the VNIC. If $k_1 = k_2 = 1$ we have a unity-gain CNIC (UCNIC).

Various other converters can be obtained by assuming different special values for the parameters A and D. The resulting converters are summarized in Table 5-1.

TABLE 5-1. VARIOUS FORMS OF CONVERTERS AND THEIR TRANSMISSION MATRIX PARAMETERS

Converter type	Transmission matrix	Circuit symbol	Characteristics
Generalized-impedance converter (GIC)	$\begin{bmatrix} A(s) & 0 \\ 0 & D(s) \end{bmatrix}$	GIC	Input impedance directly proportional to the load impedance: $$Z_{in} = \frac{A(s)}{D(s)} Z_L = K(s) \cdot Z_L$$
Voltage-inversion GIC (VGIC)	$\begin{bmatrix} -\dfrac{Z_a(s)}{Z_b(s)} & 0 \\ 0 & 1 \end{bmatrix}$	VGIC $$V_1 = -(Z_a/Z_b)V_2$$ $$I_1 = -I_2$$	Current is transmitted through network unchanged; voltage is inverted and scaled: $$Z_{in} = -\frac{Z_a}{Z_b} Z_L$$
Current-inversion GIC (CGIC)	$\begin{bmatrix} 1 & 0 \\ 0 & -\dfrac{Z_b(s)}{Z_a(s)} \end{bmatrix}$	CGIC $$V_1 = V_2$$ $$V_1 = (Z_b/Z_a)I_2$$	Voltage is transmitted through network unchanged; current is inverted and scaled: $$Z_{in} = -\frac{Z_a}{Z_b} Z_L$$
Voltage-inversion negative-impedance converter (VNIC)	$\begin{bmatrix} -k_1 & 0 \\ 0 & \dfrac{1}{k_2} \end{bmatrix}$	VNIC $$V_1 = -k_1 V_2$$ $$I_1 = -I_2/k_2$$	Voltage inverted, scaled; current scaled, not inverted: $$Z_{in} = -k_1 k_2 Z_L$$ $k_1 = k_2 = 1$: No scaling; voltage inverted (unity-gain VNIC, i.e., UVNIC)

Current-inversion negative-impedance converter (CNIC)	$\begin{bmatrix} k_1 & 0 \\ 0 & -\dfrac{1}{k_2} \end{bmatrix}$	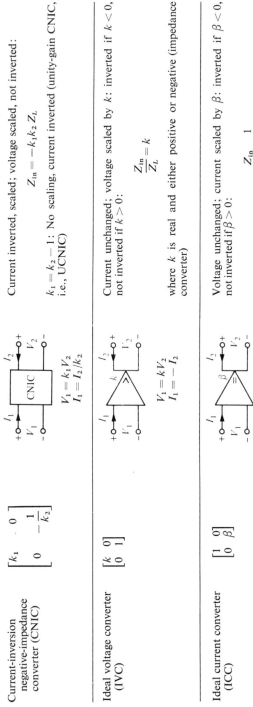 Current inverted, scaled; voltage scaled, not inverted: $$Z_{in} = -k_1 k_2 Z_L$$ $k_1 = k_2 = 1$: No scaling, current inverted (unity-gain CNIC, i.e., UCNIC)
Ideal voltage converter (IVC)	$\begin{bmatrix} k & 0 \\ 0 & 1 \end{bmatrix}$	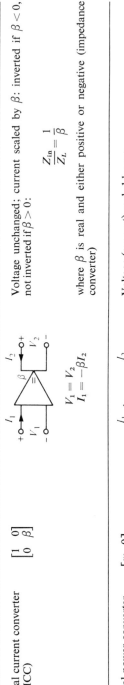 Current unchanged; voltage scaled by k: inverted if $k < 0$, not inverted if $k > 0$: $$\frac{Z_{in}}{Z_L} = k$$ where k is real and either positive or negative (impedance converter)
Ideal current converter (ICC)	$\begin{bmatrix} 1 & 0 \\ 0 & \beta \end{bmatrix}$	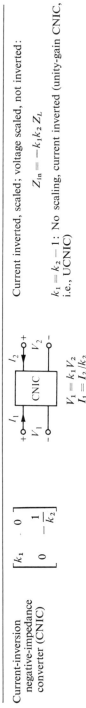 Voltage unchanged; current scaled by β: inverted if $\beta < 0$, not inverted if $\beta > 0$: $$\frac{Z_{in}}{Z_L} = \frac{1}{\beta}$$ where β is real and either positive or negative (impedance converter)
Ideal power converter (IPC)	$\begin{bmatrix} m & 0 \\ 0 & m \end{bmatrix}$	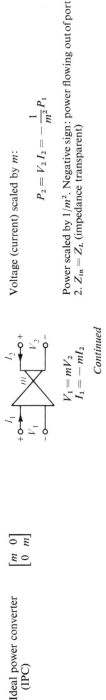 Voltage (current) scaled by m: $$P_2 = V_2 I_2 = -\frac{1}{m^2} P_1$$ Power scaled by $1/m^2$. Negative sign: power flowing out of port 2. $Z_{in} = Z_L$ (impedance transparent)

Continued

TABLE 5-1 (*Continued*)

Converter type	Transmission matrix	Circuit symbol	Characteristics
Ideal transformer (IT)	$\begin{bmatrix} n & 0 \\ 0 & \dfrac{1}{n} \end{bmatrix}$	 $V_1 = nV_2$ $I_1 = -\dfrac{1}{n}I_2$	Positive-impedance converter: $$Z_{in} = n^2 Z_L$$ Ideal transformer ($n = 1$) is ideal power converter ($m = 1$)
Complex transformer (CT)	$\begin{bmatrix} u & 0 \\ 0 & \dfrac{1}{u^*} \end{bmatrix}$	 $V_1 = uV_2$ $I_1 = -\dfrac{1}{u^*} \cdot I_2$	The IT is the special case of the CT with u real.

5.1.2 Inverters

The second basic class of networks, called *in*verters, can also be derived from the transmission matrix of the general two-port shown in Figure 5-1 when the *prin*cipal matrix diagonal equals zero:

$$A = D = 0$$
$$B, C \neq 0 \qquad (5\text{-}13)$$

(Remember that *con*verters are defined by a transmission matrix whose *non*principal diagonal equals zero.) From (5-13) inverters are characterized by the equations

$$V_1 = -BI_2$$
$$I_1 = CV_2 \qquad (5\text{-}14)$$

The input impedance follows as

$$Z_{\text{in}} = \frac{B}{C} \cdot \frac{1}{Z_L} \qquad (5\text{-}15)$$

Thus the impedance seen looking into port 1 (see Fig. 5-1) is inversely proportional to the impedance looking across port 2. Two-ports with this impedance inversion property were first characterized by Tellegen, who called them gyrators.[1] As in the case of converters, various types of inverters can be defined, depending on the identification of the B and C parameters of the corresponding transmission matrix. The most common ones (of which there are significantly less than of converters) are listed in Table 5-2. Notice that the generally accepted symbol for the gyrator is used.

5.1.3 Power Considerations

Considering the sinusoidal steady state, the total input power to the general two port of Fig. 5-1 is[2]

$$P_T = P_R + jP_I = V_1 I_1^* + V_2 I_2^* \qquad [5\text{-}16]$$

1. B. D. H. Tellegen, The gyrator, a new network element, *Philips Res. Rept.*, **3**, 81–101 (1948.)
2. Writing the voltage and current at the terminals of a one-port as the phasors $V = \hat{V}e^{j\theta}$ and $I = \hat{I}e^{j(\theta - \phi)}$ we have $P_T = P_R + jP_I = VI^* = \hat{V}\hat{I}\cos\phi + j\hat{V}\hat{I}\sin\phi$, where the asterisk denotes conjugation. P_T is often called the complex signal power, and P_R and P_I the real and imaginary signal power respectively. P_R is twice the power supplied to and dissipated by the one-port, P_I twice the amplitude of the rate at which energy is interchanged between the one-port and the external world by virtue of its reactive nature. For convenience we have not added the factor 0.5 to the terms involving signal power; it may therefore be tacitly assumed that the phasors V and I refer to rms quantities.

TABLE 5-2. COMMON FORMS OF INVERTERS AND THEIR TRANSMISSION-MATRIX PARAMETERS

Inverter type	Transmission matrix	Circuit symbol	Characteristics
Ideal gyrator (IG)	$\begin{bmatrix} 0 & \pm\dfrac{1}{g} \\ \pm g & 0 \end{bmatrix}$	$V_1 = \mp\dfrac{1}{g}I_2$ $I_1 = \pm g V_2$	The input impedance is the reciprocal of the load impedance Z_L times the gyration constant g squared: $$Z_{in} = \frac{1}{g^2 Z_L}$$
Negative-impedance inverter (NII)	$\begin{bmatrix} 0 & \pm\dfrac{1}{g} \\ \mp g & 0 \end{bmatrix}$	$V_1 = \mp\dfrac{1}{g}I_2$ $I_1 = \mp g V_2$	Negative gyrator: $$Z_n = -\frac{1}{g^2 Z_L}$$
Generalized negative impedance inverter (GNII)	$\begin{bmatrix} 0 & \pm\dfrac{1}{g(s)} \\ \mp g(s) & 0 \end{bmatrix}$	$V_1 = \mp\dfrac{1}{g(s)}I_2$ $I_1 = \mp g(s) V_2$	Negative gyrator with complex inversion ratio: $$Z_{in} = -\frac{1}{g^2(s) Z_L}$$

This quantity is in general complex. Its real part is equal to the power dissipated in the network. Therefore if

$P_R > 0$ the two-port is passive and lossy
$P_R = 0$ the two-port is passive and lossless
$P_R < 0$ the two-port is active

" Passive " means that the two-port does not supply power at the terminals. In this sense, a passive two-port can have internal active elements, without supplying any power to the external terminals from within.

For converters we have, from (5-6),

$$P_R = \text{Re } V_2 I_2^*(1 - AD^*) \qquad (5\text{-}17)$$

Thus if $AD^* = 1$, the device is a lossless passive element. The ideal transformer is such an element (see Table 5-1). The negative-impedance converters (both VNIC and CNIC) are active, since

$$P_{R_{(\text{VNIC})}}_{(\text{CNIC})} = -\text{Re } Z_L \cdot |I_2|^2 \left(1 + \frac{k_1}{k_2}\right) \qquad (5\text{-}18)$$

The ideal voltage converter (IVC), ideal current converter (ICC), and ideal power converter (IPC) are active for k, β, and $m > 1$, respectively, passive and lossless for k, β, and $m = 1$ (a degenerate case), and lossy for k, β, and $m < 1$.

A similar situation exists with the inverters. From (5-14) we have for the ideal gyrator (IG) (see Table 5-2)

$$P_{R_{(IG)}}\Big|_{BC^* = 1} = \text{Re } V_2 I_2^*(1 - BC^*) = 0 \qquad (5\text{-}19)$$

The ideal gyrator is thus passive lossless and the negative impedance inverter (NII) is active, i.e.,

$$P_{R_{(\text{NII})}}\Big|_{BC^* = -1} = \text{Re } 2V_2 I_2^* = -2 \text{ Re } Z_L|I_2|^2 < 0 \qquad (5\text{-}20)$$

5.2 NULLATORS AND NORATORS

The nullator and norator belong to a class of elements which have no conventional matrix representation. They were first introduced by Carlin as pathological networks.[3] They are useful in some analysis problems since

3. H. J. Carlin, Singular network elements, *IEEE Trans. Circuit Theory*, **CT-11**, 67–72 (1964).

certain physical elements can be approximated by combining them in appropriate configurations. The nullator is a two-terminal element which is simultaneously an open and a short circuit. It is defined by

$$V = I = 0 \qquad (5\text{-}21)$$

and represented by the circuit symbol shown in Fig. 5-2a. The norator is a two-terminal device which has the property that the voltage and current across its terminals are completely arbitrary. It is defined by

$$V = k_1 \qquad I = k_2 \qquad (5\text{-}22)$$

where k_1 and k_2 are arbitrary; it is represented by the circuit symbol shown in Fig. 5-2b. When a norator is used in a circuit, V and I take on the values needed to satisfy Kirchhoff's current and voltage laws.

Various circuit elements can be modeled by nullator–norator pairs. The most common are the ideal transistor, the operational amplifier, and circuit configurations incorporating either one. Some typical examples are listed in Table 5-3.

By inspection of the nullator–norator models shown in Table 5-3, we observe that a nullator and norator always occur in a pair, often referred to as a nullor. It can be shown that any practical or realizable network must contain the same number of nullators and norators. In order to realize the resulting network model by transistors, it is not enough for the nullators and norators to occur in pairs; each pair must also have a common node. If they do not in the original model, they very often can be rearranged to do so by using the four useful identities shown in Fig. 5-3. With these identities, nullator–norator pairs can be added to the initial circuit model in such a way as to obtain adjoining nullator–norator pairs that can be replaced by transistors. Using this technique, Mitra developed nullator–norator replacements of the four basic controlled sources.[4] These are shown in Table 5-4. Using the nullator–norator representations of the controlled sources to realize the gyrator, he was able to invent numerous transistor (or operational amplifier) gyrator realizations.[5] The same approach can be taken to develop realizations of the other network elements shown in Tables 5-1 and 5-2. In developing such realizations, it is useful to note that the dual of a nullator is a nullator, and the dual of a norator is a norator (i.e., they are self dual).

As an example of the technique used by Mitra to obtain a transistorized

4. S. K. Mitra, Nullator–norator equivalent circuits of linear active elements and their applications, *Proc. Asilomar Conf. Circuits Syst.*, Monterey, California, November 1967, pp. 267–276.
5. S. K. Mitra, Equivalent circuits of gyrators, *Electron. Lett.* **3**, 388 (1967).

FIG. 5-2. (a) Nullator. (b) Norator.

version of a controlled source from its nullator–norator representation, consider the nullator–norator version of a current-controlled current source (CCS) shown in Fig. 5-4a (see Table 5-4). This configuration is not practically realizable using transistors, since the nullator and norator have no common node. By adding the nullator–norator pair shown in Fig. 5-4b (which corresponds to an open circuit and therefore does not affect the basic network) two adjoining nullator–norator pairs are obtained, which can be transistorized as shown in Fig. 5-4c. This circuit will be recognized as a well known transistor-pair approximation of an ideal CCS (see Fig. 3-20 in Chapter 3 and the corresponding current gain (3-53)).

For convenience, the various matrix parameters of the network elements discussed above are listed in Table 5-5. For the sake of completeness this table includes the controlled sources which were already listed in Chapter 3, Table 3-6.

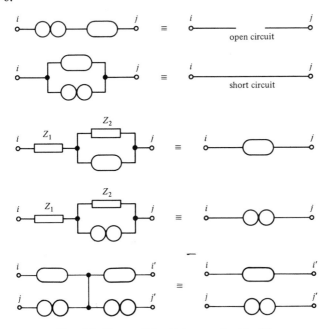

FIG. 5-3. Five useful nullator–norator identities.

TABLE 5-3. NULLATOR–NORATOR REPRESENTATIONS OF SOME BASIC TRANSISTOR AND OPERATIONAL AMPLIFIER CONFIGURATIONS

Component or circuit	Nullator–norator representation

Transistor

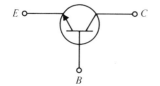

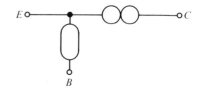

Transistor amplifier

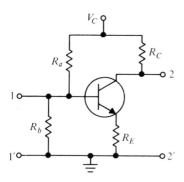

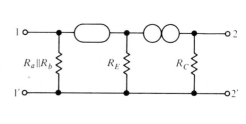

Emitter follower

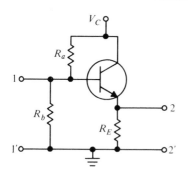

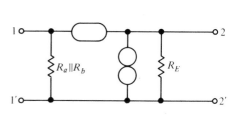

Operational amplifier (single-ended output)

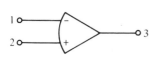

266

TABLE 5-3 (*Continued*)

Component or circuit	Nullator–norator representation

Operational amplifier (double-ended output)

Operational amplifier (inverting mode)

Operational amplifier (noninverting mode)

TABLE 5-4. NULLATOR–NORATOR REPRESENTATION OF CONTROLLED
SOURCES

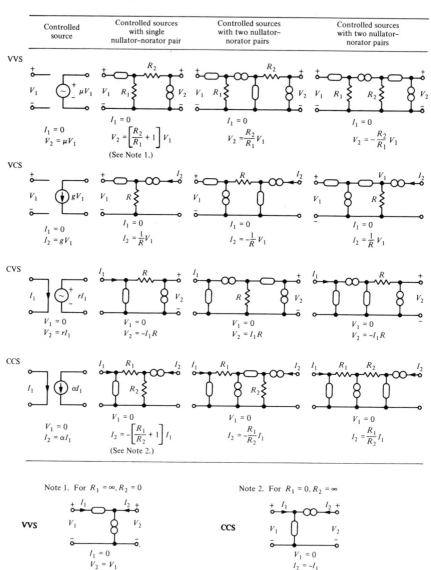

Controlled source	Controlled sources with single nullator–norator pair	Controlled sources with two nullator–norator pairs	Controlled sources with two nullator–norator pairs

VVS

$I_1 = 0$
$V_2 = \mu V_1$

$I_1 = 0$
$V_2 = \left[\dfrac{R_2}{R_1} + 1\right] V_1$
(See Note 1.)

$I_1 = 0$
$V_2 = \dfrac{R_2}{R_1} V_1$

$I_1 = 0$
$V_2 = -\dfrac{R_2}{R_1} V_1$

VCS

$I_1 = 0$
$I_2 = g V_1$

$I_1 = 0$
$I_2 = \dfrac{1}{R} V_1$

$I_1 = 0$
$I_2 = -\dfrac{1}{R} V_1$

$I_1 = 0$
$I_2 = \dfrac{1}{R} V_1$

CVS

$V_1 = 0$
$V_2 = r I_1$

$V_1 = 0$
$V_2 = -I_1 R$

$V_1 = 0$
$V_2 = I_1 R$

$V_1 = 0$
$V_2 = -I_1 R$

CCS

$V_1 = 0$
$I_2 = \alpha I_1$

$V_1 = 0$
$I_2 = -\left[\dfrac{R_1}{R_2} + 1\right] I_1$
(See Note 2.)

$V_1 = 0$
$I_2 = -\dfrac{R_1}{R_2} I_1$

$V_1 = 0$
$I_2 = \dfrac{R_1}{R_2} I_1$

Note 1. For $R_1 = \infty, R_2 = 0$

VVS

$I_1 = 0$
$V_2 = V_1$

Note 2. For $R_1 = 0, R_2 = \infty$

CCS

$V_1 = 0$
$I_2 = -I_1$

Element	$[z_{ij}]$	$[y_{ij}]$	$[h_{ij}]$	$[g_{ij}]$	$[ABCD]$
Voltage-inversion general-impedance converter (VGIC)			$\begin{bmatrix} 0 & -\dfrac{Z_a(s)}{Z_b(s)} \\ -1 & 0 \end{bmatrix}$	$\begin{bmatrix} 0 & -1 \\ -\dfrac{Z_b(s)}{Z_a(s)} & 0 \end{bmatrix}$	$\begin{bmatrix} -\dfrac{Z_a(s)}{Z_b(s)} & 0 \\ 0 & 1 \end{bmatrix}$
Current-inversion general-impedance converter (CGIC)			$\begin{bmatrix} 0 & 1 \\ \dfrac{Z_a(s)}{Z_b(s)} & 0 \end{bmatrix}$	$\begin{bmatrix} 0 & \dfrac{Z_b(s)}{Z_a(s)} \\ 1 & 0 \end{bmatrix}$	$\begin{bmatrix} 1 & 0 \\ 0 & -\dfrac{Z_b(s)}{Z_a(s)} \end{bmatrix}$
Voltage-inversion negative-impedance converter (VNIC)			$\begin{bmatrix} 0 & -k_1 \\ -k_2 & 0 \end{bmatrix}$	$\begin{bmatrix} 0 & -\dfrac{1}{k_2} \\ -\dfrac{1}{k_1} & 0 \end{bmatrix}$	$\begin{bmatrix} -k_1 & 0 \\ 0 & \dfrac{1}{k_2} \end{bmatrix}$
Current-inversion negative-impedance converter (CNIC)			$\begin{bmatrix} 0 & k_1 \\ k_2 & 0 \end{bmatrix}$	$\begin{bmatrix} 0 & \dfrac{1}{k_2} \\ \dfrac{1}{k_1} & 0 \end{bmatrix}$	$\begin{bmatrix} k_1 & 0 \\ 0 & -\dfrac{1}{k_2} \end{bmatrix}$
Ideal voltage converter (IVC)			$\begin{bmatrix} 0 & k \\ -1 & 0 \end{bmatrix}$	$\begin{bmatrix} 0 & -1 \\ \dfrac{1}{k} & 0 \end{bmatrix}$	$\begin{bmatrix} k & 0 \\ 0 & 1 \end{bmatrix}$
Ideal current converter (ICC)			$\begin{bmatrix} 0 & 1 \\ -\dfrac{1}{\beta} & 0 \end{bmatrix}$	$\begin{bmatrix} 0 & -\beta \\ 1 & 0 \end{bmatrix}$	$\begin{bmatrix} 1 & 0 \\ 0 & \beta \end{bmatrix}$

Continued

TABLE 5-5 (*Continued*)

Element	Parameters				
	$[z_{ij}]$	$[y_{ij}]$	$[h_{ij}]$	$[g_{ij}]$	$[ABCD]$
Ideal power converter (IPC)			$\begin{bmatrix} 0 & m \\ -\dfrac{1}{m} & 0 \end{bmatrix}$	$\begin{bmatrix} 0 & -m \\ \dfrac{1}{m} & 0 \end{bmatrix}$	$\begin{bmatrix} m & 0 \\ 0 & m \end{bmatrix}$
Ideal gyrator (IG)	$\begin{bmatrix} 0 & \mp\dfrac{1}{g} \\ \pm\dfrac{1}{g} & 0 \end{bmatrix}$	$\begin{bmatrix} 0 & \pm g \\ \mp g & 0 \end{bmatrix}$			$\begin{bmatrix} 0 & \pm\dfrac{1}{g} \\ \pm g & 0 \end{bmatrix}$
Negative-impedance inverter (NII)	$\begin{bmatrix} 0 & \mp\dfrac{1}{g} \\ \pm\dfrac{1}{g} & 0 \end{bmatrix}$	$\begin{bmatrix} 0 & \mp g \\ \pm g & 0 \end{bmatrix}$			$\begin{bmatrix} 0 & \pm\dfrac{1}{g} \\ \mp g & 0 \end{bmatrix}$
Generalized negative-impedance inverter (GNII)	$\begin{bmatrix} 0 & \mp\dfrac{1}{g(s)} \\ \pm\dfrac{1}{g(s)} & 0 \end{bmatrix}$	$\begin{bmatrix} 0 & \mp g(s) \\ \pm g(s) & 0 \end{bmatrix}$			$\begin{bmatrix} 0 & \pm\dfrac{1}{g(s)} \\ \mp g(s) & 0 \end{bmatrix}$

Element					
Ideal transformer (IT)	$\begin{bmatrix} \pm n & 0 \\ 0 & \pm\frac{1}{n} \end{bmatrix}$	$\begin{bmatrix} 0 & \mp\frac{1}{n} \\ \pm\frac{1}{n} & 0 \end{bmatrix}$	$\begin{bmatrix} 0 & \pm n \\ \mp n & 0 \end{bmatrix}$		
Current-controlled voltage source (CVS)	$\begin{bmatrix} 0 & 0 \\ \frac{1}{r} & 0 \end{bmatrix}$				$\begin{bmatrix} 0 & 0 \\ r & 0 \end{bmatrix}$
Voltage-controlled current source (VCS)	$\begin{bmatrix} 0 & \frac{1}{g} \\ 0 & 0 \end{bmatrix}$			$\begin{bmatrix} 0 & 0 \\ g & 0 \end{bmatrix}$	
Current-controlled current source (CCS)	$\begin{bmatrix} 0 & 0 \\ 0 & \frac{1}{\alpha} \end{bmatrix}$		$\begin{bmatrix} 0 & 0 \\ \alpha & 0 \end{bmatrix}$		
Voltage-controlled voltage source (VVS)	$\begin{bmatrix} \frac{1}{\mu} & 0 \\ 0 & 0 \end{bmatrix}$	$\begin{bmatrix} 0 & \mu \\ 0 & 0 \end{bmatrix}$			

* S. K. Mitra, *Analysis and Synthesis of Linear Active Networks*, New York: John Wiley & Sons, 1969.

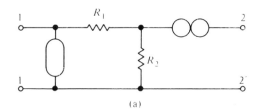

(a)

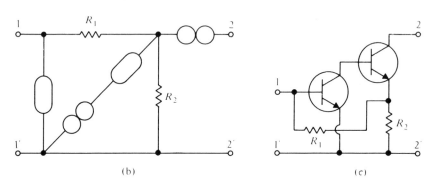

(b)   (c)

FIG. 5-4. Derivation of a transistorized current-controlled current source (CCS) from a nullator–norator CCS.

5.3 NEGATIVE IMPEDANCE CONVERTERS

The negative impedance converter itself does not have much practical significance. It has been found difficult to use even in discrete form, let alone as an integrated monolithic chip. Nevertheless it was Linvill's work on the negative impedance converter and his use of it in active filters that gave the impetus to practical active filter design in 1954.[6] In fact the negative impedance converter was really the first device considered seriously for active filter design, and for many years after Linvill's first paper the literature on active filters almost exclusively dealt with negative impedance converter configurations. Thus some elaborate theories have evolved around this device and, if nothing else, the NIC is of some academic interest. However, in recent years it has been virtually abandoned as a practical device for active filter design. Still, there is always the possibility that a useful and low-cost monolithic integrated negative impedance converter may yet appear on the scene (there are those who have argued that it holds great promise for high-frequency applications), and it will be discussed, at least in principle, here.

6. J. G. Linvill, *RC* active filters, *Proc. IRE*, **12**, 555–564 (1954).

5.3.1 The Ideal NIC

As described in Section 5.1.1, the ideal NIC is a two-port device characterized by the following network equations (see Fig. 5-5):

$$V_1 = \mp k_1 V_2$$
$$I_1 = \mp (1/k_2) I_2 \tag{5-23}$$

Instead of the transmission matrix used in Section 5.1.1, the NIC is more conveniently characterized by its h parameters. From Table 5-5 these are

$$[h] = \begin{pmatrix} 0 & h_{12} \\ h_{21} & 0 \end{pmatrix} \tag{5-24}$$

where $h_{12} = \mp k_1$ and $h_{21} = \mp k_2$.

If the upper (i.e., negative) signs are used, the device is a voltage-inversion NIC (VNIC), since the voltage is inverted from input to output and the current is not. If the lower (i.e., positive) signs are used, the device is a current-inversion NIC (CNIC), since the current is inverted from input to output while the voltage is not. The impedance looking into port 1 with a load Z_L across port 2 is, for either VNIC or CNIC,

$$Z_1 = -kZ_L \tag{5-25}$$

where

$$h_{12} h_{21} = k \tag{5-26}$$

Looking the other way (with Z_L at port 1), we have

$$Z_2 = -\frac{1}{k} Z_L \tag{5-27}$$

Thus the impedance transformation (in the general case) for one direction is the reciprocal of the transformation in the other. The constant $k = h_{12} h_{21}$ is called the gain of the NIC. In terms of k the VNIC can be defined by

$$V_1 = -kV_2$$
$$I_1 = -I_2 \tag{5-28}$$

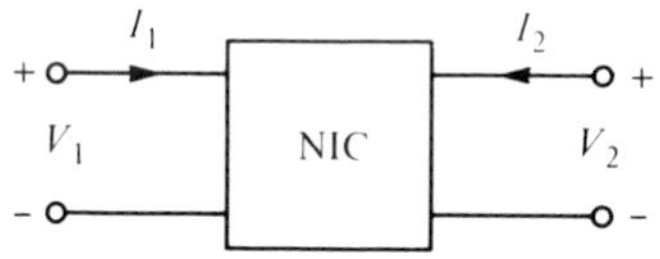

FIG. 5-5. Negative-impedance converter.

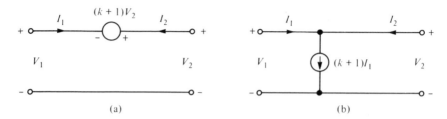

FIG. 5-6. Controlled-source representation of NICs. (a) VNIC. (b) CNIC.

and the CNIC by

$$V_1 = V_2$$

$$I_1 = \frac{1}{k} I_2 \qquad\qquad [5\text{-}29]$$

These two ideal NICs can be realized by the voltage and current sources shown in Fig. 5-6.

The NIC exhibits power gain, i.e., with

$$P_{\text{in}} = \text{Re } V_1 I_1^* \qquad\qquad (5\text{-}30)$$

we obtain

$$P_{\text{out}} = \text{Re} \left(-V_2 I_2^* \right) = \text{Re} \left(-\frac{h_{21}^*}{h_{12}} V_1 I_1^* \right) = -\frac{h_{21}}{h_{12}} P_{\text{in}} \qquad (5\text{-}31)$$

where h_{12}, h_{21} are real.

The negative sign indicates that power can flow out of both ports simultaneously; hence the NIC is an active device.

Some authors define the *ideal* NIC by the condition that $h_{12} = h_{21} = 1$. In this case the impedance transformation ratio and the power gain are both unity, and the term *unity-gain* NIC (UNIC) is often applied. Power is transmitted equally in both directions and the impedance transformation is independent of the orientation of the UNIC. Throughout this discussion "ideal" will refer to the case where $B = C = 0$ in the $[ABCD]$ matrix and will not imply that $h_{12} = h_{21} = 1$. For $h_{12} = h_{21} = 1$, UVNIC or UCNIC will be used.

5.3.2 The Nonideal NIC

The nonideal NIC has a general h matrix where $h_{12} = \mp k_1$, $h_{21} = \mp k_2$, and h_{11}, $h_{22} \neq 0$. Consider the case where $h_{22} = 0$. The impedance at port 1 with Z_L at port 2 (see Fig. 5-7a) is

$$Z_{\text{in}} = h_{11} - h_{12} h_{21} Z_L \qquad\qquad (5\text{-}32)$$

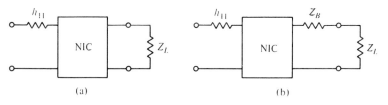

FIG. 5-7. Nonideal NIC. (a) $h_{11} \neq 0$, $h_{22} = 0$. (b) $h_{11} \neq 0$, $h_{22} = 0$, and Z_B added to Z_L.

The effect of h_{11} is therefore to reduce the negative impedance seen at port 1. Suppose an additional impedance Z_B is added in series with Z_L (see Fig. 5-7b); then we have

$$Z_{in} = h_{11} - h_{12}h_{21}(Z_L + Z_B) \qquad (5\text{-}33)$$

If

$$Z_B = \frac{h_{11}}{h_{12}h_{21}} \qquad [5\text{-}34]$$

then $Z_{in} = -h_{12}h_{21}Z_L$, which is the ideal case. Now suppose $h_{11} = 0$ and $h_{22} \neq 0$, as shown in Fig. 5-8a. Then we have

$$Z_{in} = -\frac{h_{12}h_{21}Z_L}{1 + h_{22}Z_L} \qquad (5\text{-}35)$$

The effect of h_{22} is thus to reduce the magnitude of Z_{in} (but never to make it positive). Suppose an additional impedance Z_A is connected across port 1, as shown in Fig. 5-8b. Then

$$Z_{in} = \frac{1}{\dfrac{1}{Z_A} - \dfrac{1 + h_{22}Z_L}{h_{12}h_{21}Z_L}} = \frac{h_{12}h_{21}Z_A Z_L}{h_{12}h_{21}Z_L - Z_A(1 + h_{22}Z_L)} \qquad (5\text{-}36)$$

Letting

$$Z_A = \frac{h_{12}h_{21}}{h_{22}} \qquad [5\text{-}37]$$

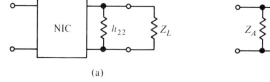

FIG. 5-8. Nonideal NIC. (a) $h_{11} = 0$, $h_{22} \neq 0$. (b) $h_{11} = 0$, $h_{22} \neq 0$, and Z_A connected across the input terminals.

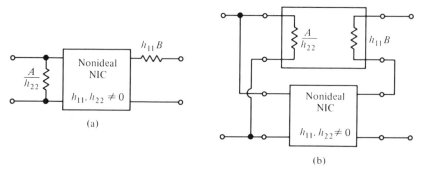

FIG. 5-9. Nonideal NIC. (a) With compensation impedances. (b) Represented as parallel–series network connection.

we obtain

$$Z_{in} = \frac{[(h_{12}h_{21})^2/h_{22}]Z_L}{h_{12}h_{21}Z_L - (h_{12}h_{21}/h_{22}) - h_{12}h_{21}Z_L} = -h_{12}h_{21}Z_L \qquad (5\text{-}38)$$

Thus with Z_A given by (5-37) we obtain the desired ideal NIC characteristics. Consequently perfect compensation is possible if either h_{11} or h_{22} is zero.

The next step is to achieve compensation in the nonideal NIC when both h_{11} and h_{22} are nonzero. From the results obtained above, the scheme shown in Fig. 5-9a seems to follow naturally. In effect, this is a parallel–series connection of two networks as shown in Fig. 5-9b. Because they are connected in this manner, we can write the g parameters (see hybrid matrix, Table 3-1) of the new network by adding the g parameters of the two individual networks:

$$[g]_{NIC} = \begin{bmatrix} \dfrac{h_{22}}{\Delta h} & -\dfrac{h_{12}}{\Delta h} \\[2ex] -\dfrac{h_{21}}{\Delta h} & \dfrac{h_{11}}{\Delta h} \end{bmatrix} \qquad (5\text{-}39)$$

$$[g]_{Comp.\,Net.} = \begin{bmatrix} \dfrac{h_{22}}{A} & 0 \\[2ex] 0 & h_{11}B \end{bmatrix} \qquad (5\text{-}40)$$

and the g parameters of the overall network are

$$[g'] = \begin{bmatrix} \dfrac{h_{22}}{\Delta h} + \dfrac{h_{22}}{A} & -\dfrac{h_{12}}{\Delta h} \\[2ex] -\dfrac{h_{12}}{\Delta h} & \dfrac{h_{11}}{\Delta h} + h_{11}B \end{bmatrix} \qquad (5\text{-}41)$$

For an ideal NIC we require that $g'_{11} = g'_{22} = 0$ (see Table 5-5); thus

$$A = h_{12} h_{21} - h_{11} h_{22} \qquad (5\text{-}42)$$

and

$$B = \frac{1}{h_{12} h_{21} - h_{11} h_{22}} \qquad (5\text{-}43)$$

Note that $AB = 1$. The $[g']$ matrix therefore becomes

$$[g'] = \begin{bmatrix} 0 & \dfrac{h_{12}}{A} \\ h_{21} B & 0 \end{bmatrix} \qquad (5\text{-}44)$$

The gain k' of the compensated NIC is now (see Table 5-5)

$$k' = h'_{12} h'_{21} = \frac{1}{g'_{12} g'_{21}} = \frac{A}{B h_{12} h_{21}} \qquad (5\text{-}45)$$

Suppose, now, that we require the gain of the compensated NIC to be unity. Assume that h_{11} and h_{22} are fixed (as they generally will be in practice) but that $h_{12} h_{21}$ can be adjusted. Then in order for k' to equal unity, we have, from (5-45),

$$h_{12} h_{21} = \frac{A}{B} = (h_{12} h_{21} - h_{11} h_{22})^2 \qquad (5\text{-}46)$$

This expression can be rewritten in a quadratic form in $(h_{12} h_{21})$;

$$(h_{12} h_{21})^2 - (1 + 2h_{11} h_{22})(h_{12} h_{21}) + (h_{11} h_{22})^2 = 0 \qquad (5\text{-}47)$$

Solving for $(h_{12} h_{21})$ we obtain

$$h_{12} h_{21} = \tfrac{1}{2}(1 + 2h_{11} h_{22} \pm \sqrt{1 + 4h_{11} h_{22}}) \qquad (5\text{-}48)$$

Rearranging in the form of (5-42) we obtain

$$h_{12} h_{21} - h_{11} h_{22} = \tfrac{1}{2} \pm \sqrt{h_{11} h_{22} + \tfrac{1}{4}} \qquad (5\text{-}49)$$

Thus with (5-42) we have

$$A = \tfrac{1}{2} + \sqrt{h_{11} h_{22} + \tfrac{1}{4}} \qquad (5\text{-}50)$$

where the positive sign is used in order to insure positive A. Thus a unity-gain NIC can be made from a nonideal NIC (assuming $h_{12} h_{21}$ in the nonideal circuit is adjustable), by the following steps:

1. Choose A such that

$$A = \tfrac{1}{2} + \sqrt{h_{11} h_{22} + \tfrac{1}{4}}$$

2. Choose B such that

$$B = \frac{1}{A}$$

3. Choose $h_{12}h_{21}$ such that

$$h_{12}h_{21} = A + h_{11}h_{22}$$

5.3.3 NIC Stability

In order to examine the stability properties of the NIC, let us consider the circuit shown in Fig. 5-10. The voltage transfer function V_2/V_s can be calculated as

$$V_2 = -I_2 Z_L = k_2 I_1 Z_L \tag{5-51}$$

$$I_1 = \frac{V_s}{Z_s - k_1 k_2 Z_L} \tag{5-52}$$

$$\frac{V_2}{V_s} = T(s) = \frac{k_2 Z_L}{Z_s - k_1 k_2 Z_L}$$

$$= \frac{k_2}{(Z_s/Z_L) - k_1 k_2} \tag{5-53}$$

$T(s)$ will have poles when $Z_s/Z_L = k_1 k_2$, where Z_s, Z_L, and $k_1 k_2$ are, in general, functions of s. If any of these poles occur on the $j\omega$ axis or in the RHP, the circuit will be unstable. Suppose, now, $Z_s = R_s$ and $Z_L = R_L$ (where R_s and R_L are real), and $k_1 k_2$ is real. If $R_s/R_L = k_1 k_2$, the circuit will be unstable. If only $k_1 k_2$ is real, there will be some combinations of Z_s and Z_L that will produce a stable circuit and some that will result in an unstable circuit. If $k_1 k_2 = 1$, interchanging Z_s and Z_L will not affect the circuit stability.

Consider now the case where Z_s and Z_L are real, but $h_{12}h_{21}$ has some time delay τ and a magnitude of unity for $s = j\omega$, i.e.,

$$h_{12}h_{21} = k_1 k_2 = e^{-s\tau} \tag{5-54}$$

where $s = \sigma + j\omega$. The poles of (5-53) will occur at the frequency s_0, where

$$R_s/R_L = e^{-s_0 \tau} = e^{-\sigma_0 \tau} e^{-j\omega_0 \tau}$$

or

$$R_s/R_L = e^{-\sigma_0 \tau}(\cos \omega_0 \tau - j \sin \omega_0 \tau) \tag{5-55}$$

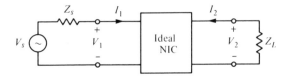

FIG. 5-10. Ideal NIC with source and load impedance.

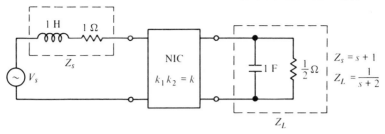

FIG. 5-11. NIC with frequency-dependent source and load impedances $Z_s(s)$ and $Z_L(s)$.

Since R_s/R_L is real and positive, we have

$$\omega_0 = \frac{2n\pi}{\tau} \qquad n = 0, \pm 1, \pm 2, \ldots \tag{5-56}$$

and

$$\sigma_0 = \frac{1}{\tau} \ln \frac{R_L}{R_s} \tag{5-57}$$

Thus, if $R_L > R_s$, $\sigma_0 > 0$ and the system is unstable. If $R_L < R_s$, $\sigma_0 < 0$ and the system is stable.[7]

As an example consider the case shown in Fig. 5-11. Here we have a case where Z_s and Z_L are complex. To determine the poles of the network set $Z_s/Z_L = k$; thus

$$s^2 + 3s + (2 - k) = 0 \tag{5-58}$$

$$s_{1,2} = -\tfrac{3}{2} \pm \tfrac{1}{2}\sqrt{1 + 4k} \tag{5-59}$$

Thus if $k \geq 2$ the system is unstable, since it will have a pole on the $j\omega$ axis or in the RHP, while for $k < 2$ the poles are in the LHP and the system is stable. Suppose Z_s and Z_L are interchanged. The equation for the poles now becomes

$$s^2 + 3s + \left(2 - \frac{1}{k}\right) = 0 \tag{5-60}$$

therefore

$$s_{1,2} = -\tfrac{3}{2} \pm \tfrac{1}{2}\sqrt{1 + 4/k} \tag{5-61}$$

Now, the system is stable for $k > \tfrac{1}{2}$ and unstable for $k \leq \tfrac{1}{2}$. Considering both cases, if $k \geq 2$ the system will be stable in one configuration but unstable if the terminations are reversed. If $k = 1$, the results are the same for either arrangement.

7. R. F. Hoskins, Stability of negative-impedance converter, *Electron. Lett.* **2**, 341 (1966).

5.3.4 NIC Realizations

We can obtain NIC configurations by first considering the basic controlled sources required for their realization. Redrawing Fig. 5-6a we obtain the VNIC as a two-port utilizing a VVS as shown in Fig. 5-12. Similarly, by redrawing Fig. 5-6b we obtain the CNIC utilizing a CCS, as shown in Fig. 5-13. Following Mitra's approach, the controlled sources can now be replaced by nullator–norator equivalents, and these then modified (by adding nullator–norator pairs to form joining pairs) to permit transistor realizations according to the equivalent circuit shown in Table 5-3.

Consider, for example, the CCS realization of the CNIC in Fig. 5-13. From Table 5-4, we can use one of the nullator–norator equivalents of a CCS, such as that shown in Fig. 5-14a. Connecting this CCS into the CNIC of Fig. 5-13, the configuration of Fig. 5-14b is obtained. This is not realizable in transistor form, however, since the nullator–norator pair has no common node. Adding the pair shown in Fig. 5-14c (which corresponds to an open circuit), and referring to Table 5-3, the transistor circuit of Fig. 5-14d is obtained. This CNIC was first described by Larky.[8a] Other similar configurations were described over a decade later.[8b] In general, such configurations can be derived by following Mitra's procedure of going from the controlled-source equivalent circuits to the corresponding nullator–norator circuits, and from these to transistor realizations. In fact, Mitra was able to derive many of the NIC transistor configurations described in the literature as well as several new ones in this way.[9] Another method of deriving new NIC configurations is to consider all possible methods of realizing the h parameters characterizing a close-to-ideal NIC. Numerous additional 2-transistor NIC-realizations have been derived by doing so systematically (i.e., morphologically).[10]

Referring to the nullator–norator version of an operational amplifier shown in Table 5-3, it is clear that adjoining nullator–norator pairs are *not* required. Thus the nullator–norator version of the CNIC shown in Fig. 5-14b can be realized directly by the operational amplifier configuration shown in Fig. 5-15a. Taking the dual of the nullator–norator configuration (i.e., VNIC) we obtain the operational amplifier realization of Fig. 5-15b.

8a. A. I. Larky, Negative-impedance converters, *IRE Trans. Circuit Theory*, CT-4, 124–131 (1957)
8b. B. R. Meyers, Some negative impedance converters, *Proc. IEEE*, **61**, 669–670 (1973).
9. Mitra's approach strongly resembles the "morphological approach" suggested by the Swiss inventor F. Zwicky. For example, by considering all possible combinations of turbine parts, and eliminating all the impractical combinations, Zwicky was able to invent an unusual number of new and useful turbine engines. He applied the same approach to a large variety of both technical and nontechnical problems; see: F. Zwicky, The morphological approach to discovery, invention, research and construction, in *New Methods of Thought and Procedure*, ed. by F. Zwicky and A. G. Wilson (New York: Springer-Verlag, 1967); see also F. Zwicky, *Discovery Invention, Research, Through the Morphological Approach* (Toronto: Macmillan, 1969).
10. C. K. Kuo and K. L. Su, Some new four-terminal NIC circuits, *IEEE Trans. Circuit Theory*, **CT-16**, 379–381 (1969).

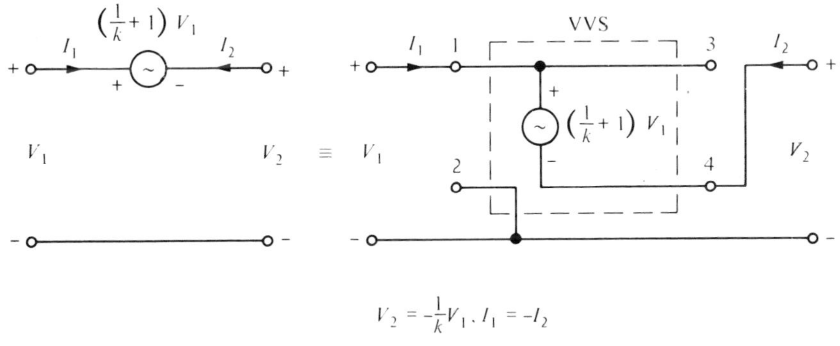

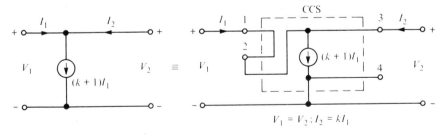

VVS
$\left(\frac{1}{k}+1\right)V_1$
I_1
I_2
V_1
V_2
1
2
3
4
$V_2 = -\frac{1}{k}V_1$, $I_1 = -I_2$

FIG. 5-12. VNIC representation using a voltage-controlled voltage source (VVS).

Many variations of NIC circuits have been suggested in the literature, but none have yet been successfully realized in monolithic integrated form so as to make them economically and commercially feasible. The main problem has been that of biasing the NIC to a given and predictable operating point. Another has been the high sensitivity of active networks using NICs. This latter point will be discussed briefly later.

5.4 GYRATORS

The significance of the gyrator in active network design has contrasted sharply with that of the negative impedance converter in the past and continues to do so in the present. After Linvill's publications in 1954 the NIC was initially considered to be one of the most useful elements for active network design; interest in it gradually declined when its attendant practical problems emerged, and attempts at realizations in integrated form have all but been abandoned completely. The gyrator, on the other hand was introduced by Tellegen as somewhat of a theoretical curiosity or even anomaly in 1948 and remained that way until Shenoi[11] published some results on a practical gyrator realization some fifteen years later. Even then it was still not taken entirely seriously

CCS
$(k+1)I_1$
I_1
I_2
V_1
V_2
1
2
3
4
$V_1 = V_2$; $I_2 = kI_1$

FIG. 5-13. CNIC representation using a current-controlled current source (CCS).

11. B. A. Shenoi, A practical realization of a gyrator circuit and *RC*-gyrator filters, *IEEE Trans. Circuit Theory*, **CT-12**, 374–380 (1965).

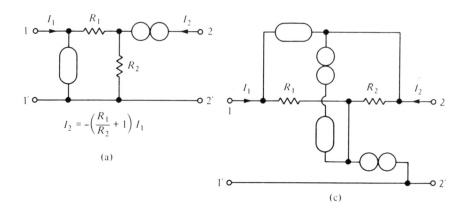

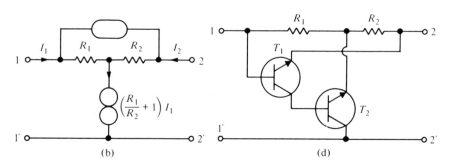

FIG. 5-14. Derivation of a transistorized CNIC from its nullator–norator equivalent.

as a viable active network element until Orchard[12] sparked a great deal of interest in gyrators with his heuristic proof that passive *LC*, and therefore also capacitor-loaded gyrator, filters are inherently superior to other realizations with respect to sensitivity. We shall discuss these matters in more detail in *Linear Integrated Networks: Design*, Chapter 1. Suffice it to say that where the significance of the NIC for active network design was initially overestimated, that of the gyrator was initially understimated and, at least at present, seems to afford a great deal more potential for practical realizability than the NIC. Whether its significance will actually provide sufficient incentive for it to become commercially available as an integrated circuit is still questionable, however, and it may be that it too, ultimately, will revert to the position of a network element with no more than circuit-theoretical interest. Presumably, however, this regression of the gyrator would be due more to economical than to technological shortcomings. More will be said about this in Chapter 7.

12. H. J. Orchard, Inductorless filters, *Electron. Lett.* **2**, 224–225 (1966).

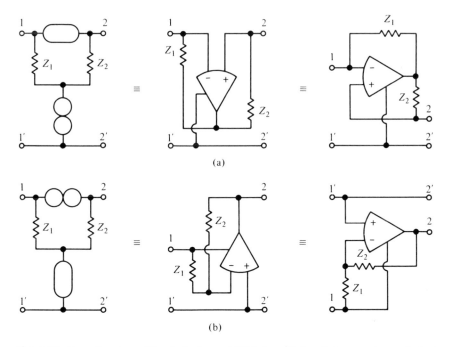

(a)

(b)

FIG. 5-15. Operational-amplifier derivations of NICs from their nullator–norator equivalent circuits. (a) CNIC: $V_1 = V_2$, $I_1 = (Z_2/Z_1)I_2$. (b) VNIC: $V_1 = -(Z_1/Z_2)V_2$, $I_1 = -I_2$.

5.4.1 The Ideal Gyrator

The ideal gyrator shown in Fig. 5-16 is a two-port network characterized by the following matrices (see Table 5-5):

$$[ABCD] = \begin{bmatrix} 0 & \dfrac{1}{g} \\ g & 0 \end{bmatrix} ; \quad [z] = \begin{bmatrix} 0 & -\dfrac{1}{g} \\ \dfrac{1}{g} & 0 \end{bmatrix} ; \quad [y] = \begin{bmatrix} 0 & g \\ -g & 0 \end{bmatrix} \quad (5-62)$$

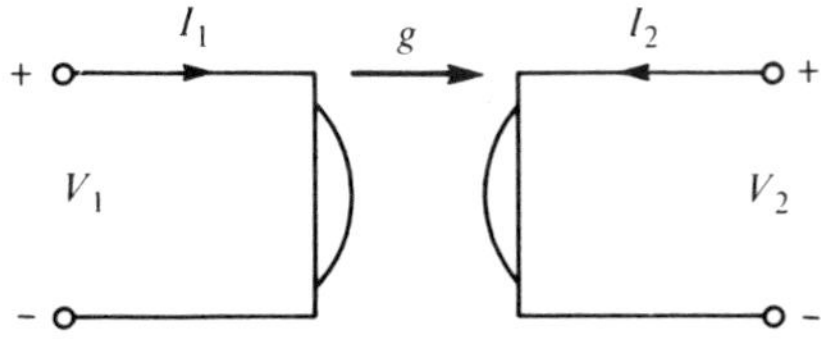

FIG. 5-16. Ideal gyrator.

Thus in terms of the y matrix parameters, which are the most convenient for gyrator characterization, we have

$$I_1 = gV_2$$
$$I_2 = -gV_1 \qquad \text{[5-63]}$$

The factor g is called the gyration conductance. The input impedance of the gyrator terminated by a load impedance Z_L is given by:

$$Z_{in} = \frac{1}{g^2 Z_L} \qquad \text{[5-64]}$$

This relationship is independent of the orientation of the gyrator. This demonstrates the bidirectional nature of the gyrator as an impedance inverter. Furthermore, since the total power into such a device is zero, i.e.,

$$P_R = \text{Re}\,(V_1 I_1^* + V_2 I_2^*) = \text{Re}\left[V_1 I_1^* - \frac{1}{g} I_1 g V_1^*\right] = 0 \qquad (5\text{-}65)$$

the ideal gyrator clearly behaves like a lossless passive element in the sinusoidal steady state.

A variation on the ideal gyrator is a two-port in which $y_{11} = y_{22} = 0$ but $y_{12} \neq -y_{21}$. A gyrator of this kind is shown in Fig. 5-17a. We now have

$$I_1 = g_1 V_2$$
$$I_2 = -g_2 V_1 \qquad \text{[5-66]}$$

where

$$g_1, g_2 > 0$$

The terminal impedances Z_1 and Z_2 (see Figs. 5-17b and 5-17c) then result as

$$Z_1 = Z_2 = \frac{1}{g_1 g_2 Z_L} \qquad \text{[5-67]}$$

As before the impedance at one port is the scaled reciprocal of the impedance at the other port and the orientation is immaterial. The total input power, however, is not zero, since we now obtain

$$P_R = \text{Re}\,(V_1 I_1^* + V_2 I_2^*) = \text{Re}\left(V_1 I_1^* - \frac{g_2}{g_1} V_1^* I_1\right) = \text{Re}\,V_1 I_1^*\left(1 - \frac{g_2}{g_1}\right) \qquad (5\text{-}68)$$

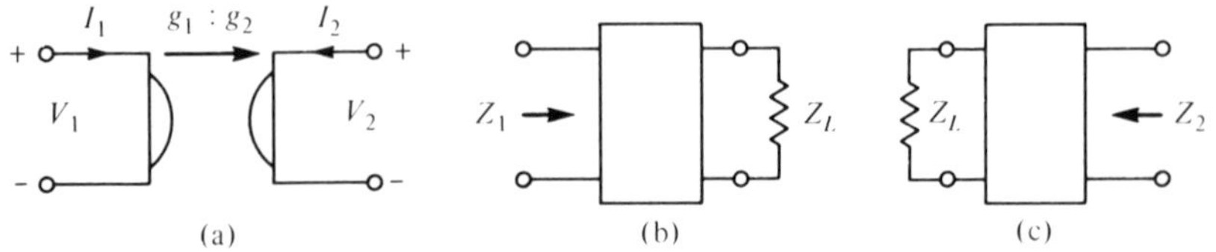

FIG. 5-17. Ideal active gyrator.

As discussed in Section 5.1.3 the gyrator now behaves like an active or a lossy passive element depending on whether g_2 is larger or smaller than g_1. In general this type of device is simply termed an *active gyrator*.[13] For the remainder of this discussion we shall consider an *ideal* gyrator as one with $y_{11} = y_{22} = 0$ and with no requirement that $y_{12} = -y_{21}$. It will be seen as we progress that in many cases the product $y_{12}y_{21}$ is significant, and not y_{12} or y_{21} individually, so that the active or passive nature of an ideal gyrator is not of much importance.

Let us now consider the gyrator realization of the floating impedance Z shown in Fig. 5-18a. Comparing the transmission matrix with that of the gyrator cascade in Fig. 5-18b we see that, in order to make them equal, we must let $g_1 = g_2 = g$ and $Y = g^2Z$ (see Fig. 5-18c).

Inductor Simulation:

There are several interesting applications of the ideal gyrator. One of these is the simulation of an inductor by the gyrator–capacitor combination shown in Fig. 5-19a. The input impedance of this configuration is given by

$$Z_{in} = \frac{1}{g_1g_2/sC} = \frac{sC}{g_1g_2} = sL_{eq} \tag{5-69}$$

where the equivalent inductor L_{eq} seen at the input terminals (see Fig. 5.19b) is given by

$$L_{eq} = \frac{C}{g_1g_2} \tag{5-70}$$

The smaller the gyration product g_1g_2 the larger the inductor appearing at the input terminals of the gyrator.

Another inductor simulation involves two gyrators and a capacitor in the configuration shown in Fig. 5-20. Lossless passive gyrators have been assumed here for convenience (i.e., $y_{12} = -y_{21} = g$).

Assuming that both gyrators are identical, the $ABCD$ matrix of the first gyrator and capacitor then results as

$$\begin{bmatrix} 0 & \frac{1}{g} \\ g & 0 \end{bmatrix}\begin{bmatrix} 1 & 0 \\ sC & 1 \end{bmatrix} = \begin{bmatrix} \frac{sC}{g} & \frac{1}{g} \\ g & 0 \end{bmatrix} \tag{5-71a}$$

and, with the second gyrator, we obtain

$$\begin{bmatrix} \frac{sC}{g} & \frac{1}{g} \\ g & 0 \end{bmatrix}\begin{bmatrix} 0 & \frac{1}{g} \\ g & 0 \end{bmatrix} = \begin{bmatrix} 1 & \frac{sC}{g^2} \\ 0 & 1 \end{bmatrix} \tag{5-71b}$$

13. T. Yanagisawa and Y. Kawashima, Active gyrator, *Electron. Lett.*, 3, 105–107 (1967).

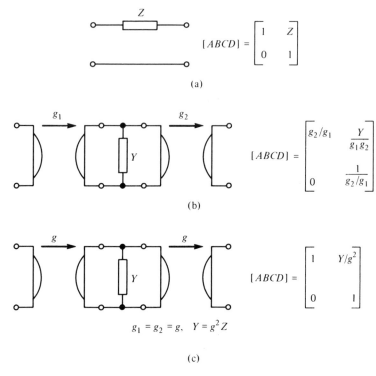

FIG. 5-18. Gyrator realization of floating impedance. (a) Floating impedance Z. (b) gyrator realization with $g_1 \neq g_2$. (c) Gyrator realization with $g_1 = g_2 = g$.

Writing out the equations represented by (5-71b) we have

$$V_1 = V_2 - \frac{sC}{g^2} I_2$$

$$I_1 = -I_2 \qquad\qquad (5\text{-}72)$$

These equations define the inductor configuration shown in Fig. 5-20. In practice, terminals 1' and 2' will generally be grounded. A floating inductance

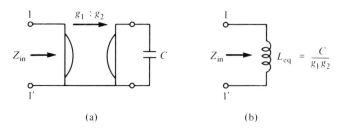

FIG. 5-19. Simulation of a grounded inductor. (a) Gyrator–capacitor combination. (b) Equivalent inductor.

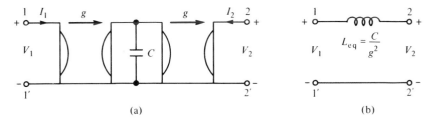

(a)   (b)

FIG. 5-20. Simulation of a floating inductor. (a) Gyrator–capacitor combination. (b) Equivalent inductor.

can therefore be realized by two grounded gyrators and a capacitor. This also follows directly from the gyrator derivation of the general floating impedance shown in Fig. 5-18.

Other Gyrator Circuit Transformations:

Suppose we connect two gyrators as shown in Fig. 5-21. Let $g_1^{(1)}$ and $g_2^{(1)}$ be the parameters of the first gyrator and $g_1^{(2)}$ and $g_2^{(2)}$ be the parameters of the second gyrator. The overall $ABCD$ matrix is

$$\begin{bmatrix} 0 & \dfrac{1}{g_2^{(1)}} \\ g_1^{(1)} & 0 \end{bmatrix} \begin{bmatrix} 0 & \dfrac{1}{g_2^{(2)}} \\ g_1^{(2)} & 0 \end{bmatrix} = \begin{bmatrix} \dfrac{g_1^{(2)}}{g_2^{(1)}} & 0 \\ 0 & \dfrac{g_1^{(1)}}{g_2^{(2)}} \end{bmatrix} \qquad (5\text{-}73)$$

If $g_1^{(i)} = g_2^{(i)} = g_i$ the matrix will become

$$[ABCD] = \begin{bmatrix} \dfrac{g_2}{g_1} & 0 \\ 0 & \dfrac{1}{g_2/g_1} \end{bmatrix} \qquad (5\text{-}74)$$

which is the matrix of an ideal transformer with a turns ratio $n = g_2/g_1$, as shown in Fig. 5-22. These and other gyrator circuit transformations are summarized in Table 5-6. Transformations of this kind are useful in deriving

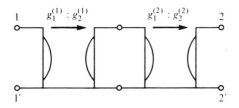

FIG. 5-21. Two active gyrators in cascade.

TABLE 5-6. GYRATOR CIRCUIT TRANSFORMATIONS

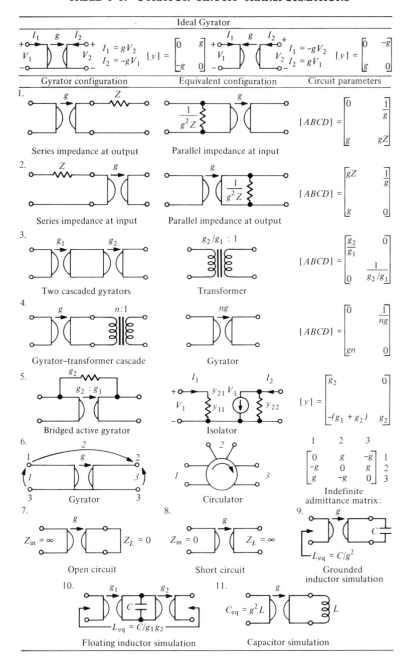

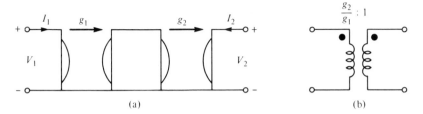

Fig. 5-22. Simulation of an ideal transformer. (a) Gyrator cascade. (b) Equivalent transformer.

the gyrator equivalent of any inductor network. For instance, using transformations 1 and 3 in the table the gyrator equivalent of a floating inductor can be obtained directly as shown in Fig. 5-23a. Likewise, the gyrator equivalent of a transformer with mutual coupling can be derived as shown in Fig. 5-23b. In precisely the same way, the gyrator equivalent networks listed in Fig. 5-24 can be obtained.

5.4.2 The Nonideal Gyrator

So far we have discussed ideal gyrators; ideal, that is, in the sense that $y_{11} = y_{22} = 0$. It is now of interest to consider the case where y_{11} and y_{22} are nonzero. We shall restrict the discussion to the case where the y parameters are real, unless otherwise stated.

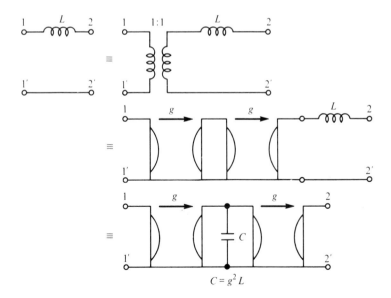

$$C = g^2 L$$

Fig. 5-23 (a). Derivation of a floating-inductor simulation by a gyrator–capacitor combination.

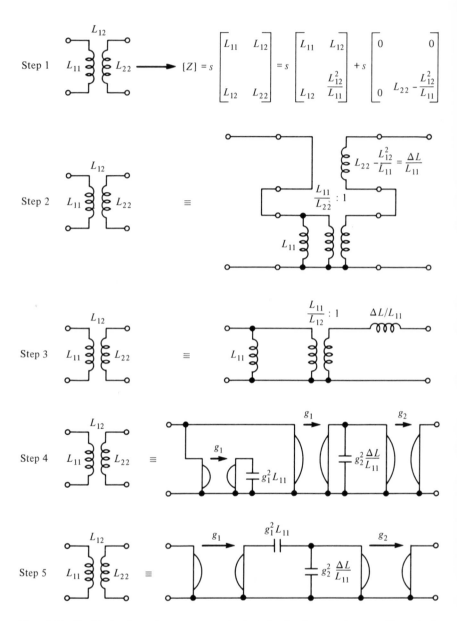

FIG. 5-23 (b). Derivation of a gyrator equivalent circuit of a transformer with mutual coupling, where $g_2/g_1 = L_{11}/L_{12}$ and $\Delta L = L_{11}L_{22} - L_{12}^2$.

Desired network   Gyrator equivalent

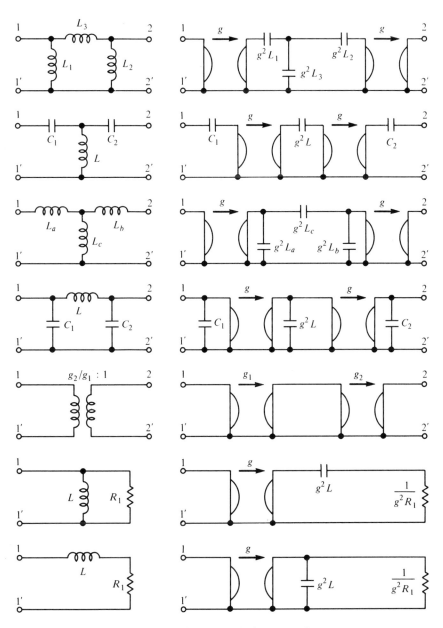

FIG. 5-24. Gyrator equivalent networks.

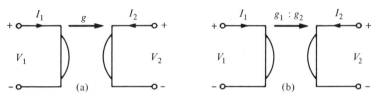

FIG. 5-25. Ideal gyrators. (a) Passive. (b) Active.

There are several ways of representing the nonideal gyrator. First, consider the passive and active ideal gyrators shown again in Fig. 5-25. The corresponding $[y]$ matrices are given by

$$[y]_{passive} = \begin{bmatrix} 0 & g \\ -g & 0 \end{bmatrix} \tag{5-75a}$$

and

$$[y]_{active} = \begin{bmatrix} 0 & g_1 \\ -g_2 & 0 \end{bmatrix} \tag{5-75b}$$

The y matrix for the nonideal gyrator can be broken up into the sum of two matrices, one of which represents an ideal passive gyrator and the other a network of admittances. The corresponding matrices are given by

$$[y] = \underbrace{\begin{bmatrix} y_{11} & g_1 \\ -g_2 & y_{22} \end{bmatrix}}_{\text{nonideal gyrator}} = \underbrace{\begin{bmatrix} 0 & \dfrac{g_1 + g_2}{2} \\ -\dfrac{g_1 + g_2}{2} & 0 \end{bmatrix}}_{\text{ideal passive gyrator}} + \underbrace{\begin{bmatrix} y_{11} & \dfrac{g_1 - g_2}{2} \\ \dfrac{g_1 - g_2}{2} & y_{22} \end{bmatrix}}_{Y \text{ network}} \tag{5-76}$$

Adding the y matrices means connecting the circuits in parallel. Thus the corresponding circuit results as shown in Fig. 5-26. A simpler representation (see Fig. 5-27) is obtained by an ideal active gyrator:

$$\begin{bmatrix} y_{11} & g_1 \\ -g_2 & y_{22} \end{bmatrix} = \underbrace{\begin{bmatrix} 0 & g_1 \\ -g_2 & 0 \end{bmatrix}}_{\substack{\text{ideal active} \\ \text{gyrator}}} + \begin{bmatrix} y_{11} & 0 \\ 0 & y_{22} \end{bmatrix} \tag{5-77}$$

Simulating a Grounded Inductor   Let us now consider the effects of the parameters y_{11} and y_{22} on the input impedance of the capacitor-terminated gyrator shown in Fig. 5-28. From Table 3-4 (in Chapter 3) we have

$$Z_{in} = \frac{y_{22} + Y_L}{\Delta y + y_{11} Y_L} \tag{5-78}$$

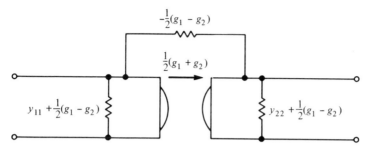

FIG. 5-26. Nonideal gyrator realized as the parallel connection of an ideal passive gyrator and an admittance network.

With $Y_L = sC$ this becomes

$$Z_{in} = \frac{y_{22} + sC}{\Delta y + y_{11}sC} \tag{5-79}$$

It is of interest to examine this impedance along the $j\omega$ axis:

$$Z(j\omega) = \frac{y_{22} + j\omega C}{\Delta y + j\omega y_{11} C} \tag{5-80}$$

Breaking $Z(j\omega)$ into its real and imaginary parts we have, with (5-77),

$$Z(j\omega) = \frac{\Delta y\, y_{22} + (\omega C)^2 y_{11}}{\Delta y^2 + (\omega C y_{11})^2} + j\omega \frac{g_1 g_2\, C}{\Delta y^2 + (\omega C y_{11})^2} \tag{5-81}$$

This expression describes a frequency dependent equivalent resistor and inductor in series (i.e., a lossy inductor) as shown in Fig. 5-28b. (The corresponding network in terms of the input admittance is shown in Fig. 5-28c). For the equivalent series resistance we have

$$R_{eq}(\omega) = \frac{\Delta y\, y_{22} + (\omega C)^2 y_{11}}{\Delta y^2 + (\omega y_{11} C)^2} \tag{5-82}$$

and for the equivalent inductor

$$L_{eq}(\omega) = \frac{g_1 g_2\, C}{\Delta y^2 + (\omega y_{11} C)^2} = \frac{g_1 g_2\, C}{(y_{11} y_{22} + g_1 g_2)^2 + (\omega y_{11} C)^2} \tag{5-83}$$

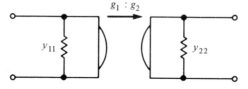

FIG. 5-27. Nonideal gyrator realized by an ideal active gyrator and two admittances.

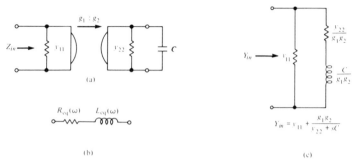

FIG. 5-28. (a) Capacitor-terminated nonideal gyrator. (b) Equivalent lossy inductor. (c) Equivalent input admittance.

The frequency dependence of R_{eq} and L_{eq} are plotted in Fig. 5-29. The equivalent inductance L_{eq} is fairly independent of frequency for $\omega \ll \Delta y / y_{11} C$:

$$L\bigg|_{\omega \ll \frac{\Delta y}{y_{11}C}} = \frac{g_1 g_2 C/(y_{11}C)^2}{\omega^2 + (\Delta y/y_{11}C)^2}$$

$$\approx \frac{g_1 g_2 C}{(y_{11}y_{22} + g_1 g_2)^2} = \frac{C}{g_1 g_2} \frac{1}{[1 + (y_{11}y_{22}/g_1 g_2)]^2} \quad (5\text{-}84)$$

A useful quantity in the description of inductors is their quality factor Q, which is defined as the ratio of reactive to resistive impedance:

$$Q = \frac{\omega L}{R} \quad (5\text{-}85)$$

Thus, for the gyrator-simulated inductor we have, from (5-82) and (5-83),

$$Q = \frac{\omega g_1 g_2 C}{\Delta y\, y_{22} + (\omega C)^2 y_{11}} \quad [5\text{-}86]$$

Note that Q is a function of frequency. As $\omega \to 0$, $Q \to 0$ and as $\omega \to \infty$, $Q \to 0$. Thus Q is a frequency function as shown in Fig. 5-30. It has a maximum value Q_m at some finite frequency ω_{Q_m}. Taking the derivative of (5-86) and setting it equal to zero we can solve for ω_{Q_m} and find

$$\omega_{Q_m} = \frac{1}{C}\sqrt{\frac{y_{22}\Delta y}{y_{11}}} \quad (5\text{-}87)$$

The resulting maximum Q is then:

$$Q_m = \frac{g_1 g_2}{2\sqrt{y_{11}y_{22}\Delta y}} \quad (5\text{-}88)$$

From (5-77) we have $\Delta y = y_{11}y_{22} + g_1 g_2$. Furthermore we may assume, for a reasonably good gyrator, that $y_{11}y_{22} \ll g_1 g_2$. Then (5-88) becomes

$$Q_m\bigg|_{y_{11}y_{22} \ll g_1 g_2} \approx \frac{1}{2}\sqrt{\frac{g_1 g_2}{y_{11}y_{22}}} \quad [5\text{-}89]$$

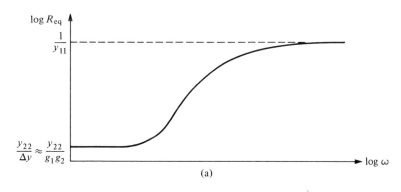

(a)

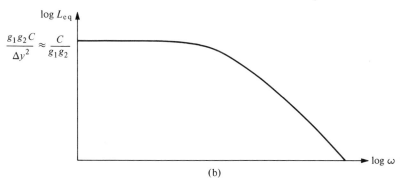

(b)

FIG. 5-29. Frequency dependence of (a) equivalent resistor $R_{eq}(\omega)$ and (b) equivalent inductor $L_{eq}(\omega)$ obtained with capacitor-loaded gyrator.

We implied previously (see (5-84)) that the upper frequency limit ω_h for L was

$$\omega_h = \frac{\Delta y}{y_{11}C} \qquad [5\text{-}90a]$$

Thus, with (5-87),

$$\omega_{Q_m} = \sqrt{\frac{\omega_h \cdot y_{22}}{C}} \qquad [5\text{-}90b]$$

which will generally be less than ω_h. Note that, in order to obtain a large inductance while using a small capacitance, $g_1 g_2$ must be as small as possible. In fact, since film and other integrated (e.g., chip) capacitors are either inherently limited in value or quite expensive when exceeding, say, 0.02 μF in value, the only economical way of simulating a large inductance by a hybrid-integrated gyrator–capacitor combination is to use a relatively low-valued capacitor and to combine it with a low-conductance gyrator. On the other hand, we see from (5-89) that the maximum Q of the inductance

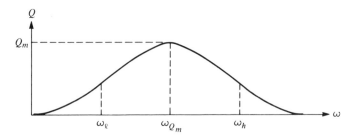

FIG. 5-30. Frequency dependence of Q of inductor simulated by capacitor-loaded gyrator.

thereby obtained is proportional to $\sqrt{g_1 g_2 / y_{11} y_{22}}$. Thus, with low conductances g_1, g_2, the input admittances y_{11}, y_{22} must be much lower in order to produce a reasonable inductor Q. Herein lies one of the problems in achieving large-valued high-Q inductances with relatively low-valued (i.e., integrable) capacitors: we require that $y_{11} y_{22} \ll g_1 g_2$ while the product $g_1 g_2$ must itself be small. In terms of the *inductance–Q product* we obtain, from (5-84) and (5-88),

$$L_{eq} \cdot Q_m = \frac{C/2}{[1 + (y_{11} y_{22} / g_1 g_2)]^{5/2} \sqrt{g_1 g_2 \, y_{11} y_{22}}}\Bigg|_{y_{11} y_{22} \ll g_1 g_2}$$

$$\approx \frac{C}{2\sqrt{g_1 g_2 \, y_{11} y_{22}}} \qquad\qquad \text{[5-91]}$$

This quantity should, of course, be as large as possible.

Frequency Dependence   In order to examine the frequency dependence of the capacitor-loaded gyrator as it is effected by frequency-dependent gyration conductances $g_1(s)$ and $g_2(s)$, we shall, for the sake of analytical simplicity, assume zero parasitic admittances, i.e., $y_{11} = y_{22} = 0$. Then from (5-70) we have

$$L_{eq}(s) = \frac{C}{g_1(s) g_2(s)} \qquad\qquad (5\text{-}92)$$

Let us assume that $g_1(s)$ and $g_2(s)$ are essentially frequency-limited by single dominant poles, i.e.,

$$g_1(s) = \frac{\omega_1}{s + \omega_1} g_1 \qquad\qquad (5\text{-}93a)$$

and

$$g_2(s) = \frac{\omega_2}{s + \omega_2} g_2 \qquad\qquad (5\text{-}93b)$$

then (5-92) becomes

$$L_{eq}(j\omega) = \frac{C}{g_1 g_2}\left[\left(1 - \frac{\omega^2}{\omega_1 \omega_2}\right) + j\omega\left(\frac{1}{\omega_1} + \frac{1}{\omega_2}\right)\right] \qquad [5\text{-}94]$$

This expression consists of a real term which decreases and becomes negative with increasing frequency, and an imaginary term. Thus, in order for the capacitor-loaded gyrator to remain useful we require that $\omega \ll \sqrt{\omega_1 \omega_2}$, or, conversely, that the corner frequencies ω_1 and ω_2 are as far away from the operative frequency range as possible. At the frequency $\omega = \sqrt{\omega_1 \omega_2}$ we have

$$L_{eq}(j\omega)\Big|_{\omega = \sqrt{\omega_1 \omega_2}} = j\,\frac{C}{g_1 g_2}\left(\sqrt{\omega_1/\omega_2} + \sqrt{\omega_2/\omega_1}\right) \qquad (5\text{-}95)$$

Thus the inductor no longer has a real component and introduces 90° of phase lead over and above that of a normal inductor.

Capacitor Loss   In general the capacitor loading a gyrator will also have some loss or dissipation associated with it, and it is of interest how this will effect the resulting simulated inductor. Consider the configuration shown in Fig. 5-31a in which an ideal gyrator is loaded by the capacitor–resistor combination Z_L. For this case, the capacitor Q is defined as

$$Q_C = \omega C R_p \qquad (5\text{-}96)$$

Calculating the input impedance of the gyrator we obtain

$$Z_{in}(j\omega) = R_{eq} + \omega L_{eq} = \frac{1}{g^2 Z_L} = \frac{1}{g^2 R_p} + j\omega\,\frac{C}{g^2} \qquad (5\text{-}97)$$

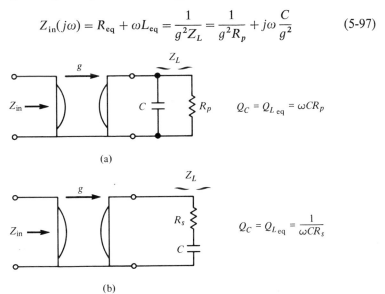

(a)

(b)

FIG. 5-31. Ideal gyrator loaded by a lossy capacitor. (a) Parallel resistor. (b) Series resistor.

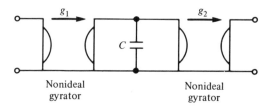

FIG. 5-32. Simulation of a floating inductor using two nonideal gyrators.

The Q of the equivalent inductor therefore results as

$$Q_{L_{eq}} = \frac{\omega L_{eq}}{R_{eq}} = \omega C R_p \tag{5-98}$$

We observe that

$$Q_{L_{eq}} = Q_C \tag{[5-99]}$$

Thus to obtain a high-Q simulated inductor we require a high-Q capacitor loading the gyrator, since the Q of the capacitor is carried over directly into the Q of the associated equivalent inductor.[14] It is an easy matter to show that for the gyrator, loaded by a capacitor whose losses can be lumped in a series resistor R_s (see Fig. 5-31b), the same Q equality results.

Simulating an Ungrounded Inductor   Let us now examine the simulation of an ungrounded inductor using two nonideal gyrators, as shown in Fig. 5-32. The $ABCD$ matrix for the overall two port is

$$\begin{bmatrix} A_1 & B_1 \\ C_1 & D_1 \end{bmatrix} \cdot \begin{bmatrix} 1 & 0 \\ sC & 1 \end{bmatrix} \cdot \begin{bmatrix} A_2 & D_2 \\ C_2 & D_2 \end{bmatrix} = \begin{bmatrix} A' & B' \\ C' & D' \end{bmatrix}$$

$$= \begin{bmatrix} A_2(A_1 + sB_1C) + B_1C_2 & B_2(A_1 + sB_1C) + B_1D_2 \\ A_2(C_1 + sD_1C) + C_2D_1 & B_2(C_1 + sD_1C) + D_1D_2 \end{bmatrix} \tag{5-100}$$

The corresponding y matrix can be obtained from Table 3-3, (Chapter 3):

$$[y'] = \begin{bmatrix} \dfrac{D'}{B'} & -\dfrac{\Delta A'}{B'} \\[2ex] -\dfrac{1}{B'} & \dfrac{A'}{B'} \end{bmatrix} \tag{5-101}$$

where A', B', C', and D' are the elements of the overall $ABCD$ matrix and $\Delta A'$ is the determinant of the $ABCD$ matrix.

---

14. In Chapters 6 and 7 we shall see that the losses in certain film capacitors may be quite high, particularly at high frequencies. They may thereby represent the main obstacle in the realization of high-Q, simulated inductors using hybrid-integrated circuit techniques.

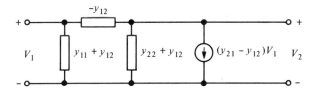

FIG. 5-33. Two-port in terms of its y parameters and a VCS.

Using one of the equivalent circuits of a two-port defined in terms of its y matrix that were derived in Chapter 3 (see Fig. 3-14a), we obtain the network shown in Fig. 5-33. The series arm is given by $-y_{12}$, which is equal to $\Delta A'/B'$. This quantity is of the form $K_1/(K_2 + sK_3)$, which represents a series RL circuit. The current source has a scale factor $y_{21} - y_{12}$, which is $(\Delta A' - 1)/B'$; this is a function of frequency. The two shunt arms are $(D' - \Delta A')/B'$ and $(A' - \Delta A')/B'$, which are both of the form $(K_1 + sK_2)/(K_3 + sK_4)$. This form corresponds to a lossy inductor and capacitor in parallel, as shown in Fig. 5-34. The corresponding conductance has the form:

$$y_s = \frac{1}{(K_3/K_1) + s(K_4/K_1)} + \frac{1}{(K_3/sK_2) + (K_4/K_2)} \qquad (5\text{-}102)$$

The equivalent circuit of Fig. 5-32 therefore has the form shown in Fig. 5-35. We can now identify the network elements of Fig. 5-35 in terms of the chain-matrix parameters of the two gyrators shown in Fig. 5-32 and obtain

$$L_1 = \frac{B_2 B_1 C}{C_1 B_2 + D_1 D_2 - \Delta A'} \qquad L_3 = \frac{B_1 B_2 C}{A_1 A_2 + C_2 B_1 - \Delta A'}$$

$$C_1 = \frac{B_2 D_1 C}{B_2 A_1 + B_1 D_2} \qquad C_3 = \frac{B_1 A_2 C}{B_2 A_1 + B_1 D_2}$$

$$R_{11} = \frac{B_2 A_1 + B_1 D_2}{C_1 B_2 + D_1 D_2 - \Delta A'} \qquad R_{31} = \frac{B_2 A_1 + B_1 D_2}{A_1 A_2 + B_1 C_2 - \Delta A'}$$

$$R_{12} = \frac{B_1}{D_1} \qquad R_{32} = \frac{B_2}{A_2}$$

$$R_2 = \frac{B_2 A_1 + B_1 D_2}{\Delta A'} \qquad K(s) = \frac{\Delta A' - 1}{B_2 A_1 + B_1 D_2 + s(B_1 B_2 C)}$$

$$L_2 = \frac{B_1 B_2 C}{\Delta A'} \qquad (5\text{-}103)$$

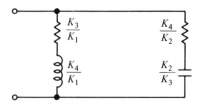

FIG. 5-34. Admittance form corresponding to (5-102).

Let us now examine some limiting cases of this network. Suppose we have two ideal gyrators with the following matrices:

$$[ABCD]_1 = \begin{bmatrix} 0 & \dfrac{1}{g_{12}} \\ g_{11} & 0 \end{bmatrix} \tag{5-104a}$$

$$[ABCD]_2 = \begin{bmatrix} 0 & \dfrac{1}{g_{22}} \\ g_{21} & 0 \end{bmatrix} \tag{5-104b}$$

The overall matrix of the two-gyrator–capacitor configuration then results as

$$[A'B'C'D'] = \begin{bmatrix} \dfrac{g_{21}}{g_{12}} & \dfrac{sC}{g_{12}g_{22}} \\ 0 & \dfrac{g_{11}}{g_{22}} \end{bmatrix} \tag{5-105}$$

and the corresponding determinant is

$$\Delta A' = \frac{g_{21}g_{11}}{g_{12}g_{22}} \tag{5-106}$$

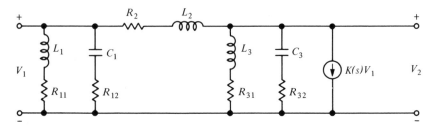

FIG. 5-35. Equivalent circuit of gyrator configuration shown in Fig. 5-32.

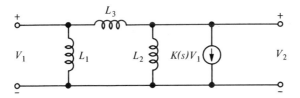

FIG. 5-36. Simplified version of Fig. 5-35.

Comparing (5-100) with (5-105) and referring to the expressions given in Fig. 5-35 it follows that for this particular case

$$C_1 = C_3 = R_{11} = R_{31} = R_2 = 0 \qquad R_{12} = R_{32} = \infty \qquad (5\text{-}107)$$

The network in Fig. 5-35 therefore simplifies to the inductive network in Fig. 5-36, where the individual inductors are now given by

$$L_1 = (C/g_{12}g_{22})\left(\frac{g_{11}}{g_{22}} - \frac{g_{21}g_{11}}{g_{12}g_{22}}\right)^{-1} = \frac{C}{g_{11}(g_{12} - g_{21})}$$

$$L_2 = (C/g_{12}g_{22})\left(\frac{g_{21}g_{11}}{g_{12}g_{22}}\right)^{-1} = \frac{C}{g_{11}g_{21}}$$

$$L_3 = (C/g_{12}g_{22})\left(\frac{g_{21}}{g_{12}} - \frac{g_{21}g_{11}}{g_{12}g_{22}}\right)^{-1} = \frac{C}{g_{21}(g_{22} - g_{11})}$$

$$K(s) = \frac{(g_{21}g_{11} - g_{12}g_{22})(g_{22}g_{12})}{sCg_{12}g_{22}} = \frac{g_{21}g_{11} - g_{12}g_{22}}{sC} \qquad (5\text{-}108)$$

The number of parasitics has been greatly reduced by eliminating the parasitics in the gyrators. If $g_{11} = g_{12}$ and $g_{21} = g_{22}$, then the parasitic current source in Fig. 5-36 disappears and we obtain the network given in Fig. 5-37. If, finally, we let $g_1 = g_2 = g$, then the ideal floating inductor shown in Fig. 5-38 results. As would be expected, its value corresponds to that derived with ideal gyrators.

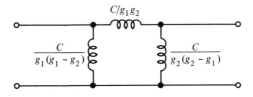

FIG. 5-37. Simplified version of Fig. 5-36.

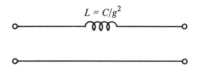

$$L = C/g^2$$

FIG. 5-38. Simplified version of Fig. 5-37.

5.4.3 Gyrator Realizations

To obtain practical gyrator realizations we shall proceed by the same systematic (i.e., morphological) method that we used for the negative impedance converter. Thus we first consider the controlled-source realizations of the ideal gyrator matrix, convert these to nullator–norator realizations, and then consider possible transistor or operational amplifier circuit equivalents by taking appropriate nullator–norator pairs.

Controlled-Source Realizations   The y matrix of an ideal gyrator can be expressed as the sum of two y matrices:

$$[y] = \begin{bmatrix} 0 & g_1 \\ -g_2 & 0 \end{bmatrix} = \begin{bmatrix} 0 & g_1 \\ 0 & 0 \end{bmatrix} + \begin{bmatrix} 0 & 0 \\ -g_2 & 0 \end{bmatrix} \qquad (5\text{-}109)$$

Letting

$$[y_A] = \begin{bmatrix} 0 & g_1 \\ 0 & 0 \end{bmatrix}; \qquad [y_B] = \begin{bmatrix} 0 & 0 \\ -g_2 & 0 \end{bmatrix} \qquad (5\text{-}110)$$

we can design a gyrator by realizing $[y_A]$ and $[y_B]$ separately and then connecting the two resulting networks in parallel. Referring to Table 5.5 we find that $[y_A]$ and $[y_B]$ represent a positive and a negative VCS, respectively, as shown in Fig. 5-39a. Connecting the two VCS in parallel (see Fig. 5-39b), we obtain the VCS representation of a gyrator, shown in Fig. 5-39c.

Similarly, by splitting up the z matrix of the gyrator into the sum of two CVSs (see Table 5-5) of which one is positive, the other negative, and connecting the two CVSs in series, one obtains the CVS gyrator representation shown in Fig. 5-40. Numerous other methods of matrix decomposition, including those involving VVS or CCS representations, are possible, but those involving VCS and CVS generally lead to more practical solutions. Indeed, the VCS realization of Fig. 5-39 has also been found to yield more practical network realizations than the CVS realization of Fig. 5-40; the lattter results in balanced instead of grounded circuit configurations.

It is not our intention here to go through the derivations of the many transistor, and amplifier gyrator realizations that follow from a morphological regrouping of nullator-norator pairs obtainable from the corresponding gyrator controlled-source representations. As Mitra has shown,[15] in so doing

15. S. K. Mitra, op. cit., see footnote 5 of this chapter.

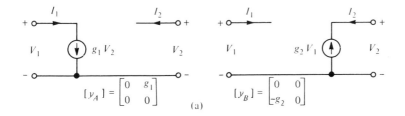

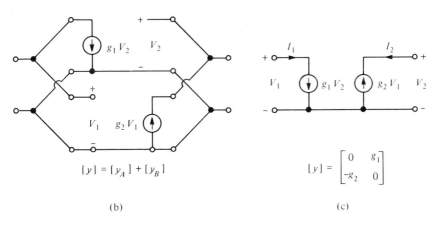

FIG. 5-39. Derivation of a gyrator as the parallel connection of a positive and a negative VCS.

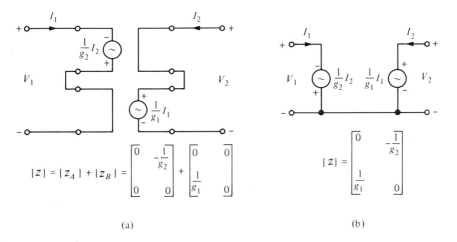

FIG. 5-40. Derivation of a gyrator as the series connection of a positive and a negative CVS.

we would almost inadvertently stumble onto the many transistor and amplifier realizations that have recently been published in the literature and, presumably, onto many more. Clearly each "realization" derived is not necessarily a practical one. It must first be investigated for realizability with respect to biasing,[16] power supply connections, and, in particular, for compatibility with integrated circuit technology. Furthermore, some of the realizable configurations may be very much more prone to parasitic effects than others. It is in this sorting process, and in the ability to detect the extent of parasitic effects, that the circuit designer has an opportunity to demonstrate how well he masters his craft.

In what follows we shall derive a few representative transistor and amplifier gyrator realizations from their nullator–norator controlled-source representations as we did for the NIC. More will be said about practical realization and design for integrated-circuit implementation in Chapter 7.

Transistor Realizations   As pointed out earlier, the VCS realization of the gyrator derived in Fig. 5-39 has been found to provide usable gyrator circuits. We require the two nullator–norator VCS representations given in Table 5-4 to provide the polarity requirements and to obtain suitable nullator–norator equivalents of Fig. 5-39b. The resulting configuration is shown in Fig. 5-41a. Designating the ith norator by N_i and the jth nullator by n_j, transistor realizations can be obtained by combining adjoining (N_i, n_j) pairs. Two three-transistor realizations are possible by grouping different nullator–norator pairs together, as shown in Figs. 5-41b and c. These circuits were first described by Sipress and Witt;[17] later the circuit in Fig. 5-41b was described in detail by Shenoi.[18] Clearly, by adding more nullator–norator pairs without altering the operating conditions (such as N_4, n_4 in Fig. 5-41a), other transistor configurations evolve.

It is not necessary, in the procedure described above, to restrict oneself to the nullator–norator representations of controlled sources given in Table 5-4. In fact, with a few basic guidelines, new nullator–norator equivalents of controlled sources can be derived quite easily. It should be clear by now that an infinite input impedance as required by an ideal voltage-controlled source must necessarily have a nullator in series with the input terminal, as shown in Fig. 5-42a. The other terminal conditions necessary for the various controlled sources are also listed in Fig. 5-42. As can be seen, *a nullator must appear at the input and a norator at the output of a controlled source.* The rectangular elements in the nullator–norator configurations can be either resistors or complementary singular elements (i.e., nullators or norators). Since these terminal conditions determine a particular controlled source,

16. Z. Galani and G. Szentirmai, DC operation of three-transistor gyrators, *IEEE Trans. Circuit Theory*, **CT-18**, 738–739 (1971).
17. J. M. Sipress and F. J. Witt, U.S. Patents Nos. 3,001,157 (September 19, 1961) and 3,120,645 (February 4, 1964).
18. B. A. Shenoi, op. cit.

modifications of a particular source, such as phase reversal, can be obtained merely by cascading a given source realization. Care must be taken, however, that in doing so an open or closed circuit is not inadvertently generated. If it is, it can be eliminated by the appropriate placement of an additional resistor.

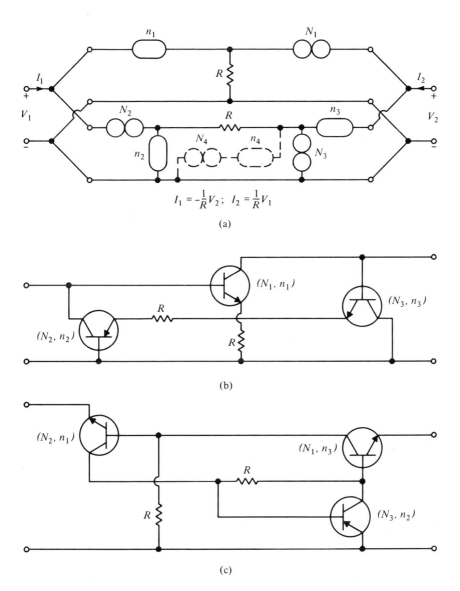

$$I_1 = -\frac{1}{R}V_2; \quad I_2 = \frac{1}{R}V_1$$

(a)

(b)

(c)

FIG. 5-41. Derivation of a transistorized gyrator. (a) Nullator–norator representation. (b) First transistor realization. (c) Second transistor realization.

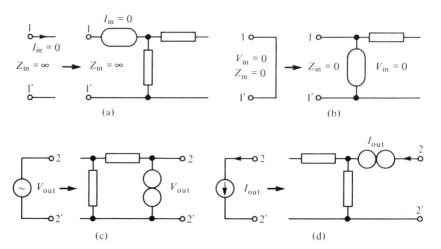

FIG. 5-42. Guidelines for the nullator–norator derivation of controlled sources. (a) Infinite input impedance. (b) Zero input impedance. (c) Voltage-source output. (d) Current-source output.

Consider, for example, the cascading of the two VCS shown in Fig. 5-43a. Although the terminal conditions of the new circuit would appear to be the right ones for a VCS, the resulting circuit is, of course, nothing of the kind, because of the resulting open circuit. By adding resistor R_0 (Fig. 5-43b) this error is eliminated and we obtain an inverting VCS by cascading two noninverting VCSs. Using this newly derived VCS as the inverting source of the gyrator in combination with the original noninverting source we obtain a new nullator–norator pair (Fig. 5-44a). Combining adjoining nullator–norator pairs, we obtain the three-transistor circuits shown in Figs. 5-44b and c. The circuit of Fig. 5-44b was described elsewhere,[19] that in Fig. 5-44c was derived by Mitra using the approach described here. The nullator–norator gyrator models of Figs. 5-41a and 5-44a, along with their transistorized versions, were recently independently advanced by Bendik.[20] Clearly more complicated circuits, and possibly improved ones, can be obtained with more complicated nullator–norator models of the controlled sources used, or by appropriately inserted additional nullator–norator pairs, such as (N_4, n_4) in Fig. 5-44a.

Operational Amplifier Realizations One way of realizing a gyrator by operational amplifiers would be to replace each nullator–norator pair by an operational amplifier. However, the configurations shown in Figs. 5-41a and 5-44a would require three operational amplifiers each, which is quite uneconomical. For this reason most operational amplifier realizations have

19. W. New and R. Newcomb, An integrable time-variable gyrator, *Proc. IEEE*, **53**, 2161–2162 (1965).
20. J. Bendik, Equivalent gyrator networks with nullators and norators, *IEEE Trans. Circuit Theory*, **CT-14**, 98 (1967).

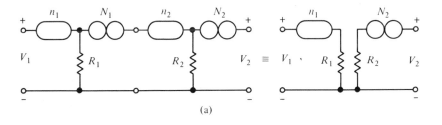

(a)

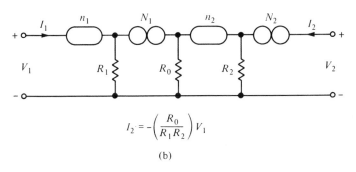

$$I_2 = -\left(\frac{R_0}{R_1 R_2}\right) V_1$$

(b)

FIG. 5-43. Cascading two VCSs to obtain a new VCS. (a) Cascade produces undesirable open circuit. (b) Addition of R_0 provides usable new VCS.

been derived by other means, resulting in configurations with only two amplifiers.

A common approach has been to start out with a general negative-impedance inverter (NII) (see Table 5-5) whose chain matrix is given by

$$[ABCD]_{\text{NII}} = \begin{bmatrix} 0 & \pm R_2 \\ \mp \dfrac{1}{R_1} & 0 \end{bmatrix} \qquad (5\text{-}111)$$

where

$$R_1 R_2 = k_2 > 0 \qquad (5\text{-}112)$$

An NII is an active two-port network which has an input impedance

$$Z_{\text{in}} = - k_2 \left(\frac{1}{Z_L}\right) \qquad (5\text{-}113)$$

when terminated by Z_L. Thus it is essentially a negative gyrator.

The reason for starting out with an NII is that a useful and simple bridge-type circuit capable of realizing it was advanced by Larky and Lundry.[21]

21. A. I. Larky, Negative-impedance converters, *IRE Trans. Circuit Theory*, **CT-4**, 124–131 (1957). W. R. Lundry, Negative-impedance circuits—some basic relations and limitations, *IRE Trans. Circuit Theory* **CT-4**, 132–139 (1957).

This circuit, also capable of realizing a VNIC or a CNIC, is shown in Fig. 5-45. The two basic NIC types and the NII result from this bridge if the impedances Z are selected as follows:

Circuit type	Z_A	Z_B	Z_C	Z_D
CNIC	R_1	R_2	∞	∞
VNIC	∞	∞	R_1	R_2
NII	R_1	∞	∞	R_2

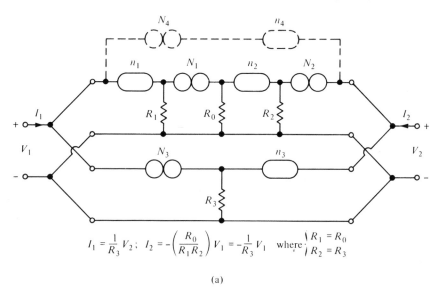

$$I_1 = \frac{1}{R_3} V_2; \quad I_2 = -\left(\frac{R_0}{R_1 R_2}\right) V_1 = -\frac{1}{R_3} V_1 \quad \text{where} \begin{cases} R_1 = R_0 \\ R_2 = R_3 \end{cases}$$

(a)

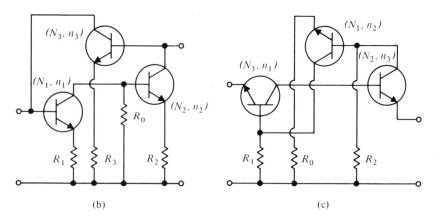

(b)   (c)

Fig. 5-44. Derivation of transistorized gyrators (b) and (c) using the newly developed VCS gyrator model (a).

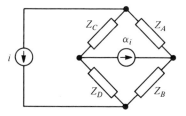

FIG. 5-45. General bridge network used to derive NIC and NII configurations.

If the $[ABCD]$ matrices of the two corresponding NICs are given as

$$[ABCD]_{\text{VNIC}} = \begin{bmatrix} -\dfrac{R_1}{R_2} & 0 \\ 0 & 1 \end{bmatrix}$$ (5-114)

and

$$[ABCD]_{\text{CNIC}} = \begin{bmatrix} 1 & 0 \\ 0 & -\dfrac{R_2}{R_1} \end{bmatrix}$$ (5-115)

then we obtain the same matrices as those given in Table 5-5 if we let $R_1/R_2 = k_1$ and $k_2 = 1$ for the VNIC and vice versa for the CNIC.

By using the equivalence between a nullator–norator pair and an infinite-gain CCS, the nullator–norator equivalent circuits for the NICs and the NII of Fig. 5-45 can be obtained as shown in Fig. 5-46. The nullator–norator CNIC and VNIC representations will be recognized as the same ones as those obtained from the controlled-source representations and used in Fig. 5-15.

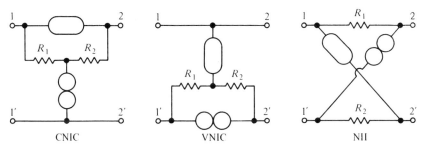

FIG. 5-46. Nullator–norator representations of the NICs and the NII obtainable from the bridge network in Fig. 5-45.

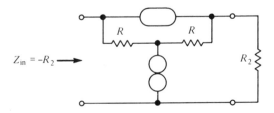

FIG. 5-47. Negative resistor realization using a resistor-terminated NIC.

Compared to the $[ABCD]$ matrix of the NII given by (5-111), that of an ideal gyrator is given by

$$[ABCD]_{\text{gyrator}} = \begin{bmatrix} 0 & \pm R_2 \\ \pm \dfrac{1}{R_1} & 0 \end{bmatrix} \qquad (5\text{-}116)$$

The NII becomes a gyrator if we let R_1 or R_2 be replaced by their negative equivalents (i.e., $-R_1$ or $-R_2$). A negative resistance can be obtained by using an NIC terminated by a positive resistance, as shown in Fig. 5-47. This is used to replace R_2 in the NII of Fig. 5-46, as shown in Fig. 5-48. Clearly two different configurations can be obtained by using the CNIC (Fig. 5-48a) or the VNIC (Fig. 5.48b). Other configurations can be obtained by interconnecting the NII with the NICs the other way round (e.g., exchanging the connecting terminals).

The nullator–norator gyrator configurations shown in Fig. 5-48 use only two nullator–norator pairs instead of the three resulting from the controlled-source derivations (e.g., Figs. 5-41 and 5-44). They are therefore more suitable

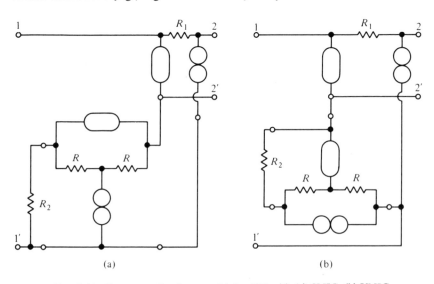

(a)   (b)

FIG. 5-48. Gyrator realizations combining NII with (a) CNIC, (b) VNIC.

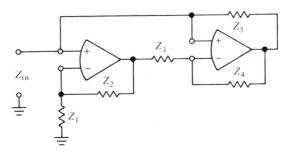

FIG. 5-49. Gyrator realization using two operational amplifiers.

for operational-amplifier implementation where a nullator–norator pair is replaced by an amplifier. Following this procedure, numerous operational amplifier realizations of gyrators have been derived by Antoniou.[22] Other operational amplifier methods are related to the nullator–norator configurations of Fig. 5-48. One with which excellent laboratory results have been obtained was reported by Riordan.[23] It is shown in Fig. 5-49. The input impedance Z_{in} is given by

$$Z_{in} = \frac{Z_1 Z_3 Z_5}{Z_2 Z_4} \qquad [5\text{-}117]$$

If either Z_2 or Z_4 is a capacitor C and all other elements are resistors R, Z_{in} becomes

$$Z_{in} = j\omega C R^2 \qquad (5\text{-}118)$$

Thus the circuit behaves like an inductance L where

$$L = CR^2 \qquad (5\text{-}119)$$

Clearly the gyration conductance g is $1/R$.

5.5 OPERATIONAL AMPLIFIERS

In Chapter 3 (Section 3.1.5) we briefly discussed networks incorporating operational amplifiers and various methods of analyzing them. It was pointed out there that the operational amplifier can be considered as a special kind of controlled source, i.e., a very-high-gain voltage-controlled voltage source (VVS). Furthermore, based on the nullator–norator equivalent representation of an operational amplifier on the one hand (see Table 5-3), and on various matrix identities on the other, we were in a position to realize

22. A. Antoniou, Realization of gyrators using operational amplifiers and their use in *RC*-active-network synthesis, *Proc. Inst. Elect. Engrs.* **116**, 1838–1850 (1969).
23. R. H. S. Riordan, Simulated inductors using differential amplifiers, *Electron Lett.*, **3**, 50–51 (1967).

some of the circuit elements discussed in this chapter by equivalent operational amplifier configurations (see, e.g., the NICs in Fig. 5-15 and the gyrator in Fig. 5-49). Thus, although we have discussed applications using them, we have not yet discussed operational amplifiers themselves, nor the numerous interesting properties they afford.

In this section we shall discuss the operational amplifier as a basic network element and demonstrate how, in its ideal form, it can be used in linear active networks to perform many useful functions. Because of its preeminence in present-day hybrid-integrated network design, which is mainly due to its availability as a low-cost monolithic integrated circuit, it will be discussed in more detail again in Chapter 7. There we shall deal with the operational amplifier in silicon monolithic integrated form and with the nonideal characteristics that result.

5.5.1 The Ideal Operational Amplifier

The ideal operational amplifier is shown in Fig. 5-50a, its equivalent circuit in Fig. 5-50b. A signal appearing at the negative terminal (V_1) is inverted at the output, a signal at the positive terminal (V_2) appears at the output with no change in sign. Hence the negative terminal is called the "inverting," the positive terminal the "noninverting" terminal. In general, the output voltage is directly proportional to the difference of the input voltages, i.e., proportional to $V_d = (V_1 - V_2)$. The constant of proportionality, $-A_0$, is the voltage gain of the amplifier; ideally, A_0 is a positive real constant.

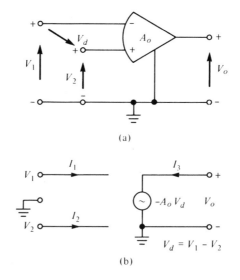

FIG. 5-50. Ideal operational amplifier. (a) Symbolic representation. (b) Equivalent circuit.

The ideal operational amplifier is characterized as follows:

1. $A_0 \to \infty$ (infinite gain)
2. $I_1 = I_2 = 0$
    $\therefore R_i = \infty$ (infinite input impedance)
3. $R_0 = 0$ (zero output impedance)
4. $V_0|_{V_d=0} = 0$ (zero offset)
5. $A_0 = $ constant   (infinite bandwidth)

Thus the ideal operational amplifier has infinite gain, input impedance, and bandwidth (properties 1, 2, and 5); it has zero output impedance (3) and zero output voltage when the voltages at both input terminals are zero (4). The latter property is referred to as that of zero offset voltage. Clearly, by itself the operational amplifier is not a very useful device, since the slightest voltage at the input will cause it to go into saturation at the output (i.e., infinite output voltage). In fact, it is only with feedback to establish finite overall voltage gain that the many useful properties of the operational amplifier become evident, as we shall see in the following applications.

The Inverting Mode   The circuit diagram of an operational amplifier in the inverting mode is shown in Fig. 5-51a, the equivalent circuit diagram in Fig. 5-51b. In this mode of operation the noninverting (positive) input terminal of the amplifier is grounded and the input signal (V_{in}) is applied to the inverting (negative) input terminal through resistor R_G. The feedback, applied through resistor R_F from the output to the inverting input terminal, is negative.

The signal processing operation performed by the circuit in Fig. 5-51 is determined by the feedback elements R_F and R_G. It is simply that of negative (i.e., inverting) voltage amplification. In order to calculate the gain of the circuit while avoiding the concept of infinite gain (which is awkward in linear network analysis) we may reason as follows. To be useful the voltages and currents of the circuit must be finite; consequently, if the circuit contains

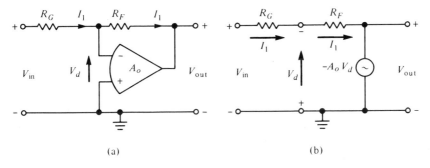

(a)   (b)

FIG. 5-51. (a) Ideal operational amplifier in the inverting feedback mode. (b) Equivalent circuit diagram.

an ideal operational amplifier with infinite gain, the potential difference between the input terminals of the amplifier must be zero. In the circuit of Fig. 5-51 the voltage V_d between the input terminals of the infinite-gain amplifier is forced to zero by the negative feedback around the amplifier. Moreover, since the input impedance of the amplifier is infinite, the input current to the amplifier is zero. Hence, for circuits using ideal operational amplifiers in which the voltages and currents are to be finite, the following two statements apply:

1. The potential difference between the amplifier input terminals is zero.
2. The current into each input terminal is zero.

These two statements enable us to analyze circuits incorporating ideal operational amplifiers without having to deal with the fact that their gain is infinitely large. Notice, however, that the first statement depends on the assumption that the output voltage of the amplifier is finite, or, in more practical terms, linearly related to the input. Whether this assumption is valid or not should be evident from the application in question.

According to the discussion presented above, the operation of the ideal amplifier in Fig. 5-51 may be described in the following terms. Because the feedback in the circuit forces the voltage V_d at the inverting input terminal to be zero at all times, there exists a virtual short circuit or virtual ground at the input to the amplifier proper. The term "virtual" is used to imply that since the feedback serves to keep the voltage V_d at zero no current actually flows through this "short." Thus, the current I_1 flowing through R_G actually continues past this virtual short through the impedance R_F. This fact, together with the fact that V_d is equal to zero, yields

$$I_1 = \frac{V_{in}}{R_G} = -\frac{V_{out}}{R_F} \qquad (5\text{-}120)$$

Therefore

$$\frac{V_{out}}{V_{in}} = \alpha_I = -\frac{R_F}{R_G} \qquad [5\text{-}121]$$

α_I is referred to as the closed-loop gain of the inverting operational amplifier. It is a negative quantity because the closed-loop amplifier reverses the sign of the input signal. Furthermore, it depends only on the resistance ratio R_F/R_G. Hence the closed-loop gain α_I is only as precise and stable as the resistance ratio itself.

If we consider the special case $R_F = R_G$, then $\alpha_I = -1$ and we obtain a so-called unity-gain inverter, i.e., a circuit used as a sign inverter with unity gain. It is sometimes used as a buffer to match a high-impedance source to a low-impedance load.

The input impedance of the amplifier in Fig. 5-51 results immediately from the fact that the input terminal of the ideal amplifier is held at virtual ground by the feedback ($V_d = 0$). Thus

$$R_{in} = \frac{V_{in}}{I_1} = R_G \qquad [5\text{-}122]$$

i.e., the input impedance depends only on the external resistor R_G.

The output impedance is defined as the impedance seen at the output of the amplifier when the input terminal is set equal to zero. By inspection of Fig. 5-51b we see that the output impedance must equal zero because of the ideal voltage source ($- A_0 V_d$) which is connected across the output terminals.

The Noninverting Mode   The circuit diagram of an ideal operational amplifier in the noninverting mode is shown in Fig. 5-52a, the equivalent circuit diagram in Fig. 5-52b. In this case the input signal is applied directly to the noninverting (positive) input terminal of the amplifier, and the feedback resistors R_F and R_G are connected between the output terminal, the inverting (negative) input terminal, and ground. Since we are considering an ideal amplifier, the input current is zero and the feedback maintains a zero potential difference between the input terminals (i.e., $V_d = 0$); the voltage V_I from the inverting terminal to ground is therefore equal to the input voltage V_{in}. Notice that V_I is not equal to zero in this case, meaning that the noninverting circuit has no virtual ground at either one of its input terminals. Since V_I is equal to V_{in}, we obtain by inspection

$$V_I = \frac{R_G}{R_F + R_G} V_{out} = V_{in} \qquad (5\text{-}123)$$

Thus

$$\frac{V_{out}}{V_{in}} = \alpha_N = \frac{R_F + R_G}{R_G} = 1 + \frac{R_F}{R_G} \qquad [5\text{-}124]$$

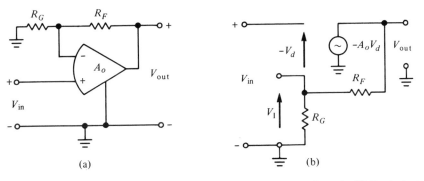

FIG. 5-52. (a) Ideal operational amplifier in the noninverting feedback mode. (b) Equivalent circuit diagram.

α_N is the closed-loop gain of the noninverting amplifier. It depends only on the resistance ratio R_F/R_G and, as in the inverting case, is only as precise and stable as this ratio. Because of the ideal voltage source across the output terminals, the output impedance of the noninverting amplifier is again zero. In contrast to the inverting amplifier mode, the input impedance to the noninverting feedback amplifier is infinite, since the input current to the ideal amplifier is zero.

5.5.2 Some Common Operational Amplifier Circuits

The operational amplifier derives its name from the fact that it may be used to accomplish a number of mathematical operations. In fact, prior to its use in integrated linear active networks, its main use was to perform these mathematical operations in analog computers. In what follows we shall consider some of the operations it is most commonly used to perform.

Voltage Follower   If we remove the resistor R_G in Fig. 5-52 (i.e., $R_G = \infty$), we obtain, from (5-124),

$$\frac{V_{\text{out}}}{V_{\text{in}}} = \alpha_N \bigg|_{R_G = \infty} = 1 \qquad [5\text{-}125]$$

This noninverting unity-gain feedback amplifier is known as a voltage follower. It, as well as the other circuits to be described here, is shown in Table 5-7. The voltage follower makes an ideal buffer between high-impedance sources and low-impedance loads because of its infinite input impedance and zero output impedance.

Summing Amplifier   The summing amplifier (see Table 5-7) is the same as the inverting feedback amplifier shown in Fig. 5-51, except that it has several input terminals. Here again the feedback forces a virtual ground to exist at the inverting input to the ideal amplifier; furthermore, the input current to the ideal amplifier is zero. Thus, the current equation for the node at the inverting input terminal gives

$$\frac{V_1}{R_1} + \frac{V_2}{R_2} + \cdots + \frac{V_n}{R_n} + \frac{V_{\text{out}}}{R_F} = 0 \qquad (5\text{-}126)$$

The output voltage, therefore, results as

$$V_{\text{out}} = -\left[\frac{R_F}{R_1} V_1 + \frac{R_F}{R_2} V_2 + \cdots + \frac{R_F}{R_n} V_n\right] \qquad [5\text{-}127]$$

It corresponds to the negative weighted sum of the input voltages. Notice that the summing operation depends exclusively on the sum of the resistance ratios.

TABLE 5-7. COMMON OPAMP CIRCUITS.

Type	Circuit	Function
Voltage follower		$V_{out} = V_{in}$
Summing amplifier		$V_{out} = -R_F \left(\dfrac{V_1}{R_1} + \dfrac{V_2}{R_2} + \cdots + \dfrac{V_n}{R_n} \right)$
Differential amplifier		$V_{out} = \dfrac{R_F + R_G}{R_G} t(s) V_2 - \dfrac{R_F}{R_G} V_1$ $V_{out} = \alpha_N t(s) V_2 + \alpha_1 V_1$
Integrator		$V_{out} = -\left(\dfrac{1}{sRC} \right) V_{in}$ $v_{out}(t) = -\dfrac{1}{RC} \int v_{in}(t)\,dt$
Differentiator		$V_{out} = -sRC V_{in}$ $v_{out} = -RC \dfrac{dv_{in}}{dt}$

This circuit is used widely to form linear combinations of various signals in analog computers. Because the addition of signals by this circuit is a result of the summing of currents at the inverting input terminal, this terminal is often called a summing node. In the ideal amplifier case, the virtual ground established at the summing node by the feedback around the amplifier prevents any interaction between the various signal sources connected to the input terminals. Each source, therefore, behaves as if it were alone.

For the same reason as in the previous cases, the output impedance of the summing amplifier is zero. The input impedance is different for each input signal. Because of the virtual ground at the summing node the input impedance seen by each signal source is simply the resistance connecting that source to the summing node.

The Differential Amplifier   The differential amplifier shown in Table 5-7 is characterized by an output voltage proportional to the difference between the two input voltages, V_1 and V_2. Assuming a network with a transfer function $t(s)$ between V_2 and the noninverting input terminal we obtain

$$V_{NI} = t(s) \cdot V_2 \qquad (5\text{-}128)$$

Similarly, by superposition, the voltage at the inverting input terminal is

$$V_I = \frac{R_F}{R_G + R_F} V_1 + \frac{R_G}{R_F + R_G} V_{out} \qquad (5\text{-}129)$$

Equations (5-128) and (5-129) are direct results of the fact that the input current to the ideal amplifier is zero. Since the potential difference between the two input terminals of the ideal amplifier is forced to zero by the feedback through R_F, V_I must equal V_{NI}. Thus setting V_I equal to V_{NI}, we obtain, with (5-128) and (5-129),

$$\frac{R_F}{R_G + R_F} V_1 + \frac{R_G}{R_F + R_G} V_{out} = t(s)V_2 \qquad (5\text{-}130)$$

Solving for the output voltage we have

$$V_{out} = \frac{R_F + R_G}{R_G} t(s)V_2 - \frac{R_F}{R_G} V_1$$

$$= \alpha_N t(s)V_2 - \frac{R_F}{R_G} V_1 \qquad [5\text{-}131]$$

For the case that $t(s)$ is equal to $R_F/(R_F + R_G)$ we obtain

$$V_{out}\Big|_{t(s) = \frac{R_F}{R_F + R_G}} = \frac{R_F}{R_G} (V_2 - V_1) \qquad [5\text{-}132]$$

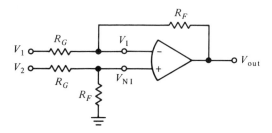

FIG. 5-53. Ideal operational amplifier in the differential mode.

In this case, the circuit subtracts one signal voltage from the other, producing an amplified version of the difference at the output.

Notice that there is no virtual ground at the input to the ideal amplifier in this circuit. The feedback in the circuit forces V_I to be equal to V_{NI}. Thus, in effect, the feedback causes a part of the input voltage V_2 to appear at the inverting input terminal of the ideal amplifier, and this voltage is coupled by R_G into the source of the input signal V_1. This interaction between the two signals may have undesirable consequences. For example, if the source of V_1 has any nonlinearity, intermodulation between the two signals will result.

It is useful to distinguish between three different modes of operation when discussing a differential amplifier. The first occurs when either V_1 or V_2 is equal to zero. In this case, we speak of the *single-ended* mode of operation. From (5-131) it follows directly that if V_1 is equal to zero the differential amplifier is operating in the noninverting mode, and if V_2 is equal to zero it is operating in the inverting mode. Referring to Fig. 5-53, the second mode of operation applies when

$$V_1 = - V_2 = V_D \qquad (5\text{-}133)$$

Thus, the two input signals are equal but have opposite polarity at every instant of time. The circuit is then said to be operating in the *differential mode*, and the input signals are called *differential-mode signals*. For this case we have, from (5-131),

$$V_{out}\bigg|_{V_1=-V_2=V_D} = \left\{ t(s) + \frac{R_F}{R_G}[1 + t(s)] \right\} V_D \qquad [5\text{-}134]$$

and for the special case that $t(s)$ is equal to $R_F/(R_F + R_G)$ we obtain, from (5-132),

$$V_{out}\bigg|_{\substack{V_1=-V_2=V_D \\ t(s)=\frac{R_F}{R_F+R_G}}} = 2\,\frac{R_F}{R_G}\,V_D \qquad [5\text{-}135]$$

Finally, the third mode of operation occurs when

$$V_1 = V_2 = V_{CM} \tag{5-136}$$

In this case, the two input signals are identical both in amplitude and phase at every instant of time and the circuit is said to be operating in the *common mode*, the input signals being called *common-mode signals*. In the general case we have, from (5-131),

$$\left. V_{out} \right|_{V_1 = V_2 = V_{CM}} = \left\{ t(s) + \frac{R_F}{R_G} [t(s) - 1] \right\} V_{CM} \tag{5-137}$$

When $t(s)$ is equal to $R_F/(R_F + R_G)$ we obtain

$$\left. V_{out} \right|_{\substack{V_1 = V_2 = V_{CM} \\ t(s) = \frac{R_F}{R_F + R_G}}} = 0 \tag{5-138}$$

Hence, for this case, the common-mode input signals produce no voltage at the output of the ideal amplifier. Incidentally, the common mode of a differential amplifier generally applies to the case when the input terminals are balanced such that (5-138) applies, i.e., when the output voltage is equal to zero. We shall see in applications discussed in later chapters that $t(s)$ is often a frequency-dependent network, dimensioned such that, at a particular frequency, $t(s)$ is equal to $R_F/(R_F + R_G)$. As a result, the particular frequency at which this condition occurs will be rejected, whereas frequencies on either side of it will appear, more or less unattenuated, at the output.

In another typical application, the differential amplifier can be used as a sensitive bridge null detector, as shown in Fig. 5-54. The bridge is balanced when $V_1 = V_2$, in which case the output voltage is zero. The two resistor pairs R and R_F are part of the bridge circuit. For example, R_F may be chosen such that the parallel combination of R_F and R is equal to R_G. When the bridge is balanced, we have common-mode voltages at the input terminals resulting in zero output voltage. When the bridge is unbalanced, the differential input signals cause an amplified voltage to appear at the output.

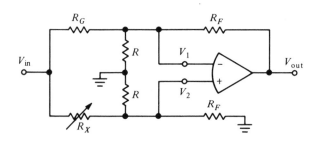

FIG. 5-54. Differential amplifier used as null detector in bridge network.

Integrator Replacing the feedback resistor R_F of the ideal amplifier in the inverting mode by a capacitor, as shown in Table 5-7, we obtain an integrator. Here again the input current to the ideal amplifier is zero and the feedback through the capacitor C forces a virtual ground to exist at the inverting input terminal. Thus, the voltage across C is simply the output voltage V_{out}. This voltage can be expressed as

$$V_{out} = -\frac{1}{C} \int_0^t I_1 \, dt + V_{out}(0) \tag{5-139}$$

Because of the virtual ground

$$I_1 = \frac{V_{in}}{R} \tag{5-140}$$

Thus,

$$V_{out} = -\frac{1}{RC} \int_0^t V_{in} \, dt + V_{out}(0) \tag{5-141}$$

The amplifier, therefore, provides an output signal which is proportional to the integral of the input voltage. Using appropriate additional circuitry it is possible to set the initial condition $V_{out}(0)$ to any desired value. Circuits of this type are commonly used for integration in analog computers.

Differentiator If the resistor R and the capacitor C of the integrator are interchanged we obtain a differentiator. Because of the virtual ground at the inverting terminal we have

$$I_1 = C \frac{dV_{in}}{dt} \tag{5-142}$$

and since

$$V_{out} = -RI_1 \tag{5-143}$$

we obtain

$$V_{out} = -RC \frac{dV_{in}}{dt} \tag{5-144}$$

Thus, the output is proportional to the time derivative of the input.

5.6 CONTROLLED SOURCES USING THE OPERATIONAL AMPLIFIER

The concept of controlled sources was introduced in Chapter 3, (see Section 3.1.5). There it was pointed out that these ideal active elements are very useful in the modeling of many electronic circuits and devices. This fact has been illustrated in the present chapter, where we have shown that such network

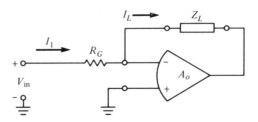

FIG. 5-55. Voltage-controlled current source using an operational amplifier: floating load.

elements as NICs or gyrators can be derived directly from their controlled-source equivalents. One reason for this is that the controlled source is the most direct circuit realization of a network element that is described in terms of matrix parameters (e.g., admittance or transmission matrix). The elements of controlled sources are, after all, generally expressed in terms of matrix parameters. Whereas controlled sources are very often used to represent a circuit in an intermediate and possibly auxiliary form that links the analytical matrix parameters of the circuit to its final realization, it is, nevertheless, often very useful to realize controlled sources themselves in circuit form. Operational amplifiers are very useful in this respect and, in fact, can be used to realize any one of the four basic controlled sources directly, as we shall see in what follows.

Consider, for example, the circuit shown in Fig. 5-55. At first glance, this circuit will be taken for an operational amplifier in the inverting mode. However, it will be noticed that the feedback resistor R_F has now been replaced by the load Z_L. Since the input current to the ideal amplifier is zero we have for the load current

$$I_L = I_1 \qquad (5\text{-}145)$$

Furthermore, since the feedback forces a virtual ground to exist at the inverting input to the ideal amplifier we have

$$I_L = \frac{V_{in}}{R_G} = gV_{in} \qquad (5\text{-}146)$$

where

$$g = \frac{1}{R_G}$$

Thus, the load current I_L is independent of the load Z_L. Seen from the terminals of Z_L, the circuit therefore appears to be an ideal voltage-controlled current source characterized by the admittance $g = 1/R_G$. Incidentally, if we consider the special case that Z_L is a capacitor, then we have the situation of an ideal current source loaded by a capacitor which, of course, is an integrator, as was discussed in the previous section.

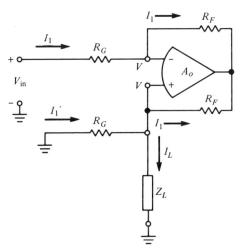

FIG. 5-56. Voltage-controlled current source using an operational amplifier: grounded load.

One of the disadvantages of the current source shown in Fig. 5-55 is that neither load terminal is grounded, although one end is at virtual ground. A voltage-controlled current source with one end of the load impedance Z_L connected to true ground is shown in Fig. 5-56. Since the input current to the ideal amplifier is zero, the same current I_1 flows in the two resistors connected to the inverting input terminal. Furthermore, since the feedback forces the potential difference between the two input terminals to zero the same voltage exists across the two feedback resistors R_F. Hence, both feedback resistors carry the same current I_1. Designating by I'_1 the current which is in the resistor R_G connected to the noninverting input terminal of the amplifier, the load current is obtained as

$$I_L = I'_1 - I_1 = -\frac{V}{R_G} - \frac{(V_{in} - V)}{R_G}$$

$$= -\frac{V_{in}}{R_G} = -gV_{in} \tag{5-147}$$

Here again, the load current is independent of the load Z_L and the circuit, seen from the load terminal, appears as an ideal voltage-controlled current source characterized by the admittance $-g = -1/R_G$. If the input voltage V_{in} and the ground connected to R_G at the noninverting terminal in Fig. 5-56 are interchanged, the only effect is to reverse the direction of the load current. Thus, the current inversion indicated in (5-147) does not occur. With the same reasoning as above, if we replace the load Z_L by a capacitor C, we have a capacitor loaded by an ideal current source or an integrator with positive or negative polarity, depending on whether the input signal is fed into the R_G at the inverting terminal or the R_G at the noninverting terminal. For

TABLE 5-8. CONTROLLED SOURCES USING THE OPAMP.

Type	Controlled source	Opamp realization

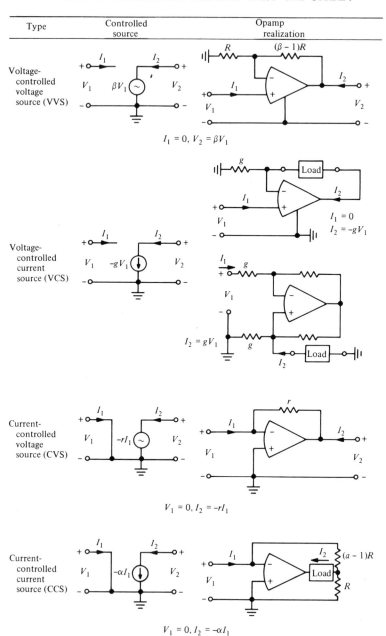

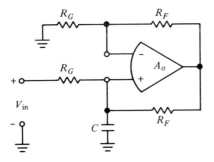

FIG. 5-57. Integrator with grounded capacitor.

example, a positive integrator with a capacitor terminal grounded is shown in Fig. 5-57.

In much the same way as was done for the voltage-controlled current source, the other three controlled sources can be derived using the ideal operational amplifier. The resulting configurations together with the defining equations have been listed in Table 5-8. Notice that the voltage-controlled voltage source is nothing but the operational amplifier in the noninverting mode (see Fig. 5-52). Likewise, the current-controlled voltage source is nothing but the operational amplifier in the inverting mode with the input resistor R_G (see Fig. 5-51) set equal to zero. The current-controlled current source in the table has a floating load, i.e., neither of its terminals is grounded. One way of obtaining a grounded terminal is by adding another operational amplifier, e.g., as described by Antoniou.[24]

5.7 SUMMARY

1. The *nonprincipal* diagonal of the transmission, or $ABCD$, matrix of an ideal *converter* is zero.

2. The input impedance Z_{in} of a converter that is loaded by an impedance Z_L is $Z_{in} = (A/D) Z_L$.

3. If $A = - k_1$ and $D = 1/k_2$ we have a voltage-inversion negative-impedance converter (VNIC); if $A = k_1$ and $D = - 1/k_2$ we have a current-inversion negative-impedance converter (CNIC). In both cases k_1 and k_2 are real constants and $Z_{in} = - k_1 k_2 Z_L$.

4. The NIC exhibits power gain. In terms of its h parameters, the output power P_{out} is related to the input power P_{in} by the relation $P_{out} = -(h_{21}/h_{12})P_{in}$.

24. A. Antoniou, Simulation of controlled sources using operational amplifiers," *IEEE Trans. Educ.* E-12, 71–74 (1969).

5. The *principal* diagonal of the transmission, or $ABCD$, matrix of an ideal *inverter* is zero.

6. The input impedance Z_{in} of an ideal inverter that is loaded by an impedance Z_L is $Z_{in} = B/CZ_L$.

7. If $B = 1/g$ and $C = g$, where g is the gyration conductance, we have an ideal gyrator. The corresponding input impedance is $Z_{in} = 1/g^2 Z_L$.

8. If $B = 1/g_1$ and $C = g_2$, the loaded ideal gyrator has the input impedance $Z_{in} = 1/g_1 g_2 Z_L$.

9. If $g_1 = g_2 = g$, the ideal gyrator behaves like a lossless passive element even though it is not reciprocal.

10. Depending on whether g_2 is larger or smaller than g_1 the ideal gyrator behaves like an active or a passive lossy device, respectively. In either case (i.e., when $g_1 \neq g_2$) we refer to an ideal *active* gyrator.

11. The input impedance of a grounded ideal gyrator loaded by a capacitor C is $Z_{in} = sC/g_1 g_2$. It therefore simulates a grounded inductance $L_{eq} = C/g_1 g_2$.

12. In order to simulate a floating inductance two grounded gyrators coupled by a grounded capacitor are required.

13. The input and output short-circuit admittances of a nonideal gyrator are not equal to zero, i.e., $y_{11} \neq 0$, $y_{22} \neq 0$.

14. The maximum Q of an inductance L_{eq} simulated by a grounded non-ideal gyrator and capacitor C is $Q_m = g_1 g_2 / 2 \sqrt{y_{11} y_{22} \Delta y}$ where $\Delta y = y_{11} y_{22} + g_1 g_2$. The corresponding inductance is frequency-dependent; its value is $L_{eq}(\omega) = g_1 g_2 C / [\Delta y^2 + (\omega y_{11} C)^2]$.

15. The nullator is a two-terminal element which is simultaneously an open and a short circuit ($V = I = 0$).

16. The norator is a two-terminal element for which the voltage and current across its terminals are arbitrary.

17. Transistors and operational amplifiers can be represented by nullator–norator pairs. The latter are often referred to as nullors.

PART II
FUNDAMENTALS OF HYBRID INTEGRATED CIRCUITS

CHAPTER

HYBRID INTEGRATED CIRCUIT TECHNOLOGY

INTRODUCTION

In Part I we attempted to cover those aspects of active network theory that are particularly useful for the design of linear active RC networks. This entailed mainly the characterization and analysis of linear active RC networks, sensitivity considerations pertinent to them, and a review of general network elements. No attention was given to physical realization.

In the present era of microelectronic systems, it very likely goes without saying that when we consider the design of linear active RC networks it is with a view to realizing them in integrated form of one kind or another. This after all is the reason for abandoning the inductor in the first place, a move that, it should be remembered, we make reluctantly: its replacement by a substitute amenable to integrated technology is by no means trivial, nor is there only one clear-cut and unique way of accomplishing it.[1] Nevertheless, linear active RC networks certainly represent one of the most practicable and economically feasible methods of designing inductorless networks, since many of them are compatible with *hybrid* integrated circuit fabrication techniques. It is with these techniques that we shall concern ourselves in this chapter. The reason that we emphasize the word " hybrid " when discussing integrated circuit technology in the present context (while the industry at large is mainly concerned with *monolithic silicon* integrated circuit techniques, at least for digital applications) will become clear after we have discussed the

1. G. S. Moschytz, Inductorless filters: a survey, *IEEE Spectrum*, August, 1970, pp. 30–36; September 1970, pp. 63–75.

329

various integrated circuit techniques available to us, and examined them with respect to their suitability for linear active network design.

With that we have arrived at the actual topic of this chapter. We shall discuss here the three main technologies of concern to us: (1) thin-film circuits, (2) thick-film circuits, and (3) monolithic silicon integrated circuits. The two first circuit types comprise passive RC components, in contrast to monolithic circuits, which are active. There is every indication that in order to design useful and economical linear active networks there is no alternative but to use *integrated* circuit techniques (as opposed, say, to discrete-component implementation); therefore the designer should have a familiarity, at least in broad terms, with the processes involved in applying them. In addition, and more important still, the designer must be well aware of the characteristics, i.e., the capabilities and the limitations, of the resulting network elements, since, as we shall see later, these characteristics should influence the approach to his design.

For the three technologies mentioned above, we shall briefly discuss the typical process steps involved in their fabrication. Our concern here will be more conceptual than detailed; as linear active network designers we are not generally required to become involved in the actual fabrication of network components and devices. A general familiarity with the fabrication processes should, however, enable us more easily to specify our components requirements in terms both understandable to the component manufacturer and reasonable with respect to the capabilities of the technology used. With that we shall be helping to bridge a gap in communications between the integrated circuit manufacturer and user, at least as far as linear networks are concerned. This gap has widened as the complexity and degree of specialization in the fields of device technology and network theory has increased.

After the process steps, we shall discuss the characteristics of the components emerging from the three integrated circuit techniques mentioned above. In doing so, we are serving the communications flow from the component manufacturer to the designer. It is vital for the network designer to know the limitations and the advantages of the various network elements at his disposal. He can then modify his design approach accordingly and create a design that is optimized with respect to the selected technology.

Linear hybrid integrated circuits have come to mean the combination of monolithic silicon devices (generally elements providing gain in one form or another) with thin- or thick-film RC components. In linear network applications the circuits must be designed such that the passive film components rather than the monolithic integrated devices determine the stability of the network parameters (e.g., resonant frequency, bandwidth, and gain). This is because film, and in particular thin-film, components of extreme stability are readily available, whereas the stability of passive components realized by monolithic integrated circuit techniques is, by comparison, poor. The resulting integrated active networks are primarily dependent on the technology used to realize the *passive* components associated with them. Circuit design techniques providing the desired high degree of independence from variations in the characteristics of the active devices are discussed in *Linear Intergrated Networks: Design.*

Because of the importance of the passive component characteristics in linear active network design, we shall, in the following, first discuss thin-film *RC* networks, since they provide components of the highest available stability. This will be followed by a discussion of thick-film components, which may be somewhat inferior in quality but require less complex processing equipment and are therefore less costly to fabricate.[2] Finally we shall discuss the characteristics of monolithic silicon devices, which are available in the form of gain blocks with wide tolerances whose network characteristics and stability must be controlled by added passive components. The fabrication of the monolithic devices is, in fact, of least interest to the network designer, since he has little, if any, access to the internal components of the device. His interest is therefore generally limited to its terminal characteristics, which are summarized in the manufacturer's specification sheets.

6.1 THIN-FILM TECHNOLOGY

The process sequence by which film material with a thickness varying between 50 Å to 20,000 Å or 0.2 to 80.0 microinches (see Table 6-1) is deposited by vacuum or vapor methods onto a substrate (e.g., glass or ceramic), in order to fabricate resistors and/or capacitors is referred to as thin-film technology. This technology is distinct from thick-film technology, which comprises the deposition by screen-and-fire methods of film material varying in thickness between 125,000 Å (0.5 mil) and 625,000 Å (2.5 mils). However, a more fundamental distinction can be made between thin and thick films that has nothing to do with the final thickness of the films.[3] The distinction is, rather, in the manner by which the film is formed. A "thin" film is one which is grown *in situ* from either a vapor or liquid (e.g., evaporation, plating). Thus the film is grown by a process involving individual atoms or molecules. This growth process is critical because it determines the properties of the film which is formed. Thick films, on the other hand, are formed using a bulk material which is screened, sprayed, spun, etc., onto a substrate. The properties of thick film are not so much dependent on the manner in which the material is applied as on the properties of the starting material and on various treatment steps after deposition. Thus, for example, because of the method in which it is formed, the thin glaze deposited on ceramic substrates (to be described later)

---

2. Any general comparison of thick- and thin-film components, be it on the basis of quality or equipment cost, is always debatable, since it depends directly on the application and the production quantities in question. More will be said on this matter later.
3. W. H. Orr, Bell Telephone Laboratories, private communication.

TABLE 6-1. CONVERSION CHART

Dimensions	Inch (in.)	Mil	Microinch	Meter (m)	Millimeter (mm)	Micron (μm)	Ångstrom (Å)
Inch (in.)	1	10^3	10^6	25.4×10^{-3}	25.4	25.4×10^3	2.54×10^8
Mil	10^{-3}	1	10^3	25.4×10^{-6}	25.4×10^{-3}	25.4	2.54×10^5
Microinch	10^{-6}	10^{-3}	1	25.4×10^{-9}	25.4×10^{-6}	25.4×10^{-3}	2.54×10^2
Meter (m)	39.4	39.4×10^3	39.4×10^6	1	10^3	10^6	10^{10}
Millimeter (mm)	39.4×10^{-3}	39.4	39.4×10^3	10^{-3}	1	10^3	10^7
Micron (μm)	39.4×10^{-6}	39.4×10^{-3}	39.4	10^{-6}	10^{-3}	1	10^4
Ångstrom (Å)	3.94×10^{-9}	3.94×10^{-6}	3.94×10^{-3}	10^{-10}	10^{-7}	10^{-4}	1

is often referred to as a "thick" film while the heavy gold plating used as conductor material is called a "thin" film. "Thin" films *tend* to be thin for many reasons, one of which is the fact that it normally takes a long time to form a thick deposit by atomistic or molecular processes. "Thick" films *tend* to be thick because it is difficult to form an extremely thin uniform layer using a bulk material. Both technologies are used to fabricate the same types of component (i.e., resistors and capacitors) but the basically different processing methods involved provide components with different characteristics. Thick-film technology will be discussed in Section 6.2.

Numerous methods of thin-film deposition have been investigated. However, the two methods that are most prominently used are cathode sputtering and vacuum evaporation, because they offer the most convenient control of the dominant process parameters and, consequently, film properties. Nickel–chromium, or nichrome, as it is commonly called,[4] tantalum nitride, and tantalum oxynitride head the list of resistive films, with aluminum and gold being used for the interconnecting conductors. Most often the deposition processes involve coating the entire substrate surface with a thin uniform layer of film, and selectively etching away the film in unwanted areas. Vacuum evaporation is used for the deposition of nickel–chromium films; cathode sputtering is normally used for refractory, that is high-melting-point (between 1500°C and 3000°C) materials such as tantalum films, since it provides a better control of the film deposition and its properties.

The manufacture of thin-film circuits requires the deposition of stable chemical compounds or elements on inert substrates. Thin-film resistors are typically of the order of a few hundred to a few thousand Ångstrom units thick and require a uniform substrate surface roughness that is small compared to this figure. In addition, the substrate must be an electrically insulating material, chemically stable to water and acids, and be able to withstand the temperatures (100°C to 450°C) either required or generated during the deposition of the films. A high-quality low-alkali-content glass has been found to meet these requirements adequately. However, to provide a more rugged package and to improve the thermal conductivity of the substrate in order to minimize hot-spot temperatures, ceramics, and for special circuits sapphire plates, are frequently used.

Materials for resistor films must adhere well to the chosen substrate, be stable, be compatible with materials used for other parts of the circuit and with photoetch techniques of pattern generation, and have as high a resistivity as possible, consistent with such other electrical requirements as a low, or at least controlled, temperature coefficient of resistance and low noise. Sputtered tantalum films and vacuum-evaporated nickel–chromium films have so far been found to meet these requirements best. This is why they are most widely

---

4. Nichrome was originally the trade name for a nickel–chromium–iron thin-film resistor material, but has since been generally adopted to designate the most commonly used thin-film resistor made of a 4:1 ratio nickel–chromium compound.

used for thin-film resistors. Of course the choice of material and technology depends on the market to be supplied. Tantalum-film resistors can be adjusted very precisely by an anodization process which at the same time produces a hard, protective oxide coating providing a considerable degree of immunity from electrochemical corrosion. There is no comparable way of providing a similar degree of protection for nickel–chromium films, and their stability is not as good. Tantalum technology is therefore preferred for the production of very stable, high-precision resistors. However, the nickel–chromium process is in some ways simpler. This is partly because, with nickel–chromium resistive films, conductive films can be deposited during the same vacuum cycle, whereas in the case of tantalum resistive films this is undesirable because of the more stringent requirements on their purity and the necessity of maintaining accurate doping levels during deposition. As a result, both sputtering (for the resistor films) and evaporation (for the conductor paths) are generally used for tantalum, which in turn increases the necessary capital investment. In addition, solutions for etching tantalum contain hydrofluoric acid, which attacks glass and glazed ceramic substrates, so that the etching process requires very careful control.

A thin-film electronic circuit requires the generation of a pattern of conductors and components. This can be achieved either in an additive or subtractive manner. In the former process, materials for the various components are deposited sequentially through a series of mechanical masks; in the latter, unwanted material is removed after deposition. It is possible to construct vacuum systems in which the mechanical masks can be changed without opening the evaporation chamber to the atmosphere and so produce all the thin-film components in a single vacuum cycle. However, mechanical masks have the disadvantage that the area covered by the mask is quite limited, since the masks have to be mechanically rigid while at the same time being thin to minimize shadow effects. A further disadvantage is the mechanical complication required to shift and accurately register a series of masks which must be held in uniform intimate contact with the substrate to obtain good edge definition.

Subtractive pattern generation has none of the disadvantages of mechanical masking, but when used with multilayer films places a selective etching requirement on the materials that can be used to form the circuit. The process is a refined version of the normal printed-wiring procedure and involves the removal of unwanted material using photoresist and -etch procedures. The advantages of the process are that only a single vacuum cycle is required to deposit most if not all of the materials[5] required for the circuit and that, in addition, large areas can be exposed and etched at a time, permitting the economies of a multiple step-and-repeated circuit pattern to be realized.

---

5. Tantalum films are normally sputtered in a separate operation.

In what follows we shall go through the basic thin-film processes described above in more detail. Because tantalum is presently the most versatile thin-film material for passive integrated circuit components we shall, in discussing thin-film processing, use tantalum technology as our main example. Its versatility comes about because tantalum film possesses the unusual advantage of being convertible to both resistors and capacitors. Besides, using tantalum to illustrate the basic thin-film processes is by no means a limitation, since the process steps required for tantalum films encompass most of the possible steps used for any other thin-film technology. Thus tantalum serves the illustrative purposes of this chapter, very well.

6.1.1 The Four Basic Processes in Thin-Film Technology

Thin-film processing can be broken down into four basic steps; (1) sputtering, (2) anodization, (3) evaporation, and (4) pattern generation. Let us look at these four basic steps in greater detail.

Sputtering Sputtering or, more accurately, glow-discharge sputtering (also sometimes called cathode sputtering or impact evaporation) is a method by which a refractory material, i.e., a material possessing a very high melting point (e.g., tantalum, tungsten, molybdenum), can be deposited in a well controlled way onto a carrier or substrate such as glass or ceramic. Figure 6-1 shows a sputtering bell jar. A small amount of argon is added to the evacuated chamber containing a tantalum cathode and the glass (or ceramic) substrate. When a negative potential is applied to the cathode, a plasma of electrons and positive argon ions forms. The ions bombard the cathode, dislodging tantalum atoms. These condense as a very adherent film on the substrate. Uniform film thicknesses are obtained over relatively large substrate areas because the sputtering source is broad and flat. Sputtering is especially suitable for refractory materials such as tantalum because it is essentially a " cold " process; despite the exceptionally low vapor pressure and high melting point of tantalum, for example, deposition is achieved with the cathode remaining at between 400°C to 500°C, which is well below the evaporation temperature of the metal. Depositing refractory materials by evaporation, on the other hand, would require extremely high temperatures to raise the vapor pressure of the refractory materials. Thus, depositing tantalum by evaporation might require temperatures in the order of 1500°C to 3000°C resulting in a possible interaction with the crucible and other adjacent materials.

One of the big advantages of the sputtering process is that various useful forms of film material can be produced simply by altering the sputtering process in appropriate ways. For example, by controlling the sputtering voltage or chamber pressure appropriately, at least three useful forms of tantalum can be produced that have a wide variety of electrical properties;

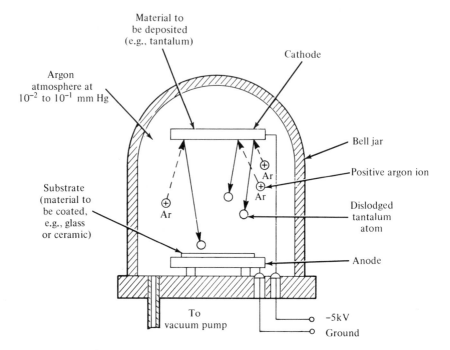

FIG. 6-1. Glow-discharge sputtering (also called cathode sputtering or impact evaporation) in a belljar.

these in turn permit a great deal of freedom in component design. One form, called beta tantalum film, is widely used because the films it forms are relatively free of stresses. Beta tantalum is a crystal phase with a tetragonal structure which is stable only at low temperatures. Consequently it is found in grown films but has never been observed in bulk tantalum. Another thin-film form, called body-centered cubic (bcc) tantalum, because its atoms are arranged in cubic cells with one atom in the center of the cell, has properties similar to those of bulk tantalum, making it most suitable for applications requiring low resistivity. Thus, for example, bcc tantalum could be used for the base electrode of thin-film capacitors, where its low series resistance means better high-frequency characteristics. Unfortunately, bcc tantalum films tend to be highly stressed, which adversely affects their adherence. Finally, a low-density form of tantalum film (obtained by sputtering at low energies) is used primarily in moderate-to-high-value resistors. Its low density and high resistivity are due to many fissures (about 40 Å wide) in its physical texture.

Both tantalum compounds and tantalum "solid solutions" are formed as thin films by a process called reactive sputtering, which consists of adding a reactive gas to the argon in the sputtering chamber. Each tantalum compound has its own unique crystal structure. Solid (or "interstitial") solutions of tantalum, on the other hand, are formed by adding such a small amount of reactive gas that the tantalum retains its original lattice structure. The gas can therefore, be thought of as having "dissolved" in the tantalum. Only the amount of reactive gas added to the vacuum chamber determines whether a compound or an interstitial solution is formed. Reactive sputtering thus allows a precise control to be maintained over such film characteristics as resistivity, temperature coefficient, and electrical stability.

Films with a wide variety of properties can be produced by altering the amount of reactive gas in solution. For example, oxygen in interstitial solution can be used to control precisely the temperature coefficient of tantalum thin films. In the extreme, adding relatively large amounts of oxygen during sputtering produces the compound tantalum pentoxide, a good insulator used for capacitor dielectrics. However, the preferred method of making tantalum pentoxide dielectrics is anodization, which we shall describe later.

In cosputtering, the tantalum cathode is partially covered with wires or strips of a second metal, such as molybdenum, tungsten, or vanadium, to form thin films of alloys. Various other alloy thin films have been studied as possible improvements over the tantalum thin film presently used in capacitors. These new alloys may some day increase the yield (the proportion manufactured that are acceptable for use) of future capacitor designs. Cosputtered tantalum–aluminum alloy films can be used to form resistors with higher sheet resistances and improved stability.

A summary of the various ways of sputtering thin-film material on a substrate using tantalum as the basic material, is given in Table 6-2.

Anodization Anodization is the process by which conductive tantalum film is converted electrolytically into the insulating compound tantalum pentoxide (i.e., Ta_2O_5). The process is generally used to stabilize and "trim," i.e., adjust to value, tantalum thin-film resistors by converting the upper layer of the conductive resistor material into tantalum oxide and thereby decreasing its thickness in a well controlled manner. A schematic of this electrolytic process is shown in Fig. 6-2. The tantalum thin film is given a positive

TABLE 6-2. TANTALUM FILMS OBTAINABLE BY CONTROLLED SPUTTERING

Manner of control	Type of film
Process control (sputtering voltage, chamber pressure, etc.)	Physical structures, e.g., β-tantalum, bcc tantalum, low-density tantalum
Codeposition (cathode material: tantalum alloy)	Alloys, e.g., tantalum–aluminum
Reactive sputtering (reactive gases added to vacuum chamber)	Compounds and interstitial solutions, e.g., nitrides, oxides, carbides, hydrides

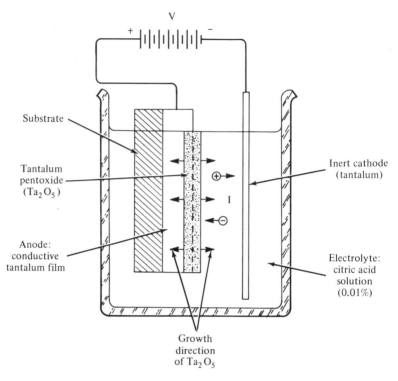

FIG. 6-2. Anodization: electrolytic conversion of conductive tantalum film to insulating tantalum pentoxide compound for resistor adjustment (trim anodization), formation of capacitor dielectric (Ta_2O_5, dielectric constant = 22), or protective coating (hard, glassy, uniform film) of thin-film devices (e.g., resistors).

potential with respect to another electrode and both are immersed in an electrolyte. As a result, an electron current flows in the external circuit and ions flow in the electrolyte. The reaction at the anode converts tantalum into tantalum pentoxide in an amount that is directly proportional to the amount of charge that has been transferred through the electrolyte. Furthermore, the voltage required to maintain a prescribed constant current is proportional to the total charge which has already passed in the solution or to the thickness of the film already formed. Since tantalum oxide is an electric insulator most of the anodic potential drop will occur across this film. The resistance of the tantalum oxide film increases as its thickness increases. Therefore, as the anodic film grows the anodizing voltage must be increased to continue the reaction at a prescribed rate. Hundreds of volts may be required to form thick anodic films. In summary, the voltage–charge dependence results from two facts: (1) it takes a certain field to pass current through the oxide already grown, so that the voltage is proportional to the thickness of the oxide; (2) the amount of oxide formed (i.e., thickness) is proportional to the charge passed,

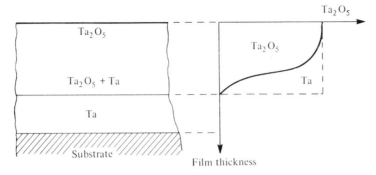

FIG. 6-3. Formation of stochiometric Ta_2O_5 and interstitial tantalum-rich Ta_2O_5 during thermal oxidation.

so that the voltage is also proportional to this charge. This can be used to control dielectric film thickness very precisely.

The tantalum pentoxide produced by anodizing forms a hard, glassy, uniform film over the resistor. This affords unusually effective protection from damage during necessary handling and from deterioration of the resistor material by heat, oxygen, or other corrosive agents. Apart from protecting thin-film resistors, anodization is mainly used to adjust their values conveniently and precisely (more will be said about trimming resistors later) and to form the dielectric material for thin-film capacitors, since tantalum pentoxide has excellent dielectric properties.

Besides being oxidized electrolytically, films can also be oxidized thermally; in this case the films are placed in an air or oxygen environment at elevated temperatures. Because of the effects of thermal inertia, precise control of this method of obtaining a reactive (i.e., oxide) film on the surface of a tantalum thin film is difficult. Thus thermal oxidation is not used to trim thin-film resistors accurately; it is used, however, to stabilize them after the anodization process. This *thermal stabilization* mechanism can be explained as follows.

Thermal stabilization is easy to understand if we consider the simplified diagram shown in Fig. 6-3. The effects of thermally stabilizing a tantalum film are shown schematically.[6] Two types of film material are thereby produced. At the top, toward the surface of the film, the pure stochiometric compound Ta_2O_5 is dominant; lower down, at the interface with the Ta, the stochiometric Ta_2O_5 is mixed with Ta atoms in an interstitial solution, resulting in so-called tantalum-rich tantalum pentoxide. The transition from the pure stochiometric Ta_2O_5 state to the tantalum-rich Ta_2O_5 state is a continuous one; it is also a stable one. Consider now the discontinuous transition of stochiometric Ta_2O_5 to Ta at the interface of an electrolytically

6. This scheme was communicated orally to the author by D. J. Sharp, Engineering Research Center, Western Electric Company, Inc., Princeton, N.J. The discussion strictly applies only to pure tantalum film. However, with other films, such as tantalum nitride (Ta_2N), the effect of thermal stabilization is very similar, if somewhat more complicated.

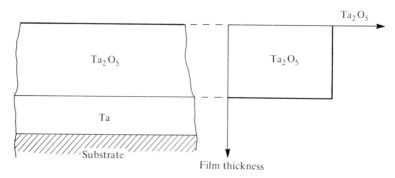

FIG. 6-4. Formation of stochiometric Ta_2O_5 resulting from anodization of Ta resistor film material.

oxidized (i.e., anodized) tantalum film as shown in Fig. 6-4. This type of discontinuous transition is unstable. By thermal oxidation (i.e., heating and aging), oxygen atoms diffuse into the tantalum at the metal–oxide interface,[7] producing oxygen-doped tantalum and tantalum-rich tantalum oxide, and the continuous transition, required for resistor stability, results. Thus, for maximum stability, thermal stabilization should always follow anodization. Because of the oxygen diffusion into the tantalum there is a small increase in the resistor value, since the conductivity of the Ta at the interface is thereby slightly reduced. Thus, to achieve a desired precision in resistance value, a final trim-anodization step may be necessary after thermal stabilization. However, the more a resistor is trim-anodized after thermal stabilization, the less stable the resistor will be. Thus, the optimum in precision and stability is achieved by anodizing close to the final value *before* thermal stabilization.[8]

Naturally, even after thermal stabilization, resistors exposed to air at room or elevated temperatures will continue to oxidize thermally at a very slow rate. This unavoidable oxidation will contribute to the overall long-term stability or aging process of the resistor. Depending on the application and the resistor in question, this may only become objectionable or even noticeable after years, if not decades. Clearly the thermal aging process will be inhibited all the more, the thicker the anodically, or thermally, grown tantalum pentoxide layer covering the tantalum resistor film material. It will be aggravated all the more, the higher the power dissipated (i.e., the greater the heat generated) in the resistor itself.

Evaporation Our third basic thin-film process, vacuum evaporation, or simply evaporation, as it is often called, is the method of thin-film deposition

---

7. Actually, oxygen diffuses into Ta if the heat treatment is in air. If the process is carried out in a vacuum, two-way diffusion takes place, in that Ta also diffuses into the oxide.
8. The increase in resistance due to thermal stabilization—typically, less than one percent—is predictable, depending on the number of hours the film is held at an elevated temperature (typically 3 to 5 hours at 250°C). It is therefore sometimes taken into account when specifying the nominal value of a resistor.

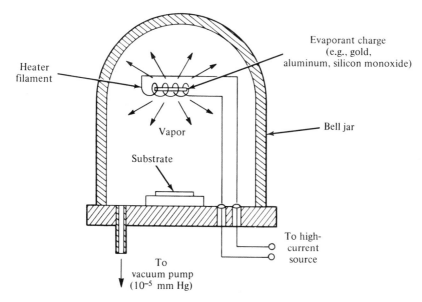

FIG. 6-5. Evaporation: deposition of thin-film conducting material and capacitor counter-electrodes by thermal activation and condensation.

commonly used for metals such as gold, copper, chromium, nickel–chromium alloys, and aluminum.[9] A schematic diagram of the vacuum evaporation process is shown in Fig. 6-5. In essence, this process consists of thermally activating an evaporant charge, and thereby transferring atoms or molecules from its surface into an extremely low-pressure gas phase. In general, the source material may be liquid or solid; in the solid case the evaporation process is often referred to as "sublimation." The source is heated until its vapor pressure reaches a high enough value for evaporation to take place at a reasonable rate. Since the substrate is considerably cooler, condensation occurs, resulting in growth of the film. Compared to sputtering, in which the deposition rates range between 50 Å to 500 Å per minute, evaporation permits much higher deposition rates of 1000 Å to 10,000 Å per minute. Thus, evaporation allows much thicker films to be deposited in a reasonable time. On the other hand, the films deposited by sputtering tend to be more uniform and, as mentioned above, permit a wider range of process control by the use of alloys and reactive gases. The vacuum required in the evaporation process is comparable to that used in sputtering (before admitting the argon), namely, on the order of 10^{-5} to 10^{-9} torr.[10] In tantalum technology, thin-

9. These metals are suitable for evaporation because at relatively low temperatures their vapor pressure is high enough for evaporation at a reasonable rate. The evaporation temperature does not always have to exceed the melting temperature. For example a sufficiently high vapor pressure for evaporation of chromium is attained at approximately 1400°C; the melting point is 1900°C. However, gold must be heated up to the same temperature (i.e., 1400°C) to obtain that same vapor pressure, but its melting point is 1063°C.

10. 1 torr (Torricelli) is equal to the pressure of 1 mm of mercury (Hg).

film conducting material and capacitor counterelectrodes are deposited by vacuum evaporation, or, as as we may also call it, thermal activation and condensation.

Pattern Generation   The fourth basic process in thin-film technology is pattern generation, the process by which a desired circuit pattern is formed on a substrate. In principle the process is similar to that used to make printed wiring circuits except that pattern lines for integrated circuits must be much finer and more precise.

There are two basic kinds of pattern generation, namely, generation by mechanical masking (additive process) or by selective removal (subtractive process). In contrast to the former method, which generally involves metal masks, the latter utilizes photolithographic techniques. A schematic of these two methods is shown in Fig. 6-6. Mechanical masking is generally used to selectively deposit film material onto a substrate; e.g., using negative masks, gold is evaporated for conducting paths and for capacitor counterelectrodes in tantalum-film technology. In general, the mechanical masks required for evaporation are made of metal, graphite, or glass. To minimize thermal expansion, graphite masks are frequently used because of their low thermal

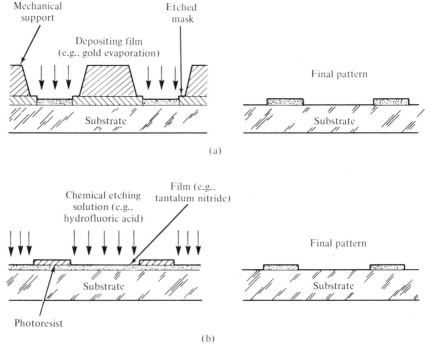

FIG. 6-6. Pattern generation. (a) Rejection masking. (b) Selective removal (photolitho-graphy).

expansion coefficient. The advantage of mechanical masks is, of course, that they can be used many times, as long as they are thoroughly cleaned at frequent intervals.

Selective removal usually involves photolithographic techniques. The film to be patterned is deposited over the entire substrate and a photosensitive "resist" then applied to the surface. The photoresist is a polymer which undergoes a chemical change when exposed to light. If it is a *positive* photoresist, the *exposed* regions can be removed in a developer solution. If it is a *negative* photoresist it is the *unexposed* regions which are removed by the developer. In either case a pattern is developed in the resist and the exposed or unexposed film on the substrate is then etched away in a suitable acid. A mild heat treatment is sometimes used to improve the etch resistance of the photoresist.

Since it is very commonly used in all types of integrated circuit technologies (including that of monolithic integrated circuits) let us illustrate the selective removal process as it may typically be used in tantalum-film technology. The first step consists of applying photoresist over the thin-film structure in a thin (1 to 2 microns) layer. Depending on the type of photoresist used, a contact print is then made for an accurate photographic positive or negative mask. In the second step the photoresist is developed, and selective chemical etching of each film, and removal of the photoresist, takes place. Typically the etchant used for tantalum is a mixture of citric and hydrofluoric acids, tantalum being resistant to most other acids. Because the photoresist mask is in intimate contact with the tantalum film it is covering (no shadow effects), this process is capable of achieving a high degree of pattern complexity and line-width precision. One of the problems that may arise with the tantalum technology, however, is that undercutting of the substrates may result because glass or glazed ceramic substrates etch more rapidly than the tantalum film itself. This is avoided by depositing a thin coating of tantalum pentoxide on the substrate before the tantalum is sputtered; compared with tantalum or glass, tantalum pentoxide is inert to hydrofluoric acid.

Both methods of pattern generation require precise photographic techniques to generate the positive or negative copy of the desired pattern. The layout of the desired circuit is first accurately cut out of a two-layer mylar material. The lower, clear mylar layer acts as a support; the upper layer, an optically opaque, generally ruby-colored mylar film is cut and peeled where the thin-film components are to be located. The resulting art master is generally 10 to 100 times larger than the final desired configuration. The cutting is customarily performed by a motor-driven precision coordinatograph, which is accurate to less than 25 microns (i.e., 1 mil). High-precision photoreduction methods are then used to obtain final positive or negative glass or film masks to the exact desired dimensions. Incidentally, most of the processes involved in the generation of thin-film circuits, from the optimization of the layout to the generation of the final masks, can be carried out by computer.

Having examined the four fundamental thin-film processes, we shall now briefly illustrate how they are used to make thin-film resistors and capacitors. Here again, we shall use tantalum technology as our prime example, because it is only with this technology that both resistors and capacitors can be obtained from the same material in thin-film form.

6.1.2 Thin-Film Integrated Components

Resistors   Figure 6-7 shows the top view and cross-section of a typical thin-film resistor. It will be clear from the figure that all four basic processes described above are necessary to generate a resistor of this kind. Tantalum nitride has been assumed as the resistor material because this is the material most commonly used in tantalum technology. Observe that the oxide layer resulting from anodization of the resistor to obtain the final desired value grows at the expense of the underlying film. It is consumption of the tantalum material itself that decreases the resistor cross section, thereby increasing the resistor value to the desired amount. Note that adjustment of a thin-film resistor by anodization, or by most other means for that matter, *increases* the resistor value. This, of course, must be taken into account in the initial design. The transition section between the film resistor and its associated conductor pattern is shown to be wider than the resistor path in Fig. 6-7. Referred to as a "dumbbell" termination, this widened section is used to

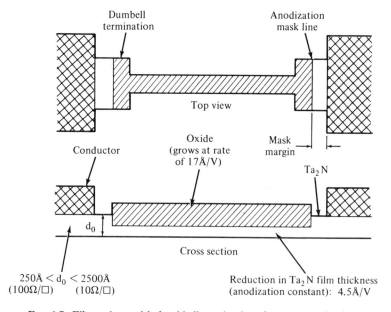

FIG. 6-7. Film resistor with dumbbell terminations for mass anodization.

minimize the effect on resistor stability of the narrow area of resistor material at each end of the resistor which must be left unanodized during processing.

Other types of tantalum film beside tantalum nitride may be used to obtain a variety of resistor properties. Tantalum nitride resistors are most commonly used because of their high stability with respect to aging and temperature. They may also be very precisely adjusted by anodization. On the other hand, by adding oxygen to the nitrogen in the reactive sputtering process ("oxygen doping"), tantalum oxynitride resistors are obtained; depending on the partial pressure of oxygen and nitrogen, a variety of temperature coefficients can be obtained. As we shall see later, this is very important for the temperature compensation of thin-film RC networks. Oxygen doping of tantalum nitride resistors is also used to provide higher-resistivity film than is obtainable with tantalum nitride alone. To provide very high specific resistivity, low-density tantalum resistors are used, as mentioned earlier.

To provide precise adjustability or tunability of resistors in a circuit, they are trim-anodized. This process is demonstrated in Fig. 6-8a. By surrounding the resistor to be anodized with a grease dam which is filled with an electrolyte (e.g., 0.01 percent citric acid) any individual resistor can be anodized by the same basic process that was illustrated in Fig. 6-2. Clearly, this method of trimming a resistor in a circuit is not very practical in the laboratory, although it can be carried out efficiently in mass production. By using a solid electrolyte, as shown in Fig. 6-8b, this problem is very effectively overcome. The solid electrolyte consists of an organic gel that can be used as a rigid carrier medium for the anodization electrolyte. The gel, when properly pattern-shaped and electrically contacted, provides a convenient tool for anodizing specific thin-film areas. If the gel is assembled into a probe as shown in the figure, a "rubber stamp" anodization technique is possible. In contrast to the method shown in Fig. 6-8a the solid electrolyte leaves a negligible residue on the resistor surface; it can therefore be used to trim a resistor during operation of the circuit.[11] Resistors can be trimmed to within fractions of a percent in this way, using the same anodization probe hundreds of times before it needs to be replaced.

In trim anodizing a thin-film resistor, the thickness of the conductive metal is uniformly reduced by the growing layer of insulating tantalum oxide. By successive anodizing and measurement of the resistor value, the trimming process can be instantaneously terminated as soon as the proper resistance value has been reached. Resistive films of tantalum, tantalum nitride, or oxygen-doped tantalum nitride are commonly made at least 5 percent lower in value than specified and are then adjusted or "trimmed" by anodizing to the precision specified. Tolerances as low as 0.01 percent can be obtained. Thus, in adjusting a tantalum thin-film resistor, which typically may be 500 Å

---

11. In AC operation, care must be taken, however, to take into account the shunting capacitance of the electrolyte. In high-precision circuits, the probe should be removed during the actual measurement.

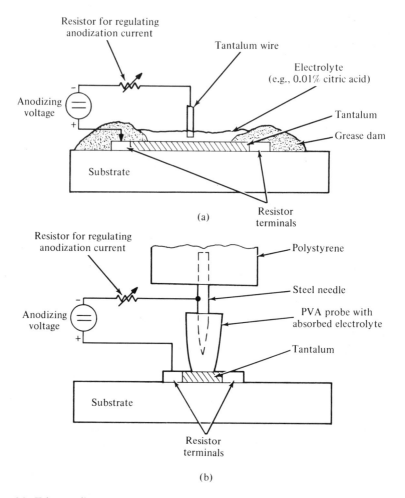

FIG. 6.8. Trim anodizing tantalum thin-film resistors. (a) Liquid electrolyte in grease dam. (b) Solid electrolyte used as anodization probe.

thick (i.e., 2 millionths of an inch), the average film thickness is controlled to within an effective tolerance of 0.05 Å.

It should be remembered that, although we used tantalum as an example for the reasons given earlier, similar process steps are required when other film materials are used, the most common of which are nickel–chromium and chromium–cobalt for thin-film resistors and aluminum or gold for interconnecting conductors. Since nickel–chromium (i.e., nichrome) film formation does not involve refractory metals, nichrome resistors are obtained by evaporation instead of sputtering. Apart from that, each of the deposition processes involves coating an entire substrate surface with a thin uniform layer of film and then selectively etching away the film in unwanted areas.

Although anodizing is the preferred method for trimming tantalum film resistors, many other methods are used for the various other film-resistor types. These methods involve changing the resistance of the film by the application of heat, by physical removal of material, or by chemical action. Heat trimming involves localized heating to adjust the resistance of the film. Heat sources include pulsed lasers, hot gas or flame, infrared lamps, and carbon arcs. Electrical trimming is very similar, in that localized heat is generated by passing an appropriate current through the resistor film. Mechanical trimming involves removal of a portion of the resistor material by milling, cutting, sandblasting, or vaporization. Equipment used in mechanical trimming can include a miniature sandblast unit, diamond scribe, electron beam, laser beam, ultrasonic scribe, arc or spark discharge probe, or simply an ink eraser. The methods most commonly used are the ultrasonic diamond scribe, the electron beam, and the laser. These are the most easily controllable methods, and the least costly. The laser, in particular, is a versatile resistor-trimming tool. The beam can be focused to a fine point (about 25 microns or 1 mil in diameter) and can burn away or vaporize portions of the resistor film. In all cases the resistance is increased by removal of resistive material. Apart from anodizing and etching, chemical trimming methods include silver painting with conductive inks and bidirectional electrochemical trimming. In the latter method hydrogen is used to reduce the resistor film (decrease resistance) or oxygen to oxidize the resistor film (increase resistance). Thus the two last-mentioned trimming methods are noteworthy in that they permit resistance to be *decreased* in value.

It is not possible, nor is it deemed necessary in the present context, to describe the large variety of trimming methods in detail. For the present suffice it to say that, one way or another, thin-film resistors can generally be fine-tuned to an accuracy equal to or better than what is possible with discrete resistors or potentiometers.

Capacitors   Figure 6-9 shows the top view and cross section of a simple thin-film capacitor. Since by far the most experience with deposited film capacitors has been obtained using tantalum films, we are again using tantalum to illustrate the fabrication of film capacitors. One of the main reasons for the use of tantalum for thin-film capacitors is that a reasonably high capacitance per unit area can be obtained with this material while at the same time, it provides high-quality resistors. Anodically grown tantalum pentoxide, which is used as the capacitor dielectric, has a dielectric constant of 25 and high dielectric strength (6.5 million volts per centimeter). Because of this high dielectric strength very thin films (i.e., less than 1000 Å) of tantalum pentoxide can be used as the dielectric.

As shown in Fig. 6-9, the simplest type of tantalum capacitor, the tantalum–metal (TM) capacitor, comprises a tantalum pentoxide dielectric sandwiched between a tantalum electrode and a counterelectrode of evaporated gold,

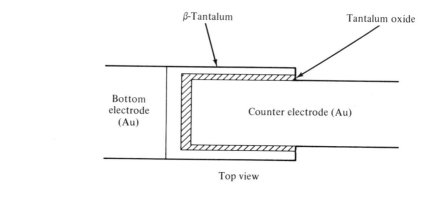

Top view

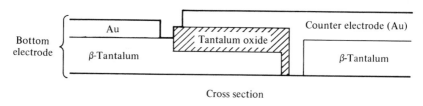

Cross section

FIG. 6-9. Thin-film capacitor using tantalum pentoxide as dielectric.

aluminum, or other metal. This type of counterelectrode adheres to the substrate well enough for some assemblies, but is not suitable for potted assemblies where the electrodes may be subjected to physical strain. Furthermore, moisture readily diffuses through the electrodes, so that, in spite of their high dielectric strength, these capacitors cannot be used in precision networks because of their high sensitivity to humidity.

This moisture sensitivity is greatly reduced by using a counterelectrode consisting of a thin layer of nichrome or titanium followed by gold or aluminum. This is called an "adherent counterelectrode," and provides capacitors for applications with stringent stability requirements. On the other hand, because of the very much improved counterelectrode adherence (the nichrome or titanium fills up cracks and crevices in the dielectric film) the dielectric strength and with it the safe working voltage of this capacitor type is decreased to about half that of the nonadherent counterelectrode type. Fortunately, though, the yield for the adherent counterelectrode type is only slightly less than for the nonadherent type.

Thin-film capacitors cannot be fine-tuned but they, as well as resistors, can be coarse-adjusted in two basically different ways. Resistors can be tapped at given intervals (e.g., in a binary series) by short-circuiting shunt paths, as shown in Fig. 6-10a. By scribing open appropriate taps mechanically (e.g.,

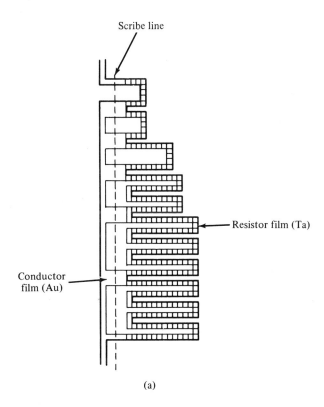

(a)

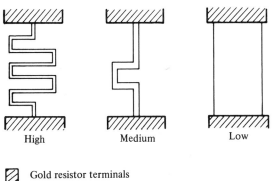

(b)

FIG. 6-10. Coarse adjustments of thin-film resistors. (a) Tapped resistor. (b) Selectively etched resistors.

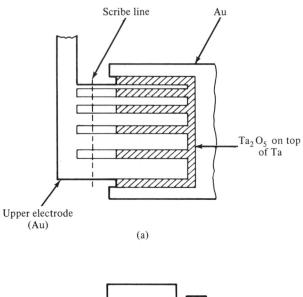

(a)

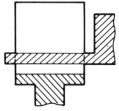

▨   Gold counterelectrode
▢   Tantalum oxide dielectric on base (tantalum) electrode
◩   Gold capacitor terminal

(b)

FIG. 6-11. Coarse adjustments of thin-film capacitors. (a) Multielectrode capacitor. (b) Capacitor with selectively etched counterelectrode.

by diamond scribing), or by laser cutting,[12] a large range of resistor values can be covered in discrete steps. Alternatively, the desired resistor pattern corresponding to a given value can be individually etched from the resistor film. An example of three different resistor values, selectively etched between the same terminal pair, as required, is shown in Fig. 6-10b. Similarly, the same two methods can be used for film capacitors. In Fig. 6-11a, a capacitor with finger electrodes is shown. By mechanical or laser scribing of appropriate

12. When using a laser it is more common to cut rectangular or L-shaped segments out of the resistor film. The resistors are then shaped appropriately, i.e., in rectangular segments rather than as meander paths. The resulting trimming process is continuous rather than in discrete steps.

fingers a wide spread of capacitor values can be obtained. Alternatively, using selective film removal, a part of the capacitor counterelectrode is etched away, leaving behind a covered area corresponding to the desired capacitance value. This is shown in Fig. 6-11b. Both methods have been used for resistors and capacitors and each has its distinct advantages. While the scribing method requires no additional etch mask and is particularly handy for laboratory use, the method using selective film removal provides circuits requiring smaller substrate area while using a basic thin-film process; it may therefore be more economical in production.

The processes discussed so far concern mainly the device engineer responsible for the fabrication of thin-film components. We shall now discuss the considerations concerning the network designer directly, namely, the characterization of those components.

6.1.3 The Characterization of Thin-Film Components

Resistors   In Fig. 6-12 we have a simplified top and cross-sectional view of a thin-film resistor. The resistor value is given by

$$R = \frac{\rho}{d} \cdot \frac{l}{w} \qquad [6\text{-}1]$$

where ρ is the specific resistivity in ohm centimers, i.e., the resistivity in bulk form, which is determined by the film material used; d is the thickness of the deposited film; l is the length of the resistor; and w is the width of the resistor. The ratio l/w is called the *aspect ratio* of the resistor and is determined exclusively by the geometry of the layout. It can be expressed as a number n of squares as indicated in the figure. Since the thickness d is constant over the total substrate, the ratio ρ/d is a constant for a given film deposition and is called the *sheet resistance* R_s. Its dimensions are ohms per square ($\Omega/\square$),

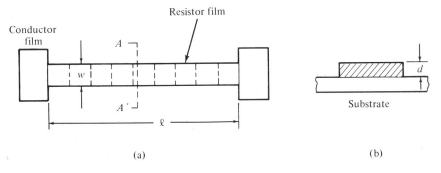

FIG. 6-12. Characterization of thin-film resistor. (a) Top view. (b) Cross section A-A'.

TABLE 6-3. SPECIFIC RESISTIVITY OF VARIOUS TANTALUM FILMS

Ta film:	bcc Ta	β-Ta	Ta_2N	Ta + O	Ta + O + Ta_2O_5
$\rho[\mu\Omega$ cm]:	24-50	180-220	240-300	40-300	250-2000

and it characterizes the specific resistivity of a given resistor film whose thickness is constant:

$$R_s = \frac{\rho}{d} \ [\Omega/\square] \qquad [6\text{-}2]$$

The sheet resistance also corresponds to the resistance of a single square of material. Referring to Fig. 6-12 this corresponds to a resistor for which $n = l/w = 1$. Therefore the total resistance of a resistor R is given by

$$R = n \cdot R_s = n \frac{\rho}{d} \qquad [6\text{-}3]$$

The specific resistivity of various tantalum film materials is listed in Table 6-3.

The two materials Ta + O and Ta + O + Ta_2O_5 represent materials obtained from reactive sputtering with oxygen, in which the amount of oxygen added to the argon is increased. As this is done, the interstitial solution Ta + O becomes a compound Ta_2O_5. The film thickness typically varies from 50 Å to 10,000 Å. Because this thickness is constant for a uniformly deposited film, resistors are generally specified in terms of sheet resistance which is a material property characterizing what is essentially a two-dimensional film. Obviously, for a given resistor film material, the lower the specified sheet resistance, the thicker the deposited film must be.

Assume, for example, that the resistor shown in Fig. 6-13 consists of tantalum nitride (Ta_2N) film and that the film thickness d is 500 Å. From Table 6-3 the specific resistivity of Ta_2N is 250 $\mu\Omega$ cm. It is convenient to convert this into ohm ångstroms; thus with the help of Table 6-1 we obtain

$$\rho = 250 \ \mu\Omega \ \text{cm} = 250 \cdot 10^{-2} \ \Omega \ \mu\text{m} = 250 \cdot 10^2 \ \Omega \ \text{Å} \qquad (6\text{-}4)$$

The corresponding sheet resistance then results as

$$R_s = \frac{\rho}{d} = 50 \ \Omega/\square \qquad (6\text{-}5)$$

Consequently, with the aspect ratio $n = 3$ the value of the resistor shown in Fig. 6-13 is 150 Ω.

It is customary for the network designer to specify a given resistor film material by its sheet resistance rather than by its specific resistivity and thickness. This leaves the question of film thickness and possible dopants (in the case of reactive sputtering) to the materials experts. Knowing the sheet resistance it is a simple matter to lay out a resistor to a given value with the

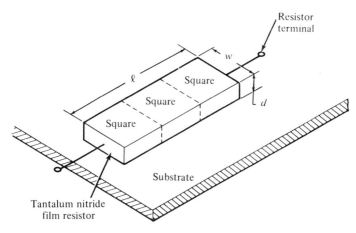

FIG. 6-13. Schematic diagram of tantalum film resistor with an aspect ratio $n = l/w = 3$.

commonly used resistor meander pattern as shown in Fig. 6-14. Such a pattern compresses a film resistor with a high square count into a configuration utilizing the substrate area more efficiently than a straight line; furthermore, a meander pattern can more readily be adapted to the available substrate area. To a good approximation, each corner of the meander pattern can be considered one-half square. Thus, assuming the resistor shown in Fig. 6-14 has a sheet resistance of 100 $\Omega/\square$, the total resistance is $[33 + (14/2)]\square$ $\times$ 100 $\Omega/\square = 4000$ Ω, where the dumbbells each contribute approximately half a square to the total resistance.

The sheet resistances of frequently used resistor films are listed in Table 6-4. Included in this Table are the temperature coefficient of resistance and the stability of the material, both of which are important to the network designer; the method of deposition is also shown.

Capacitors   Consider the schematic of a thin-film capacitor shown in Fig. 6-15. The capacitor corresponds to a planar configuration whose capacitance value is given by

$$C = A \frac{\varepsilon_r \varepsilon_0}{d} \qquad [6\text{-}6]$$

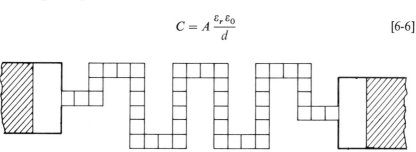

FIG. 6-14. Resistor meander pattern.

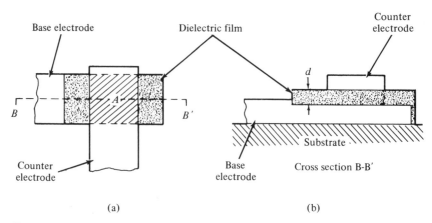

FIG. 6-15. Characterization of thin film capacitor. (a) Top view. (b) Cross section B-B'.

A is the active area of the capacitor (overlapping electrodes), it is determined by the geometry of the layout; ε_0 is the permittivity of free space, which is equal to 8.85×10^{-14} F/cm; ε_r is the dielectric constant relative to that of free space;[13] and d is the thickness of the dielectric film. This thickness is determined by the dielectric strength and, with tantalum-film capacitors, by the anodization voltage.

For tantalum pentoxide the thickness typically lies between 3000 Å and 3500 Å. As a rule of thumb, the dielectric breakdown voltage is typically equal to two-thirds the anodizing voltage.[14] The relative dielectric constant as well as various other useful parameters of film materials commonly used for film capacitors are listed in Table 6-5.

In designing thin-film capacitors, it is useful to define the term *capacitance density* C/A in farads per cm^2 or more often μF/cm^2. This is the available capacitance per unit area and characterizes the dielectric material and

TABLE 6-4. PROPERTIES OF TYPICAL THIN-FILM RESISTOR MATERIALS

Material	Property			
	Sheet resistance ($\Omega/\square$)	Temperature coefficient of resistance TCR (ppm/°C)	Stability at 150°C (%/1000 hr.)	Method of deposition
Nichrome	100–250	± 100	<0.2	evaporation
Chromium	100–500	± 300	<0.5	evaporation
Tantalum	10–100	± 200	<1.0	sputtering
Tantalum nitride	10–100	-100 ± 20	<0.1	sputtering

13. The relative dielectric constant of a vacuum is unity.
14. The recommended maximum operating voltage will, of course, be lower; it may, typically, be less than one-fifth the anodizing voltage.

TABLE 6-5. PROPERTIES OF TYPICAL THIN-FILM DIELECTRIC MATERIALS

Material	Dielectric constant ε_r	Dissipation factor tan δ'	Property Temperature coefficient of capacitance TCC (ppm/°C)	Voltage breakdown	Dielectric strength ($\times 10^6$ V/cm)	Capacitance density (μF/cm^2)	Thickness range (Å)
Silicon monoxide (SiO)	5–7	0.01–0.03	150–400	5–100	1–2	0.001–0.015	3000–40,000
Silicon dioxide (SiO$_2$)	3–4	0.004–0.04	100	50–200	3	0.002–0.02	800–10,000
Tantalum pentoxide (Ta$_2$O$_5$)	20–27	0.002–0.006	180–220	30–150	3–4	0.03–0.2	1000–6,000
Titanium dioxide (TiO$_2$)	30–100	0.01–1.0	200–800	25–90	0.3–1	0.1–1.0	1000–2,000
Aluminum oxide (Al$_2$O$_3$)	8–10	0.2–0.24	200–300	25–120	2–4	0.03–0.25	400–2,500

dielectric film thickness. Thus with (6-6) and the dielectric film thickness d given in centimeters we have

$$\frac{C}{A} = \frac{8.85 \times 10^{-8}}{d} \varepsilon_r [\mu F/cm^2] = \frac{0.0885}{d} \varepsilon_r [pF/cm^2] \qquad [6\text{-}7]$$

The capacitance density is proportional to the basic material constant ε_r and inversely proportional to the dielectric film thickness d. It characterizes the dielectric property of thin-film capacitor material, much as sheet resistance characterizes resistor film material. Typical values of capacitance density for commonly used dielectric materials are included in Table 6-5. To calculate the typical capacitance density of a tantalum pentoxide dielectric film, for example, we proceed as follows. Assume, typically, an anodization voltage of 225 V. From Fig. 6-7 we see that the oxide film grows at a rate of 17 Å/V, so that our film is 17 Å/V $\times$ 225 V = 3800 Å = 38 $\times$ 10^{-6} cm thick. The relative dielectric constant of Ta$_2$O$_5$ is typically 25. Thus from (6-7) we obtain the available capacitance density of $[(0.085 \times 25)/(38 \times 10^{-6})]$ pF/cm^2 = 0.056 μF/cm^2. The dielectric breakdown voltage, which is approximately two-thirds the anodizing voltage, will be of the order of $\frac{2}{3} \times$ 225 V = 150 V.

Another important characteristic of thin-film as well as any other capacitors is their ohmic and dielectric loss. For thin-film capacitors these losses take on well defined quantities. There are three principal sources of loss in thin-film capacitors. These are (1) dielectric loss, (2) electrode resistance, and (3) lead-in resistance. The planar geometry of the film capacitor makes the electrode and lead-in resistances important factors in the design of a capacitor pattern. On the other hand, the dielectric loss is a property of the dielectric material, and, hence, should be considered, along with other properties, when the capacitor materials are first chosen.

Capacitor losses are normally evaluated in terms of the *dissipation factor* of the capacitor. The dissipation or loss factor (tan δ) is the tangent of the loss angle δ and is equal to the inverse quality factor (Q_C) of the capacitor.[15] Defining the loss in this manner normalizes it in terms of the capacitance value.

It is of interest to consider separately the contribution that the dielectric makes to the total capacitor loss. The dielectric loss (tan δ') is thus defined as the dissipation factor of the dielectric alone. The dielectric loss is independent of the dielectric area and thickness. For most of the dielectrics used in thin-film capacitors the dielectric loss is also independent of frequency, to a first approximation. This is an important point, since it means that the high-frequency performance of a thin-film capacitor is normally not limited by the properties of the dielectric.

---

15. Sometimes the capacitor losses are also characterized by the power factor, which is the sine of the loss angle δ. For small δ, tan $\delta \approx \sin \delta \approx \delta$.

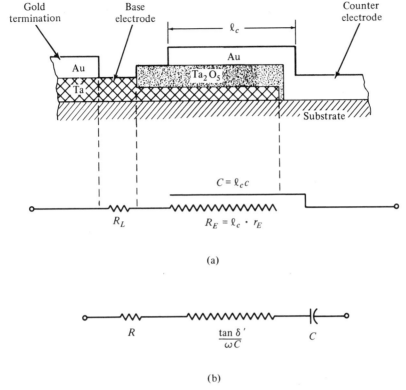

FIG. 6-16. Origin of resistive and dielectric losses in thin-film capacitor. (a) Cross-sectional view and equivalent distributed resistor–capacitor network. (b) Equivalent series RC circuit.

It is useful to represent the film capacitor by an equivalent circuit when analyzing its effect on circuit performance. Because the electrode resistance and lead-in resistance are, in effect, in series with the dielectric, it is convenient to represent the capacitor by a series equivalent circuit.

Consider the cross-sectional view of a tantalum thin-film capacitor shown with its equivalent schematic in Fig. 6-16a. It is assumed that the sheet resistance of the gold counterelectrode is negligible compared to the tantalum base electrode. This is generally true in practice because of the much lower sheet resistance of the gold counterelectrode (typically, 0.05 $\Omega/\square$) than of the tantalum base electrode (typically, 5 $\Omega/\square$). Taking the lumped equivalent of the distributed resistor–capacitor configuration shown in Fig. 6-16a we obtain the equivalent RC circuit of the thin-film capacitor, shown in Figure 6-16b. Here, R is the ohmic resistance representing the total effective series resistance provided by the electrodes (R_E) and the lead-in paths (R_L), tan δ' is the dielectric loss of the dielectric material used, and C is the value of the actual thin-film capacitance. With the lumped equivalent circuit shown in Fig. 6-16b we obtain the impedance of a thin-film capacitor as

$$Z = R + \frac{\tan \delta'}{\omega C} - \frac{j}{\omega C} \qquad [6\text{-}8]$$

We can now define an overall dissipation or loss factor $\tan \delta$ as the sum of the dielectric loss and the ohmic lead-in and electrode resistance, as follows:

$$\tan \delta = \frac{1}{Q_c} = \tan \delta' + \omega RC \qquad [6\text{-}9]$$

Clearly, the dielectric loss $\tan \delta'$, which is frequency-independent, dominates at low frequencies, whereas the resistive-loss term R dominates at high frequencies. The resistive loss R is determined by the resistivities of the electrode materials, the thickness of electrode films, and the shape of the capacitor pattern. When tantalum is used as the base electrode its relatively high sheet resistance, which is on the order of 5 $\Omega/\square$, makes the design of the capacitor pattern more critical. Thus, for high-frequency applications, three-sided capacitors, also called minimum-resistance capacitors, are very often used.

Examples of a conventional crossed-electrode capacitor pattern and of a minimum-resistance pattern are shown in Fig. 6-17. As would be expected

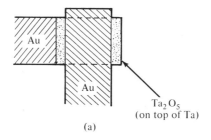

(a)

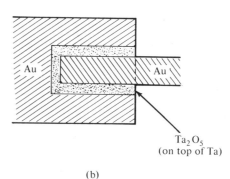

(b)

FIG. 6-17. Thin-film capacitor patterns. (a) Conventional crossed-electrode configuration. (b) Three-sided, minimum-resistance configuration.

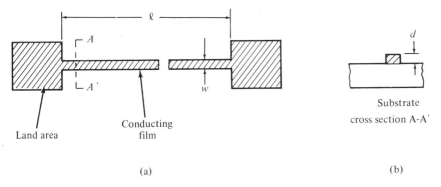

FIG. 6-18. Characterization of thin-film conductor. (a) Top view. (b) Cross section A-A'.

from the figure, the minimum-resistance capacitor decreases the ohmic resistance of the capacitor significantly (typically to less than 1 Ω). However, the corresponding geometric configuration takes up somewhat more space and may require higher-precision mask alignment than the conventional crossed-electrode pattern.

Conductors   In contrast to conventional, discrete circuits, or even those on printed-wiring boards, the conducting paths of thin-film circuits may have appreciable resistivity. Upper limits on this resistivity must therefore be guaranteed. The characterization of a typical thin-film conducting path is illustrated in Fig. 6-18. As with resistors, the sheet resistance of the film material given by (6-2) constitutes the most important characteristic and determines the applicability of the film as conductive material. It is given, for a number of commonly used film materials, in Table 6-6. In contrast to resistor film material, conductor film should, of course, have as low a sheet resistance as possible. Since the maximum film thickness of conducting material, or any other obtained by evaporation for that matter, is on the order of 10,000 Å to

TABLE 6-6. CHARACTERISTICS OF THIN FILM CONDUCTOR MATERIALS

Material	R_s for 1000 Å film ($\Omega/\square$)	Substrate adherence	Resistance to corrosion	Solderability	Bondability	Ease of evaporation
Copper	0.2	poor	poor	good	poor	good
Gold	0.27	poor	good	poor	good	good
Aluminum	0.33	fair	good	poor	poor	
Palladium	1.3	poor	good	good		poor
Palladium–Gold (50 : 50)	3	good	good	good		good
Titanium	10	good				good
Nichrome	15	good				good

20,000 Å, the minimum sheet resistance of conducting material will lie between 0.01 to 0.03 $\Omega/\square$. Besides making the conducting film sufficiently thick to minimize sheet resistance, the circuit designer still has the aspect ratio of the conducting paths as an additional degree of freedom to minimize the conducting resistance. He will therefore be careful to keep the conducting paths as wide as possible, whenever minimum-resistance conductors or interconnections are required.

Other important considerations beside the sheet resistance of conductor materials will be whether they can be easily soldered or bonded, whether their adherence to a given substrate is sufficiently good, how resistant they are to corrosion by ambient contaminants, and whether their deposition is compatible with the processes required to fabricate other film components. Some of these characteristics have been included in Table 6-6 for commonly used conductor films. Notice that, whereas gold has very low sheet resistance, it cannot be soldered reliably. If gold is combined with palladium the resulting dual conductor film can be soldered but the sheet resistance has gone up by an order of magnitude. On the other hand, gold by itself can be bonded quite easily; due to its low sheet resistance it is therefore frequently used in networks in which no soldering is required.

Substrates   There is one final component that has to be considered in a thin-film circuit, and that is the substrate material itself. The main materials used as substrates are unglazed and glazed alumina ceramic, beryllia ceramic, and Corning 7059 glass. The important factors influencing the choice of a substrate type are:

1. chemical stability,
2. surface smoothness,
3. thermal conductivity,
4. mechanical strength, and
5. cost.

Whether to use glass, ceramic, or glazed ceramic depends on which of the five characteristics mentioned above weigh most heavily in a given application. Table 6-7 gives a qualitative appraisal of the most commonly used substrate types with respect to those characteristics.

TABLE 6-7. QUALITATIVE APPRAISAL OF SUBSTRATE TYPES*

Material	Chemical stability	Surface smoothness	Thermal conductivity	Mechanical strength	Cost
Corning 7059 Glass	×	○	●	●	○
Alumina ceramic	○	●	○	○	×
Glazed alumina ceramic	×	○	×	×	×
Beryllia ceramic	○	×	○	○	●

* ○ Excellent.   × Good.   ● Fair.

As far as chemical stability is concerned the four substrate types listed are all satisfactory. All have high resistivity, high dielectric strength, and low loss tangent, and all are chemically inert. Thus, as a whole, they resist chemical action, especially during the etching of conductive and resistive films. Tantalum films present an exception; since hydrofluoric acid is required as the etchant, the glass or glazed substrates must be covered with a thin layer of tantalum oxide which, compared to glass (or glaze) is inert with respect to hydrofluoric acid.

Surface smoothness (lack of roughness) and flatness (lack of waviness) are important requirements, particularly if thin and well defined line widths of the order of 50 μm (i.e., 2 mils) are to be used. Consider, for example, the ceramic substrate shown in Fig. 6-19a. The surface is shown exaggeratedly wavy to demonstrate how the line width of a resistor, supposedly well controlled by the surface mask, is widened by the waviness of the substrate surface. The resulting resistor will have a corresponding error (in this case the value will be too low). On the other hand surface roughness can result in an unevenness in film thickness, as shown in Fig. 6-19b.

Raw ceramic materials are sometimes not smooth enough for thin-film applications using very fine line widths or requiring extreme stability. They have a surface roughness in the "as-fired" condition that is related to grain

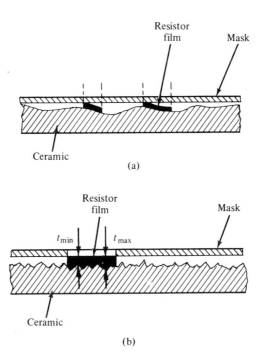

FIG. 6-19. Ceramic substrates. (a) Wavy surface. (b) Rough surface.

size and thus to the firing process. Grain size ranges from 10,000 Å to 15,000 Å. Although films deposited on such surfaces tend to replete, or smooth over, the surface roughness so that as-fired ceramic does not cause discontinuities even in films only 250 Å thick, the aging characteristics of these films are detrimentally affected. For this reason capacitors and high-stability resistors are generally deposited on glazed ceramic substrates. Glazing yields a surface comparable in smoothness to that of glass, i.e., a surface roughness of less than 1 microinch (approximately 250 Å). It should be pointed out, however, that the glazing process does leave a certain degree of surface waviness compared to the smooth surface of glass, which may in turn cause uneven line width (to a lesser degree than without glaze) due to poor mask fit.

The thermal conductivity of the substrate must be as high as possible to allow adequate dissipation of heat from the active components (e.g., beam-leaded chips bonded onto the substrate) and from the resistive components deposited on the surface. As shown in Table 6-7, the thermal conductivity of ceramics is superior to that of glass, in fact, something like twice as good (approximately 3 W/cm^2, i.e., 20 $W/in.^2$ if a heat sink is provided). Glazed ceramic is somewhat inferior to raw ceramic in this respect, because of the glass-like properties of the glaze; the thicker the glaze the poorer the thermal conductivity of glazed ceramic. However, with the very thin glaze layers generally used, the additional thermal impedance due to the glaze is a very small part of the total.

The mechanical strength of the substrates must be as high as possible for obvious reasons (e.g., to provide a rugged package and to reduce breakage in production). As would be expected, glass, being the most brittle of the substrate materials, shows up poorest in this respect. Glazed ceramic is somewhat inferior to raw ceramic due to the glasslike glaze layer on its surface. Quite apart from ruggedness of the substrates, mechanical strength gives an indication of how well external leads and appliquéed components, such as beam-leaded chips, can be bonded to the substrate surface. However, bonding problems can be avoided with glazed ceramic by using so-called spot glazing, i.e., by glazing only those areas on which thin-film components are to be deposited, or, in other words, by leaving those areas free of glaze on which chips or other appliquéed components (e.g., lead frames) are to be bonded.

Finally, in terms of the cost of the substrates, given in Table 6-7, we see that all but the beryllia ceramics are acceptable. This brings us to the beryllia ceramics themselves, which, as the reader will have noticed, are excellent in every other respect but cost. Thus, one would expect them to be used in high-precision networks where cost is not the decisive factor. This would be true, if it were not for the fact that beryllia dust is highly toxic. Handling these ceramics requires very special precautions, which make their use impractical beside being costly; they are used only when high thermal conductivity is absolutely necessary.

Summarizing our qualitative appraisal of the most frequently used substrates we see that only glazed alumina ceramics are good or excellent in every factor listed and, in fact, are equal to or better than 7059 glass in all respects except flatness and cost. Since even the cost of glazed ceramics is beginning to approach the cost of glass (which was heretofore the least expensive substrate material), their use in thin-film technology is rapidly becoming ever more widespread, whenever capacitors or resistors of extremely high stability are required; on the other hand, unglazed ceramics are perfectly satisfactory, and most commonly used, when resistor circuits and conductance paths of average quality will do.

Standard Coated Substrates   For the convenience of the user, several companies have begun to furnish substrates coated with the appropriate layers of resistor–conductor, or capacitor–conductor thin-film materials. Because of their shiny, reflecting surfaces these coated substrates are frequently referred to as "mirrors." The user fabricates resistor–conductor and capacitor–conductor networks from these substrates by subtractive etching of the appropriate films in the corresponding areas. In this way he need only invest in the masking and etching equipment without the more expensive sputtering and evaporation gear.[16]

The cross section of a typical substrate with predeposited nichrome (or chrome) resistor film and gold (or gold combined with copper) conductor film is shown in Fig. 6-20a. A corresponding mirror used in tantalum technology is shown in Fig. 6-20b. Notice the tantalum oxide underlay deposited on glass or glazed ceramic substrates for the reasons mentioned earlier. The resistor material on this substrate is tantalum nitride. With the typical film thicknesses indicated in the figure (500 Å to 1000 Å), and the specific resistivity of Ta_2N (see Table 6-3), namely $\rho = 250 \ \mu\Omega$ cm $= 2.5 \times 10^4 \ \Omega$ Å, the corresponding initial sheet resistance may typically vary between 25 $\Omega/\square$ and 50 $\Omega/\square$.

The rather complex film composite used to provide the conductor film on the coated subsrate in Fig. 6-20b is noteworthy;[17] it is a result of the properties of the individual conductor materials listed in Table 6-6. The low sheet resistance ($<0.05 \ \Omega/\square$) is provided by metals with low resistivity, such as gold or, as shown here, copper. Since neither of these materials adheres well to the substrate by itself, a thin film of more adherent material such as nichrome (as shown), chromium, titanium, or a combination of palladium over titanium is evaporated immediately before the copper or gold. Since copper is susceptible to oxidation, or other forms of corrosion, it is generally covered by a more noble metal, such as gold or palladium. A gold film is the best

---

16. However, as we shall see presently, thin-film capacitors require an evaporation step *after* the preceding etch and anodization steps, to deposit the counterelectrodes. Thus, coated substrates only obviate the need for evaporation equipment for the fabrication of thin-film resistors.
17. Recently a film composite consisting of titanium (600 Å), palladium (4000 Å), and gold (10,000 Å) has been found very useful. The thicknesses given are typical values. Here titanium serves as the adhesive layer and palladium as a diffusion barrier between the titanium and the gold conductor. In addition, if soldering is required, rhodium is spot plated so as to eliminate short circuits due to solder flow.

choice when a thermal compression bond is to be formed between the conducting film and a gold lead. If a lead is to be soldered to the conducting film, palladium is better than gold (see Table 6-6), since gold forms a brittle alloy with most solder materials. On the other hand, pure palladium is difficult to evaporate from a conventional tungsten filament; it tends to alloy with the tungsten at the evaporation temperature. For this reason a 50 : 50 palladium–gold alloy is often used as the top layer of the conducting film when solder bonds are required (see Fig. 6-20b). The gold alleviates the palladium–tungsten alloying problem and the palladium prevents gold-embrittlement of the solder joint. As an alternative, electron-beam evaporation can be used to circumvent the problem of filament attack by palladium.

The selection of a substrate from a choice of standard coated substrates of the type shown in Fig. 6-20 will be based on such properties, in the case of a resistor substrate, as the sheet resistivities of the resistor and conductor materials, and on whether appliquéed components are to be soldered or bonded onto the surface; in the case of a capacitor substrate, the selection will be based on such properties as the sheet resistivity of the base-electrodes and counterelectrodes, and on the capacitance density obtainable with a given dielectric film material and film thickness. Clearly it is advisable for the network and device designers to interact closely at this stage of the design process to ensure that optimum trade-offs are provided by a standard coated substrate to satisfy a given application. Naturally this close interaction is beneficial throughout the whole design process.

In order to minimize the interconnections required between resistor and capacitor substrates, the tantalum thin-film technology needed to combine resistors and capacitors on the same substrate has been developed. This requires two separate sputtering runs in order to deposit the most suitable tantalum film for capacitors (β-tantalum) and for resistors (e.g., tantalum oxynitride). Although the number of processing steps is thereby increased over that required for separate resistor and capacitor substrates, the economy of batch processing all the RC components in one continuous sequence normally outweighs this disadvantage. A single-substrate network may typically have the form shown, greatly simplified, in Fig. 6-21a. It is interesting to note that, in combining resistors and capacitors on the same substrate, the substrate area required is not as large as the sum of the corresponding separate resistor and capacitor substrates, and may indeed be closer to half that amount. This is because the dual-substrate approach requires additional area for conductor paths interconnecting the two separate substrates (i.e., one resistor and one capacitor substrate). On the other hand the single-substrate approach generally requires some judicious conductor routing to minimize the number of intersecting paths or crossovers. These can be easily avoided with the dual-substrate approach, particularly if the two are mounted on top of each other in sandwichlike fashion (also sometimes referred to as "piggyback assembly") as shown in Fig. 6-21b.

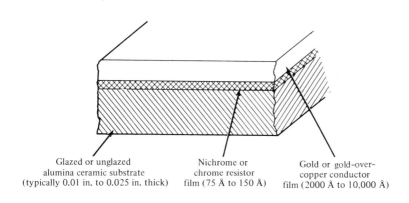

Glazed or unglazed
alumina ceramic substrate
(typically 0.01 in. to 0.025 in. thick)

Nichrome or
chrome resistor
film (75 Å to 150 Å)

Gold or gold-over-
copper conductor
film (2000 Å to 10,000 Å)

(a)

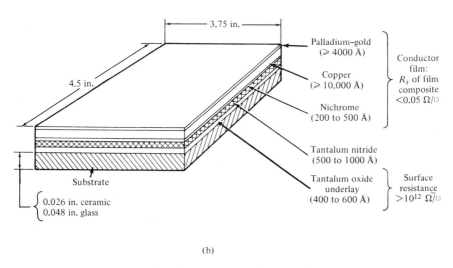

3.75 in.

4.5 in.

Palladium–gold
(⩾ 4000 Å)

Copper
(⩾ 10,000 Å)

Nichrome
(200 to 500 Å)

Conductor
film:
R_s of film
composite
$<0.05\ \Omega/\square$

Tantalum nitride
(500 to 1000 Å)

Substrate

Tantalum oxide
underlay
(400 to 600 Å)

Surface
resistance
$>10^{12}\ \Omega/\square$

0.026 in. ceramic
0.048 in. glass

(b)

FIG. 6-20. Substrates with predeposited resistor–conductor thin-film materials. (a) Nichrome technology. (b) Tantalum technology.

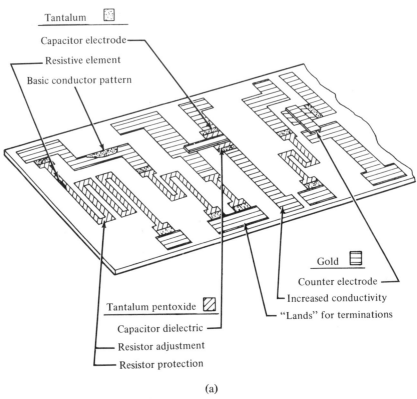

Tantalum ▢

Capacitor electrode

Resistive element

Basic conductor pattern

Gold ▤

Counter electrode

Increased conductivity

"Lands" for terminations

Tantalum pentoxide ▨

Capacitor dielectric

Resistor adjustment

Resistor protection

(a)

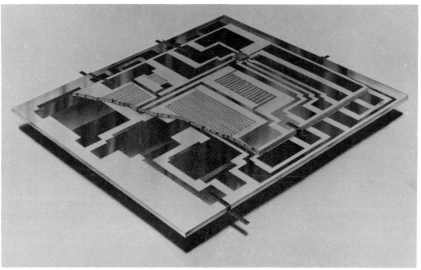

(b)

FIG. 6-21. Tantalum thin-film *RC* networks. (a) Resistors and capacitors on single substrate. (b) Separate resistor and capacitor substrates assembled in "piggy-back" fashion.

6.1.4 The Characteristics of Thin–Film Components

Having discussed in some detail the fabrication and characterization of thin-film resistors and capacitors, we come now to the most important aspect of this technology from the network designer's point of view, namely, the characteristics of the components it provides. As we shall see, this technology offers certain capabilities but also certain constraints peculiar to it, and it will be up to the network designer to utilize these peculiarities in an optimum way. He can do this only if, before beginning his design, he is thoroughly familiar with the component characteristics so that, from the outset, he can develop his design accordingly.

Resistors To appreciate the characteristics of thin-film resistors we shall start out by considering tantalum thin-film resistors (i.e., tantalum nitride, tantalum oxynitride, etc.). We do this because, as we shall see later, the quality of these thin-film resistors is superior, as a whole, to that of most other types of thin-film resistor; they therefore permit the important characteristics of thin-film resistors to be pointed out most clearly.

The characteristics of tantalum thin-film resistors are listed in Table 6-8. They are listed in two columns: the first indicates routinely available properties, the second those that could be routinely available in the future if needed, or that are already available, but at additional cost. The resistor characteristics can be summarized as follows.

Resistor Range and Spread The available resistor range using tantalum thin-film material is limited at the high and low ends. The upper limit results

TABLE 6-8. CHARACTERISTICS OF TANTALUM THIN–FILM
RESISTORS

Property	Value	
	Routine	Available*
Sheet resistance ($\Omega/\square$)	10 to 100	up to 3000
Resistor range	10 Ω to 1.0 MΩ	1 Ω to 5 MΩ
Max. resistance per substrate (MΩ)	1.0	5
Initial precision (%)	0.1	0.01
Aging: 20 yrs (%)	0.5	0.05
5 yrs (%)	0.1	0.01
Minimum line width† (μm)	50	5
Temp. coeff. (ppm/°C)	-50 to -100 ± 15	0 to -1000
Tunability (%)	$+20 \pm 0.1$	$+50 \pm 0.01$
Tracking (ppm/°C)	± 5	± 1

* Some of these characteristics are mutually exclusive.
† 25 μm $\approx$ 1 mil.

from the maximum number of squares in a given resistor pattern compatible with an acceptable yield in fabrication. The probability that a film defect will damage a resistor is greater over a longer path; at the same time, a narrower path is more vulnerable, since smaller defects are effective in damaging the resistor. If the total number of squares in all the patterns on a substrate gets too large, the substrate fabrication yield decreases. Thus, the *total* resistance on a substrate is more important than the number of resistors or the value of any one resistor.

The maximum number of squares tolerable on a substrate as a result of yield considerations depends on the minimum line width that can be reliably processed and the total substrate area required. Typically this number is on the order of 10,000. Thus an upper limit on the total resistance per substrate is:

$$R_{Tot} \text{ per substrate} < 10,000 \cdot R_s \qquad [6\text{-}10]$$

With the presently available maximum sheet resistance of tantalum film being 100 $\Omega/\square$, the total resistance per substrate is 1 MΩ. Clearly minimum line width and maximum sheet resistance depend on the state of the thin-film art and may be subject to significant improvement. It is expected that minimum line widths of 40 μm (1.5 mils) and even 25 μm (1 mil) as well as a maximum sheet resistance of between 500 $\Omega/\square$ and 1000 $\Omega/\square$ will be routinely available in the future.

The lower limit on the number of squares in any one resistor pattern, which of course determines the minimum individual resistor value compatible with an acceptable yield, is mainly determined by the ratio of the unanodized areas at the resistor terminations (see "mask margin" in top view of Fig. 6-7) to the total resistor area. Thus patterns of less than a square become difficult to fabricate within required tolerances; furthermore, depending on the percentage unanodized area, the stability of the resistor will also suffer. As the resistor becomes wider than it is long (i.e., less than a square), the small margins available between the resistor terminations and the anodizing electrolyte also make resistor adjustments more difficult.[18] Clearly the minimum permissible resistor value will depend on the specific sheet resistance being used in the design; if half a square is the practical lower dimensional limit, 100-$\Omega/\square$ film material will provide 50 Ω, 25-$\Omega/\square$ film material will provide 12.5 Ω, as the minimum resistor value. Still lower values can be obtained with parallel resistors. With 10,000 squares as the upper limit of a resistor, the maximum possible spread, assuming only two resistors on a substrate, would be on the order of 20,000:1. Practically, with five to ten resistors on a substrate, the spread between maximum and minimum resistor value should not exceed a few thousand to one. As we shall see shortly, another reason to keep the spread of resistor values on a substrate as low as possible, is to ensure their close tracking with ambient variations such as temperature.

18. If the electrolyte comes into contact with one of the conducting resistor terminations, the anodizing path is short circuited (the high resulting current may burn through the conducting material) and no anodization of the resistor film takes place.

As the sheet resistance is increased the need for fractional-square patterns to achieve low-valued resistors is accentuated. One way around this is to deposit a series of shorting sections made of a good conductor (e.g., aluminum) on top of or below the tantalum film.

Resistor Precision   The precision of initial values obtainable with tantalum thin-film resistors depends on the trimming process involved. After thermal stabilization, and with no anodization at all, the resistors may have a tolerance as wide as $\pm 15\%$. This will be due to film unevenness, line-width errors, and the like. Trimming all the resistors at once by batch anodization, i.e., dipping the complete resistor substrate into an anodizing electrolyte, tolerances no better than $\pm 5\%$ should be expected. As shown schematically in Fig. 6-22 this is the cheapest method of resistor adjustment. Initial precision between $\pm 2\%$ to $\pm 0.5\%$ requires individual resistor anodization and approximately doubles the cost. Initial precision between $\pm 0.5\%$ and $\pm 0.1\%$ is routinely available by more careful anodization but it will, in turn, increase the cost of trimming considerably. The main reason for the increased cost of trimming with higher precision requirements is that the process, as most other thin-film trimming processes, can only increase and not decrease the resistor value. Thus sheet resistances are initially produced below the desired value so that all the resistors can be trimmed to value, incuding those with an initial sheet resistance somewhat above the average for a given batch. Whereas this poses no real problem in manufacture, it does present the difficulty of not

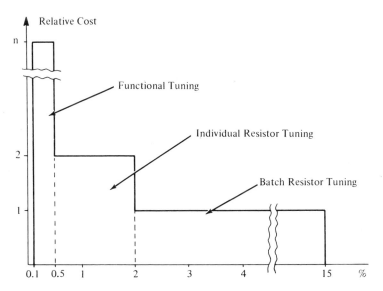

FIG. 6-22. Cost increase, in "cost units," with increased initial precision requirements of thin-film resistors.

allowing for any overshoot in the trimming process. The tighter the tolerances on a resistor of a substrate, the more likely overshoot will occur, with the result that the whole substrate may have to be discarded. Herein lies the cost-increase shown in Fig. 6-22, as the tolerances are tightened. Clearly, automation with an electrical feedback system that senses the resistor value and prevents overshoot is desirable. Very often, below $\pm 0.5\%$ tolerance requirements, it becomes more practical to trim a network functionally, than based on calculated resistor values alone. In this way the required network function, rather than the individual resistor value, is monitored, and the effect of inadvertent parasitics on the substrate is taken into account. In fact, when precisions in the order of 0.1% or better are required, parasitics must invariably be considered in the trimming process, since their contributions may cause errors of the same order of magnitude as the tolerence requirements themselves.

Resistor Stability   Closely related to the initial precision of thin-film resistors is the question of the stability to which their values can be held during life (i.e., aging). Since the anodized resistor surface is passivated by tantalum pentoxide, the stability of tantalum-based thin-film resistors is exceptionally good. Due to thermal oxidation (the main contribution to the aging process), aging will of course be accelerated if the power dissipated by the resistors and/ or the ambient temperatures in which the resistors are used is high. In active *RC* networks, it is generally the frequency function that is of interest, so that the power-handling capability, and with it the power dissipated by the resistors, may be held negligibly small. Nevertheless, if high-precision networks are required, the aging process (which is predictable provided the power dissipation and ambient temperatures can be anticipated) should be examined on an individual basis. As mentioned earlier the stability of a resistor will also depend on the percentage unanodized tantalum film due to the required mask margins at the resistor–conductor interfaces. Thus low-value resistors, particularly those less than one square in area, may be somewhat inferior in long-term stability. Typically, in active network applications, resistor aging will be well below 1% over a twenty-year life.

Resistor Tracking   Perhaps the most outstanding property of thin-film resistors compared to their discrete counterparts, and the one that may effect hybrid integrated circuit design most profoundly, is the closeness with which they as well as other components of a kind, (e.g., capacitors), track on a substrate. This is, by the way, an important characteristic of other integrated circuit technologies as well, such as thick-film or monolithic silicon integrated components. There as here, the property of component tracking has probably influenced the design of circuits, intended for these technologies, more than any other.

Tracking of resistors, as of other components, is due to individual resistors within a network being simultaneously deposited onto the substrate under precisely the same conditions. Each resistor in the network undergoes the same process, experiences the same environments, and is made from the same batch of materials. Consequently, as one would expect, the characteristics of all the resistors will very closely match and track under a common set of external stimuli. Since one important criterion for tracking is that all resistors should be subjected to the same temperature, the substrate should have good thermal conducting properties. This is particularly important if a heat-generating source (e.g., SIC-chip),[19] which would otherwise cause a thermal gradient along its surface, is located on the substrate. Consequently, alumina ceramic substrates with their superior thermal conduction compared to glass are greatly preferable when close tracking is required.

Actual data taken on production-line networks have been reported, demonstrating tracking ratios of better than 1 ppm/°C between resistors. However, this order of accuracy is only obtainable with layouts optimized to ensure tracking; the critical resistors must be physically located as close together as possible and equally distant from any heat emitting source on the substrate, such as a silicon chip. Furthermore, for optimal tracking, the values of the resistors should be as nearly equal as circuit requirements will allow, and the line widths used should also be the same.

Tantalum resistors will routinely track to within ± 5 ppm/°C of each other. Over a 60°C temperature range the resistors R_i on a substrate will therefore vary by

$$\left.\frac{\Delta R_i}{R_i}\right|_{\Delta T = 60°C} = \left.\frac{\Delta R}{R}\right|_{\Delta T = 60°C} = \pm 0.03\% \qquad (6\text{-}11)$$

This variation is often small enough to be neglected altogether.

Resistor Temperature Coefficient (TCR)   Beside tracking, another important property of thin-film resistors is how precisely their temperature coefficients can be controlled. In tantalum technology a useful degree of flexibility exists in this respect, because of the capabilities of reactive sputtering. Restricting ourselves to tantalum nitride, which has hitherto been the most commonly used material because of its overall stability, the TCR will typically be on the order of -95 ± 20 ppm/°C. By controlling the partial pressure of the nitrogen accurately during the reactive sputtering process, this value can be shifted up or down by about 50 ppm/°C. Since the TCC of tantalum thin-film capacitors is in the order of $+200$ ppm/°C, a larger negative TCR than can conveniently be obtained with tantalum nitride is required. By adding oxygen in controlled amounts to the reactive sputtering process, tantalum oxynitride with nominal TCRs of -200 ppm/°C can easily be obtained without altering any of the other crucial resistor properties significantly.

---

19. SIC = silicon integrated circuit.

Resistor Tunability   Potentiometers are, of course, not available in any form that is compatible with integrated-circuit technology, and it is important for the designer to know how much he can expect to trim his thin-film resistors by other means. Apart from the fact that the designer can only increase resistor values, thin-film resistors give him at least the flexibility he was used to with discrete components—and with a much greater accuracy and long-term stability at that. In tantalum technology, where trimming is generally done by anodizing (although laser trimming seems a feasible and even more accurate method for the near future), the percentage of fine adjustment that an assembled circuit is to undergo should be kept to a minimum; it should not exceed 20%, if heat stabilization cannot follow the trimming process. The reasons for this were given in Section 6.1.1. This limitation does not exist with laser trimming. The laser permits scribing of conductor paths and resistor trimming in one and the same step. Laser trimming is also faster and more accurately controllable than anodization. Being a noncontact process it leaves no residue on the resistor surface and permits the effects of trimming to be monitored and evaluated during the actual trimming process. In contrast, when trimming is done with a rigid electrolyte, the probe touches, and therefore shunts, the resistor in question, causing errors in measurement, and making it necessary to lift the probe off the substrate before each meter reading. On the other hand, laser implementation invariably requires more intricate and expensive equipment. Thus, the rigid-electrolyte trimming method is more convenient for experimental and laboratory use, whereas the laser should, in the long run, prove better in a production-line environment.

Now that the salient features of thin-film resistors based on the characteristics of tantalum resistors have been illustrated it is of interest to compare these with the characteristics of the other frequently used film resistors, namely, those made of nichorome. This comparison is attempted in Table 6-9. A comparison of this kind must be regarded with caution, since published data on the various types of resistor material may be based on entirely different

TABLE 6-9. A COMPARISON OF TYPICAL TANTALUM AND NICHROME RESISTOR PROPERTIES

	Resistor		
Property	Tantalum nitride	Tantalum oxynitride	Nichrome
Sheet resistance ($\Omega/\square$)	25–100	50–100	100–250
TCR (ppm/°C)	-50 to -95 ± 20	-100 to -250 ± 15	± 100
Resistor range	10 Ω to 1 MΩ	10 Ω to 1 MΩ	15–150,000 Ω
Initial precision (%)	0.1	0.1	0.5
Aging: 20 yrs (%)	0.5	0.5	
5 yrs (%)	0.1	0.1	1–5
Tracking (ppm/°C)	± 5	± 5	± 10

measuring methods and criteria. However, inasmuch as the important characteristics of thin-film resistors, such as tracking and stability, are concerned, nichrome resistors possess them sufficiently to be utilized and taken into consideration during design. In general it can be said that chromium-based technology offers higher sheet resistances and consequently better potential packing density than is presently available with tantalum technology; possibly it also offers additional advantages in applications requiring high temperature stability. Tantalum-based technology, on the other hand, allows more precise trimming and better long-term, low-temperature stability. Tantalum oxynitride affords the same advantages as tantalum nitride and in addition a choice of TCR to compensate for the TCC of tantalum thin-film capacitors. Thus nickel–chromium resistors are generally found in circuit applications where small size is a necessity but the tolerance and stability requirements are not quite so severe. In applications requiring extreme initial resistor tolerance and/or excellent aging stability, tantalum-nitride films are preferred; if controlled TCR of the order of -200 ppm/°C is required in addition, tantalum oxynitride is used.

Capacitors   Virtually the only thin-film capacitors that have been manufactured and studied in large numbers are based on tantalum technology; this technology will therefore again serve to illustrate the characteristics of thin-film capacitors in general.

There are three basic types of tantalum thin-film capacitors that can be used today. They are shown in Fig. 6-23. The first (Fig. 6-23a), the so-called TM (Tantalum–metal) capacitor, is the one used most extensively. This is the high-quality, moisture-insensitive capacitor described previously. It comprises a tantalum pentoxide dielectric sandwiched between a β-tantalum base electrode and the adherent counterelectrode. The counterelectrode consists of a thin layer (or "flash") of nichrome or titanium followed by gold or aluminum. These capacitors are used for applications with stringent stability requirements, such as generally occur in RC active networks. The TMM (Tantalum–Manganese-dioxide–Metal) capacitor shown in Fig. 6-23b is a modification of the TM type, containing a layer of semiconducting manganese dioxide (MnO_2) over the tantalum pentoxide. Because manganese dioxide is an oxygen-deficient semiconductor, it serves as an extension of the top electrode, and does not alter the capacitance of the structure. Its purpose, rather, is to provide a healing mechanism: weak spots or incipient failures in the tantalum oxide that would mean total failure in the TM type do not threaten the operation of the TMM capacitor. A third variation, shown in Fig. 6-23c, which has a low capacitance per unit area, is used for very low-value capacitors. A layer of silicon monoxide is here interposed by evaporation between the tantalum pentoxide layer and the counterelectrode. The capacitance depends almost entirely on the thickness of the silicon monoxide layer because it has a low dielectric constant (6, compared with 22 for tan-

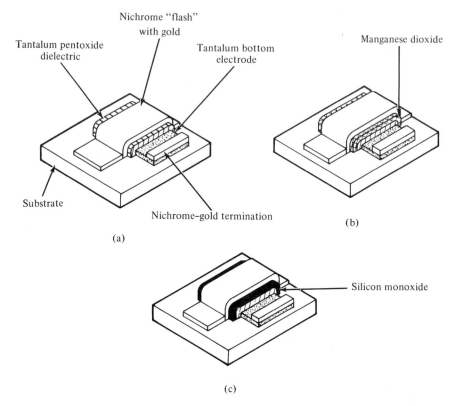

FIG. 6-23. The three basic tantalum thin-film capacitor types. (a) Adherent counterelectrode (tantalum-metal or TM) type. (b) tantalum–manganese-dioxide–metal (TMM) type. (c) duplex type incorporating silicon-monoxide–tantalum-pentoxide low-valued dielectric.

talum pentoxide). The necessary capacitance density is obtained simply by adjusting the thickness of the silicon monoxide layer. The tantalum pentoxide layer serves to increase the yield and the dielectric strength of this structure.

The characteristics of the three basic capacitor types shown in Fig. 6-23 are summarized in Table 6-10. The TM capacitor is the one most commonly used in active RC networks and we shall therefore discuss its most important characteristics in detail shortly. Considering the other two capacitor types briefly first, we notice that the anodic working voltage of the TMM capacitors is considerably higher than that of the other two types. Higher stresses and thinner films can therefore be employed with TMM capacitors, while both yield and performance are improved. For example, TMM capacitors can be made with values of up to 5 μF, which is about 100 times greater than the upper limit for the TM type. Furthermore, unlike the TM type, the TMM capacitor can be deposited on unglazed, rough-surfaced ceramic substrates. The TMM type, however, has a higher dissipation factor and temperature

TABLE 6-10. CHARACTERISTICS OF TANTALUM THIN-FILM CAPACITORS

Property	Capacitor type		
	TM (Au–NiCr/ Ta$_2$O$_5$/β-Ta)	TMM (Au–MnO$_2$/ Ta$_2$O$_5$/β-Ta)	Duplex (Au/SiO– Ta$_2$O$_5$/β-Ta)
Max C/substrate (μF)	0.08	5.0	0.1
Cap. density [μF/cm^2]	0.05–0.1	0.05–0.5	0.002–0.02
Dielectric thickness (Å)	2000–4000	400–4000	5000–15,000
Dielectric constant	21–25	21–25	5–6
Anodic* working voltage (V)	15–30	15–100	25–50
Initial precision (%)	1–5	2–10	2–10
Stability (humidity and 20-yr aging) (%)	0.2	5–10	5
Temp. coefficient (ppm/°C)	+200 ± 30	+500 to +1000	+500 to +1000
Tracking (ppm/°C)	±10		
Dielectric Loss (tan δ')	0.003–0.006	0.01–0.05	0.01–0.05

* Anodic film capacitors such as tantalum thin-film capacitors are polar in nature. The anodic voltage is the voltage in the blocking direction (base electrode positive) in contrast to the cathodic voltage. The latter is on the order of 4 to 6 V.

coefficient; both are disadvantages for the kind of applications encountered in active RC network design. The TMM capacitor may be useful, however, particularly because of the higher capacitance density and larger maximum working voltage it provides, as a power-supply-bypass capacitor, as is frequently required in active circuits using operational amplifiers.

The silicon-monoxide–tantalum-pentoxide, or duplex,[20] capacitor is useful inasmuch as its extends the capacitance range of tantalum thin films to low values (a few picofarads). The high capacitance density of the TM and TMM capacitors makes them less suitable for low-value capacitors. On the other hand, the processing involved with this capacitor type is somewhat complicated because of the silicon monoxide, which has to be evaporated onto the tantalum pentoxide before the gold counterelectrode. For this reason, it may be more practical to obtain low-value capacitors by connecting TM-type capacitors in series.

Let us now turn to the most important capacitor used in active RC networks: the TM capacitor with its β-tantalum base electrode, tantalum pentoxide dielectric layer, and nichrome–gold counterelectrode. The most important characteristics in terms of active network design, are summarized in what follows.

20. The name implies that the capacitor behaves as two capacitators in series. The capacitance value is determined principally by that of the low-valued layer; thus the range of capacitance per unit area is 0.002–0.02 μF/cm^2 which is essentially the same as that for the silicon monoxide dielectric alone.

Capacitor Range and Spread   The capacitor range available with TM capacitors is determined at the upper end by the acceptable yield, at the lower end by the acceptable capacitor tolerances. It has been found that, for a variety of reasons, very thin and very thick dielectric films have high defect densities. As would be expected, an optimum thickness range of tantalum pentoxide dielectric film therefore exists, for which the capacitor yield is at a maximum. The range is from 2500 Å to 4000 Å. Given this rather tight range of dielectric thickness, there is an upper limit on total capacitance per substrate for which acceptable yields are obtainable in manufacture. Notice that it is not the number of capacitors, nor the value of any one capacitor, that is important, but rather the total capacitance per substrate, since a defect in the area of any one capacitor will cause a defective device. It has been found that reasonable yields can be obtained if

$$C_{\text{Tot}} \text{ per substrate} < 0.08 \ \mu\text{F} \qquad (6\text{-}12)$$

The high dielectric constant of tantalum pentoxide with its consequent high capacitance density becomes a disadvantage when very low-valued capacitors are required. The area required for a capacitor of 100 pF results directly from the capacitance density of tantalum pentoxide, which is typically 0.05 $\mu\text{F}/\text{cm}^2$. Thus for 100 pF we require the capacitor area

$$A = \frac{100 \text{ pF}}{50{,}000 \text{ pF}} \text{ cm}^2 = 2 \times 10^{-3} \text{ cm}^2 = 320 \text{ sq. mils} \qquad (6\text{-}13)$$

With such a small area (namely a square approximately 18 mils on each side, which corresponds to the size of a relatively small silicon chip), a small error in defining the capacitor area results in a large relative error in capacitance. Thus, where a 100-pF capacitor may still be manufacturable with a precision of $\pm 5\%$, the tolerance on 25 pF and 10 pF will be in the order of $\pm 10\%$ and $\pm 15\%$, respectively.

Capacitor Precision   The two possible sources of capacitor error are nonuniformities in the dielectric film, which affect the capacitance density, and uncertainty in defining the active area. Dielectric errors result from variations either in the film thickness (e.g., errors in controlling the anodizing voltage) or in the dielectric constant. Area errors are due to lack of precision in delineating the capacitor boundaries; they may be caused by mask inaccuracies or misalignment. Between them, these error sources may account for a maximum of $\pm 5\%$ tolerance for capacitors on *different* substrates; typically even this error may be closer to $\pm 2\%$. The capacitors on the *same* substrate will vary by much less (on the order of $\pm 1\%$), since errors in the dielectric film formation as well as in the mask alignment affect all the capacitors in the same way. Thus, typically, the values of all capacitors on a substrate may collectively be off from the nominal value by up to $\pm 5\%$, yet by only $\pm 1\%$ of one another, as illustrated schematically in Fig. 6-24.

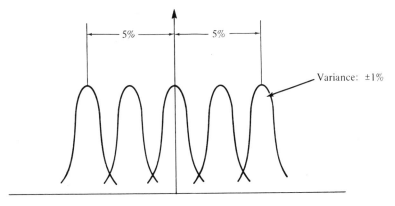

Fig. 6-24. Tolerances with respect to capacitors on different substrates ($\pm 5\%$) and with respect to capacitors on same substrate ($\pm 1\%$).

Capacitor Stability   As has been discussed earlier, the long-term stability of the adherent-counterelectrode (i.e., TM-type) capacitor is far superior to the other thin-film types; it is also more humidity-independent (see Table 6-10). This is one more reason why the TM capacitor will generally be exclusively used as the frequency-determining capacitor in active *RC* networks.

Capacitor Losses   Contrary to discrete capacitors, where the resistive and dielectric losses can generally be ignored (e.g., with dipped mica or polystyrene capacitors), this is far from the case with thin film capacitors. Whereas the resistive losses due to lead-in resistance of the electrodes, in particular of the β-tantalum base electrodes, may be reduced by special geometric configurations (e.g., the minimum-resistance configuration shown in Fig. 6-17), or by a well-conducting (e.g., aluminum) underlay, there is little that can be done to improve the inherent dielectric loss (i.e., $\tan \delta'$) of tantalum oxide. This is typically 0.003 to 0.004, compared to 0.001 or less (typically 0.0003) for a dipped mica capacitor. In many cases the dielectric loss, which can be considered an equivalent series resistance (see (6-8)), can be "tuned out" in an *RC* network; in such cases its initial value is of little concern. However, since its variation with temperature can be considerable (some measurements have indicated variations of $\pm 50\%$ over a 60°C temperature range), and few statistical data are as yet known about its overall tracking behavior on a given substrate with temperature changes, humidity, and aging, it is important to keep the effects of $\tan \delta'$ in mind and, if possible, select a circuit with most immunity to its variations. Incidentally, even if the $\tan \delta'$ of the capacitors on a substrate do track, it is unlikely that they will do so by the same amount as the resistors, particularly as the equivalent series resistor $\tan \delta'/\omega C$ is frequency-dependent. By specifying a low enough $\tan \delta'$ in critical networks (e.g., less than 0.003), the variations will very likely be negligible, particularly if the impedance level of the remaining network is not made too high, or if the capacitors themselves can be shunted by appropriate resistors.

Capacitor Tracking   For reasons similar to those given for thin-film resistors, the capacitor values on a substrate will track closely with ambient variations such as temperature and time. The reasons for this close tracking are of course simultaneous film deposition, anodization to produce the dielectric, and all the other processing steps, which will affect all capacitors on a substrate in the same way. The capacitors on a substrate can be expected to track to within ± 10 ppm/°C over the temperature range from 0°C to 60°C. Thus the total variation between the capacitors C_i on a substrate will be

$$\left.\frac{\Delta C_i}{C_i}\right|_{\Delta T = 60°C} = \left.\frac{\Delta C}{C}\right|_{\Delta T = 60°C} = \pm 0.06\% \qquad (6\text{-}14)$$

Furthermore, it is to be expected that, for the same reason that the other properties of the components on a substrate track, so does the tan δ' of the capacitors. As far as the resistive loss component is concerned (see (6-8)), this is essentially determined by the sheet resistance of the β-tantalum base electrode and will track as closely as the resistors on a substrate. This may be of interest when analyzing the losses of high-frequency capacitors for which the resistive loss is the dominant component. At voice frequencies, the resistive loss is negligible and the dielectric loss component tan δ' dominates.

Capacitor Temperature Coefficient (TCC)   The temperature coefficient of TM capacitors is determined by the TC of the dielectric, namely, tantalum pentoxide. As with the dielectric constant this is an inherent material property and is nominally $+200$ ppm/°C; actually it may vary between 170 and 230 ppm/°C.

Frequency Stability of Thin-Film *RC* Networks   We have discussed the characteristics of thin-film resistors and capacitors separately above, but in *RC* networks it is generally the *RC* product that is required to determine a frequency or time constant with specified limits on stability, temperature variation, and the like. Since the frequency f given by an *RC* product equals $(2\pi RC)^{-1}$, the frequency variation is

$$\frac{\Delta f}{f} = -\left(\frac{\Delta R}{R} + \frac{\Delta C}{C}\right) \qquad [6\text{-}15]$$

Thus, rather than inquire how accurately the resistors and capacitors of a network can be stabilized and controlled, we should be more concerned how precisely the *RC* product can be obtained and how accurately the temperature coefficients of the critical resistors and capacitors can be matched to provide stable, temperature-invariant *RC* networks.

Consider an active *RC* network whose frequencies are determined by its *RC* components. Assuming a five-year life, an ambient temperature range of 10°C to 60°C, and an ambient relative humidity varying from 20% to 95%, the overall frequency tolerance can be broken down into separate parts as shown for a typical tantalum thin-film RC network in Table 6-11. This overall tolerance includes the tolerances due to tuning, temperature, relative humidity and component aging. It also includes the effects of encapsulating the complete circuit with a silicone rubber material.[21]

The resistor film is assumed to be tantalum oxynitride, providing the optimum TCR to within ± 40 ppm/°C; the capacitor film is standard β-tantalum. By adding the median frequency changes on a worst-case basis and the distributions about the median on an rms basis, then adding the two totals on a worst-case basis, the total deviation from the design frequency is shown to be within $\pm 0.36\%$. This value and the manner of its composition (see Table 6-11) can be considered representative for tantalum thin-film RC networks.[22] The effect of increasing the ambient temperature to 75°C with other conditions remaining unchanged is also included in the table; the estimated frequency stability is seen to increase to $\pm 0.45\%$ of design value. Approximately the same increase would result if, instead of the ambient temperature being increased by 15°C, the assumed aging span had been doubled (i.e., to 10 years). Notice that the initial frequency distribution is centered slightly below the design frequency. This is done intentionally to center the total expected deviation about the design frequency.

To appreciate how the frequency stability of thin-film circuits compares with conventional, inductor–capacitor (LC) networks, we can consider two commonly used LC combinations. The frequency stability of an LC tank is given by

$$\frac{\Delta f}{f} = -\frac{1}{2}\left(\frac{\Delta L}{L} + \frac{\Delta C}{C}\right) \qquad [6\text{-}16]$$

Combining a commonly used ferrite-core inductor with a polystyrene capacitor and calculating the overall frequency deviations in terms, and for conditions, similar to those in Table 6-11, we obtain tolerances on the order of one to two percent. With a high-quality, permalloy-powder-core inductor, combined with a polystyrene capacitor in a hermetically sealed package, this tolerance can be decreased to the order of $\pm 0.35\%$. This is about the same precision obtainable with thin-film RC networks. It is therefore important to remember, when comparing the economics of active versus LC filters, that the frequency stability obtainable with tantalum thin-film circuits must be compared with the stability of high-quality, and therefore more costly, LC networks.

---

21. This coating is applied to prevent leakage currents from flowing on the bare substrate surface between closely spaced conductor paths which may be at different potentials. These currents which tend to develop in high humidity, can upset circuit performance and, in the long run, lead to catastrophic failure due to electrolytic corrosion of the film materials. The coating also provides mechanical protection and minimizes the corrosive effects of noxious environments.
22. Unpublished communications by W. H. Orr and J. S. Fisher, Bell Telephone Laboratories.

TABLE 6-11. FREQUENCY STABILITY OF TANTALUM THIN-FILM *RC* NETWORKS

Factors affecting frequency accuracy	Percentage deviations in frequency			
	ΔT: 10°C to 60°C		ΔT: 10°C to 75°C	
	Median change	Standard deviation	Median change	Distribution about median
Initial tuning and encapsulation	−0.10%	±0.10%	−0.15%	±0.10%
Relative humidity changes (20% to 95%)	±0.03%	±0.03%	±0.03%	±0.03%
Temperature changes	0	±0.18%	0	±0.20%
Aging (5 years)	0 to +0.2%	±0.10%	0 to 0.29%	±0.14%
Total	±0.13%	±0.23% (rms)	±0.18%	±0.27% (rms)
Total frequency deviation (worst case)	±0.36%		±0.45%	

6.1.5 The Design of Thin–Film *RC* Networks

In the previous sections we have considered, in detail, the design of the components constituting a typical thin-film network. We shall now conclude this discussion with a brief summary of the basic steps required to design a thin-film *RC* network, e.g., of the kind shown in Fig. 6-21a or 6-21b. These steps, broadly speaking, are the following:

1. *Draw circuit schematic.* Indicate the following:
a. Component values.
b. Component dissipation.
c. Component tolerances.
d. Critical parts of the circuit, such as conducting paths whose resistance must not exceed a maximum value (limits on minimum line width and conductor sheet resistance); high- and low-signal-level paths and high-impedance points (both of which determine the permissible lengths and proximity of conductor paths); input and output terminals; etc.
e. Resistors that are to be trimmed, and to what initial precision.

2. *Select coated substrate.* The choice of substrate will depend on:
a. Substrate material (e.g., glass, ceramic).
b. Resistor sheet resistance (depends on average resistor values).
c. Conducting material (determined by available sheet resistance).
d. Lead and chip attachment (soldering or bonding).
e. Capacitance density (depends on average capacitor values).
f. Substrate size (number of circuits per substrate).

3. *Determine layout rules.* These will depend on the technology used and on the two preceding steps. Typically, they will include:
a. Minimum distance between component and conductor or component and component (e.g., 250 to 500 μm).
b. Minimum distance required around resistors to be trimmed (e.g., for trim anodizing, 250 to 500 μm).
c. Margin to be left at substrate edges because of nonuniform surface at edges, particularly with ceramic (e.g., 1 to 1.25 mm).
d. Maximum permissible number, if any, of crossovers, feed-through holes, etc.
e. Mask registration marks.
f. Size of appliquéed devices (chip-capacitors, beam-leaded chips, lead frames, etc.).
g. Minimum resistor line width as determined by power dissipation or technology limitations (mask resolution, photographic or other optical limitations), whichever dominates.

4. *Sketch topological layout* (*no scale*). This sketch is intended to provide a rough idea of the final layout in terms of such factors as:
   a. Minimum required number of crossovers
   b. Required number of terminals and their relative location with respect to one another.

5. *Draw actual layout* (scale 20:1). This step is the crucial one for providing the actual, enlarged layout accurately, using as input all the information gathered in the preceding four steps. This involves:
   a. Determination of the individual resistor areas using the appropriate formulas for meander patterns, resistor dumbells, etc.
   b. Determination of the individual capacitor areas, taking permissible lead-in electrode resistance (cross-electrode or minimum-resistance configurations, etc.) into account.
   c. Using templates for the resistor and capacitor shapes, optimizing of layout, leaving sufficient space for conducting paths.
   d. Interconnecting components with widest possible conducting paths.

6. *Design masks*. From the comprehensive, optimized layout, derive the individual mask levels required for processing. Take heed of:
   a. Mask margins as determined by the thin-film processing to be followed.
   b. Registration method required for the individual masks.

Once again, throughout the procedure outlined above, close contact with the device-processing area must be maintained. This will prevent having to repeat some of the steps over again by getting them right the first time.

A photograph of the separate resistor and capacitor substrates as well as of the assembled unit of a hybrid integrated active network is shown in Fig. 6-25. The capacitor substrate is shown covered with a protective silicone rubber coating, as is the beam-leaded chip on the assembled device. Also shown is the final assembly, representing a dual in-line package.[23] The individual processing steps required for the fabrication of this active *RC* network are indicated in Fig. 6-26. Twelve processing steps are required for the resistor and capacitor substrates each, and five additional ones for the assembly and encapsulation of the final circuit. Most of the steps should be self-explanatory at this point. The laser separation steps indicated for the resistor and capacitor substrates refer to the fact that 20 resistor circuits and 16 capacitor circuits of the kind shown in Fig. 6-26 are simultaneously processed on resistor and capacitor master substrates, respectively. Herein lies one of the important

---

23. The dual in-line package is obtained by inserting the bent leads of the dual-substrate assembly into a plastic carrier. Note that hybrid-integrated networks based on the tantalum-thin-film. beam-leaded-chip technology do *not* require any hermetically sealed encapsulation due to the passivation of the individual components. The cost of the final product is thereby considerably reduced.

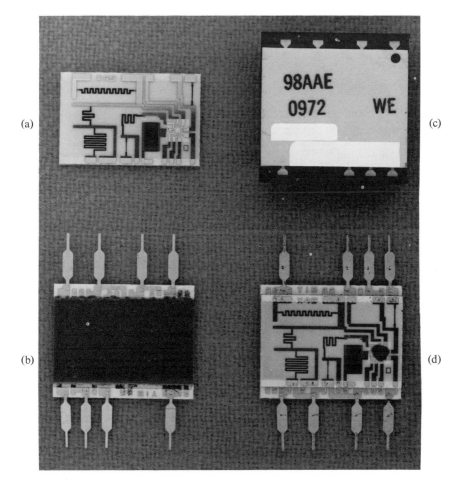

FIG. 6-25. Hybrid integrated network combining tantalum thin-film components with monolithic (beam-leaded) chip. (a) Resistor substrate. (b) Capacitor substrate coated with silicone rubber. (c) Assembly of resistor and capacitor substrates. (d) Final dual in-line package.

economies attainable with thin-film techniques as well as with most other integrated-circuit techniques. Clearly the more circuits obtainable from a master substrate the cheaper their fabrication will be. The circuit size, in turn, depends on the component density permissible with a given technology, i.e., on such factors as the minimum line width for resistor meander patterns and conductor paths, on the maximum resistor sheet resistance and capacitance density per substrate, and so on. As the technologies evolve, these factors are given most attention, since they are directly related to the overall cost of the finished product.

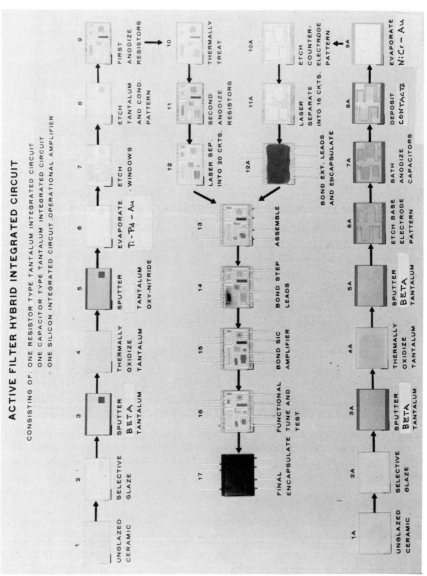

ACTIVE FILTER HYBRID INTEGRATED CIRCUIT

CONSISTING OF: ONE RESISTOR TYPE TANTALUM INTEGRATED CIRCUIT
ONE CAPACITOR TYPE TANTALUM INTEGRATED CIRCUIT
ONE SILICON INTEGRATED CIRCUIT, OPERATIONAL AMPLIFIER

1 — UNGLAZED CERAMIC

2 — SELECTIVE GLAZE

3 — SPUTTER BETA TANTALUM

4 — THERMALLY OXIDIZE TANTALUM

5 — SPUTTER TANTALUM OXY-NITRIDE

6 — EVAPORATE Ti-Pd-Au

7 — ETCH WINDOWS

8 — ETCH TANTALUM AND COND. PATTERN

9 — FIRST ANODIZE RESISTORS

10 — THERMALLY TREAT

11 — SECOND ANODIZE RESISTORS

12 — LASER SEP. INTO 20 CKTS.

13 — ASSEMBLE

14 — BOND STEP LEADS

15 — BOND SIC AMPLIFIER

16 — FUNCTIONAL TUNE AND TEST

17 — FINAL ENCAPSULATE

1A — UNGLAZED CERAMIC

2A — SELECTIVE GLAZE

3A — SPUTTER BETA TANTALUM

4A — THERMALLY OXIDIZE TANTALUM

5A — SPUTTER BETA TANTALUM

6A — ETCH BASE ELECTRODE PATTERN

7A — BATH ANODIZE CAPACITORS

8A — DEPOSIT CONTACTS

9A — EVAPORATE NiCr-Au

10A — ETCH COUNTER- ELECTRODE PATTERN

11A — LASER SEPARATE INTO 16 CKTS.

12A — BOND EXT. LEADS AND ENCAPSULATE

Fig. 6-26. Processing steps involved in fabrication of circuit shown in Fig. 6-25.

6.2 THICK-FILM TECHNOLOGY

The term "thick film" implies that the films used in this technology are thicker than those used in thin-film technology. Indeed, while thicknesses on the order of 50 Å to 20,000 Å are typical in the latter, thicknesses on the order of 125,000 Å (0.5 mil or 12.5 μm) to 625,000 Å (2.5 mils or 62.5 μm) are typical in the former. For this reason, while the thicknesses of thin films are characterized in Ångstrom units, those of thick films are generally characterized in mils or microns. However, as each technology progresses, some thin films are becoming thicker, and vice versa, and there are products of the two technologies available today whose film thicknesses are comparable. Because of this overlap in thickness ranges, a clearer distinction between the two technologies can be maintained by referring to the radically different processes used in each. Where thin films are deposited by vacuum or vapor methods, thick films are deposited by screen-and-fire methods, which involve totally different equipment and materials. Naturally the characteristics of the resulting components differ accordingly.

It will not be necessary, in the following, to go into the same depths of processing and characterization as we did for thin-film technology. For one thing, the terms characterizing thick-film components (e.g., sheet resistance, capacitance density, etc.) are the same; for another the processing itself is simpler. Furthermore, the characteristics peculiar to thin-film components generally apply to thick-film components as well—except, perhaps less decisively. Thus, thick-film components track on a substrate, but not as closely as thin-film components, nor is their stability as good. In fact most characteristics obtainable with thick-film components can be obtained with thin-film components as well—and often better; the reverse is obviously not true. Thus we can already state that thick-film components can be considered for medium-quality networks (in terms of the effects of temperature, aging, humidity, etc); high-quality networks will generally be better served with thin-film components.

Assuming that the reader has familiarized himself with the basic concepts of film technology in the preceding sections we shall now, somewhat briefly discuss the pertinent features of thick-film technology.

6.2.1 The Two Basic Processes in Thick-Film Technology

Thick-film processing consists of two basic processes, namely, silk screening and firing. As we shall see, these processes are much simpler than those involved in thin-film processing. Consequently, the required equipment is much cheaper as, of course, is the resulting product. This inherent cost difference is mitigated somewhat if the total production is large and, in some cases, it can be offset by the higher packing density provided by thin-film technology.

Silk Screening   Silk screening is a technique that has been borrowed from the graphic arts industry. In fact it is completely analogous to the conventional silk-screen printing process used for posters today, and many centuries earlier by the Chinese for their works of art, except that the "silk screen" is now replaced by a fine-mesh (150–350 wires/in.) stainless-steel screen. The mesh is stretched on an aluminum frame and the screen coated with a photosensitive emulsion or resist, which is exposed through a photographic mask and developed. This leaves the screen clear where the thick film is to be deposited and blocked by the fixed emulsion elsewhere. The screen is then placed on a substrate and carefully aligned. Circuit components are deposited on the substrate through a screening process for which a wide variety of tools are available. They are all based on the same principle, involving a squeegee driven across the patterned screen at a constant rate. The squeegee forces a paste through the holes left open in the screen areas where material is to be imprinted on the substrate surface. The paste used depends on whether resistors or conducting paths are being deposited. Separate screens are used for the conductor and resistor areas. Capacitors have been fabricated by these techniques as well, but the capabilities of thick-film capacitors have not progressed nearly as far yet as those of resistors and conductor paths. For this reason, appliquéed chip capacitors are still used most widely with this technology.

The pastes (also called liquid films or inks) used for the screening process have certain properties in common, whether they are conductive, resistive or dielectric. They consist of fine particles suspended in an organic binder. The binder provides the proper silk-screening characteristics, i.e., the necessary viscosity for screening and the lubrication required for the squeegee. It also holds the circuit element in place on the substrate prior to firing. The solids in the paste determine the electrical properties of the component being deposited. A wide variety of metallurgical systems are available for these purposes. They are principally based on mixed powders consisting of noble metals (e.g., platinum, palladium, gold, and silver) and glass or ceramic combinations.

Firing   After the thick-film paste has been silk-screened onto a ceramic substrate, the liquids of the paste dry out and the firing process begins. This generally takes place in a continuous, zoned conveyor furnace or kiln. There are usually four to eight separately controlled zones, with highest-temperature firing first so that later firing will have minimal effects on previously fired patterns. The prescribed firing-temperature profiles for the various thick-film materials determine the actual screening and firing sequence; the material with the highest firing temperature, which is generally the conductor material, is screened and fired first, followed by the material with the next highest temperature, which may be the resistor material, and so on. Typically the furnace temperatures required for firing vary between 500°C to 1000°C. During the firing process the organic binders of the thick-film paste are burned out, or, more specifically, oxidized and vaporized. The remaining paste material, consisting of the metal particles spread uniformly throughout and fired with the glass or ceramic particles into a frit, fuses with the substrate, thereby becoming a permanent part of the overall ceramic structure.

6.2.2. Thick–Film Integrated Components

Thick-film components differ mainly in the pastes, or inks, used in their fabrication. The processes involved, consisting of screening and firing, are basically the same.

Resistors   Thick-film resistor inks can be divided into three classes:

1. Cermets[24] (a mixture of precious metals or precious metal oxides and a glass frit) in an organic material added to render the ink suitable for screen deposition.
2. A precious-metal film suspended in an organic material.
3. Carbon film. This film is of little importance in modern hybrid circuits and will not be discussed further here.

The thickness needed for metal-film resistors is less than that required for cermet resistors, but an underglaze, such as is used with thin-film circuits, is needed to provide a smooth surface; see Fig. 6-27. A low-melting-point overglaze is subsequently applied to the metal film to provide a hermetic seal which protects the metal film from the environment. Resistivity and temperature coefficient of the metal-film resistor are not affected by the choice of substrate material, because of the underglaze. Sheet resistances of from 5 to 10 $\Omega/\square$ are available in metal-film compositions.

Cermet resistors are probably the most widely used of the thick-film type. They offer a wider range of resistivities than metal-film resistors. They are less sensitive to substrate surface roughness, so that special surface preparation such as underglazing is unnecessary. Usually an overglaze is provided for maximum stability and isolation from the environment. Thick-film resistors with this overglaze are rugged structures that are extremely resistant to physical abrasion and damage. Consequently, substrates can safely be stacked for automated handling during manufacture.

---

24. The popularly used term "cermet" (for "ceramic–metallic" composition) was originally an industry trademark.

The most commonly used cermet inks are palladium–silver–glass or palladiumoxide and glass compositions. Several resistive inks are available commercially with sheet resistances varying from 5 to 100 $\Omega/\square$. The resistance is determined by the percentage composition of glass.

Typically, thick-film resistors are between 10 and 30 μm (i.e., between 0.5 and 1.5 mils) thick. Firing temperatures are between 700°C and 800°C. Firing is the critical step in making a thick-film resistor, since under or overfiring causes resistance drift to increase with time or the temperature coefficient to increase. The as-fired resistor tolerance is between 10% and 20%. Usually, however, the resistor is purposely made lower in value than desired and is trimmed to the desired value by removing resistor material. Trimming is generally accomplished by air–abrasive techniques. A fine, abrasive powder, air-propelled through a jet nozzle, is directed at the edge of the resistor, removing some of the material to adjust the resistor to a higher resistance value. Trimming is usually done automatically. Tolerances to less than 1% percent are possible, with a range between 2% and 5% being commonly achieved.

The temperature coefficient of thick-film cermet resistors is, to some degree, a function of the substrate characteristics, but in general lies between -100 and $+300$ ppm/°C. The nominal TCR is determined by the particular material used, but it can vary from lot to lot for a given material.

Conductors   Conductors for hybrid thick-film application are fabricated from metal–glass compositions. The conductor formulations consist of finely divided suspensions of either palladium–silver, platinum–silver, or platinum–gold powder plus a glass frit in an organic vehicle which renders the paste (or ink) suitable for screening. As with the resistor inks, the organic material burns off during the firing process and the glass powder or frit binds the metal–glass composition to the substrate. The inks are sufficiently low in resistivity to be suitable for interconnections between thick-film and discrete appliquéed components. The added-on discrete components can be soldered, welded, or ultrasonically bonded to the conductors without difficulty.

Firing temperatures range from 500°C to 1000°C depending on the ink used; the timing is set to optimize the conductivity, adhesion, and solderability of the conductor.

Because of its versatility and ease of processing, the most commonly used conductive ink is the gold–platinum composition. Generally the conductor is screened and fired first, then the resistor, at a somewhat lower temperature. Resistivity is approximately 0.01 to 0.1 $\Omega/\square$ per mil thickness (approximately 1.3×10^{-5} Ω cm for a conductor thickness of 10 μm, i.e., 0.5 mil), and may be reduced further by oversoldering the conductor. Conductor line widths of 100 μm (approx. 4.5 mils) are possible, but 200- to 250-μm (8- to 10-mil) widths are more easily fabricated. Typical conductor thickness is 10 μm

(0.5 mils). As might be expected, conductor geometry has a significant effect on the stray capacitance and inductance at high frequencies. However, reasonable care in circuit layout should limit problems from this cause to second-order effects. If necessary, stray capacity can be minimized by using a substrate with the lowest possible dielectric constant, e.g., beryllia ($\varepsilon_r = 6.4$) rather than alumina ($\varepsilon_r = 9.2$).

When separated by a layer of glaze, screened conductors form conductor crossovers, thus eliminating the need for double-sided boards or lead interconnections, which are fairly common in printed-circuit techniques. These crossovers typically have a stray capacitance of 0.25 to 1.0 pF.

Capacitors   Screened and fired capacitors require two conductive layers on the substrate, separated by a dielectric film. Values up to 0.01 μF can be produced practically in this way. However, capacitor inks are still in the development stage and the current tendency is to meet any capacitor needs in a thickfilm circuit by attaching discrete chip capacitors to the substrate. The cost of these readily available chip capacitors is low enough that it is seldom, at present, more economical to use the printing technique. As we shall see presently highly rugged chip capacitors are available whose electrical characteristics compare favorably even with those of high-quality discrete capacitors.

6.2.3 The Characteristics of Thick–Film Components

In this section we shall attempt to summarize those characteristics of thick-film components that may be of interest to the active network designer.

Resistors   Some of the relevant characteristics of thick-film resistors are listed in Table 6-12. They are summarized in what follows.

TABLE 6-12. CHARACTERISTICS OF TYPICAL THICK–FILM
RESISTORS

Characteristic	Resistor type	
	Metal	Cermet
Sheet resistance ($\Omega/\square$)	5000–10,000	100–100,000
Minimum line width (μm)	125	125
Resistor spread (for one ink)	10 : 1	10 : 1
Initial precision (%)	±15 to ±20	±15 to ±20
Tracking of initial values (%)	±2.5	±2.5
Trimming accuracy (%)	±1.0	±1.0
Temperature coefficient (ppm/°C) (depends on sheet resistivity)	±500	±100 to ±500

Resistor Range and Spread   Using different inks to cover different resistor ranges, a very wide range of sheet resistances can be obtained. Ranges of the order of 10 $\Omega/\square$ to 100 $\Omega/\square$ are often mentioned for cermet materials; metal films typically cover an order of magnitude less. However, the fact that these ranges can only be attained using more than one ink, therefore requiring a corresponding number of screening and firing runs, is not always pointed out. Actually, a realistic maximum resistor spread that can be covered with a single ink is ten to one; e.g., 1.5 kΩ to 15 kΩ. Typically, at least two (often three and sometimes even four) separate resistor ink patterns have to be printed separately, all adding to cost. As a rule of thumb a resistor spread of 10:1 can be obtained with one ink, a spread of 100:1 with two inks. The absolute resistor values should, if possible, be between 100 Ω and 100 kΩ.

Resistor Precision   Many variables affect the initial, untrimmed values of screened- and fired thick-film resistors. Among the geometric factors affecting the initial precision are substrate uniformity, screen mesh and emulsion, location of resistors on substrate, and shape and size of resistors. Chemical factors include firing profile, atmosphere and contamination, chemical composition, termination materials, and substrate composition. The most important variable, however, is the print thickness; the highest percentage of resistor variations can be accounted for by thickness effects.

Under optimum conditions it may be possible to produce resistors, as fired, to $\pm 10\%$. A more reasonable expectation, however, is $\pm 20\%$. If resistors with 10% or lower tolerances are to be made consistently, the resistors must be trimmed. Air abrasion is one accepted method, providing high-speed automatic trimming readily. Of the many other methods of trimming that have been studied, tried, and perhaps used (e.g., electrolyte, RF discharge, arc, ultrasonic, laser, etc.) the laser looks particularly encouraging. As with thin films, laser-trimmed thick-film resistors can attain stability equal to, or even better than, that of air-abrasion-trimmed resistors.[25] Two of the reasons for this superiority seem to be the absence of dust and the confined area involved in laser trimming. Whether trimming is accomplished by air abrasion (often referred to as "sandblasting") or laser, resistor deposition is deliberately designed to be 15% to 20% low.

Although trimming accuracies of 0.1% and even less are possible under highly controlled conditions, $\pm 2.5\%$ is very much more typical; in fact a practical production limit is $\pm 1\%$. Obviously the wider the permissible tolerances, the less costly the trimming process. In order to reduce the cost of thick-film assemblies, the trend is to design as far as possible with 20% resistors and to allow for functional trimming with a few selected resistors to achieve the DC and AC performance required. Thus the circuit is initially designed in such a way that one or at most two resistors are critical. After final assembly of the circuit, including the appliquéed components such as

25. Provided that microcracks in the resistor material, which sometimes occur adjacent to the laser cut, can be avoided.

active devices, capacitor chips, etc., the circuit is made operative and the critical resistors are adjusted until the desired output is observed. Beside saving on trimming costs, this has the advantage that active devices with relatively wide tolerances can be employed; the initial errors they, as well as residual parasitics of the assembly, cause can be "tuned out" very easily.

Whereas the initial precision of thick-film resistors may be considerably poorer (e.g., 20%) than is usually encountered with discrete resistors (e.g., 5% to 10%), they do have the useful property that the ratio of resistance values on a given substrate, that is, the initial tracking tolerances, are much closer than the absolute values. Thus the tolerance distribution will behave similarly to that shown in Fig. 6-24. Tracking tolerances of 2.5% without trimming, as compared to 15% to 20% absolute, may be considered realizable production targets.

Designing to wide absolute, but close tracking tolerances is a new concept for the circuit designer, who is used to specifying tight individual resistance tolerances. He must now try to work for the widest possible absolute tolerances, at the same time taking advantage of the much closer tracking tolerances, to achieve his results. For example, in discrete designs, if the designer wants to attenuate a voltage with an accuracy of $\pm 10\%$ or less he specifies two 5% resistors for the corresponding voltage divider. In thick-film there is no need to specify 5% resistors. It is more economical to leave the absolute tolerance at 15% or even 20% and to specify an overall tracking ratio within $\pm 10\%$. Thus, absolute tolerances on resistors should be as wide as possible, and the design built around tight resistor ratios. One should, however, take care to specify close tracking tolerances only for resistances of similar value, i.e., values which may be fabricated with one ink. Tracking tolerances will not hold as well for extreme values of resistors, where different inks must be used.

The preceding comments hold as well for thick-film as for thin-film components. In fact, avoiding close tolerance specifications of resistances comes as a byproduct of either thin- or thick-film technology. In either case it is possible to incorporate the film equivalent of a preset potentiometer fairly economically, provided that there are not too many of them. In commercial discrete designs, the designer will try, as far as possible, to avoid specifying a resistor to be selected or adjusted on test. In film circuitry the technique is used widely, particularly since the capability of fine adjustment by resistor trimming exists.

Resistor Stability   Failure in a thick-film resistor may be understood to occur when the resistance variation under specified ambient conditions exceeds an arbitrarily defined value. Some manufacturers prefer to speak of the resistance variation resulting from "all causes" (i.e., TCR *and* permanent drift or aging), while others choose to consider the resistor variation resulting from the TCR separately. The all-causes resistor variation is probably more realistic, so it will be used here in discussing stability.

Like conductivity itself, the drift mechanism in thick-film resistors is not too well understood. Moreover, it probably differs considerably for the chemically different resistor systems used. In the palladium–silver resistor system, which is the oldest and best known, it is possible to observe at least three different drift-controlling effects: temperature, current, and voltage gradient. With great oversimplification it can be said that for sheet resistances below 1 $\Omega/\square$ the effect of temperature appears to dominate; between 1 and 15 $\Omega/\square$ current dominates, and above 15 $\Omega/\square$ the voltage gradient dominates.

It is known, for example, that over a rather wide resistivity span, increasing the resistor temperature (without power) increases the drift in a positive direction (increasing R). It is also known that when power is increased in an environment where temperature is not permitted to rise, the drift is increased, but in a negative direction (decreasing R). Thus, in a given practical case, temperature and power effects tend to compensate one another and the resistors under load are actually more stable than those without load.

The stability claimed by different thick-film-resistor manufacturers varies appreciably. Typically, though, the all-causes resistance variations may be broken down as follows: initial tolerance (after trimming) 0.1 % to 1 %; TCR effect, 0.5 % to 5 %, and permanent resistance variation 0.5 % to 3 %. Thus the total variation may be between 1 % and 10 %, approximately.

In general it can be said that with respect to stability thin films are superior to thick films. Typically, thick films are quoted at a maximum change of 1 % to 2 % per 1000 hours at 100°C for rated load. Rated load is usually 3 W/cm^2 (i.e., 20 W/in.^{2}) of film. Under similar stress conditions, tantalum nitride on glazed ceramic has a maximum expected change of 0.5 %. Under less stress, the difference in load life between thick films and thin films is even greater.

Because the screen-and-fire operation is far less expensive than the processes required in thin-film technology, thick-film resistors are useful wherever tolerances and load life variations of 2 % to 5 % are acceptable, and where TCR is of no great importance. However, since stability (i.e., component variation) is of extreme importance in linear active network design, whereas power-handling capability, and even resistor spread, are much less so, thin-film resistors (in combination with thin-film or chip capacitors) will very likely remain the preferred components for high-quality linear active networks. For low-Q circuits and circuits in which relatively wide frequency variations can be tolerated, thick-film microcircuitry, with its special advantage of wide range of sheet resistances and, above all, low cost, should be given serious consideration.

Capacitors   Thick-film capacitors are not nearly as common as thick-film resistors, although they appear to offer some promise for the future. For the time being, they still have the following drawbacks:

1. For acceptable yields, capacitance values should be limited to less than 1000 pF.
2. Capacitance tolerances are greater than $\pm 20\%$. This tolerance may be acceptable for coupling and bypass applications, but for precision networks it will very likely be unacceptable.
3. The temperature coefficients are very large (between 500 and 3000 ppm/°C).
4. Large-valued capacitors for coupling and bypass applications (where drawbacks 2 and 3 do not constitute a severe problem) require large areas on the substrate.

Nevertheless, dielectric film materials have become available that provide improved temperature coefficients and higher dielectric constants. Furthermore, the capability of air-abrasively trimming capacitors to tolerances of about 2% already exists. However, in general, deposited capacitors are still more costly and require more space than discrete chip capacitors. The latter are readily available in a large range of values and are used widely, both in combination with thick- and thin- (e.g., nichrome) film resistors.

Ceramic chip capacitors, or monolithic ceramic capacitors as they are also called, are made by screening noble-metal electrodes on strips of green unfired ceramic, stacking the layers, and firing at approximately 1300°C to fuse the whole into a solid, monolithic unit, as shown in Fig. 6-28. The units are finished off by metallizing the two ends where alternate electrodes protrude. Usually these metallized ends are presoldered, so that the chips can be simply attached to hybrid circuit conductors by placing them down and reflow soldering. Welding and ultrasonic bonding can also be used, but at increased cost. Individual dielectric layers may be as thin as 25 μm, and stacks of up to 80 or 100 layers may be obtained. Chip capacitors also serve as crossovers, since conductors may be run underneath them. Silver–palladium or gold electrodes are used to solder the chip capacitors to the conductor pads. Typical electrical characteristics of available chip capacitors are shown in Table 6-13.

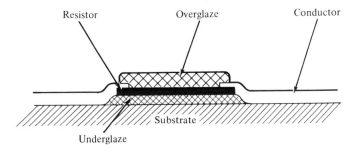

FIG. 6-27. Metal-film resistor on ceramic structure.

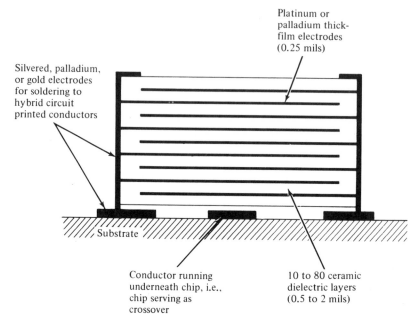

Platinum or
palladium thick-
film electrodes
(0.25 mils)

Silvered, palladium,
or gold electrodes
for soldering to
hybrid circuit
printed conductors

Conductor running
underneath chip, i.e.,
chip serving as
crossover

10 to 80 ceramic
dielectric layers
(0.5 to 2 mils)

Substrate

FIG. 6-28. Monolithic ceramic chip capacitor.

The quality of available chip capacitors surpasses that of thick-film capacitors in range of available values, in long-term stability, and in temperature coefficient. On the other hand, chip capacitors are not, strictly speaking, integrated components; they will not, therefore, track as well as their integrated counterparts in terms of initial values or with temperature. Furthermore, they are not as compatible with batch processing methods. Thus, assuming that thick-film capacitors will ultimately be available with the same qualities and high yields as chip capacitors, they will presumably provide a cost advantage in the long run.

TABLE 6-13. CHARACTERISTICS OF TYPICAL
CERAMIC CHIP CAPACITORS

Capacitance range:	10 pF to 0.1 μF
Capacitance tolerance:	0.5 % available
Temperature coefficient:	NPO ± 30 ppm/°C
Aging:	less than 0.1 % available
Self-resonant frequency:	>2 GHz
Working voltage (DC)	25–500 V depending on C value
Quality Factor (Q):	10,000 at 1 MHz

6.3 MONOLITHIC[26] SILICON INTEGRATED CIRCUIT TECHNOLOGY

In the context of linear active networks we need concern ourselves only with *linear* silicon integrated circuit (SIC) technology. The most important product of this technology has been the operational amplifier, which we shall discuss in more detail in Chapter 7. Here we shall briefly review the processes involved in SIC technology and the characteristics of the resulting components.

6.3.1 The Three Basic Processes in SIC Technology

Epitaxy[27]   Epitaxial growth is a chemical reaction in which silicon is precipitated from a gaseous solution, or vapor phase, and grows in a very precise manner, i.e., as a thin film of single-crystal (monocrystalline) silicon, on the surface of a silicon wafer placed in the solution. To form the epitaxial layer, the silicon wafer, or slice, is heated by RF energy in a special furnace, called an epitaxial reactor, in which vapors containing silicon and the desired dopants are passed over the slice. Silicon and dopant atoms from the vapor skid about on the surface of the growing epitaxial film until they find a correct position in the lattice and become fastened into the growing structure by interatomic forces.

The important features of the epitaxial process are as follows:

1. The atoms of the newly grown layer are arranged in single-crystal fashion on the single-crystal substrate. Moreover, the lattic structure of the newly grown layer is an exact extension of the substrate crystal structure. This is illustrated in Fig. 6-29.

2. In contrast to the procedure used for producing bulk semiconductor crystals in which a single crystal is grown from the liquid phase, the epitaxial process involves growth from the gas phase and no portion of the system is at a temperature anywhere near the melting point of the material.

3. Unlike diffusion, epitaxial growth forms regions of very uniform resistivity.

4. Epitaxial layers can be grown over previously diffused surfaces and may be of the same or the opposite conductivity type as the substrate wafer.

5. The impurity concentrations in the epitaxial layer can be controlled within wide limits and complex impurity profiles may be grown.

26. From the greek "monos" = single and "lithos" = stone.
27. From the greek "epi" = upon and "teinein" = arrange.

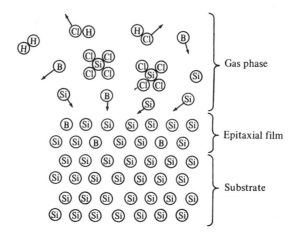

FIG. 6-29. Growth of an epitaxial silicon film, with dopants (boron), from the gas phase onto a single-crystal silicon substrate.

Diffusion   The diffusion process generally takes place in two steps. First a high concentration of dopant atoms, or "carriers," is brought in contact with the surface of a silicon slice for a relatively short time. This is called the "predeposition" and is intended to bring the proper number of carriers onto the surface of the slice. In the second step the dopant source is removed and, under the influence of high temperature, the dopant atoms on the slice surface penetrate the slice, i.e., they are dissolved in the still-solid silicon crystal according to well established physical laws. The depth of penetration and the concentration within the slice depend on time, temperature, the kind of dopant, and its original concentration. In general the dopant concentration has a maximum value near the surface of the solid slice, falling off in a Gaussian or complementary-error-function fashion within the slice itself. If the diffusion cycle were to proceed for a sufficiently long time, the dopant would become uniformly distributed throughout the wafer. In practice, however, it is desired to form junctions in the wafer, and the diffusion cycle is terminated long before this type of saturation occurs; the graded distribution of the dopants is thereby locked in.

The important features of the diffusion process are as follows:

1. It permits dopants to be introduced into a crystal without destroying the crystal (as opposed to alloying, for example, which does destroy the crystal).

2. Most of the circuit elements (transistors, resistors, etc.) on a silicon integrated circuit slice are formed by a number of diffusion steps. Thus the same slice area must often be subjected to diffusion a number of times.

3. To prevent a continuation of diffusion by previously diffused dopants when exposed to high temperatures a second time, dopants with differing diffusion constants can be selected. Slowly diffusing dopants (i.e., dopants

possessing small diffusion constants) are thereby used for the first and deepest diffusion. Subsequent diffusions can then be carried out with faster diffusing dopants at a lower temperature and for a shorter time so that they have little effect on the previously formed regions.

4. The diffusion constants of certain dopants commonly used with silicon are much greater in silicon than in silicon dioxide (SiO_2). For this reason SiO_2 layers are widely used as "masks" in diffusion procedures, where the SiO_2 layer itself is produced by diffusion in a diffusion furnace maintained at very high temperatures (between 900°C and 1300°C). This key process in SIC technology is commonly referred to as *oxidation*. As the oxide layer to be used as a mask in subsequent diffusion steps forms, the rate of growth decreases with time because of the time necessary for the oxygen to diffuse through the already formed SiO_2 so that the reaction can continue. Thus, the thicker, the layer, the slower the process becomes.

5. Regardless of the diffusion technique, there is a limit to the maximum concentration of dopant that can diffuse into silicon. This maximum concentration depends on the "solid solubility" of the dopant, a property limiting the number of atoms that can be dissolved per unit volume of silicon. Solid solubility is a function of temperature.

6. The dopant diffuses sideways into the slice as well as perpendicularly. Diffusion into the slice sideways under an oxide layer means that the *p-n* junctions are never exposed to the contaminating atmosphere (see Fig. 6-30). Depending on the process and the material, sideways diffusion may proceed at anywhere between the same rate as, to less than half, the rate of perpendicular diffusion.

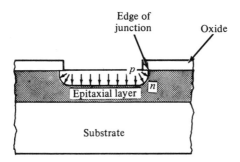

FIG. 6-30. Diffusion into slice is sideways as well as vertical; thus edge of *pn* junction is under oxide.

Photolithography   This process is essentially the same as that described in the discussion of thin-film technology. Thus, for a silicon dioxide layer to be patterned, for example, it is first covered with a film of photoresist. A photographic mask (generally a chrome pattern on glass) with the desired circuit pattern is then placed over the slice. On exposure to ultraviolet light, the portions of photoresist under the transparent areas of the mask become polymerized (i.e., hardened). Subsequently, the unpolymerized areas are dissolved away. The portions of the oxide layer not protected by the polymerized photoresist are then removed by dipping the slice into an etching solution. Finally, after the processing in question, the remainder of the photoresist (i.e., the polymerized area) is dissolved with another solution.

6.3.2 The Six–Mask Planar Epitaxial Process

The planar process, which is the foundation of SIC technology, is based on the fact that layer upon layer of differently doped semiconductor material can be formed from the same side of a silicon slice. Previously, devices had been manufactured by working from both sides of a slice; the individual processes were difficult to control accurately in three dimensions and thin, fragile slices were needed. With the planar process it is possible to realize linear integrated silicon devices by a six-mask sequence of the epitaxial, diffusion, and oxidation fabrication steps that were described in the previous section. This six-mask planar epitaxial process, as it is often called, has become the standard process for high-volume production because it has been found to be the lowest in cost.

A cross section of some of the basic components (i.e., pinch resistor, *n-p-n* transistor, lateral *p-n-p* transistor, and standard resistor) which can be made with the six-mask planar epitaxial process is shown in Fig. 6-31. Referring to this figure, the process can be broken down into the following basic steps.

1. *Buried Layer* (*Mask 1*). Slice manufacture starts with a high-resistivity *p*-type substrate, the silicon itself. In order to reduce the collector resistance of the planar transistors to be generated, *n*-type islands (referred to as n^+ regions because of their high dopant concentration which, in turn, guarantees low resistivity of the material) are diffused into the high-resistivity *p*-type substrate at each collector location. This so-called "buried layer" serves to reduce the effects of the parasitic vertical *p-n-p* transistor consisting of the base, collector, and substrate layers (see Section 6.3.3) as well as to short-circuit the collector series resistance, thereby reducing parasitic capacitance.

2. *Epitaxial Growth.* A high-resistivity (2 to 4 Ω cm for standard linear ICs) *n*-type layer is epitaxially grown over the entire surface of the slice. A *p-n* junction is thereby created across the entire slice; this is used to isolate the circuit elements from each other. The epitaxially grown *n*-type material forms the collector of the *n-p-n* transistors and the base of the lateral *p-n-p*'s.

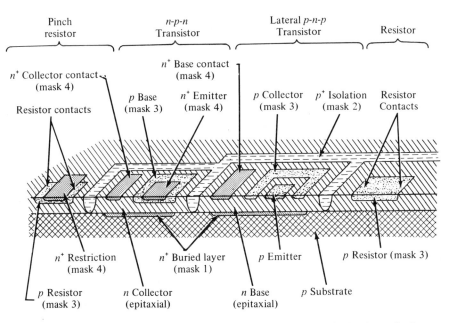

FIG. 6-31. Some of the integrated components that can be made with the six-mask planar epitaxial process include pinch resistors, *n-p-n*, and lateral *p-n-p* transistors.

3. *Isolation Diffusion* (*Mask 2*). During this and all subsequent diffusions, an oxide layer, generated by oxidation on the slice surface prior to the actual diffusion, acts as a mask against the dopant vapor. With this mask the epitaxially grown *n*-type material is separated into islands with a p^+ (i.e., high-dopant-concentration) isolation diffusion. This isolation diffusion converts the *n*-type epitaxial layer into *p*-type material in thin strips surrounding the circuit elements. At a temperature of about 1200°C, the slice is exposed to a gas containing *p*-type dopants until the new *p*-type regions join the underlying *p* substrate, thus surrounding the elements with a *p-n* junction and isolating them from each other electrically.

4. *Base Diffusion* (*Mask 3*). After the isolation diffusion, the oxide layer is formed once more and the slice covered with photoresist, as it was prior to step (3). The new layer of photoresist is exposed through a mask which outlines all the areas to be diffused with a *p*-type dopant to form the transistor base regions, the diode electrodes (e.g., anodes,[28] if formed from *n-p-n* transistors), and the resistors. The base diffusion is also applied to the isolation regions (which are also *p*-type material) as indicated in Fig. 6-32, in order to prevent surface leakage across the isolation barriers and thereby further

28. The term "anode" implies that the current flow is from the *p* base to the *n* emitter. Actually, nearly all the current (except for a small base drive) flows from the collector *across* the base to the emitter.

improve circuit performance. The unexposed portions are then removed and the unprotected oxide etched away. In a furnace, the slice is then exposed to a gas containing *p*-type dopant and the relatively shallow *p*-type base diffusion takes place.

5. Emitter Diffusion (Mask 4). The oxide layer is formed once again, and the slice covered with photoresist. The fourth mask is placed on top and the photoresist is exposed, outlining all the *n*-type regions, including transistor emitters, diode electrodes (e.g., cathodes if formed from *n-p-n* transistors) and collector contact regions. The emitter diffusion, which follows the removal of photoresist and oxide, produces an n^+-type region of low resistivity and still shallower depth than in the base diffusion. This diffusion basically completes the integrated circuit structure, which is left entirely covered with insulating silicon dioxide. The two final steps involve making contact to the individual devices and interconnecting them.

6. Contact Windows (Mask 5). Since the entire slice surface is once again covered with an oxide layer, a new mask is required to outline the areas through which contact will be made to all the devices. After exposure and removal of unexposed photoresist, the oxide is removed in those areas.

7. Interconnect Metallization (Mask 6). Once the contact windows through the isolating oxide layer have been generated in the preceding step, a thin film of metal, usually aluminum, is deposited on the slice. More photoresist is spread on top of this film and exposed through a mask which contains the interconnection patterns. After the unexposed portions of the resist have been removed, the regions of the metallic film not needed for the interconnection of the devices are etched away. The silicon dioxide layers on top of the silicon structure insulate the interconnection paths from the underlying devices. Large areas of metal are left at the periphery of the circuit as bonding pads for lead wires.

The circuits are now complete and can be tested on the slice. To mount the chips in individual packages, break lines are scribed between the circuits with a diamond tool and the individual chips broken apart.

If beam leads are used instead of bonded lead wires, three steps in addition to step (7) are required, after the contact windows in step 6 have been generated. They involve the fabrication of the gold beams used to bond the finished chip to a substrate.

In the standard six-mask planar epitaxial process described above, the selection of resistivities and diffusion schedules is centered around making high-quality *n-p-n* transistors. In fact, it is possible to make these devices with characteristics comparable to almost any discrete transistor. On the other hand, the performance of lateral[29] *p-n-p* transistors is inferior to that

29. Beside parasitic, or substrate *p-n-p* transistors (see Section 6.3.3) these are the only *p-n-p* transistors which can be made without process or design complications.

TABLE 6-14. CHARACTERISTICS OF TYPICAL SILICON
INTEGRATED SHEET RESISTANCES

	Characteristic		
Sheet resistance	Typical value ($\Omega/\square$)	Tolerance (%)	Temperature coefficient (ppm/°C)
Base diffusion	140	± 20	2000
Emitter diffusion	2.5	± 30	100
Epitaxial layer	200	± 50	2000
Deposited resistance layer	50–1000	± 5	100
Pinch resistor	5000–30,000	$+100, -50$	3000–5000

of their discrete counterparts. For one thing, conventional lateral *p-n-p* transistors have a very much lower current gain (typically between 5 and 50 depending on processing). Also, the usable bandwidth is very much narrower, since serious phase shift begins to appear at around 2 to 6 MHz. Consequently, care must be exercised when designing lateral *p-n-p*'s into a circuit.

Resistors are normally made using the base diffusion which, typically has a sheet resistance of about 140 $\Omega/\square$, although this may vary between 100 to 200 $\Omega/\square$ depending on the manufacturer (see Table 6-14). To obtain high resistivity, pinch resistors with a typical sheet resistance of about 15 k$\Omega/\square$ can be made; they allow the use of high-value resistors while taking up very little chip area. Unfortunately, they exhibit low breakdown voltages (typically about 5 V) and an FET-like nonlinear voltage characteristic. In addition, control of the resistance value is poor.

6.3.3 The Characteristics of Silicon Integrated Circuit Components

Resistors   Silicon integrated resistors can be made by either diffusion or deposition. Diffused resistors are formed at the same time as transistors, diodes, and diffused capacitors; no additional processing steps are necessary. However, diffused resistors exhibit a large variation of resistance with temperature, and the available range of resistance is limited. Deposited-film resistors, on the other hand, require additional processing steps; therefore they are more expensive. However, they offer a much wider range of resistance, a lower temperature coefficient of resistance, and lower stray capacitance.

A comparison, from a network designer's point of view, between the characteristics of diffused silicon and tantalum thin-film resistors is given in Table 6-15. This comparison shows that it is only reasonable to use silicon diffused resistors if medium values with wide tolerances and high temperature coefficients can be tolerated. Somewhat lower tolerances and smaller temperature coefficients (± 1000 ppm/°C) can be obtained with deposited resistors.

TABLE 6-15. CHARACTERISTICS OF TYPICAL DIFFUSED SILICON AND
THIN-FILM TANTALUM RESISTORS

	Resistor type		
	Diffused silicon		
Property	Type 1**	Type 2	Tantalum thin-film
Sheet resistance ($\Omega/\square$)	100–200		100 max.*
			3000†
Minimum line width (μm)	10 ($\pm 20\%$ tolerance)		50
	5 ($\pm 30\%$ tolerance)		
Maximum resistance (kΩ)	20	60	<1000
Minimum resistance (Ω)	50	20	10
Initial precision (%)	± 20	± 40	standard: ± 1.0
			available: ± 0.01
Temp. coeff. (ppm/°C)	+1500 to +2000		-50 to -100 ± 15*‡
Tracking (ppm/°C)	± 50		± 5
Cost of real estate ($/cm²)	8.00		0.15

* Ta_2N.
† Low density.
‡ Available in prescribed values.
** Types 1 and 2 represent the material characteristics of various manufacturers.

Compared to the precision of initial resistor values, matching and temperature
tracking between identical resistors in close proximity on a chip is very good.
As with the other integrated circuit techniques discussed earlier, this property
can be exploited by designing circuits such that their performance depends
on the ratios of similar resistors, rather than on absolute values. In fact,
in some cases, the ratios may be very much more precise and predictable than
they would be for discrete components. One reason for this is the uniformity
of the temperature on a chip. Since all elements are within a short distance
of each other, and since they are all part of a monolithic chip of fairly high
thermal conductivity, they are likely to be at very nearly the same temperature during operation, regardless of ambient temperature gradients.[30] If the
temperature of the chip changes, the temperature of all elements on the chip
change similarly and the ratios of the element values tend to stay the same.
Because of this close thermal tracking effect, silicon integrated circuits may
actually be more stable than their discrete counterparts.

Capacitors   Two basic types of capacitor are available to the silicon-
integrated-circuit designer, namely, the thin-film integrated capacitor and
the junction capacitor. Both consist of two low-resistance layers (" plates "),
separated by a carrier-free region. In a thin-film capacitor, one of the plates is

30. It should be pointed out, however, that in very high-gain devices such as operational amplifiers,
thermal feedback can be troublesome nevertheless. Consider the fact that the temperature
variation of V_{BE}, the base emitter diode, is -2.4 mV/°C. Thus for a 0.1°C mismatch,
ΔV_{BE} will be -0.24 mV. With a device gain of, say, 50,000 this error in V_{BE} may be amplified,
in the open-loop condition, producing an error voltage at the output of 12 V.

formed by deposition of a metal film, and the carrier-free region is formed by a dielectric material. In junction capacitors, both plates are formed by diffusion of low-resistance layers (of opposite dopant type), and the carrier-free region results from the depletion of charges at the p-n junction. In either case, the maximum value of available capacitors is very limted, i.e., on the order of 50 to 100 pF. Furthermore, fabrication may require additional processing steps, increasing costs and decreasing yield. Typically, tolerances of diffused capacitors are on the order of $\pm 25\%$, those of the deposited -or thin-film capacitors are on the order of $\pm 20\%$, and matching of identical capacitors in close proximity may be on the order of 3% to 5%. Because of the very limited values and the wide tolerances, the only capacitors generally included on a silicon monolithic chip are low-value, frequency-compensating capacitors used in operational amplifiers. Capacitors on a silicon monolithic chip are not used as components of frequency-selective active networks, nor, for that matter, are resistors that are fabricated on a chip.

Transistors   Although we are not here directly concerned with the design of silicon integrated circuits, it is of interest, in order to appreciate their performance, to know some of the noteworthy characteristics obtainable with integrated n-p-n transistors (n-p-n diodes), as well as some of the constraints imposed by lateral p-n-p transistors. Typical characteristics of these devices are listed in Table 6-16. It will be observed that, in contrast to silicon integra-

TABLE 6-16. CHARACTERISTICS OF TYPICAL INTEGRATED TRANSISTORS AND DIODES

Characteristic	Symbol	Typical value	Tolerance	Temperature coefficient
n-p-n transistors and diodes				
Current amplification factor	h_{FE}	75–300	$+50, -30\%$	$+0.5\%/°C$
Tracking of h_{FE}	Δh_{FE}		$\pm 10\%$	$+10.5\%/°C$
Base-emitter diode forward voltage drop at $I_E \approx 1$ mA	V_{BE}	0.7 V	$\pm 6\%(\pm 40\ mV)$	$-2.4\ mV/°C$
Tracking of V_{BE}	ΔV_{BE}		± 4 mV	$\pm 10\ \mu V/°C$
Base-emitter diode reverse breakdown voltage	BV_{EBO}	6–8 V	$\pm 5\%$	$+3\ mV/°C$
Collector-base breakdown voltage	BV_{CBO}	<45 V	$\pm 30\%$	
Collector-substrate breakdown voltage	BV_{CS}	<60 V		
Lateral p-n-p transistors				
Current amplification factor	h_{FE}	5–50	$+100\%, -50\%$	$+0.5\%/°C$
Base-collector and base-emitter breakdown voltage	BV_{CBO}, BV_{EBO}	<45 V		

ted resistors or capacitors, n-p-n transistors have characteristics at least as good as, and in some respect better than, those of their discrete counterparts. This is a result of the fact that the planar process now used to fabricate silicon integrated circuits was originally developed for discrete transistors; the integration of transistors therefore requires fewer compromises than that of any other components.

Although the integrated transistor is almost identical in performance with a discrete transistor, there are some differences. Unlike its discrete counterpart, the contact to the collector region of an integrated transistor is made through the top surface, not through the substrate (see Fig. 6-32), since the substrate must be electrically isolated from the collector region. However, the current path between the collector contact and the actual collector–base junction is through a narrow region of high-resistivity n-type epitaxial material. Thus, the collector series resistance of integrated bipolar transistors is not negligible. This may be a disadvantage when low saturation voltages or large current-driving capabilities are required. The buried layer, that is the n^+-type region diffused in the collector area, as described in Section 6.3.2, solves this problem to a large extent, by shorting out the collector resistance. The collector series resistance can be decreased even more by adding a processing step between the isolation and base diffusions. A so-called "deep-collector" diffusion path, i.e., an n^+ path joining the deep collector with the collector contact, can be introduced. Clearly this means an added processing step and therefore added cost. The deep collector is not generally used in operational amplifier design, because the series collector resistance can be decreased sufficiently by the buried-collector layer alone.

The fact that an integrated transistor is isolated from the other components by a p-n junction introduces effects not found in discrete transistors. The junction is between the collector and the substrate, and when it is reverse-biased (as it must be to isolate the device), a capacitance is created between these two regions which degrades the performance of the transistor at high frequencies. In this respect, then, the integrated n-p-n transistor may well be somewhat inferior to its discrete counterpart. In addition, there is some leakage current between the collector and the substrate which can become significant in low-current applications. Furthermore, the isolation junction creates a parasitic transistor. Just as the top three layers of the structure in Fig. 6-32 form an n-p-n transistor, the bottom three layers form a parasitic p-n-p transistor. Under certain voltage conditions, when the collector and the substrate are negative with respect to the base of the n-p-n transistor, the parasitic transistor is in the active region and can conduct significant current between base and substrate. In linear circuits however, this condition can be avoided relatively easily, and the buried layer further reduces the effects of this transistor.

A very noteworthy characteristic of integrated transistors is the very close matching of V_{BE} (base-emitter forward voltage) between identical transistors

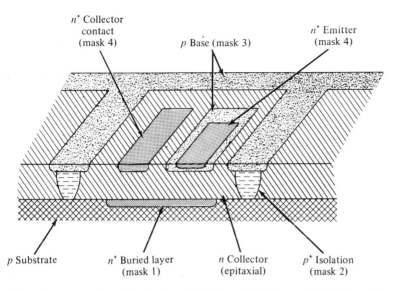

n^+ Collector contact (mask 4)

p Base (mask 3)

n^+ Emitter (mask 4)

p Substrate

n^+ Buried layer (mask 1)

n Collector (epitaxial)

p^+ Isolation (mask 2)

FIG. 6-32. An n-p-n transistor showing base diffusion applied also to the isolation regions to prevent surface leakage.

in close proximity. The base-emitter forward voltage of a transistor is typically on the order of 700 mV, yet the base-emitter voltages of two identical transistors in close proximity will not differ by more than 2 mV or so, which is less than 0.3%. Thus V_{BE} will track to within 10 μV/°C with temperature, or by 0.6 mV over a temperature range of 60°C. As we shall see in the next chapter, this outstanding feature of integrated transistors can be utilized to great advantage in designing linear active networks such as operational amplifiers. Notice, however, that unlike the V_{BE} of transistors in close proximity, the current gains h_{FE} do not necessarily track particularly well.

6.3.4 Three Basic Rules for Silicon Integrated Circuit Design

The reader may well be somewhat disheartened when he compares the characteristics of silicon integrated circuit components, as described in the preceding section, with those of discrete components, to which he has long become accustomed. The limitations and constraints imposed by silicon integrated components are indeed numerous. To name the most serious ones, resistor tolerances and temperature coefficients are large, the range of permissible capacitor values is very limited, inductors are forbidden altogether, p-n-p transistors perform considerably worse than their discrete counterparts, and zener diodes are only realizable in the form of base-emitter breakdown diodes or multiples thereof, and they introduce appreciable internal circuit noise at that.

Fortunately, this is not the whole story, for if it were, silicon technology would be unacceptable in linear circuits for all but the very simplest of applications. We know, of course, that this is not so at all, and that on the the contrary, countless electronic systems in industrial, military, and commercial fields exist today, with highly sophisticated linear microcircuits that exceed even the most optimistic predictions of less than a decade ago. However, it is well to remember that this success only materialized after linear integrated circuit design was approached as a new discipline, obeying rules that differ significantly from those used for discrete-component circuit design.[31] The rules are few and simple, yet unconventional. In essence, they can be summarized by three points.

1. Capacitors and large-valued resistors should be eliminated or their number at least minimized, even if it takes several more diodes and transistors to do so.

The justification for this rule is not hard to find. It is fairly accurate to assume that the basic cost of processing a silicon slice is constant over a wide range of circuit (i.e., chip) size and complexity. Thus the smaller the individual chip, the greater the number of chips per slice and the lower the cost of each individual chip. Furthermore, the smaller the chip size, the lower the probability of a random defect occurring within the chip area, causing it to be a reject. Consequently, cost is a direct function of chip area. Within this framework we must now consider the relative sizes of resistors, capacitors, diodes, and transistors. If all of these components are made with the same diffusion process, the only factor that determines their relative cost is their area. Based on this criterion, capacitors are the most expensive, while transistors and diodes are the lowest-cost devices. If we consider the case that not diffused, but deposited resistors and capacitors are used, then this conclusion is only reinforced, since these devices require additional processing steps.

2. The utmost use should be made of the matched characteristics of components (e.g., resistors, transistors, and diodes) on a chip, and the close tracking of these components with ambient variations.

The reasons for this have been discussed earlier. The disadvantage of wide parameter tolerances can be overcome by the advantage of close parameter matching. Where production variations in sheet resistivity may, for example, cause 15% to 20% tolerances in the absolute value of resistors on a chip, the facts that they are processed in the same diffusion furnace and located within a few thousands of an inch of each other will cause them to maintain their resistance ratio to within 2% to 3%. If they are located close together on the chip their temperature coefficients will be nearly identical and they will track

31. R. J. Widlar, Some circuit design techniques for linear integrated circuits, *IEEE Trans. Circuit Theory*, **CT-12**, 586–590 (1965). Design techniques for monolithic operational amplifiers, *IEEE J. Solid-State Circuits*, **SC-4**, 184–191 (1969).

closely with temperature. The same applies to diode and transistor parameters; these are matched automatically within a monolithic chip to a much greater accuracy than is obtainable with discrete devices—even after the latter have gone through expensive testing and sorting processes. Not only does this initial matching come free with integrated circuits, but the planar process also insures matched aging characteristics over the life of the equipment.

   3. For economical feasibility, as high a degree of process standardization as possible should be achieved.

The concept of circuit standardization is vital to the economical success of linear integrated circuits and introduces design guidelines that differ fundamentally from those pertaining to conventional circuit design. While monolithic linear integrated circuits can be produced very inexpensively in high volume, there are minimum costs that, if amortized over a small volume, make the individual circuits too expensive. Thus integrated circuits designed with the fewest number of components to satisfy a particular requirement may be very much less economical than more complex designs that cover a larger variety of applications, thereby increasing the total production volume. High production volumes can also be obtained by taking the same approach with linear circuits that has proven to be highly successful with integrated digital circuits, i.e., to design linear networks of a given complexity using multipurpose building blocks of a lesser complexity; the latter can be used in a large variety of applications and are thereby required in large quantities.

CHAPTER

7

SILICON INTEGRATED DEVICES FOR LINEAR ACTIVE NETWORK DESIGN

INTRODUCTION

In the preceding chapter we acquainted ourselves with the technologies used to fabricate linear integrated circuits. More important still, we familiarized ourselves, as designers of linear hybrid integrated networks, with the characteristics, constraints, peculiarities, and advantages of components and circuits resulting from these technologies. Thus we are now in a position to discuss some of the basic circuits that have been developed for these technologies and to appreciate the main concepts that were used in their design. The examples discussed will demonstrate how closely related a successful approach to circuit design is, and must be, to the technology used for its implementation. Proven circuit designs of an established technology cannot simply be forced onto new, emerging technologies; they must be appropriately modified and adapted to each new technology, to fully benefit from the characteristics it affords. Only in this way can a combination of circuit design and technology that is optimum with respect to performance and cost be guaranteed.

7.1 BUILDING BLOCKS FOR SILICON INTEGRATED CIRCUIT DESIGN

We saw in the preceding chapter that to optimize a circuit for silicon monolithic implementation one must take the following features of that technology into consideration:

1. Capacitors and high-valued resistors take up more chip area than transistors and diodes, and their number should be minimized, even if it takes more transistors and diodes to do so.
2. Tolerances on absolute component values tend to be high (e.g., $\pm 15\%$ to $\pm 20\%$ for resistors).
3. Tracking of component values on a chip is much closer than would be possible with discrete components (e.g., resistors to within a few percent, V_{BE} of transistors with equal geometry and operating conditions to within a few millivolts).
4. High quality *n-p-n* transistors are obtainable on a chip; lateral *p-n-p*'s requiring no extra processing are limited both in current gain and frequency response.

As we shall see in the following examples, the building blocks most commonly used in silicon integrated circuit designs rely heavily on the tracking capability of resistors and transistors on a chip; *p-n-p* transistors are used only where necessary for biasing purposes, and where their limited frequency response has no adverse effects.

7.1.1 The Differential Amplifier

The circuit diagram of a differential amplifier is shown in Fig. 7-1. This circuit is also used in discrete-component circuits, where it is sometimes referred to as a "long-tailed pair." The performance of the circuit depends mainly on how closely the two transistors T_1 and T_2 can be matched. Where this requirement was one of the major problems in discrete circuits, it comes

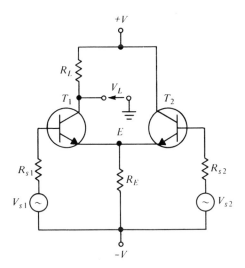

FIG. 7-1. Differential amplifier; V_{s1}, V_{s2}: signal sources; R_{s1}, R_{s2}: source impedances.

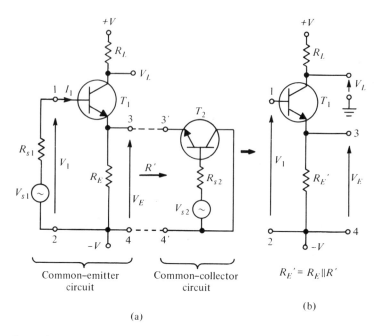

FIG. 7-2. Equivalent diagram of differential amplifier. (a) Parallel connection of common-emitter and common-collector transistor stages. (b) Equivalent common-emitter stage.

free on a silicon chip. In discrete circuits, matched transistors had to be selected by an expensive testing and sorting process and, if possible, the two matched transistors were mounted with a common heat sink. On a monolithic chip, the two transistors are matched automatically to a greater accuracy than was ever possible with discrete transistors, in spite of the precautions taken with the latter. Furthermore the planar process insures matched transistor stability over life.

To understand the operation of the differential amplifier, it is helpful to interpret transistor T_1 as a common-emitter amplifier with a voltage-variable, current-feedback resistor R_E' at its emitter, as shown in Fig. 7-2. The current feedback combination R_E' consists of the actual current-feedback resistor R_E in parallel with T_2 in the common-collector mode. Using the low-frequency hybrid transistor equivalent circuit shown in Fig. 7-3, the equivalent circuit in Fig. 7-2 permits a simple analysis of the differential amplifier.

Input Impedance   We shall make the assumptions here that the h parameters of transistors T_1 and T_2 are identical and that

$$|h_{re} V_c| \ll |h_{ie} I_b| \tag{7-1a}$$

$$\frac{1}{h_{oe}} \gg h_{ie} \tag{7-1b}$$

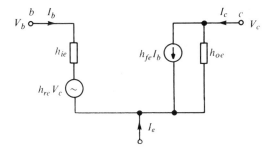

FIG. 7-3. Low-frequency hybrid transistor equivalent circuit.

These assumptions are valid for a monolithic chip at low frequencies. Referring to Fig. 7-2 we now obtain the following expression for the input impedance of the differential amplifier:

$$R_{in} = \frac{V_1}{I_1} = h_{ie} + R'_E(1 + h_{fe}) \qquad [7\text{-}2]$$

where

$$R'_E = R_E \| R' \qquad (7\text{-}3)$$

$$R' = \frac{V_{s1}}{V_{s1} - V_{s2}} (R'_c + R'_{s1}) - R'_{s1} \qquad (7\text{-}4)$$

$$R'_c = \frac{h_{ie} + R_{s2}}{1 + h_{fe}} \qquad (7\text{-}5)$$

$$R'_{s1} = R_E \| \frac{h_{ie} + R_{s1}}{1 + h_{fe}} \qquad (7\text{-}6)$$

As we shall see below, beside close matching of T_1 and T_2, another important requirement for high performance of the differential amplifier is that the common-emitter resistor R_E be as large as possible; it should, in fact, simulate an ideal current source (i.e., $R_E \to \infty$). Thus

$$R_E \gg \frac{h_{ie} + R_{s1}}{1 + h_{fe}} \qquad (7\text{-}7)$$

and therefore

$$R'_{s1} \approx \frac{h_{ie} + R_{s1}}{1 + h_{fe}} \qquad (7\text{-}8)$$

Notice that R'_E is a variable emitter impedance whose value depends on the two input signals V_{s1} and V_{s2}. It consists of the large current-feedback resistor R_E shunted by the input impedance R' of the common-collector transistor T_2.

The amount by which R_E is shunted by R' depends on the level and phase of the input signals V_{s1} and V_{s2}. Thus the current feedback R'_E of T_1 is regulated by the input signals as illustrated by the following two extreme cases.

1. Single-ended input: $V_{s1} \neq 0$, $V_{s2} = 0$. According to (7-4) we have

$$R' = R'_c = \frac{h_{ie} + R_{s2}}{1 + h_{fe}} \tag{7-9}$$

Since $R_E \gg R'_c$ it follows that

$$R_{in} \approx 2h_{ie} + R_{s2} \tag{7-10}$$

The input impedance depends on the source impedance R_{s2}. As we shall see presently, R_{s2} should be as small as possible to give large signal amplification and to minimize offset voltage at the output. Whether the term $2h_{ie}$ or R_{s2} dominates in (7-10) therefore depends on the collector current of T_1 and T_2 as well as on the value of R_{s2} for a given application.

2. Common-mode signal: $V_{s1} = V_{s2}$. For this case R' is infinitely large according to (7-4). Thus

$$R' \gg R_E$$

We can therefore neglect R' in (7-3) and the large current-feedback resistor R_E remains:

$$R'_E \approx R_E$$

Therefore, with (7-2) we have

$$R_{in} \approx h_{ie} + R_E(1 + h_{fe}) \tag{7-11}$$

Considering the features of silicon integrated circuits listed earlier, it is clearly unwise to design the differential amplifier with a large-valued resistor R_E which should function as an ideal current source. Instead, the high collector impedance of the common-emitter transistor T_3 shown in Fig. 7-4 can be used, where the collector of T_3 is connected to the point E in Fig. 7-1. In terms of the h parameters of T_3, R_E is then given by

$$R_E = \frac{1}{h_{oe_3}} + \frac{R_e}{h_{ie_3} + R_e}\left(h_{ie_3} + \frac{h_{fe_3}}{h_{oe_3}}\right) \tag{7-12}$$

This will be on the order of several hundred kilohms for typical h-parameter values. Substituting (7-12) into (7-11) the resulting input impedance will be in the order of several megohms.

Gain   As described above, the differential amplifier may be interpreted as a common-emitter amplifier, incorporating current feedback R'_E, which is dependent on the input signals. From Fig. 7-2 the gain therefore results as

$$G = \frac{V_L}{V_{s1}} = \frac{h_{fe}R_L}{R_{s1} + h_{ie} + R'_E(1 + h_{fe})} \tag{7-13}$$

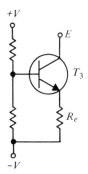

+V

−V

·Fig. 7-4. Current source consisting of common-emitter stage with current feedback.

Considering the same two extreme cases as we did for the input impedance, the effect of current feedback on the gain, regulated by the input signals, can be shown.

1. Single-ended input: $V_{s1} \neq 0$, $V_{s2} = 0$. The feedback is very small, the gain therefore high. With (7-9) we have $R' = R'_c$. Since $R_E \gg R'_c$ it therefore follows that

$$R'_E \approx R'_c$$

and, with (7-5),

$$G_{SE} \approx \frac{h_{fe} R_L}{R_{s1} + R_{s2} + 2h_{ie}} \qquad [7\text{-}14]$$

With typical *h*-parameter values, and a load resistor $R_L = 5$ kΩ, the gain may range from 20 to 30 dB.

2. Common-mode signal: $V_{s1} = V_{s2}$. Here we have $R'_E \approx R_E$; therefore

$$G_{CM} \approx \frac{h_{fe} R_L}{R_{s1} + h_{ie} + R_E(1 + h_{fe})} \qquad [7\text{-}15]$$

The feedback is very large in this case, the gain correspondingly low. With the same numerical values as in case 1, the attenuation of the input signals will be over 60 dB.

In some cases, the output signal of the differential amplifier is taken between the collectors of T_1 and T_2 (see Fig. 7-1). In this case, transistors T_1 and T_2 have equal load resistors R_L in their collectors, and a balanced, or push-pull, output signal, amplified by twice the amount indicated by (7-14) and (7-15), results.

In the discrete version of the differential amplifier, the biasing of T_1, T_2, and T_3 will be obtained by resistor voltage dividers at the base of each transistor. To insure equal collector currents in T_1 and T_2, a low-valued variable potentiometer is connected between the emitters of T_1 and T_2, the potentiometer tap being connected to R_E (or the collector of T_3). The voltage dividers at the bases of T_1 and T_2 have a high impedance level to maintain a high input impedance; they are, of course, as incompatible with monolithic silicon design as is the potentiometer at the emitters of T_1 and T_2. By using a current source that has been developed especially for this technology, these problems can be elegantly overcome.

7.1.2 Current Sources

The simplest version of a monolithic current source[1] is shown in Fig. 7-5a. This is a good example of a circuit that would be quite impractical and expensive in discrete-component design. It is naturally suited for silicon integrated circuit design, because it relies on close initial matching between transistors T_1 and T_2, and subsequent close tracking with temperature. For all practical purposes T_1 and T_2 can be considered identical, as long as their geometries on the chip are the same and they are placed close enough to one another that the temperature difference between them can be neglected.

Assuming, then, that T_1 and T_2 are identical, it follows that the emitter current of T_2 is equal to that of T_1, namely, $I_1 - I_{B2}$. The collector current is then

$$I_2 = \frac{h_{fe_2}}{h_{fe_1}}(I_1 - 2I_{B2}) \tag{7-16}$$

Assuming, furthermore, that the current gains of T_1 and T_2 are matched and reasonably large, we obtain

$$I_2 \approx I_1 \tag{7-17}$$

If the base-emitter voltage is neglible compared to the supply voltage (or if a diode is connected in series with R_2) we obtain the DC collector voltage of T_2 as

$$V_{c2} = V - I_2 R_2 = V\left(1 - \frac{R_2}{R_1}\right) \tag{7-18}$$

The output voltage depends on the *ratio of two resistors* rather than on their absolute value; this is as it should be, in a monolithic silicon integrated circuit. Furthermore, if I_2 should be desired other than equal to I_1, this can be obtained by dimensioning the emitter base junction of T_2 accordingly. If, say, I_2 should be twice as large as I_1, the emitter of T_2 would be made twice the size of the emitter of T_1. On the other hand if I_2 is to be very small (e.g., microamperes), this cannot be obtained by reducing the size of the T_2 emitter.

1. R. J. Widlar, Some circuit design techniques for linear integrated circuits, *IEEE Trans. Circuit Theory*, **CT-12**, 586–590 (1965).

Instead, T_1 and T_2 are designed equal, and a resistor R_e is inserted in series with the emitter of T_2, as shown in Fig. 7-5b. In order to start out with as low a current I_1 as possible, R_1 is made as large as is practical on a silicon chip. With the minimum possible value of I_1 thus established, the value of R_e required for a specified current I_2 then follows:

$$R_e = \frac{kT}{qI_2} \ln \frac{I_1}{I_2} \qquad (7\text{-}19)$$

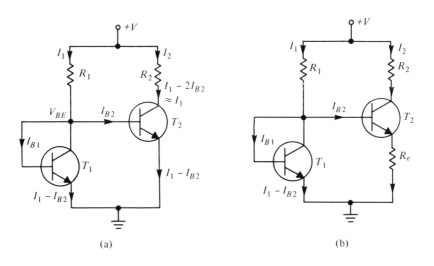

(a)   (b)

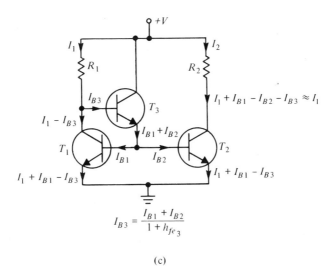

(c)

FIG. 7-5. Various forms of constant-current sources for silicon integrated circuit design.

where k is the Boltzmann constant, T is the absolute temperature of the junctions, and q is the magnitude of the electronic charge. At room temperature (i.e., 25°C), $kT/q \approx 26$ mV.

The current sources of Fig. 7-5 have been widely used in linear silicon integrated circuits. To overcome some of the errors inherent in the two-transistor circuits, somewhat more elaborate versions have been developed by adding one or more transistors to the circuit. In some precision applications, the fact that I_2 is smaller by $2I_{B2}$ than I_1 can be corrected by adding an emitter follower between the collector and base of T_1, as shown in Fig. 7-5c. As in a and b, the emitter junctions of T_1 and T_2 are connected in parallel, and the corresponding emitter currents can be considered equal because of matched characteristics. The output current of T_2 is $I + I_{B1} - I_{B2} - I_{B3}$, and since T_1 and T_2 are matched, I_{B1} and I_{B2} will cancel in this expression. If the current gains are large, the remaining error term I_{B3} will be very small and the output current will be virtually identical to the control current I_1.

Numerous other modifications of the prototype current source of Fig. 7-5a have been suggested, which correct the output current inaccuracy, increase the output impedance, correct for increases of h_{fe_2} (of T_2 in Fig. 7-5a) with increasing collector voltage due to base-width modulation, and so on. However, the basic idea of matching the V_{BE}'s of two virtually identical transistors is common to them all, and any additional transistors generally correct for second-order effects.

7.1.3 Push–Pull–to–Single–Ended Converters

Very often it is desirable to convert the balanced output signal of a push-pull amplifier into a single-ended output signal. One way of achieving this is with the circuit shown in Fig. 7-6. T_1 and T_2 are the transistors of the differential amplifier (see Fig. 7-1). T_3 and T_4 are matched transistors placed close to one another on the chip. Their bases are biased from a common voltage point (i.e., through R_3) through matched resistors (the load resistors of the differential stage, R_1 and R_2). Thus, when the collector currents of the differential stage are equal, the collector currents of T_3 and T_4 will also be equal, and the second stage will be DC-balanced. To obtain the single-ended output signal from the differential voltage across the collectors of T_1 and T_2, transistor T_3 functions as a unity-gain amplifier that inverts the output of T_1 and combines the resulting signal with the output of T_2 at the base of T_4. As shown in Fig. 7-6, the full gain of the differential stage is thereby obtained at the base of T_4 and appears, greatly amplified by the gain of T_4, as a single-ended output signal at the collector T_4. Thus, with the configuration shown in Fig. 7-6, T_1 and T_2 form a first, generally medium-gain, differential-amplifier stage and T_4 a single-ended, high-gain second stage. T_3 provides

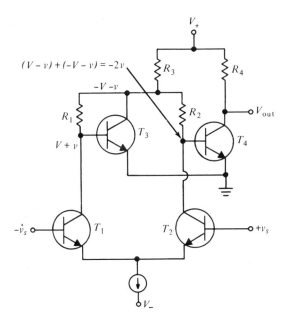

FIG. 7-6. Push-pull-to-single-ended converter providing DC biasing and high-gain stage (T_4).

the balanced DC biasing that enables the push-pull-to-single-ended conversion to be performed on a DC-coupled basis. Incidentally, by making R_3 equal to R_4, an additional useful feature of this circuit is obtained, i.e., that under balanced conditions the single-ended output of T_4 is insensitive to changes in positive supply voltage. If the positive supply voltage varies, the collector currents of both T_3 and T_4 will vary in such a way that the voltage at the collector of T_4 remains constant. Furthermore, by keeping the collector resistors R_1 and R_2 of the differential pair relatively small (e.g., on the order of 2 kΩ), the second amplifier stage (T_4) is essentially voltage-driven. Thus, the gain is not greatly affected by the current gain of T_4 and is relatively constant over a wide range of operating temperatures.

Another example of a push-pull-to-single-ended signal converter is shown in Fig. 7-7. T_1 and T_2 are the transistors of the differential pair and T_3 and T_4 are emitter followers minimizing the load on the collectors of the differential pair. T_5 and T_6 constitute a current source of the kind shown in Fig. 7-5. Assuming that the individual transistor pairs are closely matched (i.e., T_1 and T_2, T_3 and T_4, T_5 and T_6), as are the two resistors R, then it can be seen by inspection that the voltage drop across both resistors R is $V + v - 2V_{BE}$, where V is the DC or common-mode voltage component, v is the differential-mode, or push-pull component constituting the desired signal, and V_{BE} is a base-to-emitter voltage drop. If the current source, formed by T_5 and T_6, is

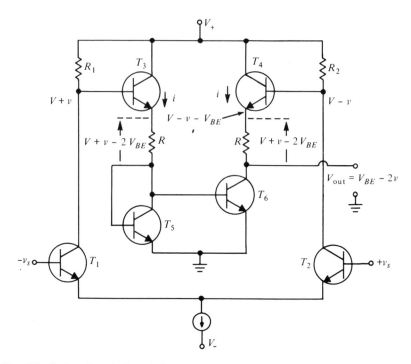

FIG. 7-7. Push-pull-to-single-ended converter comprising emitter followers (T_3 and T_4) and current sources (T_5 and T_6).

ideal then the output voltage appearing between the collector and the emitter (i.e., ground) of T_6 is $V_{BE} - 2v$.

Notice that, in both push-pull-to-single-ended converters discussed, the common-mode components of the input voltage V cancel; they contribute nothing to the output voltage. In both cases this has the desirable consequence that changes in the common-mode input voltage due to changes in temperature or power-supply voltage do not affect the output voltage. Furthermore, the full differential input signal $2v$ appears at the output of each converter. It is interesting to note that in the scheme shown in Fig. 7-7 the DC level of the signal is shifted by one base-to-emitter voltage drop above ground (or, more generally, above the negative power-supply terminal). This is just the right DC level to be used as the input to a common-emitter amplifier stage such as T_4 in Fig. 7-6.

Numerous other schemes for push-pull-to-single-ended conversion have been used. For example, to avoid using the resistors R in Fig. 7-7 (which may become large when a correspondingly large DC-level shift is required) circuits combining lateral p-n-p transistors with n-p-n's are often used. Account is thereby taken of the fact that individual resistors on a monolithic chip should not exceed 10 kΩ or so, and also of the fact that, for large shifts in DC level, the power dissipated in the resistors may be substantial, on the order of 100 mW. On the other hand, the lateral *p-n-p* devices, being located in the signal path, will cause the performance of the converter to deteriorate at high frequencies.

7.1.4 Voltage–Level Shifters

In discrete-component circuit design, coupling of successive gain stages is generally carried out AC-wise using coupling capacitors; if direct coupling is required, this can conveniently be achieved using complementary transistors. With silicon integrated circuits, AC coupling is virtually impossible because of the coupling capacitors required. On the other hand, with only lateral *p-n-p* transistors readily available that do not match the performance of *n-p-n* transistors, complementary transistor design is also quite impractical. Thus in monolithic design *n-p-n* transistor gain stages are generally cascaded, resulting in the need to shift the DC voltage levels within a circuit in order to obtain a desired output voltage swing. Naturally the shift in voltage level should occur with minimum AC signal attenuation.

A voltage-level shifter commonly used in monolithic circuits is shown in Fig. 7-8a. It consists of an emitter follower, providing a low-impedance source,

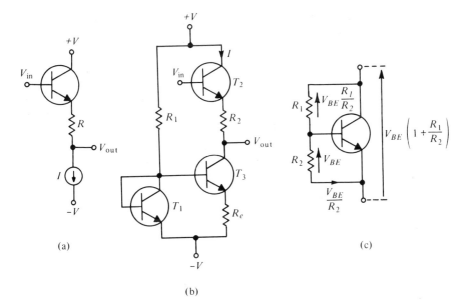

FIG. 7-8. Various forms of voltage-level-shifting networks for silicon integrated circuit design

a direct-current source I, and a diffused-silicon resistor R. The current I flows through R causing a corresponding voltage drop across it. As long as the load on the output terminal is small compared to R, there is little signal attenuation across R, even though the shift in DC voltage level may be relatively large. With the current source of Fig. 7-5b the level shifter of Fig. 7-8a takes on the form shown in Fig. 7-8b. Thus, assuming that the currents in T_2 and T_3 are the same (this implies sufficiently high current gains), we have

$$I = \frac{V_{BE_1} - V_{BE_3}}{R_e} \tag{7-20}$$

where V_{BE_1} and V_{BE_3} are the base-emitter drops of T_1 and T_3, respectively. Since the collector impedance of T_3 is very large, the AC signal from the emitter of T_2 to the output terminal suffers negligible attenuation. Due to the DC current drawn by T_3, the DC voltage level at the output is more negative with respect to the input by an amount ΔV, where

$$\Delta V = V_{BE_2} + (V_{BE_1} - V_{BE_3}) \frac{R_2}{R_e} \tag{7-21}$$

The level shifting scheme of Fig. 7-8b has several disadvantages. Firstly, the resistor R_2, as a diffused silicon resistor, will be restricted in value to a few tens of kilohms. Therefore, when a large voltage-level shift is required the power dissipated across R_2 may be considerable. Secondly, although R_2 is limited in value, it will still cause the level-shift stage to have a relatively high output impedance (on the order of a few tens of kilohms). Therefore its high-frequency performance will be limited by the capacitive loading at the output

To cope with these disadvantages, numerous modifications of the basic level-shifting principle using a current source have been suggested. Instead of the level-shifting resistor R_2, a string of diodes has been used in some applications. In others, in which lateral p-n-p transistors have been combined with n-p-n transistors, the power dissipation problem is improved but not the frequency response. One simple method with some merit is shown in Fig. 7-8c. The current through R_2 is V_{BE}/R_2. If the current into the base of the transistor is negligibly small, this same current flows through R_1. Thus the total voltage drop ΔV across the level shifter is

$$\Delta V = V_{BE} \left(1 + \frac{R_1}{R_2} \right) \tag{7-22}$$

Notice that the shift in level depends on the ratio of two resistors; as we know, this is a desirable feature in integrated circuit design.

7.2 THE OPERATIONAL AMPLIFIER

In two previous chapters various aspects of the ideal operational amplifier in linear active network design were discussed. In Chapter 3 methods of analyzing active networks incorporating operational amplifiers were dealt with, while in Chapter 5 the versatility of the operational amplifier as a basic network element was demonstrated. It is because of the extraordinary versatility of the operational amplifier, which guarantees its usage in a range of applications encompassing far more than active filter networks, that so much effort has been invested in the semiconductor industry to provide low-cost, high-quality operational amplifiers in monolithic integrated form. Indeed, monolithic integrated operational amplifiers had reached extremely low costs (compared to that of their discrete counterparts) long before they were used in active filters on a large scale.

In the following, the monolithic integrated operational amplifier will be discussed as a building block for linear active network design. We shall show how the building blocks for silicon integrated circuit design dealt with in Section 7-1 figure heavily in their design, and shall consider the advantages and shortcomings, compared to the ideal, of the resulting device. The two-port characteristics, gain, frequency stability, and sensitivity of the monolithic operational amplifier will be considered in some detail, and error sources inherent in this network element will be pointed out.

7.2.1 The Silicon Integrated Operational Amplifier

The functional block diagram of a typical silicon integrated operational amplifier is shown in Fig. 7-9a, a simplified circuit equivalent in Fig. 7-9b. The first stage (box *1*) is a differential amplifier; it is used for the input stage because of its high common-mode rejection and its low drift properties. In some cases, two or even three differential amplifier stages are cascaded to give the required open-loop agin of the amplifier. The current source at the two emitters are realized by circuits of the type discussed in Section 7.1.2. Following the differential amplifier(s), a push-pull to single-ended converter (box *2*) feeds into a high-gain amplifier stage (*3*) which may be no more complex than a single common-emitter transistor stage. The latter may not be required if more than one differential pair is used at the input. Somewhere along the cascade of DC-coupled gain stages, the DC level of the signal must be shifted so as to permit a sufficiently large voltage swing without distortion and to ensure an output DC level that is nominally at ground potential. The latter requirement corresponds to the zero-offset property of an operational amplifier, i.e., the output voltage (both AC and DC) must equal zero with respect to ground when the input terminals are at ground potential.

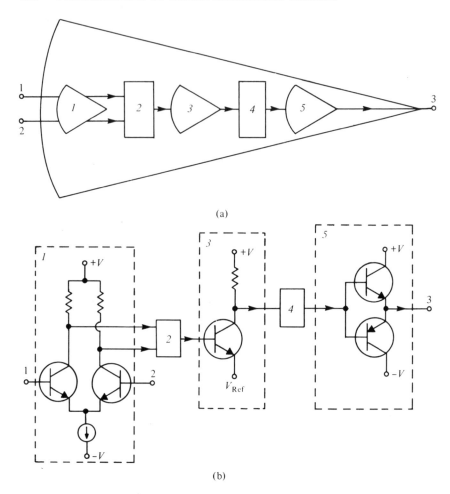

(a)

(b)

FIG. 7-9. Typical silicon integrated operational amplifier. (a) Functional block diagram. (b) Simplified equivalent circuit diagram.

In Fig. 7-9 the voltage-level shifter is indicated by box *4*, which is connected between the high-gain stage and the output stage, here shown as a Class B complementary emitter follower. The latter provides a low-output impedance and low stand-by power drain; it is also capable of swinging the output voltage positively or negatively by an amount almost equal to the power-supply voltage.

Using the integrated–circuit building blocks described in Section 7.1, a practical realization of the rudimentary circuit shown in Fig. 7-9b is illustrated in Fig. 7-10. Transistors T_2 and T_3 will be recognized as the input differential pair, and the T_1, T_9 combination as an emitter current source of the kind shown in Fig. 7-5b. T_4 constitutes a push-pull-to-single-ended converter

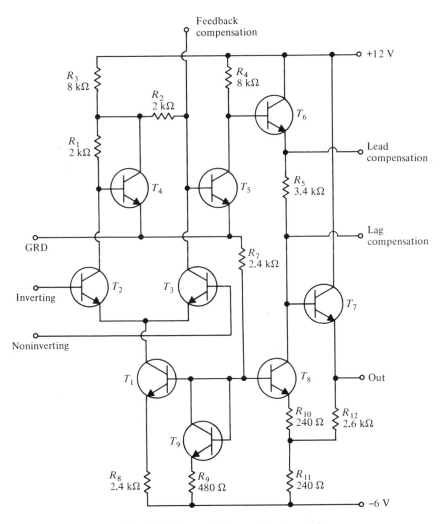

FIG. 7-10. Type μA702 operational amplifier.

combined with the high-gain stage T_5. This combination was discussed in Section 7.1.3 (see Fig. 7-6). T_6, T_8, and T_9 provide level shifting from the collector of T_5, which is at several volts positive with respect to ground, to approximately one base-to-emitter voltage drop (0.7 V) above ground. This combination was discussed in Section 7.1.4 (see Fig. 7-8b). Notice that the diode-connected transistor T_9 is used both for the input current source (with T_1) and for the level-shifting process. The output stage (T_7) is a simple emitter follower (rather than a Class B push-pull stage of the kind shown in Fig. 7-9b) in order to avoid using a lateral p-n-p transistor in the circuit. The trade-off here is between amplifier bandwidth and power consumption; the lateral *p-n-p* transistor, being in the signal path, would limit the bandwidth of the amplifier but permit Class B operation at the output, thereby conserving power. The single-stage *n-p-n* emitter-follower output stage T_7 has a wider bandwidth at the cost of dissipating more power. Notice that the load resistor of T_7 is fed back to the emitter of T_8, giving a controlled amount of positive feedback. In this way T_8, in combination with T_7, actually provides some gain, in addition to shifting the signal level in combination with T_6 and T_9. This feedback also enhances the available output swing so that it is nearly equal to the supply voltages.

The circuit diagram in Fig. 7-10 represents the Fairchild μA702 type, which was one of the first commercially available monolithic silicon operational amplifiers. It demonstrates clearly that the efficient design of a linear integrated circuit requires more than simply the interconnection of silicon integrated building blocks as was done in Fig. 7-9. Thus some, of the transistors are shared between functions, e.g., T_4, which provides DC biasing as well as push-pull-to-single-ended conversion; similarly T_9 is shared by the input current source, the voltage-level shifter, and the output feedback stage; and T_8, which is involved in the DC-level-shifting operation, is simultaneously a part of the output feedback stage. These are relatively simple, and perhaps obvious, examples, yet they paved the way for a considerable degree of ingenuity in linear-circuit design, that has resulted in silicon integrated operational amplifiers of extraordinary complexity and outstanding performance. Functional sharing of individual transistors, total exploitation of the characteristics afforded by silicon integrated components, and, perhaps most important, a complete disregard for conventional discrete-component design rules resulted in a new approach to circuit design that is deliberately adapted to the peculiarities of silicon integrated components.[2] The operational amplifiers that have subsequently emerged far exceed the first 702 type in performance, yet cost no more.

An example of one of the many sophisticated successors of the 702 amplifier, the Fairchild μA741, is shown in Fig. 7-11a. Notice that while the 702 had only 9 transistors the 741 has 20; the resistor count is the same in both. Lateral *p-n-p* transistors are used liberally in the 741, an indication of the improved performance of these transistors, although the bandwidth of the 741 is still narrower than that of the earlier 702, which contained only *n-p-n* transistors. On the other hand, just about every other property is far superior.

As with the 702, the basic rules used in designing the 741 device are the same as those listed in Chapter 6 (Section 6.3.4) and the basic design concept the same as that indicated by Fig. 7-9. Due to the added circuit complexity, however, which is incorporated to overcome the second-order imperfections occurring in simpler circuits, it may be difficult to recognize the basic building blocks used. For this reason a simplified diagram of the amplifier, showing only its essential functions as far as the signal-transmission path is concerned,

2. R. J. Widlar, op. cit.

is shown in Fig. 7-11b. Briefly, transistors T_1 through T_7 form the differential-input stage with level shifting and push-pull-to-single-ended conversion. The resistors in the emitter circuits of T_5 and T_6 permit the output offset voltage to be nulled with the aid of external potentiometers or trimmed film resistors. Transistors T_8 and T_9 (see Fig. 7-11a) form a Darlington pair in a common-emitter amplifier similar to the stage 3 shown in Fig. 7-9b, except that the load resistor in Fig. 7-9b is here replaced by a current source T_{18}. Transistors T_{10} and T_{11} form a complementary Class B output stage like the one in Fig. 7-9b, stage 5.

The transistors shown in Fig. 7-11b make up the complete signal-transmission path through the μA741 amplifier; the remaining transistors provide bias and short-circuit protection for the output stage. Referring to Fig. 7-11a, transistor T_{12} is a level shifter, as shown in Fig. 7-8c; it provides a turn-on bias for T_{10} and T_{11} and thereby eliminates the dead zone, associated with Class B or Class C operation, in the output stage. Transistors T_{13} through T_{16} form two current sources of the types shown in Fig. 7-5. The collector current for T_{15} is controlled by the current through the bias resistor R_b, and

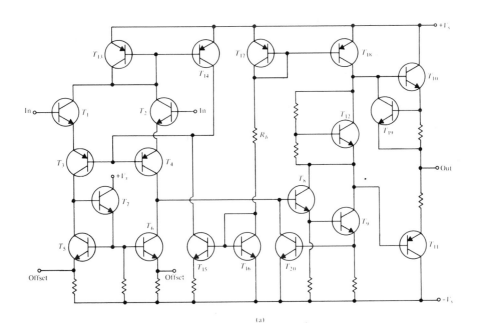

(a)

FIG. 7-11. Type μA741 Operational amplifier. (a) Detailed.

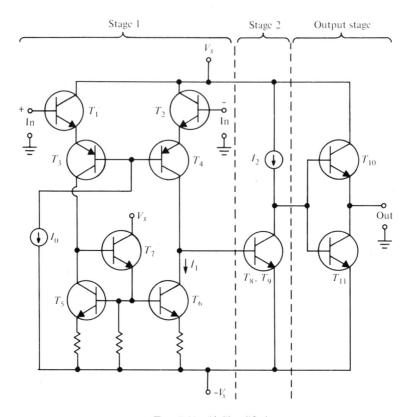

FIG. 7.11. (b) Simplified.

the collector current for T_{14} by the common-mode current in the differential amplifier. The difference between these collector currents provides the base currents for T_3 and T_4. If these base currents are small, then the common-mode current in the differential amplifier adjusts itself so that the collector current of T_{14} is approximately equal to the collector current of T_{15}. Thus the common-mode bias current for the differential amplifier is controlled by the current through the bias resistor R_b.

The current through R_b also controls the bias current for the Darlington common-emitter stage, T_8 and T_9, because it determines the collector current for T_{18}. Transistors T_{17} and T_{18} form a current source that serves as the collector load for the Darlington pair. This current-source load permits the Darlington pair to develop a high voltage gain.

Transistor T_{19} protects power transistor T_{10} from damage by excessive load currents. When the current through the resistor in series with the emitter of T_{10} develops a voltage drop greater than about 0.6 V, T_{19} turns on and diverts current away from the base of T_{10}. This action limits the current through T_{10}. Similarly, T_{20} provides overload protection for T_{11}.

7.2.2 DC Characterization of the Nonideal Operational Amplifier [*]

Two-Port Characteristics The equivalent-circuit model of a nonideal operational amplifier is shown in Fig. 7-12, where R_i and R_o are the open-loop input and output impedances, respectively, and R_b the intrinsic feedback impedance from the output to one of the input terminals. A_o is the open-loop differential voltage gain of the operational amplifier. Actually, these parameters may be frequency-dependent, but for the present we shall consider them to be constant. V_i is the voltage at the inverting input terminal, V_n the voltage at the noninverting input terminal of the amplifier.

In Fig. 7-13, a differential-input operational amplifier is shown in the inverting open-loop mode. The corresponding transmission matrix is

$$
\begin{bmatrix} V_i \\ I_i \end{bmatrix} = \begin{bmatrix} \dfrac{R_b + R_o}{R_o - A_o R_b} & \dfrac{R_o R_b}{R_o - A_o R_b} \\[2em] \dfrac{R_b + R_o + R_i(A_o + 1)}{R_i(R_o - A_o R_b)} & \dfrac{R_o}{R_i} \cdot \dfrac{R_b + R_i}{R_o - A_o R_b} \end{bmatrix} \begin{bmatrix} V_o \\ -I_o \end{bmatrix} \tag{7-23}
$$

Generally the intrinsic feedback impedance R_b is very large and can be assumed to approach infinity. This assumption will be made from here onwards. The transmission matrix given by (7-23) then simplifies to

$$
\begin{bmatrix} V_i \\ I_i \end{bmatrix} = \begin{bmatrix} -\dfrac{1}{A_o} & -\dfrac{R_o}{A_o} \\[2em] -\dfrac{1}{A_o R_i} & -\dfrac{R_o}{A_o R_i} \end{bmatrix} \begin{bmatrix} V_o \\ -I_o \end{bmatrix} \tag{7-24}
$$

Typical values for the most important characteristics of nonideal, commercially available operational amplifiers are summarized in Table 7-1.

The Inverting Feedback Mode Adding an external feedback resistor R_F and an input resistor R_G to the configuration shown in Fig. 7-13 results

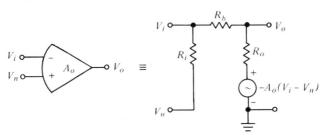

FIG. 7-12. Equivalent circuit of nonideal operational amplifier.

[*] The following discussion is based on: G. S. Moschytz, The operational amplifier in linear active networks, IEEE Spectrum, Vol. 7, 42–50(1970).

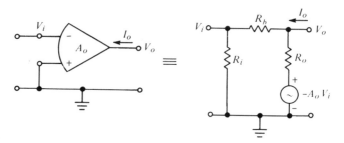

F$_{\text{IG}}$. 7-13. Equivalent circuit of inverting operational amplifier without feedback.

in the inverting feedback mode of the operational amplifier, as shown in Fig. 7-14a. The transmission matrix for this configuration is given by

$$
\begin{bmatrix} V_{in} \\ I_{in} \end{bmatrix}_{\substack{\text{inverting} \\ \text{mode}}} = \begin{bmatrix} \dfrac{R_i(R_F + R_o) + R_G[R_F + R_o + R_i(A_o + 1)]}{R_i(R_o - A_o R_F)} & \dfrac{R_i R_o R_F + R_o R_G R_F + R_o R_G R_i}{R_i(R_o - A_o R_F)} \\[3ex] \dfrac{R_F + R_o + R_i(A_o + 1)}{R_i(R_o - A_o R_F)} & \dfrac{R_o(R_F + R_i)}{R_i(R_o - A_o R_F)} \end{bmatrix} \begin{bmatrix} V_{out} \\ I_{out} \end{bmatrix} \quad (7\text{-}25)
$$

In general, a resistor R_n is required at the noninverting input terminal of the operational amplifier (see Fig. 7-14b) to minimize the output offset voltage. The transmission matrix for the resulting configuration then becomes

$$
\begin{bmatrix} V_{in} \\ I_{in} \end{bmatrix}_{\substack{\text{inverting} \\ \text{mode}}} = \begin{bmatrix} \dfrac{(R_F + R_o)(R_i + R_n) + R_G(R_F + R_o + R_i + R_n) + A_o R_G R_i}{R_o(R_i + R_n) - A_o R_i R_F} & \dfrac{R_o[R_G(R_F + R_i + R_n) + R_F(R_i + R_n)]}{R_o(R_i + R_n) - A_o R_i R_F} \\[3ex] \dfrac{R_F + R_o + R_n + R_i(A_o + 1)}{R_o(R_i + R_n) - A_o R_i R_F} & \dfrac{R_o(R_F + R_i + R_n)}{R_o(R_i + R_n) - A_o R_i R_F} \end{bmatrix} \begin{bmatrix} V_{out} \\ I_{out} \end{bmatrix} \quad (7\text{-}26)
$$

The Noninverting Feedback Mode The noninverting feedback configuration is shown in Fig. 7-15a. The same assumption is made as with the inverting mode, namely, that the internal feedback resistor R_b is very large and tends to infinity. The amplifier is assumed driven from a voltage source or a source

TABLE 7-1. TYPICAL PROPERTIES OF THE NONIDEAL OPERATIONAL AMPLIFIER

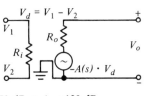

DC open–loop gain	$70\ \text{dB} < A_o < 120\ \text{dB}$
Input impedance	$20\ \text{k}\Omega < R_i < 4\ \text{M}\Omega$
	$(20\ \text{nA} < I_i < 5\ \mu\text{A})$
Output impedance	$0.8\ \Omega < R_o < 2000\ \Omega$
Offset voltage	$0.6\ \text{mV} < (V_o/A_o)|_{Vd=0} < 5\ \text{mV}$
Gain–bandwidth Product	$1\ \text{MHz} < (A_o \times \text{BW}_{-3\text{dB}}) < 500\ \text{MHz}$
(see diagram below)	

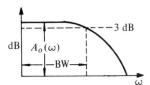

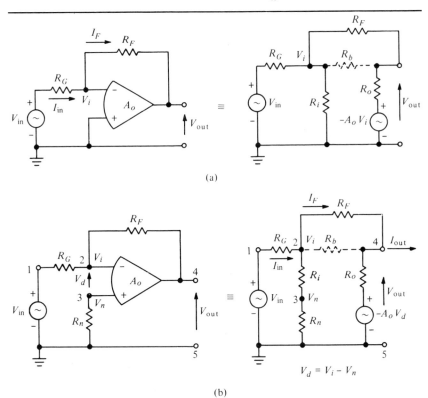

(a)

(b)

FIG. 7-14. Equivalent circuit of operational amplifier in inverting feedback mode. (a) Noninverting terminal grounded. (b) DC-offset resistor R_n added in series with noninverting terminal.

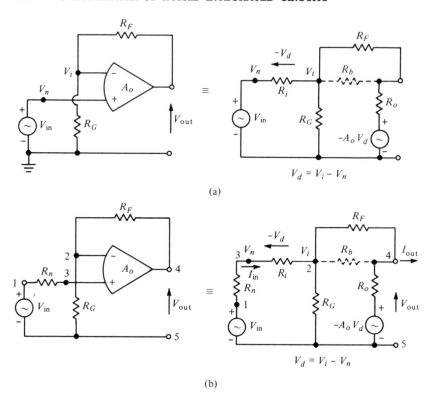

FIG. 7-15. Equivalent circuit of operational amplifier in noninverting feedback mode. (a) With source impedance zero. (b) With finite source impedance R_n for DC-offset compensation.

with negligibly small output impedance so that a resistor R_n must be supplied to minimize the output offset voltage as shown in Fig. 7-15b. By direct analysis of this equivalent circuit diagram, the transmission matrix is obtained as follows:

$$
\begin{bmatrix} V_{in} \\ I_{in} \end{bmatrix}_{\substack{\text{noninverting} \\ \text{mode}}} = \begin{bmatrix} \dfrac{(R_i + R_n)(R_G + R_F + R_o) + R_G(R_F + R_o) + A_o R_i R_G}{R_G R_o + A_o R_i (R_G + R_F)} & \dfrac{R_o[(R_i + R_n)(R_F + R_G) + R_F R_G]}{R_G R_o + A_o R_i (R_G + R_F)} \\[2em] \dfrac{R_G + R_F + R_o}{R_G R_o + A_o R_i (R_G + R_F)} & \dfrac{R_o(R_F + R_G)}{R_G R_o + A_o R_i (R_G + R_F)} \end{bmatrix} \begin{bmatrix} V_{out} \\ I_{out} \end{bmatrix} \quad (7\text{-}27)
$$

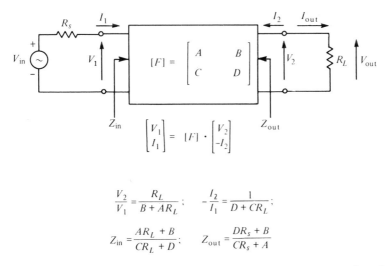

$$\frac{V_2}{V_1} = \frac{R_L}{B + AR_L} ; \qquad -\frac{I_2}{I_1} = \frac{1}{D + CR_L} ;$$

$$Z_{in} = \frac{AR_L + B}{CR_L + D} ; \qquad Z_{out} = \frac{DR_s + B}{CR_s + A}$$

FIG. 7-16. Linear two-port given by its transmission matrix. R_s = source impedance; R_L = load impedance.

Closed-Loop Gain In terms of the transmission-matrix parameters of the two-port shown in Fig. 7-16, the voltage and current gains, input impedance, and output impedance of the operational amplifier can be assumed to be independent of both the source impedance R_s and the load impedance R_L. The former can be included in R_G or R_n (see Figs. 7-14 and 7-15, respectively); the latter can be considered much larger than the amplifier output impedance and, therefore, negligible. Thus[3]

$$G = \frac{V_{out}}{V_{in}} \bigg|_{I_{out} = 0} = \frac{1}{\bar{A}} \qquad (7\text{-}28)$$

$$-\frac{I_2}{I_1} \bigg|_{V_2 = 0} = \frac{1}{\bar{D}} \qquad (7\text{-}29)$$

$$Z_{in} = \frac{V_{in}}{I_1} \bigg|_{I_{out} = 0} = \frac{\bar{A}}{\bar{C}} \qquad (7\text{-}30)$$

$$Z_{out} = \frac{V_2}{I_2} \bigg|_{V_1 = 0} = \frac{\bar{B}}{\bar{A}} \qquad (7\text{-}31)$$

It follows from (7-28) that the inverse of the first terms of the transmission matrices given by (7-26) and (7-27) are the closed-loop gain G of the operational amplifier in the inverting and noninverting feedback modes, respectively.

3. A bar is used over the transmission matrix parameters in this chapter, in order to prevent confusing the transmission matrix parameter A with the finite open-loop operational amplifier gain A. The latter, as opposed to A_0, will be defined shortly.

Designating the inverting mode by a subscript I, the noninverting mode by a subscript N, we obtain

$$G_I = \frac{V_{out}}{V_{in}}\bigg|_{I_{out}=0} = \frac{R_o(R_i + R_n) - A_o R_i R_F}{(R_F + R_o)(R_i + R_n) + R_G(R_F + R_o + R_i + R_n) + A_o R_G R_i}$$

[7-32]

and

$$G_N = \frac{V_{out}}{V_{in}}\bigg|_{I_{out}=0} = \frac{R_G R_o + A_o R_i(R_G + R_F)}{(R_i + R_n)(R_G + R_F + R_o) + R_G(R_F + R_o) + A_o R_i R_G}$$

[7-33]

These are the general expressions for the closed-loop DC voltage gain of the nonideal operational amplifier in the inverting and noninverting mode. The closed-loop gains for the two modes of operation, using ideal operational amplifiers, result when we set

$$A_o = \infty$$
$$R_i = \infty \qquad (7\text{-}34)$$
$$R_o = 0$$

Letting α_I and α_N, respectively, designate the closed-loop gains of the inverting and noninverting modes under the ideal conditions characterized by (7-34), we obtain from (7-32)

$$\alpha_I = G_I\bigg|_{\substack{A_o=\infty \\ R_i=\infty \\ R_o=0}} = -\frac{R_F}{R_G} \qquad [7\text{-}35]$$

and, from (7-33),

$$\alpha_N = G_N\bigg|_{\substack{A_o=\infty \\ R_i=\infty \\ R_o=0}} = \frac{R_G + R_F}{R_G} = 1 + \frac{R_F}{R_G} \qquad [7\text{-}36]$$

Feedback Representation   It is useful to consider the operational amplifier in the inverting or noninverting mode as a feedback amplifier. The corresponding signal-flow graph is shown in Fig. 7-17a, and the transmission function is given by

$$G = \frac{V_{out}}{V_{in}} = \alpha \frac{A\beta}{1 + A\beta} \qquad [7\text{-}37]$$

α is the closed-loop gain, using an ideal operational amplifier, as given by (7-35) and (7-36), A is the forward gain, and β is the feedback factor obtained when using a nonideal operational amplifier. From the essential signal-flow graph shown in Fig. 7-17b (notice the absence of a leakage path) it follows

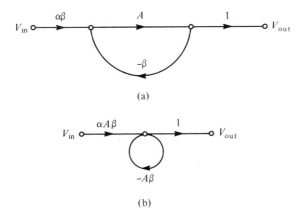

FIG. 7-17. Operational amplifier in feedback (inverting or noninverting) mode .(a) Signal-flow graph. (b) Essential signal-flow graph. $\alpha =$ ideal closed-loop gain; $A =$ forward gain; $\beta =$ feedback factor.

that the product $A\beta$ is the loop gain of the feedback configuration. The representation given by (7-37) as well as the signal-flow graphs in Fig. 7-17 are valid both for the inverting and noninverting mode, even though the actual expressions for α, A, and β for the two modes of operation are different. To keep them apart, we shall continue to use the subscripts I and N for the inverting and noninverting modes, respectively, as we did in (7-35) and (7-36).

The ratio of *actual* to *idealized* closed-loop gain follows from (7-37):

$$\frac{G}{\alpha} = \frac{1}{1 + (1/A\beta)} \tag{7-38}$$

As we shall see shortly, $A\beta$, the amplifier-loop gain, must be much larger than unity for gain stability. Thus, (7-38) can be expanded in a Taylor series as follows:

$$\frac{G}{\alpha} = 1 - \frac{1}{A\beta} + \left(\frac{1}{A\beta}\right)^2 - \left(\frac{1}{A\beta}\right)^3 + \cdots = 1 - E \tag{7-39}$$

where E is the gain error. In general, E can be approximated by the linear term in (7-39).

In order to appreciate the difference between the actual and the idealized closed-loop gains we must calculate the error term E. For this we need to know the A, and β terms for the inverting and noninverting feedback modes.

The Inverting Feedback Mode

The Forward Gain A_I

The forward gain in the inverting mode corresponds to the output voltage when a unit-voltage signal is applied to the input of the operational amplifier (terminals 2 and 5 in Fig. 7-14b) with the input signal source V_{in} set equal to zero. Thus, by inspection of Fig. 7-14b we have

$$A_I = A_o \cdot \frac{R_i}{R_i + R_n} \cdot \frac{R_F}{R_F + R_o} - \frac{R_o}{R_F + R_o} \tag{7-40}$$

or, rewritten,

$$A_I = A_o \frac{1 - (1/A_o)(R_o/R_F)[1 + (R_n/R_i)]}{[1 + (R_o/R_F)][1 + (R_n/R_i)]} \tag{7-41}$$

It is useful, now, to introduce the following substitutions:

$$\beta_o = \frac{R_G}{R_F + R_G} = \frac{1}{1 + (R_F/R_G)} \tag{7-42}$$

$$\varepsilon_i = \frac{R_F}{R_i} \tag{7-43}$$

$$\varepsilon_o = \frac{R_o}{R_F} \tag{7-44}$$

Furthermore, we must examine the external resistor R_n more carefully. This resistor is added to the noninverting input terminal of the amplifier, in both the inverting and noninverting modes, in order to minimize DC offset at the amplifier output. The value of R_n results from the following considerations. The bias currents at each input terminal of the operational amplifier (i.e. the input terminals of the differential gain stage) generate a DC error, or offset voltage, at the output if the DC paths to ground at each input are not equal. Thus, for minimum offset voltage at the output, R_n must equal the parallel combination of R_G with the series combination of R_F and R_o. (The load impedance can generally be neglected compared with the small value of R_o.) Thus

$$R_n = \frac{R_G(R_F + R_o)}{R_G + R_F + R_o} \tag{7-45}$$

Letting

$$R_p = \frac{R_G R_F}{R_G + R_F} = \beta_o R_F \tag{7-46}$$

and considering that R_o is generally negligibly small compared with R_F, we have

$$R_n \approx R_p = R_G \| R_F = \beta_o R_F \tag{7-47}$$

Returning to (7-41) and applying the substitutions introduced above, we obtain

$$A_i = A_o \frac{1 - (\varepsilon_o/A_o)(1 + \beta_o \varepsilon_i)}{(1 + \varepsilon_o)(1 + \beta_o \varepsilon_i)} \tag{7-48}$$

It is well at this point to look at some numerical values for the parameters occurring in (7-48). From Table 7-1 we can assume the following values for a typical[4] voiceband operational amplifier

$$(A_o)_{min} = 10,000$$
$$(R_i)_{min} = 50 \text{ k}\Omega \qquad (7\text{-}49)$$
$$(R_o)_{max} = 100 \text{ }\Omega$$

Assuming a minimum closed-loop gain of unity,[5] the limits on β_o are

$$0 \leq \beta_o \leq 0.5 \qquad (7\text{-}50)$$

A typical feedback resistor might be

$$R_F = 10 \text{ k}\Omega \qquad (7\text{-}51a)$$

Considering two extreme cases for the ideal closed-loop gain, namely, $\alpha_I = -100$ and $\alpha_I = -1$ (where $\alpha_I = -R_F/R_G$), we obtain for the terms in (7-48)

	$\alpha_I = -100$ $(R_F = 100 \text{ } R_G)$	$\alpha_I = -1$ $(R_F = R_G)$
$\beta_o = [1 + (R_F/R_G)]^{-1}$	0.01	0.5
$1/A_o$	10^{-4}	10^{-4}
ε_i	0.2	0.2
$\beta_o \varepsilon_i$	0.002	0.1
ε_o	0.01	0.01
ε_o/A_o	10^{-6}	10^{-6}
$A_o \beta_o$	100	5000
$\varepsilon_o/A_o(1 + \beta_o \varepsilon_i)$	1.002×10^{-6}	1.1×10^{-6}

$$(7\text{-}51b)$$

Clearly the following inequality holds:

$$\varepsilon_o/A_o(1 + \beta_o \varepsilon_i) \ll 1 \qquad (7\text{-}52)$$

and the term $\varepsilon_o/A_o(1 + \beta_o \varepsilon_i)$ can be neglected in (7-48) (ε_o and $1/A_o$ are very small numbers, their product even more so). Thus (7-48) can be simplified as

$$A_I \approx \frac{A_o}{(1 + \varepsilon_o)(1 + \beta_o \varepsilon_i)} \qquad [7\text{-}53a]$$

With $\varepsilon_o \ll 1$, this simplifies to

$$A_I \Big|_{\varepsilon_o \ll 1} \approx \frac{A_o}{1 + \beta_o \varepsilon_i} \qquad [7\text{-}53b]$$

4. Compared to the limits given in Table 7-1 and, indeed, to most inexpensive, commercially available amplifiers, the values chosen here are deliberately pessimistic. This is in order to lend credence to the approximations that we shall assume for the general case, based on the numerical quantities derived for this special case.
5. From Eq. (7.36) it is clear that unity gain is actually the lowest gain possible in the noninverting mode. In the inverting mode, a closed-loop gain of less than unity can be obtained [with $R_F < R_G$ in (7-35)] but is rarely used.

For example, with the values given by (7-51), we have

$$A_I\Big|_{\alpha_I = -100} \approx \frac{A_o}{1.01 \times 1.002} \approx 0.99 A_o;$$

$$A_I\Big|_{\alpha_I = -1} \approx \frac{A_o}{1.01 \times 1.1} \approx 0.91 A_o$$

Thus, the forward gain is actually less by about 10% for the low-closed-loop-gain case than it is for the high-closed-loop-gain case.

The Feedback Factor β_I

The feedback factor in the inverting mode corresponds to the voltage fraction at the input terminals 2 and 5 of Fig. 7-14b when a unity-voltage signal is applied across the output terminals of the amplifier and the input signal V_{in} is short-circuited. By inspection of Fig. 7-14b we obtain

$$\beta_I = \frac{(R_i + R_n)\|R_G}{(R_i + R_n)\|R_G + R_F} = \frac{R_G(R_i + R_n)}{R_G(R_i + R_n) + R_F(R_i + R_n + R_G)} \qquad (7\text{-}54)$$

With (7-42) and (7-46) this becomes

$$\beta_I = \frac{\beta_o}{1 + [R_p/(R_i + R_n)]} \qquad (7\text{-}55)$$

Notice that for the ideal operational amplifier β equals β_o, and that it is only the finite input impedance R_i which prevents this from being so, i.e.,

$$\beta_I\Big|_{R_i \to \infty} = \beta_o = \frac{R_G}{R_F + R_G} \qquad (7\text{-}56)$$

Thus the resistance ratio β_o represents the feedback factor of the ideal operational amplifier, where the amplifier need be ideal only in as far as input impedance is concerned. With (7-43) and (7-44), (7-55) becomes

$$\beta_I = \beta_o \frac{1 + \beta_o \varepsilon_i}{1 + 2\beta_o \varepsilon_i} \qquad [7\text{-}57]$$

With the values given by (7-51b) we have

$$\beta_I\Big|_{\substack{\alpha_I = -100 \\ \beta_o = 0.01}} = \frac{1.002}{1.004} \beta_o \approx \beta_o = 0.01$$

and

$$\beta_I\Big|_{\substack{\alpha_I = -1 \\ \beta_o = 0.5}} \approx \frac{1.1}{1.2} \beta_o = 0.92 \beta_o = 0.46$$

As we would expect, the feedback factor β_I is significantly larger for a gain of -1 than for a gain of -100.

The Closed-Loop Gain G_I

From (7-37) the closed-loop gain of the amplifier in the inverting mode is

$$G_I = \alpha_I \cdot \frac{A_I \beta_I}{1 + A_I \beta_I} \tag{7-58}$$

With (7-53) and (7-57) and loop gain results as

$$A_I \beta_I = \frac{A_o \beta_o}{(1 + \varepsilon_o)(1 + 2\beta_o \varepsilon_i)} \tag{7-59a}$$

Thus the finite input and output impedance contained in ε_i and ε_o, respectively, has the effect of reducing the loop gain $A_o \beta_o$. When $\varepsilon_o \ll 1$,

$$A_I \beta_I \Big|_{\varepsilon_o \ll 1} \approx \frac{A_o \beta_o}{1 + 2\beta_o \varepsilon_i} \tag{7-59b}$$

and, with (7-35), the closed-loop gain becomes

$$G_I = -\frac{R_F}{R_G} \cdot \frac{A_o \beta_o}{(1 + \varepsilon_o)(1 + 2\beta_o \varepsilon_i) + A_o \beta_o} \tag{7-60}$$

When $\varepsilon_o \ll 1$,

$$G_I \Big|_{\varepsilon_o \ll 1} \approx -\frac{R_F}{R_G} \cdot \frac{A_o \beta_o}{1 + 2\beta_o \varepsilon_i + A_o \beta_o} \tag{7-61}$$

Using the numerical values of (7-49) and (7-51) we obtain

$$G_I \Big|_{\alpha_I = -100} = -100 \cdot \frac{100}{1.01 \times 1.004 + 100} \approx 99$$

and

$$\tag{7-62}$$

$$G_I \Big|_{\alpha_I = -1} = -1 \cdot \frac{5000}{1.01 \times 1.2 + 5000} \approx -1$$

Thus for a gain of -100 the *actual* closed-loop gain is off by approximately 1% from the value obtainable with an *ideal* operational amplifier; for $\alpha_I = -1$, the deviation of the actual from the ideal value is indiscernable.

The Gain Error E_I

From (7-39) it follows that

$$\frac{G_I}{\alpha_I} = 1 - \frac{1}{A_I \beta_I} + \left(\frac{1}{A_I \beta_I}\right)^2 - \left(\frac{1}{A_I \beta_I}\right)^3 + \cdots = 1 - E_I \tag{7-63}$$

and from (7-59) we have

$$\frac{1}{A_I \beta_I} = \frac{(1 + \varepsilon_o)(1 + 2\beta_o \varepsilon_i)}{A_o \beta_o} \tag{7-64}$$

With the numerical values of (7-49) and (7-51) we have

$$\left. \frac{1}{A_I \beta_I} \right|_{\alpha_I = -100} = \frac{1.01 \times 1.004}{100} \approx 10^{-2}$$

and $\hspace{12cm}$ (7-65)

$$\left. \frac{1}{A_I \beta_I} \right|_{\alpha_I = -1} = \frac{1.01 \times 1.2}{5000} = 2.4 \times 10^{-4}$$

Thus, as α_I decreases, $1/A_I \beta_I$ decreases approximately in proportion to it.

Having obtained $1/A_I \beta_I$ it is a simple matter to obtain the remaining error terms of the series given by (7-63). Generally, however, it will suffice to consider only the linear term, so that the percentage error term E_I is approximately given by

$$E_I[\%] \approx \frac{100}{A_I \beta_I} \approx \frac{100}{A_o \beta_o} (1 + \varepsilon_o)(1 + 2\beta_o \varepsilon_i) \tag{7-66}$$

A glance at (7-65) shows that for our typical numerical example this approximation is valid for the $\alpha_I = -100$ case as it is for $\alpha_I = -1$. In fact, both the numerical gain errors obtained by (7-65) correspond closely to the values obtained by (7-62).

In terms of the resistor ratio

$$-\alpha_I = \frac{R_F}{R_G} \tag{7-67}$$

which is the ideal closed-loop voltage gain of the inverting operational amplifier, (7-66) becomes

$$E_I[\%] \approx \frac{100}{A_o} \left[2\frac{R_F + R_o}{R_i} + \left(1 + \frac{R_o}{R_F}\right)(1 - \alpha_I) \right] \tag{7-68}$$

In this expression A_o, R_i, and R_o are determined by the particular operational amplifier used, and α_I is determined by the required closed-loop gain. Thus, the only parameter remaining to be chosen is the feedback resistor R_F. To see what influence this resistor has on the gain error E_I, (7-68) is plotted as a function of R_F and the parameter α_I in Fig. 7-18. The plots are for typical values of two basically different amplifiers, namely a wideband and a voice-band type, as listed in Table 7-2. It is clear from the curves shown that there is a value of R_F for each curve for which E_I has a minimum. Setting the

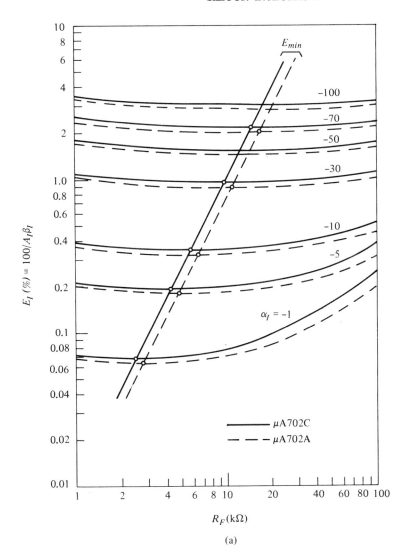

FIG. 7-18. Gain error E_I for the operational amplifiers characterized in Table 7-2. (a) Wideband, low-gain type.

derivative of E_I with respect to R_F equal to zero and solving for R_F this value is obtained as[6]

$$R_F\bigg|_{E_{Imin}} = R_e\sqrt{1 - \alpha_I} \qquad (7\text{-}69)$$

where

$$R_e = \sqrt{R_i R_o/2} \qquad (7\text{-}70)$$

6. Since α_I is negative (see (7-35)) the value under the square root is always positive.

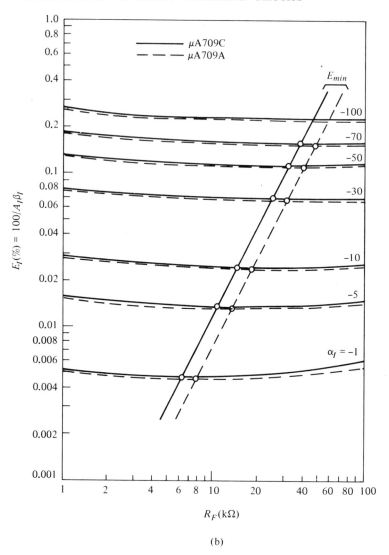

(b)

FIG. 7-18. (b) Voiceband, high-gain type.

Thus, the value of R_F required to minimize E_I increases with closed-loop gain and is proportional to the geometric product of the input and output impedances of the operational amplifier. The values of R_F for minimum E_I are indicated in Fig. 7-18. Equation (7-69) is plotted in Fig. 7-19 for the "typical" amplifier values listed in Table 7-2. The optimum values of R_F for the two gain values assumed in our numerical example (see (7-49) and (7-51)) are

$$R_F|_{\alpha_I=-100} = 10\sqrt{2.5 \times 10^6} = 15.8 \text{ k}\Omega$$

and

$$R_F\big|_{a_I = -1} = \sqrt{5 \times 10^6} = 2.2 \text{ k}\Omega$$

It is evident from Fig. 7-18 that the minima of E_I are only evident for low values of closed-loop gain. In fact, for gain values above about 10 (i.e., $|\alpha_I| > 10$) the curves become increasingly independent of R_F. Since E_I increases linearly with closed-loop gain and is virtually negligible at gain values below about 10 for most available operational amplifiers, the choice of R_F may, in most low-closed-loop-gain cases, be determined by other considerations than those intended to minimize E_I. One such consideration is very often the minimization of temperature drift of the offset voltage at the output of the amplifier. Nevertheless, in cases of medium gain and critical stability one might do well to consider (7-69) as a guideline for the selection of R_F. Fortunately the resulting values are relatively low and thus also satisfy one of the requirements for minimum offset drift. However, there is no point in realizing the value given by (7-69) too accurately, as R_o and R_i are subject to the relatively wide tolerances typical for silicon integrated circuits.

It is clear from (7-68) that E_I increases with R_o and decreases with R_i and A_o. Thus the commonly known fact is reiterated that the higher A_o and R_i and the lower R_o and α_I, the closer the actual closed-loop gain G_I of the amplifier will be to the ideal value α_I.

TABLE 7-2. TYPICAL *DC* CHARACTERISTICS OF TWO OPERATIONAL AMPLIFIERS

	Wideband amplifier					
	μA702A			μA702C		
Characteristic	Min. (−55°C–125°C)	Typical (25°C)	Max. (25°C)	Min. (0°C–70°C)	Typical (25°C)	Max. (25°C)
A_o	2000	3600		1500	3400	
R_i[kΩ]	6.0	40.0		6.0	32.0	
R_o[kΩ]		0.2	0.5		0.2	0.6
	Voiceband amplifier					
	μA709A			μA709C		
Characteristic	Min. (−55°C–125°C)	Typical (25°C)	Max. (25°C)	Min. (0°C–70°C)	Typical (25°C)	Max. (25°C)
A_o	25,000	45,000		12,000	45,000	
R_i[kΩ]	40	400.0		35.0	250.0	
R_o[kΩ]		0.15	0.15		0.15	0.15

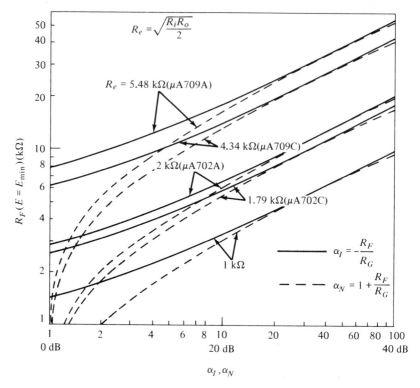

FIG. 7-19. Value of feedback resistor R_F providing minimum gain error for amplifier types characterized in Table 7-2.

The two amplifiers for which E_I is plotted in Figure 7-18b differ only in the value of their input impedances. For α_I values above 10 this difference (almost a factor of two) has negligible effect on E_I. For lower values of closed-loop gain, the difference does become noticeable, particularly for relatively large values of R_F.

In order to get a rapid estimate of the gain error E_I, it is useful to consider only the error introduced by finite gain of the amplifier while assuming an infinite input and a zero output impedance. From (7-66) we then obtain

$$E_I[\%] \approx \frac{100}{A_o \beta_o} \tag{7-71}$$

which for convenience is plotted in Fig. 7-20.

The Noninverting Feedback Mode

The Forward Gain A_N

The forward gain in the noninverting mode corresponds to the output voltage when a unity voltage is generated across the amplifier input (terminals 2 and 3

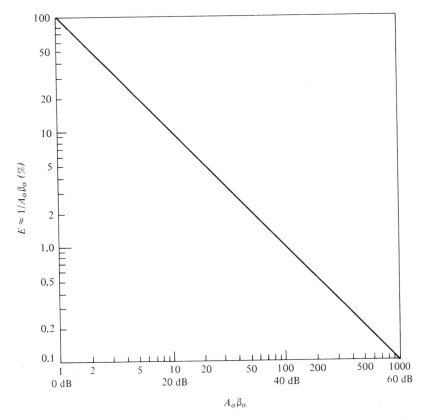

Fig. 7-20. Approximate gain error E for amplifiers nonideal only with respect to open-loop gain A_0.

in Fig. 7-15b) and the signal source V_{in} is removed (i.e., open circuit at the input). By inspection of Fig. 7-21, in which a current source $1/R_i$ is placed across terminals 3 and 5 (to generate 1 V across R_i) and the signal source is removed, we obtain

$$A_N = \frac{R_G}{R_i} \cdot \frac{R_o}{(R_G + R_F + R_o)} + A_o \frac{R_G + R_F}{R_G + R_F + R_o} \qquad (7\text{-}72)$$

Using the terms β_o, ε_i, ε_o, and R_p as defined earlier, and recognizing that (7-47) must be valid here as well, we see that (7-72) becomes

$$A_N = A_o \frac{1 + (\varepsilon_o/A_o)\beta_o \varepsilon_i}{1 + \varepsilon_o(1 - \beta_o)} \qquad [7\text{-}73]$$

With (7-52) this simplifies to

$$A_N \approx \frac{A_o}{1 + \varepsilon_o'} \qquad [7\text{-}74]$$

where

$$\varepsilon_o' = \varepsilon_o(1 - \beta_o) \tag{7-75}$$

As we shall see, ε_o' appears in place of ε_o for the expressions of the non-inverting feedback mode. Since $0 < \beta_o < 1$ it follows that

$$\varepsilon_o' < \varepsilon_o \ll 1 \tag{7-76}$$

Assuming the operational amplifier characteristics given by (7-49), and a feedback resistor $R_F = 10$ kΩ, most of the numerical values given in (7-51b) will be valid for the noninverting mode except those involving α_N and β_o. This is because the ratio R_F/R_G must take on different values to obtain the same value for α_N as for α_I. Thus

	$\alpha_N = 100$ $(R_F = 99\ R_G)$	$\alpha_N = 1$ $(R_G = \infty)$	
$\beta_o = [1 + (R_F/R_G)]^{-1}$	0.01	1.0	
$\beta_o \varepsilon_i$	0.002	0.2	(7-77)
$A_o \beta_o$	100	10,000	

and we have

$$A_N\Big|_{\alpha_N = 100} \approx A_N\Big|_{\alpha_N = 1} \approx A_o$$

The Feedback Factor β_N

The feedback factor in the noninverting mode corresponds to the voltage fraction appearing across the input terminals 2 and 3 (see Fig. 7-15b) when a unity voltage signal is applied across the output terminals and the input

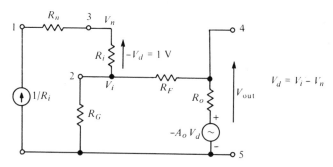

FIG. 7-21. Equivalent circuit of operational amplifier in noninverting feedback mode with current source $1/R_i$ impressed across input impedance R_i to provide 1 V across input terminals.

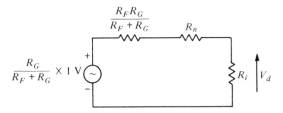

FIG. 7-22. Thévenin equivalent diagram for the calculation of the feedback factor β_N.

signal source V_{in} is short-circuited. By inspection of the resulting Thévenin equivalent network (shown in Fig. 7-22), the feedback factor results as

$$\beta_N = \frac{R_G}{R_G + R_F} \cdot \frac{R_i}{R_i + R_n + R_G \| R_F} = \frac{R_G R_i}{(R_i + R_n)(R_G + R_F) + R_G R_F} \quad (7\text{-}78)$$

Substituting the terms β_o, ε_i, ε_o, and R_p and utilizing (7-47) we obtain

$$\beta_N = \frac{\beta_o}{1 + 2\beta_o \varepsilon_i} \quad [7\text{-}79]$$

Notice that for the amplifier with infinite input impedance (i.e., $\varepsilon_i = 0$), $\beta_N = \beta_o$. When R_G is removed from the noninverting feedback mode (i.e., $R_G = \infty$) the amplifier is operating with 100% feedback and $\alpha_N = \beta_o = 1$. With the values given by (7-49) and (7-77) we have

$$\beta_N \bigg|_{\substack{\alpha_N = 100 \\ \beta_o = 0.01}} = \frac{\beta_o}{1.004} \approx \beta_o = 0.01; \qquad \beta_N \bigg|_{\substack{\alpha_N = 1 \\ \beta_o = 1}} = \frac{\beta_o}{1.4} \approx 0.715$$

Note that although β_N for unity gain is less than the ideal value (i.e., 1) by more than 25%, it is still appreciably higher than β_I at unity gain.

The Closed-Loop Gain G_N

The closed-loop gain of the amplifier in the noninverting mode is

$$G_N = \alpha_N \frac{A_N \beta_N}{1 + A_N \beta_N} \quad (7\text{-}80)$$

With (7-74) and (7-79) the loop gain results as

$$A_N \beta_N = \frac{A_o \beta_o}{(1 + \varepsilon_o')(1 + 2\beta_o \varepsilon_i)} \quad (7\text{-}81a)$$

Notice the similarity to (7-59a) except that ε_o is replaced by ε_o'. When $\varepsilon_o \ll 1$ we have

$$A_N \beta_N \bigg|_{\varepsilon_o' \ll 1} = A_I \beta_I \bigg|_{\varepsilon_o \ll 1} = \frac{A_o \beta_o}{1 + 2\beta_o \varepsilon_i} \quad (7\text{-}81b)$$

With the numerical example of (7-49) we have

$$A_N \beta_N \Big|_{\substack{\alpha_N = 100 \\ \beta_0 = 0.01}} = 100; \qquad A_N \beta_N \Big|_{\substack{\alpha_N = 1 \\ \beta_N = 1}} = 7.15 \times 10^3 \qquad (7\text{-}82)$$

Now, calculating the closed-loop gain we obtain

$$G_N = \left(1 + \frac{R_F}{R_G}\right) \frac{A_o \beta_o}{(1 + \varepsilon'_o)(1 + 2\beta_o \varepsilon_i) + A_o \beta_o} \qquad [7\text{-}83]$$

When $\varepsilon'_o \ll 1$

$$G_N \Big|_{\varepsilon_o' \ll 1} \approx \left(1 + \frac{R_F}{R_G}\right) \frac{A_o \beta_o}{1 + 2\beta_o \varepsilon_i + A_o \beta_o} \qquad [7\text{-}84]$$

With (7-49), (7-51) and (7-77) we have

$$G_N \Big|_{\substack{\alpha_N = 100 \\ \beta_N = 0.01}} \approx 0.99 \alpha_N = 99; \qquad G_N \Big|_{\substack{\alpha_N = 1 \\ \beta_N = 1}} \approx \alpha_N = 1 \qquad (7\text{-}85)$$

Notice that the values obtained here are the same as the corresponding ones obtained in the inverting mode (see (7-62)).

The Gain Error E_N

The ratio of actual to idealized closed-loop gain is again given in the form of (7-39), and because of the close analogy between the inverting and non-inverting cases we can immediately say

$$E_N[\%] \approx \frac{100}{A_N \beta_N} \qquad (7\text{-}86)$$

With (7-81a) we therefore have

$$E_N[\%] \approx \frac{(1 + \varepsilon'_o)(1 + 2\beta_o \varepsilon_i)}{A_o \beta_o} 100 \qquad [7\text{-}87]$$

Considering the numerical example of (7-49) and using (7-86) we therefore obtain

$$E_N \Big|_{\alpha_N = 100} \approx 1\%; \qquad E_N \Big|_{\alpha_N = 1} \approx 0.014\%$$

These values are consistent with the corresponding closed-loop gain values obtained in (7-85).

Comparing (7-66) and (7-87) while recalling (7-76), we have

$$\frac{E_N}{E_I} \approx \frac{1 + \varepsilon'_o}{1 + \varepsilon_o} \leq 1 \qquad [7\text{-}88]$$

that is to say, the noninverting gain error is always less than or equal to the inverting gain error. To illustrate this, the gain error E is plotted as a

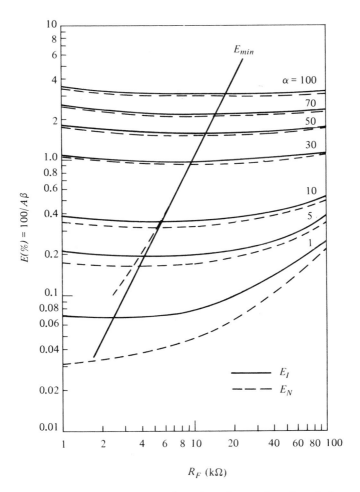

FIG. 7-23. Gain error E for inverting and noninverting modes of typical operational amplifier.

function of R_F and α for the inverting and noninverting modes of a typical operational amplifier in Fig. 7-23.

In terms of the resistance ratio

$$\alpha_N = \frac{R_G + R_F}{R_G} = \frac{1}{\beta_o} \tag{7-89}$$

which determines the ideal closed-loop voltage gain of the noninverting operational amplifier, (7-87) becomes

$$E_N[\%] \approx \frac{100}{A_o}\left[2\,\frac{(R_F + R_o)}{R_i} - \frac{R_o}{R_F} + \alpha_N\left(1 + \frac{R_o}{R_F}\right) - \frac{2}{\alpha_N}\frac{R_o}{R_i}\right] \tag{7-90}$$

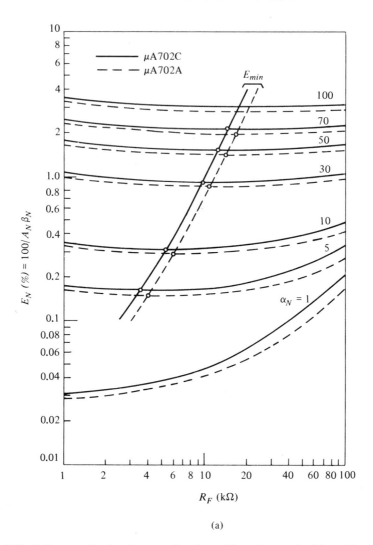

(a)

Fig. 7-24. Gain error E_N for the operational amplifiers characterized in Table 7-2. (a) Wideband, low-gain type.

This expression is plotted in Fig. 7-24 as a function of R_F and α_N for the same amplifiers (see Table 7-2) as were used to plot E_I as a function of α_I in Fig. 7-18. As with the inverting case, there is a value of R_F for a given gain α_N for which the gain error E_N has a minimum. It can be obtained by taking the derivative of (7-90) with respect to R_F. This yields the expression

$$R_F \bigg|_{E_N \text{ min}} = R_e \sqrt{\alpha_N - 1} \tag{7-91}$$

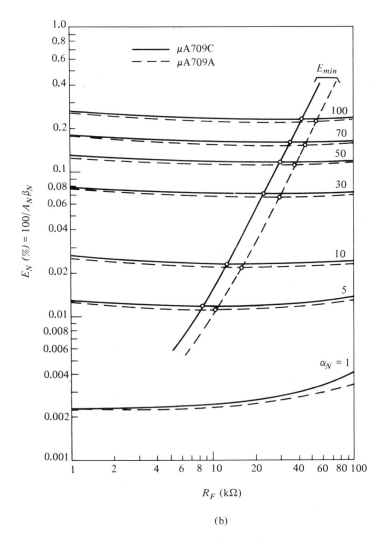

FIG. 7-24. (b) Voiceband, high-gain type.

where R_e is given by (7-70). Equation (7-91) is plotted in Fig. 7-19 for the amplifiers listed in Table 7-2. Notice that for low gains the optimum resistance R_F is very much smaller in the noninverting case than in the inverting case. For high gains the optimum values converge. For $\alpha_N = 100$, for example, the optimum resistor value is the same as for $\alpha_I = -100$.

As indicated by (7-88) the gain error E_N is somewhat less than the error E_I because ε_o, or the output impedance R_o, appears in the noninverting case reduced by the factor $1 - \beta_o$. Thus such expressions for the noninverting mode as (7-90) or (7-87) can be obtained from the corresponding expres-

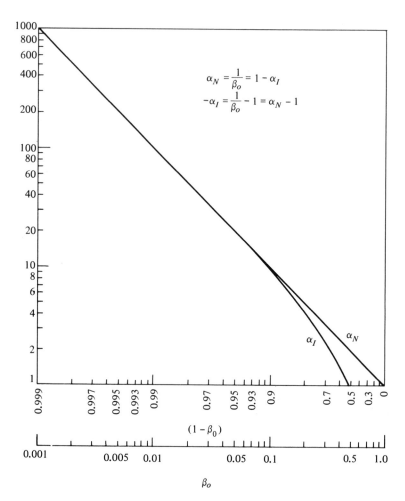

$$\alpha_N = \frac{1}{\beta_o} = 1 - \alpha_I$$

$$-\alpha_I = \frac{1}{\beta_o} - 1 = \alpha_N - 1$$

FIG. 7-25. Graph of relationships among $-\alpha_I$, α_N, β_o and $1 - \beta_o$.

sions for the inverting mode (i.e., (7-88) and (7-66)) simply by substituting ε_o by $\varepsilon_o' = \varepsilon_o (1 - \beta_o)$. With (7-35) and (7-36) it follows that $(1 - \beta_o) = - \alpha_I/(1 - \alpha_I) = 1 - (1/\alpha_N)$. For convenience this relationship between α_I, α_N, and $1 - \beta_o$ is shown graphically as a function of β_o in Fig. 7-25.

The gain error for the inverting and noninverting amplifier modes for a given closed-loop gain would be identical were it not for the effect of the output impedance R_o and for the fact that, for a given closed-loop gain, β_o will be different for the two modes of operation.[7] To illustrate the difference

---

7. This is particularly true for low closed-loop gains. Thus, for example, for unity gain $1/A_o \beta_o$ equals $2/A_o$ for the inverting mode and $1/A_o$ for the noninverting mode.

in gain error for the two cases, the values corresponding to an R_F value of 5 kΩ and 50 kΩ, respectively, have been extrapolated from Figs. 7-18 and 7-24 and plotted in Fig. 7-26. As is to be expected, E_I is only noticeably larger than E_N for low closed-loop gain values (below about 10). This is because ε_o' approaches ε_o and β_o differs ever less in the two modes of operation as the closed-loop gain increases. Furthermore, since the minima of E_I and E_N occurring at low gain values require low values of R_F (see (7-69) and (7-91)), the error values for low gains are smaller with $R_F = 5$ kΩ than with $R_F = 50$ kΩ. (Although not noticeable in the figures, there is a crossover point at gain values on the order of 60, above which the 50-kΩ curves are below those corresponding to 5 kΩ.)

As pointed out earlier, the parameters of the two amplifiers for which the error curves are shown in Figs. 7-18b and 7-24b differ only in input impedance (see "typical" values for μA709A and μA709C in Table 7-2). Similarly, E_I and E_N for the same amplifier differ only because of the difference in the equivalent output impedance (i.e., R_o and $R_o(1 - \beta_o)$). Thus differences in E_I and E_N due to differences in input and output impedances manifest themselves only at relatively low gains. At low gain values, however, the gain errors E_I and E_N are generally so small that these differences are of little consequence. At higher gain values, E_I and E_N both become significantly larger, but there the difference between them disappears. *Thus both at low and high gains, the gain errors E_I and E_N can generally be considered equal for a given value of closed-loop gain.*

Input Impedance In terms of the chain matrix elements $\bar{A}$, $\bar{B}$, $\bar{C}$, and $\bar{D}$ and a load impedance R_L, the input impedance of a two-port is given by

$$Z_{in} = \frac{\bar{A}R_L + \bar{B}}{\bar{C}R_L + \bar{D}} \tag{7-92}$$

Assuming an open circuit, or merely a large load impedance compared with the output impedance of the two-port,[8] (7-92) simplifies to

$$Z_{in} = \frac{\bar{A}}{\bar{C}} \tag{7-93}$$

The Inverting Feedback Mode Referring to Fig. 7-14b and the matrix elements given by (7-26) we obtain

$$(Z_{in})_I = \frac{V_{in}}{I_{in}}\bigg|_{I_{out}=0} = R_G + \frac{R_P}{1 + A_o B_o} \tag{7-94}$$

8. A low output impedance is characteristic of the operational amplifier, so that this assumption can generally be made. The limit on the permitted loading is determined by the current-driving capabilities of the output stage of a given amplifier.

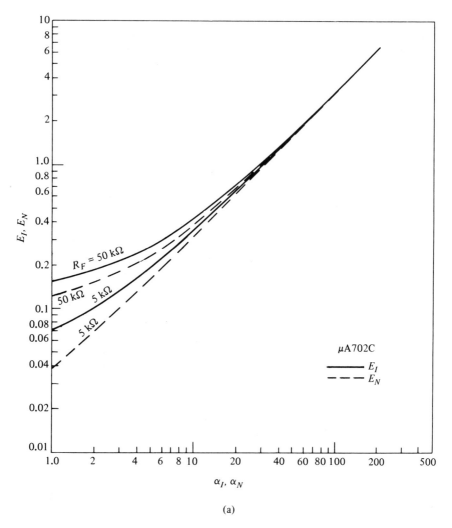

Fig. 7-26. Gain error E_I and E_N for two different feedback resistors R_F. (a) Wideband, low-gain type.

where

$$R_P = (R_F + R_o)\|(R_i + R_n)$$

$$= \frac{(R_F + R_o)(R_i + R_n)}{R_F + R_o + R_i + R_n} \tag{7-95}$$

and

$$B_o = \frac{R_i}{R_F + R_o + R_i + R_n} \tag{7-96}$$

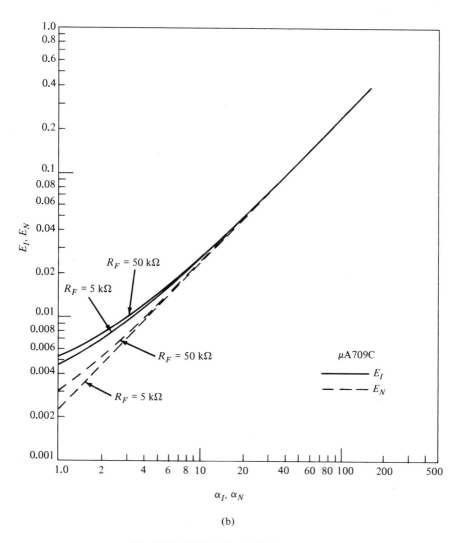

FIG. 7-26. (b) Voiceband, high-gain type.

With (7-35), (7-43), (7-44), (7-46), and (7-47), (7-94) becomes

$$(Z_{in})_I = R_G(1 + \eta) \qquad [7\text{-}97]$$

where

$$\eta = -\alpha_I \frac{(1 + \beta_o \varepsilon_i)(1 + \varepsilon_o)}{1 + \varepsilon_i(1 + \beta_o + \varepsilon_o) + A_o} \qquad [7\text{-}98]$$

To gain some insight into the quantity η it is evaluated for the numerical values given by (7-49) and (7-51). We obtain

$$\eta\Big|_{\alpha_I=-100} = -\alpha_I \frac{1.01}{1.2 + A_o} \approx -1.01 \frac{\alpha_I}{A_o}$$

$$\eta\Big|_{\alpha_I=-1} = -\alpha_I \frac{1.1}{1.3 + A_o} \approx -1.1 \frac{\alpha_I}{A_o}$$

In general

$$(1 + \beta_o \varepsilon_i)(1 + \varepsilon_o) \approx 1 \tag{7-99}$$

and

$$1 + \varepsilon_i(1 + \beta_o + \varepsilon_o) \ll A_o \tag{7-100}$$

These inequalities are also evident from the preceding numerical example. Thus:

$$\eta \approx -\frac{\alpha_I}{A_o} = \frac{1 - \beta_o}{A_o \beta_o} \tag{7-101}$$

and

$$(Z_{\text{in}})_I \approx R_G\left(1 - \frac{\alpha_I}{A_o}\right) \tag{7-102}$$

The input impedance to an inverting operational amplifier is therefore proportional to the ratio of ideal closed-loop gain to open-loop gain, both of which gains are readily accessible quantities. With the numerical values listed above, this approximation gives

$$Z_{\text{in}}\Big|_{\alpha_I=-100} \approx 1.01 R_G; \qquad Z_{\text{in}}\Big|_{\alpha_I=-1} \approx R_G \tag{7-103}$$

This is accurate to within less than 1%, which should be sufficient for most practical purposes.

The Noninverting Feedback Mode   Since (7-92) is valid again here, we have, referring to Fig. 7-15b and using the matrix elements given in (7-27),

$$(Z_{\text{in}})_N = \frac{V_{\text{in}}}{I_{\text{in}}}\Big|_{I_{\text{out}}=0} = R_i + R_n + R_G \frac{R_F + R_o + A_o R_i}{R_G + R_F + R_o} \tag{7-104}$$

With (7-43), (7-44), (7-46), (7-47), and (7-81a), (7-104) can be simplified as follows:

$$(Z_{\text{in}})_N = R_i(1 + \beta_o \varepsilon_i + A_N \beta_N \zeta) \tag{7-105}$$

where

$$\zeta = \left(1 + \frac{\varepsilon_i(1 + \varepsilon_o)}{A_o}\right)(1 + 2\beta_o \varepsilon_i) \tag{7-106}$$

Using the numerical values in (7-49) and (7-51) we obtain, for the terms in (7-105),

$$Z_{in}|_{\alpha_N=100} = R_i(1.002 + 1.004A_N\beta_N) = 101.4R_i \approx 5 \text{ M}\Omega$$

$$Z_{in}|_{\alpha_N=1} = R_i(1.2 + 1.4A_N\beta_N) = 10^4 R_i \approx 500 \text{ M}\Omega \qquad (7\text{-}107)$$

In general

$$\frac{\varepsilon_i(1 + \varepsilon_o)}{A_o} \ll 1 \qquad (7\text{-}108)$$

and

$$\zeta \approx 1 + 2\beta_o\varepsilon_i \qquad (7\text{-}109)$$

Since

$$2\beta_o\varepsilon_i \ll 1 \qquad (7\text{-}110)$$

for α_N values larger than unity (see (7-107)), we can set

$$\zeta|_{\alpha_N>1} \approx 1 \qquad (7\text{-}111)$$

Furthermore, even for $\alpha_N = 1$, $\beta_o\varepsilon_i$ is insignificant compared to $A_N\beta_N$, so that we obtain

$$(Z_{in})_N \approx (1 + A_N\beta_N)R_i \approx A_N\beta_N R_i \qquad [7\text{-}112]$$

or, in terms of the gain error as given by (7-86),

$$(Z_{in})_N \approx 100 \cdot \frac{R_i}{E_N[\%]} \qquad [7\text{-}113]$$

This expression shows the input impedance to be directly proportional to the loop gain or inversely proportional to the gain error. With the numerical values for $A_N\beta_N$ obtained earlier (see (7-82)) we obtain

$$Z_{in}\bigg|_{\alpha_N=100} \approx 5 \text{ M}\Omega; \qquad Z_{in}\bigg|_{\alpha_N=1} \approx 358 \text{ M}\Omega.$$

Compared to the detailed calculation based on (7-105), the approximation given by (7-113) should be accurate enough for most practical purposes. It can be obtained directly for the amplifiers concerned by using gain-error curves of the kind shown in Fig. 7-24. Notice that for α_N near unity, greater accuracy is obtained with $\zeta = 1 + 2\beta_o\varepsilon_i$, in which case $(Z_{in})_N \approx (1 + 2\beta_o\varepsilon_i) A_N\beta_N R_i$.

Output Impedance The output impedance of a two-port in terms of the chain matrix elements and a source impedance R_s is given by

$$Z_{out} = \frac{\bar{D}R_s + \bar{B}}{\bar{C}R_s + \bar{A}} \qquad (7\text{-}114)$$

Since we can always assume that R_s is included in the resistor R_G (see Fig. 7-14b), the output impedance for an operational amplifier in the inverting mode is given by

$$Z_{out} = \frac{\bar{B}}{\bar{A}} \tag{7-115}$$

Inverting Feedback Mode With the matrix elements of (7-26), (7-115) becomes

$$(Z_{out})_I = R_o \frac{R_G(R_F + R_i + R_n) + R_F(R_i + R_n)}{(R_F + R_o)(R_i + R_n) + R_G(R_F + R_o + R_i + R_n) + A_o R_G R_i} \tag{7-116}$$

With (7-43), (7-44), (7-46), (7-47), and (7-59a), (7-116) can be simplified as follows:

$$(Z_{out})_I = \frac{R_o}{A_I \beta_I (1 + \varepsilon_o)(1 + \xi)} \tag{7-117}$$

where

$$\xi = \frac{1}{A_I \beta_I} \left(1 - \beta_o \frac{\varepsilon_o}{1 + \varepsilon_o} \cdot \frac{1 + \beta_o \varepsilon_i}{1 + 2\beta_o \varepsilon_i} \right) \tag{7-118}$$

Substituting the typical numbers of our numerical example (see (7-49)) in (7-118) it follows that

$$\beta_o \cdot \frac{\varepsilon_o}{1 + \varepsilon_o} \cdot \frac{1 + \beta_o \varepsilon_i}{1 + 2\beta_o \varepsilon_i} \ll 1 \tag{7-119}$$

so that

$$\xi \approx \frac{1}{A_I \beta_I} \tag{7-120}$$

Since $\varepsilon_o \ll 1$, (7-117) simplifies to

$$(Z_{out})_I \approx \frac{R_o}{A_I \beta_I} = \frac{E_I [\%]}{100} \cdot R_o \tag{7-121}$$

Using the numerical values in (7-49) and (7-65) we obtain

$$Z_{out} \bigg|_{\alpha_I = -100} \approx 100 \frac{1.01 \times 1.004}{100} \approx 1 \, \Omega$$

$$Z_{out} \bigg|_{\alpha_I = -1} \approx 100 \times 2.4 \times 10^{-4} \approx 0.024 \, \Omega$$

Thus the output impedance is directly proportional to the gain error E_I; even for relatively large values of α_N, Z_{out} is typically on the order of 1 Ω or less.

Noninverting Feedback Mode   The output impedance for the noninverting operational amplifier is also given by (7-115), since the source impedance R_s of (7-114) can be included in R_n (see Fig. 7-15b). With the matrix elements of (7-27) we therefore obtain

$$(Z_{out})_N = R_o \frac{(R_i + R_n)(R_F + R_G) + R_F R_G}{(R_i + R_n)(R_G + R_F + R_o) + R_G(R_F + R_o) + A_o R_i R_G} \tag{7-122}$$

With (7-43), (7-44), (7-46), (7-47), and (7-81a), (7-122) becomes

$$(Z_{out})_N = \frac{R_o}{A_N \beta_N (1 + \varepsilon_o')(1 + \psi)} \tag{7-123}$$

where ε_o' is given by (7-76) and

$$\psi = \frac{1}{A_N \beta_N} \left(1 + \frac{\beta_o \varepsilon_o}{1 + \varepsilon_o(1 - \beta_o)} \cdot \frac{\beta_o \varepsilon_i}{1 + 2\beta_o \varepsilon_i} \right) \tag{7-124}$$

Substituting the typical numerical values of (7-49) and (7-51a) into (7-124) we find that in general

$$\frac{\beta_o \varepsilon_o}{1 + \varepsilon_o(1 - \beta_o)} \cdot \frac{\beta_o \varepsilon_i}{1 + 2\beta_o \varepsilon_i} \ll 1 \tag{7-125}$$

Thus

$$1 + \psi \approx 1 + \frac{1}{A_N \beta_N} \approx 1 \tag{7-126}$$

Since

$$\varepsilon_o' = \varepsilon_o(1 - \beta_o) \ll 1 \tag{7-127}$$

(7-123) can be simplified as follows:

$$(Z_{out})_N \approx \frac{R_o}{A_N \beta_N} = \frac{E_N [\%]}{100} \cdot R_o \tag{7-128}$$

This expression corresponds exactly to that for the inverting case given by (7-121), expect that $A_I \beta_I$ must be replaced by $A_N \beta_N$. With the numerical quantities given for (7-81b) we have

$$Z_{out}\bigg|_{\alpha_N = 100} \approx 1\,\Omega; \qquad Z_{out}\bigg|_{\alpha_N = 1} \approx \frac{100}{7.15 \times 10^3} \approx 0.014\,\Omega$$

Compared with the detailed calculation based on (7-123) the approximation given by (7-128) is accurate enough for most practical purposes. It shows the output impedance to be directly proportional to the gain error E_N and to be less than or equal in value to the output impedance obtained in the inverting mode.

Summary of DC Characteristics   A summary of the DC characteristics of operational amplifiers expressed in terms of the parameters discussed above is given in Table 7-3.

TABLE 7-3. OPERATIONAL AMPLIFIER DC CHARACTERISTICS*

Characteristic	Inverting Mode	Noninverting Mode
1. Ideal closed-loop gain	$\alpha_I = -R_F/R_G$	$\alpha_N = (R_F + R_G)/R_G$
2. Forward gain	$A_I \approx \dfrac{A_o}{(1+\varepsilon_o)(1+\beta_o\varepsilon_i)}$	$A_N \approx \dfrac{A_o}{1+\varepsilon_o'}$
3. Feedback factor	$\beta_I = \beta_o\,\dfrac{1+\beta_o\varepsilon_i}{1+2\beta_o\varepsilon_i}$	$\beta_N = \dfrac{\beta_o}{1+2\beta_o\varepsilon_i}$
4. Loop gain	$A_I\beta_I \approx \dfrac{A_o\beta_o}{(1+\varepsilon_o)(1+2\beta_o\varepsilon_i)}$	$A_N\beta_N \approx \dfrac{A_o\beta_o}{(1+\varepsilon_o')(1+2\beta_o\varepsilon_i)}$
5. Gain error	$E_I = \dfrac{1}{A_I\beta_I} - \left(\dfrac{1}{A_I\beta_I}\right)^2 + \cdots \approx \dfrac{1}{A_I\beta_I}$	$E_N = \dfrac{1}{A_N\beta_N} - \left(\dfrac{1}{A_N\beta_N}\right)^2 + \cdots \approx \dfrac{1}{A_N\beta_N}$
6. Closed-loop gain	$G_I = \alpha_I\,\dfrac{A_I\beta_I}{1+A_I\beta_I} \approx \alpha_I(1-E_I)$	$G_N = \alpha_N\,\dfrac{A_N\beta_N}{1+A_N\beta_N} \approx \alpha_N(1-E_N)$
7. Feedback resistor for minimum gain error	$R_F\vert_{E_I\ \min} = [\tfrac{1}{2}R_i R_o(1-\alpha_I)]^{1/2}$	$R_F\vert_{E_N\ \min} = [\tfrac{1}{2}R_i R_o(\alpha_n - 1)]^{1/2}$
8. Input impedance	$(Z_{\text{in}})_I \approx R_G\left(1+\dfrac{\alpha_I}{A_o}\right)$	$(Z_{\text{in}})_N \approx (1+2\beta_o\varepsilon_i)A_N\beta_N R_i \approx A_N\beta_N R_i$
9. Output impedance	$(Z_{\text{out}})_I \approx \dfrac{R_o}{A_I\beta_I}$	$(Z_{\text{out}})_N \approx \dfrac{R_o}{A_N\beta_N}$

* Throughout the table (see Figs. 7-14b and 7-15b), $\beta_o = R_G/(R_F + R_G)$; $\varepsilon_i = R_F/R_i$; $\varepsilon_o = R_o/R_F$; and $\varepsilon_o' = \varepsilon_o(1-\beta_o)$.

7.2.3 AC Characterisation of the Nonideal Operational Amplifier

In the preceding discussion A_o, the open-loop differential voltage gain of the amplifier, was considered to be constant. In reality, however, it is frequency-dependent; in fact, it can be approximated by a rational function of the following form:

$$A_o(s) = \frac{A_o}{(s+\omega_1)(s+\omega_2)(s+\omega_3)} \qquad (7\text{-}129)$$

This representation corresponds to the Bode plot of the open-loop gain, where the poles ω_1, ω_2, and ω_3 are the corner frequencies in the frequency band of interest (from zero frequency to unity-gain crossover frequency), and the slope increases by 6 dB per octave at each corner. A typical representation of the Bode plot for the open-loop gain corresponding to (7-129) and the closed-loop gain corresponding to (7-37) is shown in Fig. 7-27. This representation shows the open-loop gain A, the closed-loop gain G, and the loop gain $A\beta$. As long as the loop gain $A\beta$ is large, the closed-loop gain is approximately equal to α—i.e., the error E discussed earlier is negligibly small. It is clear from

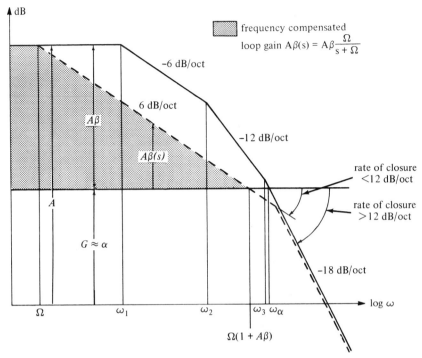

dB

frequency compensated
loop gain $A\beta(s) = A\beta\dfrac{\Omega}{s + \Omega}$

−6 dB/oct

6 dB/oct

$A\beta$

−12 dB/oct

$A\beta(s)$

rate of closure
<12 dB/oct

A

rate of closure
>12 dB/oct

$G \approx \alpha$

−18 dB/oct

log ω

Ω ω_1 ω_2 ω_3 ω_α

$\Omega(1 + A\beta)$

FIG. 7-27. Bode plot of open-loop gain and closed-loop gain for typical operational amplifier.

this figure, however, that the loop gain is now frequency-dependent and, in fact decreases to 0 dB at the frequency ω_α at which the open-loop and the closed-loop responses intersect. To satisfy Bode's criterion for absolute stability of a feedback amplifier,[9] the frequency response of the operational amplifier must generally be limited still further by external frequency-stabilizing RC circuits. This is shown for a typical situation by the broken lines in Fig. 7-27. The loop gain has thereby been modified to roll off at a − 6 dB/octave slope at the frequency Ω, which is lower than the first natural-break frequency (ω_1) of the amplifier.

Numerous methods of incorporating frequency-stabilizing RC networks to ensure the required 6-dB/octave rate of closure are available in the literature. One approach is to modify the frequency response of the open-loop gain by incorporating lead–lag networks in the forward path of the amplifier. Another is to modify the frequency response of the closed-loop gain by incorporating frequency-dependent networks in the feedback network. Either way, the

9. According to Bode's stability criterion, the rate of closure between the open-loop and the closed-loop frequency response must be less than 12 dB per octave. To guarantee sufficient phase margin the rate of closure is usually chosen to be 6 dB per octave. More will be said about this in our discussion of frequency compensation in Section 7.2.5.

resulting loop gain generally has a single pole[10] (on the negative real axis) and can be expressed in the following form:

$$A\beta(s) = A\beta \frac{\Omega}{s + \Omega} \qquad [7\text{-}130]$$

The corresponding loop-gain magnitude thus becomes

$$|A\beta(j\omega)| = A\beta \cdot F_\omega \qquad (7\text{-}131)$$

where

$$F_\omega = \frac{1}{\sqrt{1 + (\omega/\Omega)^2}} \qquad (7\text{-}132)$$

F_ω takes into account the frequency characteristics of the loop gain (in this case, a single pole) and must, of course, be appropriately modified if another pole–zero configuration is used.

Substituting (7-130) into (7-37) we obtain, for the closed-loop gain,

$$G = \alpha \cdot \frac{A\beta\Omega}{s + \Omega(1 + A\beta)} \qquad [7\text{-}133]$$

where, for a given value of Ω, the cutoff frequency is proportional to the loop gain $A\beta$, as shown in Fig. 7-27. Note that for equal ideal gains, i.e., $|\alpha_I| = \alpha_N = \alpha$, and for $\beta = \beta_0$, the cut-off frequency in the inverting mode is $\alpha/(1 + \alpha)$ times *smaller* than in the noninverting mode. In particular, *for unity gain, the bandwidth for the inverting mode is only half that of the noninverting mode.*

The gain error is also modified now by F_ω and, according to either (7-66) or (7-86), its magnitude can be approximated by

$$E \approx (A\beta F_\omega)^{-1} \qquad [7\text{-}134]$$

where A and β are A_I and β_I or A_N and β_N, as the case may be, and F_ω is the correcting factor taking into account the loop-gain frequency response. Assuming the most common case, given by (7-132), the correction factor $1/F_\omega$ is plotted in Fig. 7-28. It must also be considered for all other amplifier parameters that depend on loop gain, such as the input and output impedances[11] (see Table 7-3). More will be said about obtaining the required frequency compensation and rolloff for stable operation in Section 7.2.5.

10. The number of loop-gain poles can be any number N as long as there are $N-1$ zeros such that the rolloff in the vicinity of unity crossover is -6 dB per octave. Whereas multiple poles and zeros widen the loop-gain bandwidth, the compensating networks generating them depend on the value of loop gain available. They are therefore useful only when an amplifier is required to provide a predetermined and constant amount of gain. When the required gain is variable, a compensating network is used to generate a single pole such that the -6-dB/octave rolloff covers at least the entire range of required gain anticipated. To provide adequate phase margin, it extends either from the minimum loop gain required for gain stability or, more generally, from DC open-loop gain to unity closed-loop gain. The latter is the most commonly employed frequency-compensating scheme and has been assumed in the text.
11. K. Soundararajan and K. Ramakrishna, Characteristics of Nonideal Operational Amplifiers, IEEE Trans. Circuits and Systems, CT-21, 69-75 (1974).

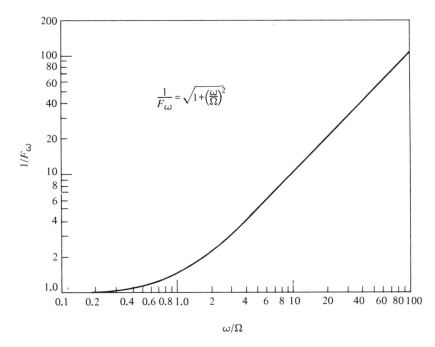

$$\frac{1}{F_\omega} = \sqrt{1+\left(\frac{\omega}{\Omega}\right)^2}$$

FIG. 7-28. Loop-gain correction factor $1/F_\omega$ for the most common method of frequency compensation.

7.2.4 Gain Sensitivity and Stability

One of the main reasons for the popularity of operational amplifiers in active network synthesis is that because of their high open-loop gain they can provide highly stable closed-loop gain. As will be shown in what follows, the degree of gain stability is mainly dependent on the amount of available loop gain and on the stability of the resistor feedback network used.

From Table 7-3 (lines 4 and 6) and with the substitutions $\alpha_I = 1 - (1/\beta_o)$ and $\alpha_N = 1/\beta_o$ we obtain for the closed-loop gains of the inverting and non-inverting operational amplifier, respectively,

$$G_I = \frac{A_o\beta_o - A_o}{1 + \varepsilon_o + 2\beta_o\varepsilon_i + 2\beta_o\varepsilon_i\varepsilon_o + A_o\beta_o} \qquad (7\text{-}135)$$

and

$$G_N = \frac{A_o}{1 + \varepsilon_o + 2\beta_o\varepsilon_i + 2\beta_o\varepsilon_i\varepsilon_o - \varepsilon_o\beta_o - 2\beta_o^2\varepsilon_i\varepsilon_o + A_o\beta_o} \qquad (7\text{-}136)$$

In terms of the network sensitivity function, the relative variation in closed-loop gain with small variations in the parameters contained in either (7-135) or (7-136) is given by

$$\frac{dG}{G} = S_{A_o}^G \frac{dA_o}{A_o} + S_{\beta_o}^G \frac{d\beta_o}{\beta_o} + S_{\varepsilon_i}^G \frac{d\varepsilon_i}{\varepsilon_i} + S_{\varepsilon_o}^G \frac{d\varepsilon_o}{\varepsilon_o}$$ [7-137]

The corresponding sensitivity functions have been calculated and are listed in Table 7-4. Assuming that $A\beta \gg 1$ and considering that $A_o \gg 1$ and that $\varepsilon_o' < \varepsilon_o \ll 1$, the resulting functions can be greatly simplified as shown in the Table.

In practice it is useful to take finite differentials rather than derivatives when evaluating (7-137). A linear relationship between the closed-loop gain G and the parameters in (7-137) is thereby assumed. This is acceptable as a first approximation as long as the differentials remain small. They are obtained as follows: $\Delta A_o/A_o$ includes the relative tolerance and drift (e.g , with temperature) of the open-loop gain. It is generally specified by the manufacturer. $\Delta\beta_o/\beta_o$ can be calculated directly from the definition in (7-42). One obtains

$$\frac{\Delta\beta_o}{\beta_o} = (1 - \beta_o) \cdot \frac{(\Delta R_G/R_G) - (\Delta R_F/R_F)}{1 + [\Delta(R_G + R_F)/(R_G + R_F)]}$$ (7-138)

The term $[(\Delta R_G/R_G) - (\Delta R_F/R_F)]$ is a measure of the tracking capability of the external feedback resistors. For integrated circuits, and in particular for tantalum thin-film resistors, it can be made extremely small—about ± 5 ppm/°C. $\Delta\varepsilon_i/\varepsilon_i$ follows from (7-43):

$$\frac{\Delta\varepsilon_i}{\varepsilon_i} = -\frac{(\Delta R_i/R_i) - (\Delta R_F/R_F)}{1 + (\Delta R_i/R_i)}$$ (7-139)

In general the resistor R_F is external to the operational amplifier, whereas R_i is a part of it. R_i is thus subject to the relatively wide tolerances and drift common with silicon integrated resistors—typically of the order of $\pm 20\%$. In contrast, the external resistor R_F can be chosen to be arbitrarily stable (typically varying by less than 1% with temperature, humidity, and aging). Thus in general

$$\frac{\Delta R_F}{R_F} \ll \frac{\Delta R_i}{R_i}$$ (7-140)

so that

$$\frac{\Delta\varepsilon_i}{\varepsilon_i} = -\frac{\Delta R_i/R_i}{1 + (\Delta R_i/R_i)}$$ (7-141)

$\Delta\varepsilon_o/\varepsilon_o$ follows from (7-44):

$$\frac{\Delta\varepsilon_o}{\varepsilon_o} = \frac{(\Delta R_o/R_o) - (\Delta R_F/R_F)}{1 + (\Delta R_F/R_F)}$$ (7-142)

TABLE 7-4. GAIN SENSITIVITY FUNCTIONS

Inverting Mode	Noninverting Mode

1. $$S_{Ao}^{GI} = \frac{1}{1+A_I\beta_I} \approx \frac{1}{A_I\beta_I}$$

2. $$S_{\beta o}^{GI} = \frac{1}{1-\beta_o} \cdot \frac{\dfrac{1}{\beta_o} + \dfrac{1}{A_I\beta_I}\dfrac{1+2\varepsilon_i}{1+2\beta_o\varepsilon_i}}{1+(1/A_I\beta_I)} \approx -\frac{1}{1-\beta_o}$$

3. $$S_{\varepsilon i}^{GI} = -\frac{2\beta_o\varepsilon_i}{1+2\beta_o\varepsilon_i} \cdot \frac{1}{1+A_I\beta_I} \approx -\frac{2\beta_o\varepsilon_i}{1+2\beta_o\varepsilon_i} \cdot \frac{1}{A_I\beta_I}$$

4. $$S_{\varepsilon o}^{GI} = -\frac{\varepsilon_o}{1+\varepsilon_o} \cdot \frac{1}{1+A_I\beta_I} \approx -\frac{\varepsilon_o}{A_I\beta_I}$$

1. $$S_{Ao}^{GN} = \frac{1}{1+A_N\beta_N} \approx \frac{1}{A_N\beta_N}$$

2. $$S_{\beta o}^{GN} = -\frac{1 + \dfrac{1}{A_N\beta_N}\dfrac{2\varepsilon_i\beta_o}{1+2\varepsilon_i\beta_o}\dfrac{1+\varepsilon_o(1-2\beta_o-0.5\varepsilon_i)}{1+\varepsilon_o(1-\beta_o)}}{1+(1/A_N\beta_N)} \approx -1$$

3. $$S_{\varepsilon i}^{GN} = -\frac{2\beta_o\varepsilon_i}{1+2\beta_o\varepsilon_i} \cdot \frac{1}{1+A_N\beta_N} \approx -\frac{2\beta_o\varepsilon_i}{1+2\beta_o\varepsilon_i} \cdot \frac{1}{A_N\beta_N}$$

4. $$S_{\varepsilon o}^{GN} = -\frac{\varepsilon_o'}{1+\varepsilon_o'} \cdot \frac{1}{1+A_N\beta_N} \approx -\frac{\varepsilon_o'}{A_N\beta_N}$$

Since R_o is also a silicon integrated resistor, being a part of the operational amplifier, it can be assumed that

$$\frac{\Delta R_F}{R_F} \ll \frac{\Delta R_o}{R_o} \tag{7-143}$$

and thus

$$\frac{\Delta \varepsilon_o}{\varepsilon_o} = \frac{\Delta R_o}{R_o} \tag{7-144}$$

Combining (7-138), (7-141), and (7-144) with the functions in Table 7-4, the expressions for gain stability listed in Table 7-5, line 1, are obtained. For most commercially available operational amplifiers

$$\frac{\varepsilon_o'}{1 + \varepsilon_o'} \frac{\Delta R_o}{R_o} < \frac{\varepsilon_o}{1 + \varepsilon_o} \frac{\Delta R_o}{R_o} \ll \frac{\Delta A_o}{A_o} \tag{7-145}$$

Furthermore, if temperature-stable external feedback resistors are used,

$$\frac{\Delta R_G}{R_G} \approx \frac{\Delta R_F}{R_F} \approx 0 \tag{7-146}$$

Similarly, if integrated resistors are used that track closely with ambient variations, then

$$\frac{\Delta R_G}{R_G} - \frac{\Delta R_F}{R_F} \approx 0 \tag{7-147}$$

Under the circumstances specified by (7-145) and (7-146) or (7-147) the expressions for gain stability can be simplified to those given in line 2 of Table 7-5. Notice that in this case the expressions obtained for $\Delta G/G$ are basically the same for the inverting and the noninverting feedback modes, i.e., of the form

$$\left(\frac{\Delta G}{G}\right)_{I,N} = \left(\frac{1}{AB}\right)_{I,N} \frac{\Delta A'}{A'} \tag{7-148}$$

where

$$\frac{\Delta A'}{A'} = \frac{\Delta A_o}{A_o} + \frac{2\beta_o \varepsilon_i}{1 + 2\beta_o \varepsilon_i} \cdot \frac{\Delta R_i/R_i}{1 + (\Delta R_i/R_i)} \tag{7-149}$$

$\Delta A'/A'$ takes into account the inherent variation of the operational amplifier, i.e., the variation of the open-loop gain and of the input impedance. This quantity is formally the same for the inverting and noninverting feedback modes.[12] On the other hand the subscripts I and N in (7-148) indicate that

12. Note, however, that for the same closed-loop gain, β_o will take on different values in the inverting and noninverting cases.

TABLE 7-5. GAIN STABILITY

	Inverting Mode	Noninverting Mode

1.*

Inverting Mode:

$$\frac{\Delta G_I}{G_I} \approx \frac{1}{A_I\beta_I}\left[\frac{\Delta A_o}{A_o} + \frac{2\beta_o\varepsilon_i}{1+2\beta_o\varepsilon_i}\cdot\frac{\Delta R_i/R_i}{1+(\Delta R_i/R_i)}\right] - \frac{\varepsilon_o}{1+\varepsilon_o}\cdot\frac{\Delta R_o}{R_o}\right] - \frac{(\Delta R_G/R_G)-(\Delta R_F/R_F)}{1+[\Delta(R_G+R_F)/(R_G+R_F)]}$$

Noninverting Mode:

$$\frac{\Delta G_N}{G_N} \approx \frac{1}{A_N\beta_N}\left[\frac{\Delta A_o}{A_o} + \frac{2\beta_o\varepsilon_i}{1+2\beta_o\varepsilon_i}\cdot\frac{\Delta R_i/R_i}{1+(\Delta R_i/R_i)}\right] - \frac{\varepsilon_o'}{1+\varepsilon_o'}\cdot\frac{\Delta R_o}{R_o}\right] - \frac{(\Delta R_G/R_G)-(\Delta R_F/R_F)}{1+[\Delta(R_G+R_F)/(R_G+R_F)]}(1-\beta_o)$$

2.†

Inverting Mode:

$$\frac{\Delta G_I}{G_I} \approx \frac{1}{A_I\beta_I}\left[\frac{\Delta A_o}{A_o} + \frac{2\beta_o\varepsilon_i}{1+2\beta_o\varepsilon_i}\cdot\frac{\Delta R_i/R_i}{1+(\Delta R_i/R_i)}\right]$$

Noninverting Mode:

$$\frac{\Delta G_N}{G_N} \approx \frac{1}{A_N\beta_N}\left[\frac{\Delta A_o}{A_o} + \frac{2\beta_o\varepsilon_i}{1+2\beta_o\varepsilon_i}\cdot\frac{\Delta R_i/R_i}{1+(\Delta R_i/R_i)}\right]$$

* Includes second-order effects.
† First-order approximation.

$\Delta G_I/G_I$ and $\Delta G_N/G_N$ are obtained by multiplying $\Delta A'/A'$ by $1/A_I\beta_I$ and $1/A_N\beta_N$, respectively.

Referring to the simplified drawing in Fig. 7-29a, we note that A is the open-loop gain, $G \approx 1/\beta$ the closed-loop gain, and $A\beta$ the loop gain of an operational amplifier. This is true to within a small correction for both the inverting and noninverting feedback modes. Thus, assuming that $A_I\beta_I \approx A_N\beta_N = A\beta$ (which a comparison of (7-59a) and (7-81a) shows to be almost accurately true), we have from (7-148)

$$\frac{\Delta G}{G} = \frac{1}{A\beta}\frac{\Delta A'}{A'} = \frac{G}{A}\cdot\frac{\Delta A'}{A'} \qquad\qquad [7\text{-}150]$$

Remember that $\Delta A'/A' \neq \Delta A/A$ because, according to (7-149), it includes, beside the variation in open-loop gain, the variation in input impedance as well. In any event, (7-150) indicates that the variation in closed-loop gain

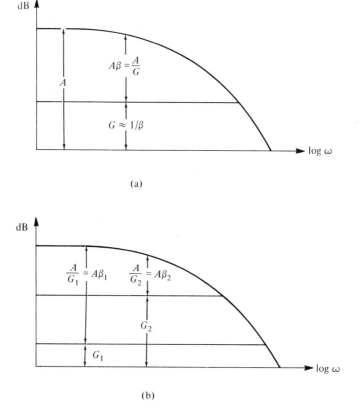

(a)

(b)

FIG. 7-29. Gain stabilization: (a) Relationships among open-loop gain A, closed-loop gain G, and loop gain $A\beta$. (b) Comparison of two closed-loop gains G_1 and G_2.

is inversely proportional to the open-loop gain of the amplifier. For a unity-gain amplifier, (7-150) simplifies to

$$\left.\frac{\Delta G}{G}\right|_{G=1} = \frac{1}{A} \cdot \frac{\Delta A'}{A'} \tag{7-151}$$

Referring to Fig. 7-29b we may now compare the relative variation of two closed-loop gain values obtained with the same amplifier. From (7-150) it follows that

$$\frac{\Delta G_2}{G_2} = \frac{G_2}{G_1} \cdot \frac{\Delta G_1}{G_1} \tag{7-152}$$

Thus, *the variation in closed-loop gain is directly proportional to the closed-loop gain itself.*

A NUMERICAL EXAMPLE: Let us now return to our numerical example, characterized by (7-49), to evaluate the variation of closed-loop gain in the inverting and noninverting modes of a typical operational amplifier. Beside the information in (7-49) we must know something about the expected variations $\Delta A_o/A_o$ and $\Delta R_i/R_i$. Typically, A_o will decrease with increasing temperature; e.g., over a temperature range of 0°C to 60°C it may decrease by up to 50%. Thus we shall assume that

$$\left.\frac{\Delta A_o}{A_o}\right|_{\Delta T = 60°C} = -50\% \tag{7-153a}$$

The input resistance R_i will have the temperature coefficient of a silicon resistor or of the emitter–base junctions of the input transistors.[13] In either case it will be positive and quite large (see Table 6-14 in Chapter 6). Thus from 0°C to 60°C we may find that typically

$$\left.\frac{\Delta R_i}{R_i}\right|_{\Delta T = 60°C} = 30\% \tag{7-153b}$$

and therefore

$$\frac{\Delta R_i/R_i}{1 + (\Delta R_i/R_i)} = 23\% \tag{7-153c}$$

For the inverting case we find from (7-51b)

	$\alpha_I = -100$ $(R_F = 100R_G)$	$\alpha_I = -1$ $(R_F = R_G)$
$2\beta_o \varepsilon_i$	0.004	0.2
$2\beta_o \varepsilon_i/(1 + 2\beta_o \varepsilon_i)$	≈ 0.004	0.167

$$\tag{7-154}$$

---

13. With the equivalent-T diagram of the transistor (see Fig. 3-19) and assuming n emitter–base junctions in series (e.g., for a Darlington input pair, $n = 2$) we have $R_i \approx nr_e\beta$ where β is the current gain. Thus $R_i \approx n(kT/qI_c)\beta = n(k/q)(T/I_b)$. The variation of I_b is specified by the manufacturer, whereas the variation of the ambient temperature (in °K) depends on the application.

and for the noninverting case from (7-77)

	$\alpha_N = 100$ $(R_F = 99R_G)$	$\alpha_N = 1$ $(R_G = \infty)$	
$2\beta_o \varepsilon_i$	0.004	0.4	
$2\beta_o \varepsilon_i/(1 + 2\beta_o \varepsilon_i)$	≈ 0.004	0.286	(7-155)

Thus with (7-149) and (7-153) we obtain

$$\left.\frac{\Delta A'}{A'}\right|_{\alpha_I = -100} = \left.\frac{\Delta A'}{A'}\right|_{\alpha_N = 100} = -50\% + 0.092\% \approx -50\% \quad (7\text{-}156a)$$

$$\left.\frac{\Delta A'}{A'}\right|_{\alpha_I = -1} = -50\% + 3.8\% \approx -46\% \quad (7\text{-}156b)$$

and

$$\left.\frac{\Delta A'}{A'}\right|_{\alpha_N = 1} = -50\% + 6.6\% \approx -43\% \quad (7\text{-}156c)$$

Notice that the contribution to $\Delta A'/A'$ from $\Delta R_i/R_i$ is negligibly small for high gain values in both the inverting and the noninverting modes. For low gains it becomes larger, in particular for the noninverting case, but still insignificantly so because of the coefficient $2\beta_o \varepsilon_i/(1 + 2\beta_o \varepsilon_i)$, which is typically much less than unity. Oddly enough, due to the opposite temperature coefficients of $\Delta A_o/A_o$ and $\Delta R_i/R_i$, a larger contribution of $\Delta R_i/R_i$ would actually help to reduce $\Delta A'/A'$.

With the values obtained in (7-156) inserted in (7-148), and using the values obtained in (7-65) and (7-82) we obtain

$$\left.\frac{\Delta G}{G}\right|_{\alpha_I = -100} = \left.\frac{\Delta G}{G}\right|_{\alpha_N = 100} \approx -0.5\% \quad (7\text{-}157a)$$

$$\left.\frac{\Delta G}{G}\right|_{\alpha_I = -1} \approx -0.011\% \quad (7\text{-}157b)$$

and

$$\left.\frac{\Delta G}{G}\right|_{\alpha_N = 1} \approx -0.006\% \quad (7\text{-}157c)$$

Due to the much larger loop gain $A\beta$ at low closed-loop gain values, the relative variation in closed-loop gain $\Delta G/G$ is much smaller than for high closed-loop gains. Because of the typically small contributions of $\Delta R_i/R_i$ to $\Delta A'/A'$ it should not be surprising that in practice the approximation

$$\frac{\Delta A'}{A'} \approx \frac{\Delta A_o}{A_o} \quad (7\text{-}158)$$

is usually made and $\Delta R_i/R_i$ ignored altogether. Because of the commonly occurring opposite signs of $\Delta A_o/A_o$ and $\Delta R_i/R_i$ this approximation is generally a worst case; the effect of $\Delta R_i/R_i$ will, if anything, decrease $\Delta A'/A'$. Thus it is common to approximate (7-150) by

$$\frac{\Delta G}{G} \approx \frac{1}{A\beta} \cdot \frac{\Delta A_o}{A_o} \approx \frac{G}{A_o} \cdot \frac{\Delta A_o}{A_o} \qquad [7\text{-}159]$$

This is valid for both the inverting and the noninverting modes. Since $G \approx \alpha$ we then obtain for the numerical example given by (7-49) and (7-153)[14]

$$\left.\frac{\Delta G}{G}\right|_{\alpha_I = -100} = \left.\frac{\Delta G}{G}\right|_{\alpha_N = 100} \approx \frac{100}{10,000}(-50\%) = -0.5\% \qquad (7\text{-}160a)$$

and

$$\left.\frac{\Delta G}{G}\right|_{\alpha_I = -1} = \left.\frac{\Delta G}{G}\right|_{\alpha_N = 1} \approx \frac{1}{10,000}(-50\%) = -0.005\% \qquad (7\text{-}160b)$$

Notice the close agreement between (7-160a) and the more accurately calculated quantity in (7-157a); it occurs at high gains where the gain variation is largest. At low gains the variation becomes so small that, in spite of the less good agreement between (7-160b) and (7-157b) and (7-157c), we can conclude here too that, for most practical purposes, (7-159) approximates $\Delta G/G$ quite well enough.

So far, in calculating $\Delta G/G$ we have concentrated on the expression given in line 2 of Table 7-5, in which variations of the output impedance R_o and of the external feedback resistors R_F and R_G have been ignored. While the first omission concerning R_o will be valid in the majority of cases, this may not be true of the variations of the feedback resistors. Let us therefore consider the term (see (7-138))

$$\Delta \beta_o/\beta_o = \frac{(\Delta R_G/R_G) - (\Delta R_F/R_F)}{1 + [\Delta(R_G + R_F)/(R_G + R_F)]}(1 - \beta_o) \qquad (7\text{-}161)$$

which takes these variations into account. The term $1 - \beta_o$ on the right-hand side occurs only in the noninverting mode, thereby making $\Delta \beta_o/\beta_o$ dependent on the closed-loop gain α_N: for unity gain (i.e., α_N equal to unity), $1 - \beta_o = 0$ and the effect of the feedback resistors R_F and R_G is eliminated (R_G is not in the circuit for this case); for high gains (e.g., $\alpha_N = 100$), β_o approaches zero and the effect of the feedback-resistor variations becomes as noticeable as in the inverting case.

The term $(\Delta R_G/R_G) - (\Delta R_F/R_F)$ depends directly on the tracking of the feedback resistors. For thin-film resistors, for example, this will be on the order of ± 5 ppm/°C or, over a 60°C temperature range, 0.03%. The term

---

14. Notice that in (7-150) and (7-159) only $|G|$ or $|\alpha|$ are used without any regard for sign.

$\Delta(R_G + R_F)/(R_G + R_F)$ gives the variation of the *sum* of the feedback resistors; this will depend on their temperature coefficients. Pursuing our example of thin-film resistors, and assuming, for example, tantalum oxynitride resistors whose TCR is controlled to be -200 ppm/°C (to match the TCC of tantalum thin-film capacitors) we obtain, over a 60°C temperature range, -1.2%. Thus for the term in (7-161) we obtain $0.03\%/0.988 = 0.0306\%$. For low closed-loop gains this is comparable in value to the contributions of the operational amplifier itself (see (7-157)). However, in the noninverting mode the additional coefficient $1 - \beta_o$ becomes very small at those low closed-loop gains (e.g., zero for unity gain), so that the contribution of the feedback resistors will be significantly decreased. In the inverting mode, the variations of the feedback resistors should, however, be taken into account.

Returning again to the noninverting mode, with a closed-loop gain of 2 (i.e., $\alpha_N = 2$ or $\beta_o = 0.5$), the contribution of the feedback resistors is only halved and should, in general, be considered. We must remember, also that the values given in (7-157b) and (7-157c) are directly proportional to the closed-loop gain (see (7-159)) and therefore also double when $G \approx \alpha = 2$. In the present example this makes the contribution to $\Delta G/G$ from the amplifier -0.012% (see 7-157c), and from the feedback resistors $\pm 0.0156\%$; in other words, the effect of each is about the same. As the closed-loop gain gets closer to unity in the noninverting mode, the coefficient $1 - \beta_o$ approaches zero rapidly, and the feedback-resistor variations become negligibly small. This represents a strong case for keeping the noninverting gain as close to unity as possible when used in active networks.

The Effect of Frequency-Compensation Networks One additional contribution to $\Delta G/G$, which was not included in the sum of variations in (7-137), should still be considered. It is the effect of variations in the frequency-compensation network used to provide the -6 dB/octave rate of closure mentioned in Section 7.2.3.

The effect of the frequency-compensation networks is only included at this point because many of the influences on $\Delta G/G$ included in (7-137) have been shown to add only second-order contributions to $\Delta G/G$. We can therefore now start out[15] with the following reasonably accurate, very much simplified, expression for closed loop gain:

$$G = \alpha \cdot \frac{A_o \beta_o}{1 + s\tau + A_o \beta_o} \qquad (7\text{-}162)$$

This expression results directly from (7-133) by substituting the ideal open-loop gain A_o for A_I or A_N, the ideal feedback factor β_o for β_I or β_N, and $1/\tau$ for the break frequency Ω of the frequency-compensated operational amplifier (see Fig. 7-27). Since β_o and α are directly related, we have from (7-162)

---

15. The following discussion is based on: D. R. Means, Compensation for temperature dependence of operational amplifier gain, *IEEE J. Solid-State Circuits*, SC-7, 507–509 (1972).

$$\frac{\Delta G}{G} = S^G_{\beta_o} \frac{\Delta \beta_o}{\beta_o} + S^G_{A_o} \frac{\Delta A_o}{A_o} + S^G_\tau \frac{\Delta \tau}{\tau} \qquad [7\text{-}163]$$

For the inverting case, where $\beta_o = 1/(1 - \alpha_I)$, we have

$$S^{G_I}_{\beta_o} = \frac{\beta_o}{\beta_o - 1} \left(\frac{1 + s\tau + A_o}{1 + s\tau + A_o \beta_o} \right) \qquad (7\text{-}164a)$$

$$S^{G_I}_{A_o} = \frac{1 + s\tau}{1 + s\tau + A_o \beta_o} \qquad (7\text{-}164b)$$

$$S^{G_I}_\tau = - \frac{s\tau}{1 + s\tau + A_o \beta_o} \qquad (7\text{-}164c)$$

The last two terms are small for low frequencies (i.e., $s \to 0$); they increase (i.e. approach unity) at high frequencies, where there is less feedback to stabilize the gain. The first term is less frequency-dependent, varying as it does between $1/(\beta_o - 1)$ at low frequencies and $\beta_o/(\beta_o - 1)$ at high frequencies. It will therefore be left out of the following considerations and we are left with

$$\frac{\Delta G_I}{G_I} \approx \frac{1}{1 + s\tau + A_o \beta_o} \left[(1 + s\tau) \frac{\Delta A_o}{A_o} - s\tau \frac{\Delta \tau}{\tau} \right] \qquad [7\text{-}165]$$

The phase angle corresponding to (7-162) is[16]

$$\phi = \tan^{-1} \frac{\omega \tau}{1 + A_o \beta_o} \qquad [7\text{-}166]$$

With a variation in A_o and τ the phase varies as follows:

$$\phi + \Delta\phi = \tan^{-1} \frac{\omega \tau [1 + (\Delta \tau / \tau)]}{1 + A_o \beta_o [1 + (\Delta A_o / A_o)]} \approx \tan^{-1} \frac{\omega \tau [1 + (\Delta \tau / \tau)]}{A_o \beta_o [1 + (\Delta A_o / A_o)]} \qquad (7\text{-}167)$$

For $A_o \beta_o \gg 1$ and a small phase angle $\Delta\phi$ we therefore obtain

$$\Delta\phi \approx \tan^{-1} \frac{\omega \tau}{A_o \beta_o} \left(\frac{\Delta \tau}{\tau} - \frac{\Delta A_o}{A_o} \right) \qquad [7\text{-}168]$$

Inspection of (7-165) and (7-168) shows that one very effective method of stabilizing the gain and phase of an operational amplifier is to let

$$\frac{\Delta \tau}{\tau} = \frac{\Delta A_0}{A_o} \qquad [7\text{-}169]$$

Then (7-165) becomes

$$\left. \frac{\Delta G_I}{G_I} \right|_{\Delta A_o / A_o = \Delta \tau / \tau} = \frac{1}{1 + s\tau + A_o \beta_o} \cdot \frac{\Delta A_o}{A_o} \qquad [7\text{-}170]$$

16. Notice that there is no subscript on ϕ since we obtain the same value for the inverting case as for the noninverting case.

Notice that $\Delta G_I/G_I$ now no longer increases with frequency as was the case in (7-165), but, in fact, decreases. Furthermore, with (7-169), $\Delta\phi$ given by (7-168) goes to zero. Thus (7-169) is a very powerful method of stabilizing an operational amplifier with respect to gain and phase, in particular when the variation in phase is more objectionable than the variation in gain. In practice this is very often the case.

For the noninverting case (for which $\alpha_N = 1/\beta_o$), the terms in (7-163) are

$$S_{\beta_o}^{G_N} = \frac{A_o\beta_o}{1 + s\tau + A_o\beta_o} \qquad (7\text{-}171a)$$

$$S_{A_o}^{G_N} = \frac{1 + s\tau}{1 + s\tau + A_o\beta_o} \qquad (7\text{-}171b)$$

$$S_{\tau}^{G_N} = \frac{-s\tau}{1 + s\tau + A_o\beta_o} \qquad (7\text{-}171c)$$

Here the first term decreases at high frequencies, while the last two increase. All three are frequency-dependent and must be considered. Thus we have

$$\frac{\Delta G_N}{G_N} = \frac{1}{1 + s\tau + A_o\beta_o}\left[A_o\beta_o\frac{\Delta\beta_o}{\beta_o} + \frac{\Delta A_o}{A_o} + s\tau\left(\frac{\Delta A_o}{A_o} - \frac{\Delta\tau}{\tau}\right)\right] \qquad [7\text{-}172]$$

At low frequencies, i.e., as s approaches zero, the last term in (7-172) disappears. However, at high frequencies it becomes the dominant term. Here again it can be eliminated by adhering to (7-169), in which case we obtain

$$\left.\frac{\Delta G_N}{G_N}\right|_{\Delta A_o/A_o = \Delta\tau/\tau} = \frac{1}{1 + s\tau + A_o\beta_o}\left(A_o\beta_o \cdot \frac{\Delta\beta_o}{\beta_o} + \frac{\Delta A_o}{A_o}\right) \qquad [7\text{-}173]$$

As in the inverting case (see (7-170)), $\Delta G_N/G_N$ now decreases with increasing frequency. Since the phase shift is given by (7-166), the phase variation given by (7-168) also disappears. Thus it is useful to implement (7-169) in both the inverting and noninverting cases, in order to stabilize the corresponding gain and phase characteristics.

In the following section we shall discuss commonly used frequency-compensation networks for operational amplifiers. The term τ used above refers to the RC time constant of such networks. After having discussed these networks, we shall briefly consider one of the most commonly used frequency-compensating methods and how to design the corresponding network so that the resulting variation $\Delta\tau/\tau$ can be made equal to $\Delta A_o/A_o$, as required by (7-169).

7.2.5 Frequency Compensation

It was pointed out under 7.2.3 that in order for an operational amplifier in either the inverting or the noninverting feedback mode to be stable, the rate of closure (see Fig. 7-27) between the Bode plots of the open-loop frequency response $A(\omega)$ and the closed-loop response $G(\omega)$ must be less than 12 dB/octave. This is known as Bode's criterion for feedback-system stability.

The derivation of Bode's criterion follows directly from the general feedback representation of the operational amplifier as discussed in that part of Section 7.2.2, concerned with feedback representation. The closed-loop gain function $\alpha[A\beta/(1 + A\beta)]$ corresponds to a single-loop signal-flow graph, and the gain of the loop (i.e., loop gain) has the value $-A\beta$. The poles of the closed-loop system are the roots of the characteristic polynomial

$$1 + A\beta = 0 \qquad (7\text{-}174)$$

As in any other feedback system, the location in the s plane of the roots of the characteristic equation determine the stability and the dynamic behavior of the operational amplifier with feedback. For stability these roots must be on the left of the imaginary axis in the s plane. In terms' of the Nyquist stability criterion, this means that the polar plot of the loop gain $A\beta$ may not encircle the -1 point (assuming that A and β are stable on their own), or that the phase angle of $A\beta$ must be smaller than $180°$, when $A\beta$ has unity gain.

Bode showed that in a gain vs. frequency plot in which both gain and frequency are logarithmic (i.e., gain in decibels, frequency as log ω), a first-order system will have a magnitude curve with a slope of 6 dB/octave and a phase angle of $90°$ at high frequencies; similarly a second-order system with a slope of 12 dB/octave will have a phase angle of $180°$ associated with it. A general expression relating the slope in decibels/octave and the phase angle consequently has the form

$$\text{slope [dB/octave]} = \frac{\text{phase angle}}{180°} \cdot 12 \text{ dB/octave} \qquad [7\text{-}175]$$

Applying this relationship to the loop gain $A\beta$ and to Nyquist's criterion mentioned above we obtain the rate-of-closure, or stability criterion: the frequency response of the loop gain $A\beta$ must have a slope less than 12 dB/octave at the zero-crossing point. Thus, in Fig. 7-30, the frequency responses corresponding to the loop gains $(A\beta)_1$ and $(A\beta)_2$ are both stable because, at the frequency ω_o at which the loop gain of either one is zero (zero-crossing frequency), the rate of closure between the (compensated) open-loop and closed-loop gains is 6 dB/octave, i.e., less than 12 dB/octave. The advantage of $(A\beta)_2$ over $(A\beta)_1$ is obvious; the loop gain is larger over a wider frequency range and therefore, according to Section 7.2.4, the gain error is smaller and the closed-loop gain more stable.

In practice it is necessary to provide a margin of safety by which a feedback amplifier avoids the condition of instability. This is generally expressed in terms of the phase margin ϕ_m defined as

$$\phi_m = 180 - \phi_0 \qquad (7\text{-}176)$$

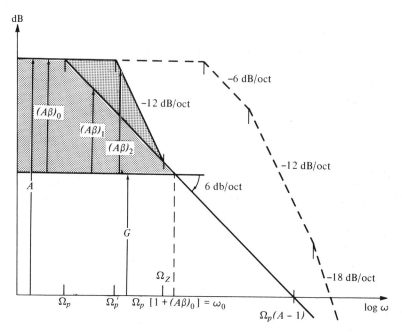

FIG. 7-30. Frequency responses corresponding to two methods of frequency compensation.

where ϕ_0 is the phase shift around the feedback loop at the unity-gain frequency f_0. The phase margin must be converted into an equivalent slope in decibels/octave by using (7-175). Thus for a rate of closure of 12 dB/octave the phase margin is 0°, for 10 dB/octave it is approximately 30°, and for 6 dB/octave it is 90°. In feedback circuits using monolithic operational amplifiers it is generally recommended to have a phase margin between 40° and 45° to ensure stability in the face of production-line spreads, temperature variations, and power-supply drift.

Let us now consider methods of frequency-compensating operational amplifiers such that (1) they are stable (i.e., rate of closure less than 12 dB/octave) and (2) the loop gain is as large as possible over a maximum frequency range.

We recall (see Fig. 7-17) that the loop gain $A\beta$ clearly characterizes the feedback loop associated with an operational amplifier in either of the two feedback modes (i.e., inverting and noninverting); A constitutes the forward path of this feedback loop, β the feedback path. Thus in shaping the frequency response of the loop gain for stability, the frequency response of either A or β can be correspondingly modified by compensation networks. These two possibilities will be briefly discussed in what follows.

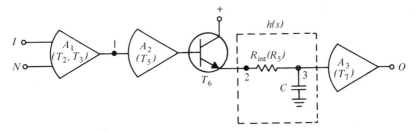

FIG. 7-31. Simplified version of the operational amplifier diagram of Fig. 7-10.

Frequency Compensation of the Forward Path For the sake of describing this compensation method consider the diagram in Fig. 7-31. We have here a simplified diagram of the operational amplifier shown in Fig. 7-10. The gain stages A_1, A_2, and A_3 correspond to the gains supplied by the transistors T_2 and T_3, T_5, and T_7, respectively. T_6 is an emitter follower connected to resistor R_5, which is here referred to as R_{int} to denote that it is internal to the amplifier chip. Terminals 2 and 3 of R_{int} are available as external leads to which external, frequency-compensating components (e.g., capacitor C in Fig. 7-31) can be attached. Designating the frequency response of the resulting RC compensation network (of which only R_5 is on the silicon chip) by $h(s)$ we have for the forward path

$$A(s) = A_1 A_2 A_3 h(s) \tag{7-177}$$

We assume now that the product $A_1 A_2 A_3$ is itself frequency-dependent, specifically of the form given by (7-129) and plotted in Figs. 7-27 and 7-30; thus

$$A_1 A_2 A_3 = A_o(s) = \frac{A_o}{(s + \omega_1)(s + \omega_2)(s + \omega_3)} \tag{7-178}$$

Based on the operational amplifier model in Fig. 7-31 we can now distinguish between the following methods of frequency compensating the forward path.

Lag Compensation In its simplest form, lag compensation is illustrated by the RC low-pass section designated $h(s)$ in Fig. 7-31. The transfer function of $h(s)$ is

$$h(s) = \frac{V_3}{V_2} = \frac{\Omega_p}{s + \Omega_p} = \frac{1/CR_{int}}{s + (1/CR_{int})} \tag{7-179}$$

The effect of $h(s)$ is shown in line 1 in Table 7-6. Note that the break frequency Ω_p of $h(s)$ must be chosen such that $A(s)$ intersects the 0-dB axis *below* the first break frequency ω_1 of $A_o(s)$. Thus the unity-gain bandwidth in this case is less than ω_1.

TABLE 7-6.   FREQUENCY COMPENSATION SCHEMES

I. Compensation of the forward path $A_o(s)$

Compensation network	Compensated open-loop gain $A(s)$	Open-loop-gain frequency response

1.

$$A_1 A_2 A_3 \frac{\Omega_p}{s + \Omega_p}$$

$$\Omega_p = 1/CR_{int}$$

2.

$$A_1 A_2 A_3 h_o \frac{s + \Omega_z}{s + \Omega_p}$$

$$h_o = \frac{R}{R_{int} + R}$$

$$\Omega_z = 1/CR$$

$$\Omega_p = 1/C(R + R_{int})$$

3.

$$A_1 A_2 A_3 \frac{s + \Omega_z}{s + \Omega_p}$$

$$\Omega_z = 1/CR_{int}$$

$$\Omega_p = 1/C(R_{int} \| R)$$

$$= (R_{int} + R)/CR_{int}R$$

4.

$$A_1 A_3 \cdot \frac{\Omega_o}{s + \Omega_p}$$

$$\Omega_o = 1/C_F R_s$$

$$\Omega_p = 1/a_2 C_F R_s$$

$$a_2 = A_2(0)$$

TABLE 7-6 (*continued*)

I. Compensation of the forward path $A_o(s)$

Compensation network	Compensated open-loop gain $A(s)$	Open-loop-gain frequency response
	$A_1 A_3 \dfrac{s + \Omega_z}{s + \Omega_p}$ $\Omega_z = 1/C_F R_F$ $\Omega_p = 1/a_2 C_F R_s$	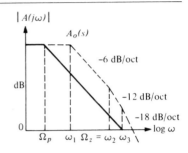

II. Compensation of the feedback path $\beta(s)$

Amplifier configuration	Loop gain $A\beta(s)$	Loop-gain frequency response $|A\beta(j\omega)|$
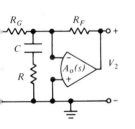	$A_o(s) \cdot \beta_o' \dfrac{s + \Omega_z}{s + \Omega_p}$ $\beta_o' = \dfrac{R_G \parallel R}{R_G \parallel R + R_F} \; ; \; \Omega_z = 1/CR$ $\Omega_p = 1/C(R + R_F \parallel R); \Omega_p < \Omega_z$	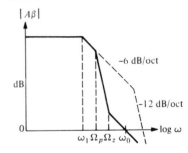
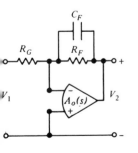	$A_o(s) \cdot \dfrac{s + \Omega_z}{s + \Omega_p}$ $\Omega_z = 1/C_F R_F$ $\Omega_p = 1/C_F(R_F \parallel R_G)$ $\Omega_p > \Omega_z$	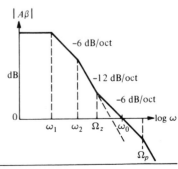

The useful bandwidth can be extended by adding the resistor R in series with C, as shown in line 2 of Table 7-6. This adds a zero Ω_z to $A(s)$, since $h(s)$ now has the form

$$h(s) = h_o \frac{s + \Omega_z}{s + \Omega_p} = \frac{R}{R + R_{int}} \cdot \frac{s + (1/\ R)}{s + [1/C(R + R_{int})]} \qquad (7\text{-}180)$$

By selecting Ω_z equal to ω_1 of $A(s)$, the unity-gain bandwidth can be extended almost up to ω_2.

Lag compensation, while easy to implement, has several disadvantages. First, the resulting frequency-compensated gain is sensitive to changes in A_o, the DC gain of the amplifier. Variations of A_o by factors of 2 to 3 are not uncommon. Assuming a constant external capacitor C and a $\pm 20\%$ variation of R_S (this is an internal silicon-diffused resistor), the pole frequency Ω_p, and with it the intersection of $A(s)$ with the unity-gain axis (i.e., the crossover frequency) can vary an octave on either side of the design value. An octave increase of the crossover frequency could cause an instability due to an excessive rate of closure at unity gain. Second, suitable internal high-impedance points are hard to find in most integrated operational amplifiers because of the difficulty of fabricating large resistors. Consequently the capacitor C required to obtain a desired ω_p inevitably becomes large. The large capacitor, in turn, causes another problem. The inability of the amplifier to charge and discharge the capacitor at high frequencies leads to a reduced peak-to-peak output voltage at high frequencies; this is referred to as a reduced slew rate and will be discussed in more detail in Section 7.2.7.

Lead Compensation   Adding capacitor C in parallel with R_{int} and a resistor to ground we obtain the lead network shown in line 3 of Table 7-6. Notice that, in contrast to the situation with the lag network of line 2, here the zero occurs before the pole, i.e., $\Omega_z < \Omega_p$. In fact, by selecting the external components accordingly, the pole frequency can be made much larger than the zero. Thus, if Ω_z is made to equal ω_2 and $\Omega_p \gg \Omega_z$, the pole at ω_2 can be cancelled and the -6 db/octave slope extended out to ω_3. In combination with separate lag compensation this, in turn, leads to an increase in the unity gain bandwidth. By itself, lead compensation is only useful in extending the bandwidth of relatively high closed-loop gains up to the break frequency ω_3.

Single-Pole Feedback Compensation   To understand feedback compensation we return to our operational amplifier model in Fig. 7-31. This time, however, we shall add no external components to form a frequency function $h(s)$. Instead, we shall examine how, typically, the poles of $A_o(s)$ as given by (7-178) are generated within the amplifier and, in so doing, will find a method of relocating them in a more desirable way.

The model in Fig. 7-31 is fairly typical of silicon integrated operational amplifiers in that it consists of two or three gain stages in cascade. Generally, one stage, say the second, has the highest gain in the amplifier. The only capacitors present in the circuit are parasitic ones, in particular those inherent in the transistors and those from the collectors to the substrate. Assuming, then, that the second stage (T_5) has the largest gain, then it is essentially the Miller effect on its collector-to-base capacitance (C_{cb}) that produces the largest effective capacitor in the circuit. This capacitance and the effective resistance at the base of T_5 are primarily responsible for the first pole of $A_o(s)$ at ω_1. Assuming that the input differential stage A_1 has the next highest gain, the second pole at ω_2 is generated mainly by this stage and the third pole ω_3 by the third stage. The fourth pole and any others occurring at still higher frequencies are not of the Miller-effect type. To determine which circuit components generate these additional poles, detailed information concerning the parasitic capacitors of the integrated-circuit chip are necessary. However, for compensation purposes, only the first three poles are generally of importance. Whether, indeed, the gain distribution between stages is as assumed above or otherwise is immaterial to our discussion. The fact is that the largest gain stage will be primarily responsible for the first amplifier pole ω_1 and it is this pole that dominates the frequency behavior of the amplifier (The "dominant" pole frequency).

In what follows we shall continue our assumption that T_5 in Fig. 7-31 provides the highest gain and produces the lowest pole of the operational amplifier. If, then, we connect a capacitor C_F in parallel with the collector-to-base capacitance C_{cb} (Fig. 7-32) then the Miller effect will shift the dominant pole at ω_1 to the lower frequency Ω_p. Thus Ω_p becomes the new dominant pole frequency in place of the original dominant pole frequency ω_1. This is seen immediately from Fig. 7-32, where straightforward analysis gives us a transfer function of the form

$$A_2(s) = \frac{V_2}{V_1} = \frac{\Omega_o}{s + \Omega_p} = \frac{1/_F R_s}{s + (1/a_2 C_F R_s)} \qquad (7\text{-}181)$$

where a_2 is the DC gain, R_s is the internal load resistor of the preceding amplifier stage, and C_F is the feedback capacitor. We have made the assumption here that C_{cb} of the transistor is included in the value for C_F and that $R_s \ll (h_{ie})_{T_5}$. Notice that (7-181) has the same form as (7-179) for the lag network, except that the capacitor is now multiplied by a_2. Thus, for a given Ω_p, the required capacitance C may easily be two orders of magnitude smaller than in the equivalent lag compensation case.

As with lag compensation, a zero can be added simply with feedback compensation in order to extend the usable bandwidth. This is achieved by including a resistor R_F in series with C_F. The resulting transfer function is

$$A_2(s) = (s + \omega_1) \cdot \frac{s + \Omega_z}{s + \Omega_p} \approx (s + \omega_1) \frac{s + (1/C_F R_F)}{s + (1/a_2 C_F R_s)} \qquad (7\text{-}182)$$

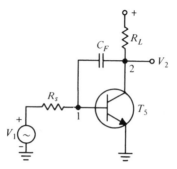

FIG. 7-32. Transistor gain stage with Miller capacitor C_F.

The zero frequency is approximated here, since it is actually reduced somewhat by g_m/C_F, where g_m is the transconductance of the transistor. For all practical purposes this reduction is negligible, however. The term $s + \omega_1$ appears in (7-182) to account for the fact that the pole at ω_1 has been shifted to Ω_p. The effect of this feedback compensation as well as that involving only C_F is shown in lines 4 and 5 of Table 7-6. Notice that with this feedback scheme the largest unity-gain bandwidth so far is achievable. Even with only C_F the bandwidth can extend up to ω_2, since the first -6 dB/octave slope is simply shifted down to Ω_p. Including R_F, the increase in bandwidth is even larger. By shifting ω_1 down to ω_p and letting $\Omega_z = \omega_2$, the unity-gain bandwidth can be extended up to ω_3.

The feedback-compensation method has several other advantages. First, since Ω_p depends on a_2 (which is the largest gain stage of the amplifier), feedback compensation is not as sensitive to variations in the overall open-loop gain as are the other methods. This lack of sensitivity will be explained in more detail in the next section. Second, as mentioned before, since C_F is multiplied by a_2, a much smaller capacitor can be used to obtain a given break at Ω_p. The smaller capacitor allows a full voltage swing to much higher frequencies, i.e., it improves the slew rate appreciably. Finally, the smaller capacitor is easier to produce in any integrated technology and actually becomes small enough in many cases (e.g., typically 30 pF) to fabricate on a silicon chip.

It should be clear that those amplifiers in which both lead and feedback compensation can be combined provide the widest bandwidths.

Double-Pole, Single-Zero Feedback Compensation   In all of the compensation schemes discussed so far, we have obtained amplifier rolloffs decreasing monotonically at -6 dB/octave. We have mentioned before that this is a safe (from a frequency-stability point of view) but by no means optimum method of frequency compensation. Clearly a double pole (see

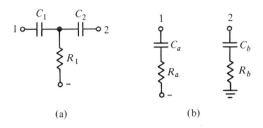

FIG. 7-33. *RC* networks for double-pole, single-zero frequency compensation.

Ω_p' in Fig. 7-30) providing an initial 12 dB/octave rolloff, and a subsequent zero (e.g., Ω_z) shortly before the unity-loop-gain crossover frequency (e.g., ω_0) gives higher loop gain over a wider frequency range while still guaranteeing a 6 dB/octave rate of closure. The desirability of high loop gain with regard to gain stability and other important amplifier characteristics should require no further elaboration at this point.

A double-pole, single-zero compensation scheme providing a loop-gain characteristic such as $(A\beta)_2$ in Fig. 7-30 can be obtained by connecting a *T* network as shown in Fig. 7-33a or a dual-lead network as shown in Fig. 7-33b either between terminals 1 and 2 in the gain stage shown in Fig. 7-32 or between terminal 1 of that stage and the output of the amplifier. R_1 of the *T* network and the terminal of the first lead combination $R_a C_a$ are preferably connected to the negative power-supply terminal rather than to ground in order to reduce power-supply noise and ripple feeding through to the output of the amplifier. By contrast the second lead combination $(R_b C_b)$ should be grounded, since the output is referenced to ground and there is virtually no gain between terminal 2 and the output.

Frequency Compensation of the Feedback Path   We remember that, in order to insure stability, the loop gain $A\beta$ of an operational amplifier must fulfill Bode's criterion, which guarantees that the rate of closure between the open-loop gain $A(s)$ and the closed-loop gain $G(s)$ must be less than 12 dB/octave. In the previous section, in which we compensated the forward path, or, more precisely, the open-loop gain $A_o(s)$ in the forward path, we assumed that the closed-loop gain $G(s)$ was independent of frequency and modified the frequency response of $A_o(s)$ so as to roll off at a gradual incline of less than 12 dB/octave with respect to $G(s)$. Instead of decreasing (or flattening) the rolloff of $A_o(s)$, the stability criterion can be equally well fulfilled by *increasing* the rolloff of $G(s)$ (Fig. 7-34). By introducing poles Ω_1 and Ω_2 in the closed-loop gain $G(s)$, while leaving the inherent rolloff of the open-loop gain $A_o(s)$ unchanged, the rate of closure is decreased to less than 12 dB/octave (actually to 6 dB/octave in Fig. 7-34) and the amplifier is stable.

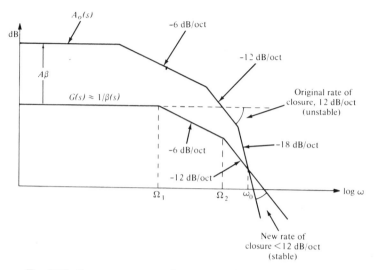

FIG. 7-34. Frequency response by compensation of the feedback path.

Let us now examine how poles can most conveniently be introduced in the closed-loop gain $G(s)$ to satisfy Bode's criterion. For sufficiently large loop gain, i.e., $A\beta \gg 1$, we find (see (7-150))

$$\frac{G}{A} \approx \frac{1}{\text{Loop gain}} = \frac{1}{A\beta} \qquad (7\text{-}183a)$$

or

$$G(s) \approx \frac{1}{\beta(s)} \qquad (7\text{-}183b)$$

i.e., the closed-loop gain is the inverse of the feedback transfer function. Consequently, poles or pole–zero combinations introduced in the closed-loop gain correspond to zeros or zero–pole combinations in the feedback path. In fact, to establish the desired pole–zero pattern in $G(s)$, it is simpler to start out with the feedback path, based on the following equivalent statement of Bode's criterion:

$$[\text{slope } A\beta]_{\omega_0} = [\text{slope } A + \text{slope } \beta]_{\omega_0}$$
$$= [\text{slope } A - \text{slope } (1/\beta)]_{\omega_0} < 12 \text{ dB/octave} \qquad (7\text{-}184)$$

where ω_0 is the crossover frequency (see Fig. 7-34). Clearly, the amount by which the slope of $A\beta$ at ω_0 is less than 12 dB/octave corresponds to the phase margin ϕ_m expressed in dB/octave. Notice that the smaller the loop gain $A\beta$ (i.e., the larger the closed-loop gain) is required to be, the easier it is to compensate an operational amplifier because the open-loop rolloff of $A_o(s)$ at the intersection with $G(s)$ will be flatter. This is true for any kind

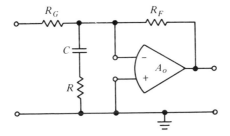

FIG. 7-35. Lag compensation of the feedback path.

of frequency compensation and implies that low-gain, or, in the extreme, unity-gain amplifiers require the most stringent measures of frequency compensation. Furthermore, because the frequency-compensating capacitors are then largest, the slew rate is correspondingly small.

In what follows we shall consider various means of frequency-compensating the feedback path $\beta(s)$ so as to satisfy Bode's criterion as expressed by (7-184).

Lag Compensation Lag-compensation of the feedback path takes on the form shown in Fig. 7-35. Resistors R_F and R_G pertain to the regular feedback network, R and C exclusively to the lag compensation. The feedback transfer function is

$$\beta(s) = \beta_o' \cdot \frac{s + \Omega_z}{s + \Omega_p} = \frac{R_G'}{R_G' + R_F} \cdot \frac{s + (1/CR)}{s + [1/C(R + R_F')]} \qquad (7\text{-}185)$$

where $R_G' = R_G \| R$; $R_F' = R_F \| R$. Since we are trying to satisfy Bode's criterion as expressed by (7-184), it is most convenient, in evaluating (7-185), to consider its effect on the loop gain $A\beta$. A typical plot is shown in Fig. 7-36. The loop-gain characteristic for the amplifier without the compensation network is given by the dashed line. By adding the compensating network, the slope of the loop-gain frequency-response curve is increased at frequencies between Ω_p and Ω_z, as shown by the solid line. Although the compensating network reduces the unity-gain frequency of the loop-gain response, it can be designed so as not to affect the phase shift at high frequencies appreciably. This compensation scheme as well as the one following has been included in Table 7-6.

Lead Compensation An operational amplifier with lead compensation in the feedback network is shown in line 7 of Table 7-6. Insofar as the loop gain is concerned, the feedback capacitor C_F together with the feedback resistors R_F and R_G act as a phase-lead network. The resulting loop-gain characteristics show that the zero Ω_z, which occurs before the pole Ω_p, decreases to -6 dB/octave, the natural -12 dB/octave slope taking place

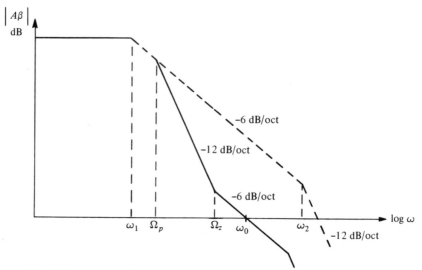

FIG. 7-36. Loop-gain frequency response, compensated in the feedback path.

at the amplifier break frequency ω_2. A stable -6 dB/octave rate of closure at the crossover frequency ω_0 is thereby obtained.

It should be clear to the reader that instead of considering the loop gain in lines 6 and 7 of Table 7-6 to demonstrate the method by which frequency compensation of the feedback path stabilizes an operational amplifier, we could consider the closed-loop gain $G(s)$. After the $\beta(s)$ functions have been inverted to obtain the corresponding $G(s)$ functions, it follows immediately that phase-lag and -lead compensations of the feedback path result in the stable closed-loop gain characteristics shown in Fig. 7-37. (Notice that Ω_p and Ω_z correspond to the expressions given in Table 7-6). In principle, phase-lag networks are used to increase the slope of the loop-gain characteristic at low frequencies without appreciably affecting the loop phase angle at frequencies in the vicinity of the unity-loop-gain crossover frequency. By contrast, phase-lead networks are used to reduce the slope of the loop-gain characteristic at frequencies in the vicinity of the unity-loop-gain crossover frequency, thereby reducing the loop phase shift at those (high) frequencies. Both these and the preceding compensation methods can be optimized for a given operational amplifier to provide the maximum loop gain over the widest possible frequency range compatible with an acceptable phase margin.

7.2.6 Matching $\Delta A_o/A_o$ with $\Delta\tau/\tau$

In Section 7.2.4 we found that a very effective method of minimizing variations of the closed-loop gain $\Delta G/G$ as well as of the phase $\Delta\phi$ is to let $\Delta\tau/\tau = \Delta A_o/A_o$ (see (7-169)), where G is given by (7-162). The frequency $\Omega = 1/\tau$ is the corner frequency of the frequency-compensated open-loop gain as given by (7-130)

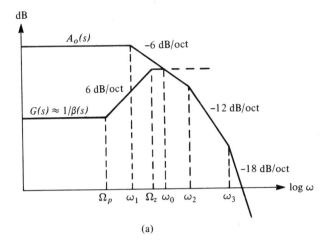

(a)

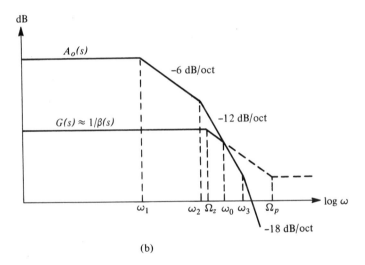

(b)

FIG. 7-37. Frequency compensation of the feedback path. (a) Lag compensation. (b) Lead compensation.

and shown in Fig. 7-27. Of the various methods of frequency compensation discussed in Section 7.2.5, the most commonly used is single-pole feedback compensation in the forward path. In essence it consists of introducing a feedback capacitor around the highest-gain stage of an amplifier. The Miller effect multiplies the capacitor by the gain of that stage, providing a large effective capacitor whose actual value is small.

Let us examine the specific case shown in line 4 of Table 7-6.[17] The open-loop gain is given by

$$A(s) = A_1 A_3 \frac{a_2}{1 + s\tau} \qquad (7\text{-}186)$$

where

$$\tau = a_2 R_s C_F \qquad (7\text{-}187)$$

R_s is the internal load resistor of the preceding amplifier stage and C_F is the feedback capacitor including the effect of the collector-base capacitance of the transistor providing the DC gain a_2. From (7-187) it follows that

$$\frac{\Delta\tau}{\tau} = \frac{\Delta R_s}{R_s} + \frac{\Delta C_F}{C_F} + \frac{\Delta a_2}{a_2} \qquad (7\text{-}188)$$

With (7-169) we therefore require that

$$\frac{\Delta R_s}{R_s} + \frac{\Delta C_F}{C_F} + \frac{\Delta a_2}{a_2} = \frac{\Delta A_o}{A_o} \qquad (7\text{-}189)$$

The term $\Delta a_2/a_2$, the gain variation of the highest-gain stage, depends on the design of the particular amplifier under consideration. Assume, for example, that A_o comprises, in equal parts, the gain a_2 and the gain provided by the remaining one or two stages. Assuming, furthermore, that both parts vary equally, then

$$\frac{\Delta a_2}{a_2} \approx \frac{1}{2} \frac{\Delta A_o}{A_o} \qquad (7\text{-}190)$$

$\Delta R_s/R_s$ is the variation of an internal silicon-chip resistance that cannot be modified in any way. C_F, on the other hand, is an external capacitor that can be selected to satisfy our condition (7-169). Thus with (7-189) and (7-190) we select C_F such that its temperature coefficient is given by

$$\frac{\Delta C_F}{C_F} = \frac{1}{2} \frac{\Delta A_o}{A_o} - \frac{\Delta R_s}{R_s} \qquad (7\text{-}191)$$

Using the typical values given by (7-153a) and (7-153b) for $\Delta A_o/A_o$ and $\Delta R_s/R_s$, respectively, we find

$$\frac{1}{2} \frac{\Delta A_o}{A_o} = \frac{\Delta a_2}{a_2} = -4000 \text{ ppm/}^{\circ}\text{C}$$

$$\frac{\Delta R_s}{R_s} = 5000 \text{ ppm/}^{\circ}\text{C}$$

and the required TCC of C_F is

$$\frac{\Delta C_F}{C_F} = -9000 \text{ ppm/}^{\circ}\text{C}$$

Actually the temperature coefficient assumed for A_o is the worst-case value (corresponding to -50% change over 60°C) and the TCR of R_s is also rather high. In practice a TCC of -4000 ppm/°C to -5000 ppm/°C may be more accurate and will certainly show significant improvement in gain and phase stability compared to a capacitor with zero (i.e., NPO) or very small TCC. To obtain more accurate numbers for the TCC of C_F, the specification sheets of the operational amplifier should be referred to, with the possible addition of some measurements of the temperature dependence of the break frequency of a compensated amplifier. When making these test measurements, NPO capacitors should be used for the compensation. This will provide information on both $\Delta A_o / A_o$ and $\Delta R_s / R_s$ with negligible effects from the compensation capacitors.

If the gain stage a_2 considered above dominates significantly over the other stages, or if, in fact, the capacitor C_F is fed back around the whole amplifier stage, then we can assume that

$$\frac{\Delta a_2}{a_2} = \frac{\Delta A_o}{A_o} \qquad (7\text{-}192)$$

and require a corresponding increase of the temperature coefficient of C_F. Exactly the same procedures as those carried out here can be used to obtain the desirable temperature coefficients of the external components in the other single-pole compensation schemes listed in Table 7-6. For pole–zero schemes, or those employing double poles, the analysis becomes somewhat more cumbersome, but the basic ideas outlined here remain the same.

7.2.7 Error Sources in Nonideal Operational Amplifiers

The error sources in commercially available integrated operational amplifiers are amply discussed in the trade literature. Inasmuch as they have a direct bearing on the design of active networks (e.g., on impedance levels or on the circuit layouts of hybrid integrated circuits with a view to minimizing critical parasitic capacitance) they will be discussed here briefly.

Offset Voltage The ideal operational amplifier has zero offset, i.e., the output voltage is zero for zero input voltage. In practice this will never be exactly true and there will always be a finite voltage, generally only several millivolts, at the output even if the input terminals are grounded. One obvious reason for this follows directly from Fig. 7-38a. In order that the opamp may function properly, a bias current i_1 and i_2 must enter the input terminals. These will generally be the base currents of the input differential-gain stage. Because of differences between the base-emitter characteristics of the input transistors, both the input currents and voltages will differ. If the input currents are high or the source resistors (i.e., R_n and R_G in parallel with R_F)

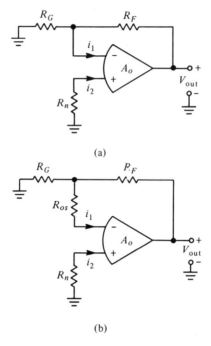

(a)

(b)

FIG. 7-38. Offset voltage minimization.

are large and unequal, the voltage at the input terminals will be nonzero and an offset voltage at the output will result.

Because the output offset voltage depends on the gain, offset is generally referred to the input terminals. To keep the offset voltage as small as possible, first the values of the source resistors R_n and $R_G \| R_F$ should be reasonably low (e.g., not higher than several kilohms), and second R_n should be selected such that

$$R_n = \frac{R_G R_F}{R_G + R_F} \qquad [7\text{-}193]$$

This expression was used in the previous analysis of operational amplifiers (see (7-47) in Section 7.2.2). In order to permit offset, gain, and source impedances to be independent of one another, particularly in the non-inverting mode, when R_n may be determined by other factors, a resistor R_{os} may be inserted in the circuit as shown in Fig. 7-38b. Then R_{os} must have the value

$$R_{os} = R_n - \frac{R_G R_F}{R_G + R_F} \qquad [7\text{-}194]$$

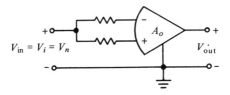

FIG. 7-39. Common-mode operation.

Clearly this is only useful if R_n is required to be large and two correspondingly large resistors R_G and R_F are to be avoided in place of only one, namely, R_{os}. The latter consideration is important when one is using thin-film resistors, which should not exceed a maximum upper limit per substrate in total value. (Typically the upper limit may be 0.5–1 MΩ.)

Common-Mode Gain   The common-mode gain is defined as the ratio of the output voltage V_{out} to the input voltages V_i and V_n when V_i and V_n are identical in amplitude and phase. An example of common-mode operation is shown in Fig. 7-39. Whereas ideally the common-mode gain is zero (infinite attenuation), in practice, of course, it is not. Thus, the designer must settle for a common-mode gain that is, at best, very much smaller than the open-loop gain $A(s)$. In the following we shall examine the conditions that are required in order for this to be true.

An examination of Figs. 7-14b and 7-15b shows that, in either the inverting or the noninverting configuration, the differential gain acts on the *difference* between the voltages V_i and V_n. On the other hand, the common-mode gain acts on those parts of V_i and V_n that are *in phase and identical in magnitude*. That is, the common-mode gain acts on the smaller of the two in-phase signals V_i and V_n. In the inverting configuration V_i is greater than V_n, in the noninverting configuration it is less than V_n. These considerations are reflected by the output-voltage equations when the effects of the common-mode gain (CMG) are included. With $V_d = V_i - V_n$ we have

$$(V_{out})_I = -A_o V_d - \text{CMG} \cdot V_n$$
$$= A_o(V_n - V_i) - \text{CMG} \cdot V_n \qquad (7\text{-}195)$$

and

$$(V_{out})_N = A_o(V_n - V_i) - \text{CMG} \cdot V_i \qquad (7\text{-}196)$$

Using the closed-loop gain formulas of (7-32) and (7-33) and assuming that the output impedance R_o is negligibly small we obtain the following closed-loop gain expressions for operational amplifiers with nonzero common-mode gain:

$$G_I' = \frac{-A_o R_i R_F - CMG_I R_n R_F}{R_G(R_F + R_i + R_n) + R_F(R_i + R_n) + A_o R_i R_G + R_n R_G CMG_I}$$

(7-197)

and

$$G_N' = \frac{A_o R_i(R_G + R_F) - CMG_N R_G R_F}{(R_G + R_F)(R_n + R_i) + R_G R_F + A_o R_i R_G + CMG_N R_G(R_i + R_n)}$$

(7-198)

Now, in order for the common-mode gain CMG to be negligible compared to the open-loop gain A_o, the following constraints must hold: for the inverting case,

$$CMG_I \ll A_o \frac{R_i}{R_n}$$

[7-199]

and for the noninverting case

$$CMG_N \ll A_o \frac{R_i}{R_i + R_n}$$

[7-200]

Notice that, like the offset considerations discussed above, the conditions in (7-199) and (7-200) favor a source resistance R_n that is as small as possible. Furthermore, the inequality (7-199) shows that the gain of an inverting configuration is not affected by the common-mode gain as long as the input impedance R_i is sufficiently large. However, even when R_i is very large in the noninverting configuration, the gain is still dependent on the common-mode gain, provided the open-loop gain A_o is finite. This is an important distinction between inverting and noninverting operation.

Often, the common-mode rejection ratio (CMR) rather than the CMG is specified. It is defined as

$$CMR = \frac{\text{Differential gain } (A_o)}{CMG}$$

This value should, of course, be as large as possible. Expressing the CMR in terms of the inequalities given above, we obtain the following conditions:

$$CMR_I \gg \frac{R_n}{R_i}$$

[7-201]

and

$$CMR_N \gg \frac{R_i + R_n}{R_i}$$

[7-202]

These inequalities do not place very stringent restrictions on the common-mode rejection.

Finally, one important consideration is the maximum input voltage that can be tolerated in common-mode operation. This is generally referred to as the common-mode voltage range. Depending on the design of the amplifier, this may be determined by the DC-bias conditions of the transistors in the input differential amplifier, in which case the total common-mode voltage range may be of the order of only 1 or 2 V. The circuit shown in Fig. 7-10 belongs to this class. By contrast, the input stage of the circuit shown in Fig. 7-11a is designed such that the input voltage swing in common-mode operation equals the (symmetrical) supply-voltage values (i.e., $\pm V_s$) to within less than a volt or so.

Parasitic Input Capacitance One error source that can be particularly damaging, since it can cause amplifier instability, is parasitic capacitance at the input terminals of the operational amplifier. Since the operational amplifier providing low gain in the noninverting mode is quite vulnerable to this parasitic effect, we shall discuss it in some detail in the following.

Assuming unity gain, which in this situation is the worst case, we have from (7-72) and (7-78)

$$A_N(s)\Big|_{\alpha_N = 1} = A_{N_1}(s) = \frac{R_o}{R_i} + A(s) \tag{7-203}$$

and

$$\beta_N\Big|_{\alpha_N = 1} = \beta_{N_1} = \frac{R_i}{R_i + R_n + R_F} \tag{7-204}$$

We use $A(s)$ instead of $A_o(s)$ in (7-203) to indicate that the open-loop gain is assumed to be frequency-compensated for a -6 dB/octave slope. Thus

$$A(s) = A_o \frac{\Omega}{s + \Omega} \tag{7-205}$$

where Ω is the compensated break frequency of the amplifier.

With (7-203) we find that

$$A_{N_1}(s) = \frac{N_{A_1}}{D_{A_1}} = \frac{R_o}{R_i} \cdot \frac{s + \Omega[A_o(R_i/R_o) + 1]}{s + \Omega} \approx A_o \frac{\Omega}{s + \Omega} = A(s) \tag{7-206}$$

Consider now, the case that the resistors R_j (where $j = i, n, F$) in (7-204) are replaced by frequency-dependent inpedances Z_j consisting of a resistor R_j in parallel with a capacitor C_j:

$$Z_j = \frac{1}{C_j} \frac{1}{s + \Omega_j} \qquad j = i, n, F \tag{7-207}$$

where

$$\Omega_j = \frac{1}{R_j C_j} \qquad j = i, n, F \tag{7-208}$$

and the C_j are parasitic capacitances. Then (7-204) becomes

$$\beta_{N_1}(s) = \frac{N_{\beta_1}}{D_{\beta_1}}$$

$$= \frac{1/C_i(s + \Omega_i)}{1/C_i(s + \Omega_i) + 1/C_n(s + \Omega_n) + 1/C_F(s + \Omega_F)}$$

$$= \frac{C_n C_F(s + \Omega_n)(s + \Omega_F)}{C_n C_F(s + \Omega_n)(s + \Omega_F) + C_i C_F(s + \Omega_i)(s + \Omega_F) + C_i C_n(s + \Omega_i)(s + \Omega_n)}$$

$$(7\text{-}209)$$

With $G = \alpha[A\beta/(1 + A\beta)]$, (7-206), and (7-209) we have

$$G_N(s)\bigg|_{\alpha_N = 1} = G_{N_1}(s) = \frac{N_{G_1}}{D_{G_1}} = \frac{A_{N_1}(s)\beta_{N_1}(s)}{1 + A_{N_1}(s)\beta_{N_1}(s)} = \frac{N_{A_1}N_{\beta_1}}{N_{A_1}N_{\beta_1} + D_{A_1}D_{\beta_1}} \quad [7\text{-}210]$$

where

$$N_{G_1} = A_o C_n C_F(s + \Omega_n)(s + \Omega_F)\Omega \tag{7-211a}$$

and

$$D_{G_1} = N_{G_1} + (s + \Omega)\{C_n C_F(s + \Omega_n)(s + \Omega_F) + C_i(s + \Omega_i)[C_F(s + \Omega_F) + C_n(s + \Omega_n)]\} \tag{7-211b}$$

Since this expression is quite complicated let us first look at the simple but no less important case shown in Fig. 7-40, where the parasitic input capacitance C_i is in parallel with R_i, but the other impedances are purely resistive, i.e., $C_n = C_F = 0$, $Z_n = R_n$, $Z_F = R_F$, and $Z_i = 1/C_i(s + \Omega_i)$. Then $A_{N_1}(s)$ is given by (7-206) and

$$\beta_{N_1}(s)\bigg|_{\substack{C_n = C_F = 0 \\ C_i \neq 0}} = \frac{\Omega_{nFi}}{s + \Omega_{pi}} \tag{7-212}$$

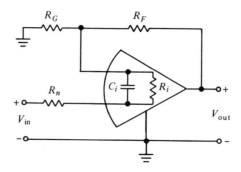

FIG. 7-40. Parasitic input capacitance.

where

$$\Omega_{nFi} = \frac{1}{(R_n + R_F)C_i} \tag{7-213a}$$

and

$$\Omega_{p_i} = \Omega_{nFi} + \Omega_i = \frac{1}{C_i[R_i\|(R_n + R_F)]} \tag{7-213b}$$

Note that

$$\Omega_{nFi} < \Omega_{p_i} \tag{7-213c}$$

so that $\beta_{N_1}(0) = \Omega_{nFi}/\Omega_{p_i}$ is less than unity (as it must be, since $\beta_{N_1}(0)$ is also given by (7-204)). The loop gain

$$A_{N_1}(s)\beta_{N_1}(s) = A_o \frac{\Omega}{s + \Omega} \cdot \frac{\Omega_{nFi}}{s + \Omega_{p_i}} \tag{7-214}$$

is plotted in Fig. 7-41. Observe that because of the additional pole Ω_{p_i} the operational amplifier will be unstable because the loop-gain rate of closure is 12 dB/octave. To avoid this problem we evidently must require that

$$\Omega_{p_i} > \omega_c \tag{7-215}$$

where ω_c is the unity-gain crossover frequency of the open-loop gain $A_{N_1}(s)$. With (7-213b) this implies that

$$R_n + R_F < \frac{R_i}{(\omega_c/\Omega_i) - 1} = \frac{1/C_i}{\omega_c - \Omega_i} \tag{7-216}$$

This inequality sets an upper limit either on the maximum impedance level of the external resistors R_n and R_F or on the maximum time constant of the input impedance, $R_i C_i = 1/\Omega_i$. Clearly excessive parasitic capacitance in parallel with C_i can cause the inequality to be violated. As a result the operational amplifier will be instable when used in the low-gain, noninverting feedback mode. Consider, for example, the case that $R_i = 50$ kΩ, $R_n + R_F = 20$ kΩ and $\omega_c = 2\pi \times 10$ MHz. Then Ω_i must be larger than $\omega_c/3.5$ and C_i must be less than 1.1 pF. A larger input resistance R_i, or lower-valued resistors R_n and R_F, permit the value of C_i to be correspondingly larger.

In the unity-gain noninverting mode, R_F is required only for offset considerations; R_F must equal R_n. If, however, a resistor R_{os} is used for offset, and a low-valued resistor R_F is AC-coupled in parallel with R_{os}, as shown in Fig. 7-42, then both the offset and stability conditions discussed above can be satisfied simultaneously. The capacitor C must be large enough to have no effect at the frequencies of operation. For $1/CR_F$ to be much lower than the break frequency Ω we require that

$$C \gg 1/\Omega R_F \tag{7-217}$$

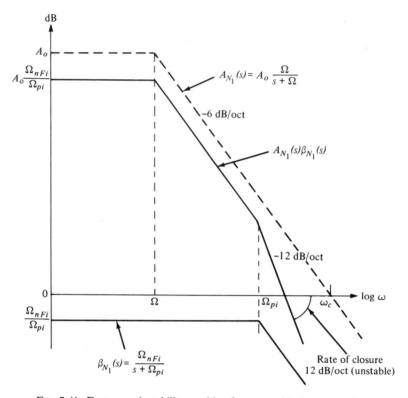

FIG. 7-41. Frequency instability resulting from parasitic input capacitance.

This inequality may be inconvenient, since it requires a large capacitor. This may be avoided by means of the configuration shown in Fig. 7-43. Here $C_n = 0$, $Z_i = 1/C_i(s + \Omega_i)$ and $Z_F = 1/C_F(s + \Omega_F)$. Identifying this case by a prime on $\beta_{N_1}(s)$, (7-209) becomes

$$\beta'_{N_1}(s) = \frac{C_F(s + \Omega_F)}{C_F(s + \Omega_F) + C_i(s + \Omega_i) + R_n C_i C_F(s + \Omega_i)(s + \Omega_F)} \tag{7-218}$$

We retain sufficient degrees of freedom if we let $\Omega_F = \Omega_i$ where $C_F \neq C_i$. Then

$$\beta'_{N_1}(s) = \frac{\Omega_{ni}}{s + \Omega'_{pi}} \tag{7-219}$$

where

$$\Omega_{ni} = \frac{1}{R_n C_i} \tag{7-220a}$$

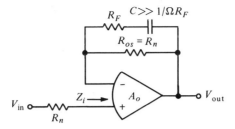

FIG. 7-42. Selection of R_F to overcome parasitic input capacitance.

and

$$\Omega'_{p_i} = \Omega_i \left(1 + \frac{R_F}{R_n} \cdot \frac{C_i + C_F}{C_i}\right) \qquad (7\text{-}220b)$$

Since $\Omega_i = \Omega_F$, it is easy to show that

$$\Omega_{n_i} < \Omega'_{p_i} \qquad (7\text{-}220c)$$

Plotting the loop gain $A_{N_1}(s)\beta'_{N_1}(s)$ we obtain a plot similar to the one shown in Fig. 7-41 and therefore require, for stability, that

$$\Omega'_{p_i} = \Omega_i \left(1 + \frac{R_F}{R_n} \frac{C_i + C_F}{C_i}\right) > \omega_c \qquad [7\text{-}221a]$$

or

$$\frac{R_F}{R_n} \cdot \frac{C_i + C_F}{C_i} > \frac{\omega_c}{\Omega_i} - 1 \qquad [7\text{-}221b]$$

We observe that in this case we can select R_F to meet offset conditions, (i.e., $R_F = R_n$) and select C_F such that (7-221b) is satisfied. Clearly, C_F should not be selected larger than necessary, in order not to reduce the loop-gain bandwidth unnecessarily. For an example similar to the one given above, in which $R_i = 50$ kΩ, $R_F = R_n$ (notice that the actual value is not required here), and $C_i = 1$ pF, we require a value of C_F larger than 1.5 pF to satisfy (7-221b). The parasitic capacitance alone, say between conductor paths on a film layout, may be sufficient to satisfy this requirement.

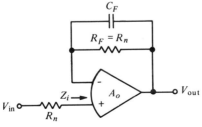

FIG. 7-43. Selection of C_F to overcome parasitic input capacitance.

In summary, the noninverting unity-gain amplifier is particularly sensitive to parasitic capacitance C_i between its input terminals. If C_i is sufficiently large to violate (7-216) instability (e.g., parasitic oscillations) will be the result. To counteract this, the configuration in Fig. 7-43 (or also that in Fig. 7-42) can be used; it comprises a capacitor C_F between the inverting and output terminals. Frequently there is a sufficiently large *parasitic* capacitance C_F to cancel the harmful effects of the parasitic capacitance C_i between the input terminals. Naturally, in an integrated (e.g., film) circuit layout, attention must be paid to the contrasting effects of these two parasitic capacitances, particularly when one is using circuits with unity- or low-gain noninverting amplifiers. The effect of parasitic capacitance between any other pair of amplifier terminals can be derived directly from (7-210) following the same procedure as was used above.

Slew Rate   Capacitive loading, either at internal points or at the output of an operational amplifier, causes slew-rate limiting, i.e., limiting to the rate at which the capacitor or capacitors in question can be charged. The slew rate depends on (1) the amount of current available to charge a capacitor, (2) the capacitor value itself, and (3) the rate at which the capacitor is to be charged. Thus, with a maximum available current i_{max} and a given capacitance C, the maximum attainable slew rate SR is

$$SR_{max} = \left(\frac{dv}{dt}\right)_{max} = \frac{i_{max}}{C} \qquad (7\text{-}222)$$

Note that the maximum slew rate may be different for positive and negative going signals.

In a slew-rate-limited circuit the signal waveforms entering the amplifier will be distorted at the output; a square waveform will become trapezoidal, a sinusoidal waveform triangular, and so on. If we consider the common case of a sinusoidal signal

$$V = \hat{V} \sin \omega t \qquad (7\text{-}223)$$

then the rate of change is

$$\frac{dv}{dt} = \omega \hat{V} \cos \omega t \qquad (7\text{-}224)$$

This is a sinusoidal function of time whose peak instantaneous value is

$$SR_{max} = \left(\frac{dv}{dt}\right)_{max} = \omega \hat{V} \qquad (7\text{-}225)$$

Thus the onset of waveform distortion with sinusoidal signals occurs when $\omega \hat{V}$ is equal to the maximum slew rate specified for the amplifier. The maximum amplitude of a sinusoidal voltage that can be handled by an amplifier without introducing distortion due to slew-rate limiting is therefore *a function of the voltage frequency*. This follows directly from (7-225):

$$\hat{V}_{max} = SR_{max}/\omega \qquad \text{[7-226]}$$

The higher the frequency of the input sinusoidal voltage, the more its amplitude must be limited to avoid distortion by slew-rate limiting.

An operational amplifier invariably has an intrinsic maximum slew rate. This is generally specified for given operating conditions, in particular for given capacitive loading and frequency compensation. The frequency-compensating networks load internal points of the amplifier capacitively, thereby directly affecting the slew-rate capabilities of the amplifier. As we saw in the discussion on frequency compensation, the capacitor values required to stabilize the amplifier increase with the amount of negative feedback around the amplifier. This, in turn, increases as the closed-loop gain decreases. Consequently, unity-gain compensation usually yields the poorest slew rate, i.e., the lowest frequency range of large signal swing.

One way of overcoming this problem is to reduce the negative feedback to a value corresponding to a higher closed-loop gain, while leaving the actual closed-loop gain unchanged.[18] This can be done with the circuit shown in Fig. 7-44. Assuming that the input impedance to the amplifier is much larger than R_g, the voltage transfer function results in a straightforward manner. Letting

$$R'_G = R_G \| R_g \qquad (7\text{-}227a)$$

and

$$\beta'_o = \frac{R'_G}{R'_G + R_F} \qquad (7\text{-}227b)$$

we obtain, by analogy to (7-37),

$$G'_I = -\frac{R_F}{R_G} \cdot \frac{A_o \beta'_o}{1 + A_o \beta'_o}\bigg|_{A_o\beta_o' \gg 1} \approx -\frac{R_F}{R_G} \qquad (7\text{-}228)$$

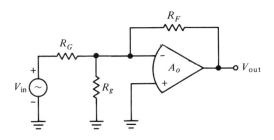

FIG. 7-44. Inverting opamp with improved slew rate.

18. D. R. Kesner, Simple technique extends op amp slew rate, *EDN Mag.* June 1, 1970, pp. 46–50.

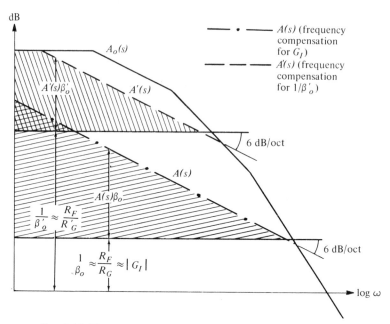

FIG. 7-45. Frequency response of opamp with improved slew rate.

Thus, for sufficiently large loop gain $A_o \beta'_o$, G'_I is approximately equal to the closed-loop gain that would be obtained without the additional resistor R_g. Since

$$R'_G = \frac{R_g}{R_G + R_g} R_G < R_G \qquad (7\text{-}229)$$

the feedback factor β'_o is smaller than the feedback factor without R_g, namely $\beta_o = R_G/(R_F + R_G)$. Consequently the loop gain $A_o \beta'_o$ is smaller than the original loop gain $A_o \beta_o$. It is not decreased as much as might at first be expected, however. Consider, for example, the situation shown in Fig. 7-45. $A(s)$ is the frequency-compensated open-loop gain necessary for the inverted feedback factor $1/\beta_o$, which is approximately equal to the closed-loop gain $|G_I|$. If the inverted feedback factor is increased to $1/\beta'_o$, the frequency compensation can be correspondingly lighter (providing a larger slew rate, which, it will be remembered, is our objective), and the open-loop gain characteristic $A'(s)$ results. Notice, that because of the lighter compensation, the loop gain $A'(s)\beta'_o$ is not reduced nearly as much, compared to $A(s)\beta_o$, as it would be if $A(s)$ would have to be used with $1/\beta'_o$. (In the latter case, the cross-hatched area in Fig. 7-45 would represent the available loop gain.) Naturally, some decrease in loop gain does result, however, and it will depend on the application to what extent this reduction can be tolerated.

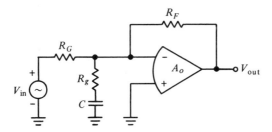

FIG. 7-46. Inverting opamp with improved slew rate and DC offset voltage.

One noticeable disadvantage in the configuration of Fig. 7-44 is the increased offset voltage at the amplifier output, compared to the circuit without R_g. Whatever offset voltage is present at the inverting terminal will be multiplied not by R_F/R_G but by R_F/R_G'. If, for example $R_F = R_G = 100$ kΩ and $R_g = 1$kΩ, any offset voltage at the input will be multiplied at the output by approximately 100. This can be avoided rather easily, however, by connecting a large capacitor C in series with R_g, as shown in Fig. 7-46. This circuit will have the same amount of DC offset voltage at the output as the circuit with R_g removed. Care should be taken to select C large enough that the frequency $f = 1/2\pi R_g C$ is small, e.g., a decade below the loop-gain unity-crossover frequency.

A strategy similar to that above for improving slew rate is possible for the noninverting mode, as shown in Fig. 7-47a. However, since the insertion of attenuation in the feedback path (i.e., R_g) in the noninverting mode increases the gain while improving the slew rate, a corresponding attenuator (i.e., R) must be added in the forward path to obtain, say, unity gain. To avoid the offset problem both resistors are DC-isolated by the capacitors C_g and C. Unity gain is obtained if $R_n = R_F$, $R_g = R$, and $C_g = C$. To reduce the component count the grounded ends of C_g and C can be tied together, as shown in Fig. 7-47b.

Noise We call noise all those signals appearing at the output terminal of an operational amplifier that are neither related to the signals at the input terminals, nor predictable from an accurate knowledge of the closed-loop transfer function of the circuit of which the operational amplifier is a part. These unwanted signals may be introduced into a circuit by the amplifier itself, by the components used in the feedback loop or in the rest of the circuit, or by the power supply; they may also be coupled or induced into the input, the output, the ground-return, or the measurement circuit from nearby, or in some cases quite distant, sources.

The measurement of the noise of an active network containing an operational amplifier is best performed at the output terminals of the amplifier. Thus, in deriving an equivalent circuit for an amplifier with noise, the noise source

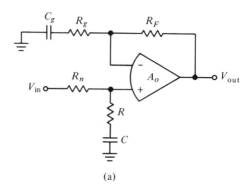

(a)

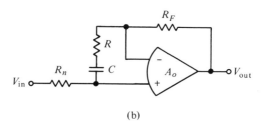

(b)

FIG. 7-47. Noninverting opamp. (a) With improved slew rate. (b) With reduced component count.

may be placed at the output of the amplifier, as shown in the circuit in Fig. 7-48. R_F, R_G, and R_n are the familiar external resistors of the amplifier feedback network and R_i and R_o are the internal input and output resistances of the amplifier, respectively. V_N is the noise voltage which would appear at the amplifier output terminals if no feedback were present and if the external resistors were noiseless. From Fig. 7-48 we obtain the voltage transfer ratio of the output voltage corresponding to the noise-voltage source as follows:[19]

$$\frac{V_{out}}{V_N} = \frac{(R_n + R_i)(R_F + R_G) + R_F R_G}{(R_n + R_i)(R_F + R_G + R_o) + R_G(R_F + R_o) + A_o R_i R_G} \tag{7-230}$$

In general, it is more convenient to refer the noise-voltage source to one of the input terminals. Placing the noise-voltage source at the inverting input terminal we obtain a voltage transfer function V_{out}/V_{N-} corresponding to the inverting transfer function G_I as given by (7-32). Comparing that expression with (7-230), i.e., setting the two output voltages equal, we obtain an expression for the equivalent input noise-voltage source V_{N-} in terms of the output noise-voltage source:

$$V_{N-} = V_N \frac{(R_n + R_i)(R_F + R_G) + R_F R_G}{R_o(R_n + R_i) - A_o R_i R_F} \tag{7-231}$$

19. D. R. Means, unpublished memorandum, Bell Telephone Laboratories.

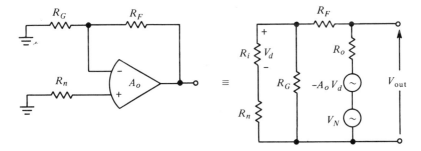

FIG. 7-48. Noise model of operational amplifier.

In precisely the same way, we can place the noise-voltage source at the non-inverting input terminal. We then obtain the voltage transfer function V_{out}/V_{N+} corresponding to G_N as given by (7-33). Solving for V_{N+}, which is the noise-voltage source at the noninverting input terminal, in terms of V_N the noise voltage source at the output terminal, we obtain

$$V_{N+} = V_N \frac{(R_n + R_i)(R_F + R_G) + R_F R_G}{R_o R_G + A_o R_i(R_F + R_G)} \tag{7-232}$$

From (7-231) and (7-232) we see that, if the operational amplifier is ideal with respect to its input and output impedances, then we obtain the following simplified expressions for the equivalent noise-voltage sources at the inverting and the noninverting input terminals, respectively:

$$V_{N-} = -\frac{V_N}{A_o}\left(1 + \frac{R_G}{R_F}\right) \tag{7-233}$$

and

$$V_{N+} = V_N/A_o \tag{7-234}$$

From these two expressions we learn that for an operational amplifier with high input and low output impedance, V_{N-} depends on external circuit components whereas V_{N+} does not.[20] For this reason, the equivalent noise-voltage source is more conveniently referred to the noninverting-amplifier input terminal, as shown in Fig. 7-49 than to the inverting terminal. To complete the noise characterization of an operational amplifier, two equivalent noise-current sources must also be included at the input terminals, as shown in the same figure.

As noted above, the noise source referred to the noninverting input terminal of the operational amplifier is independent of the closed-loop gain

---

20. This does not mean that the noise caused by V_{N+} and appearing at the output is independent of closed-loop gain. It simply means that the equivalent noise-voltage source referred to the noninverting input terminal is independent of closed-loop gain. The resulting output noise as given by (7-230), of course, depends on the closed-loop gain.

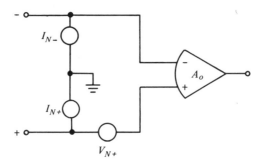

FIG. 7-49. Operational amplifier with noise model referred to input.

of the circuit as determined by R_F and R_G. Therefore, the noise at the output is directly proportional to the closed-loop gain or, equivalently, inversely proportional to the amount of feedback. Thus, if we measure the total output noise of an operational amplifier connected as shown in Fig. 7-50a we obtain plots of the kind shown in Fig. 7-50b. Here the output noise voltage has been plotted as a function of gain for three different impedance levels. For a given impedance level the noise increases linearly with gain, except at low noise levels, where it tails off to a constant value because of the noise introduced by the measuring equipment itself.

The output noise also increases with the impedance level of the feedback network. This is due to two factors. First, above a certain impedance level, known as the *characteristic noise impedance*, the total output noise is due primarily to the noise-current sources rather than to the noise-voltage source. Therefore, at impedance levels above the characteristic noise impedance, the total output noise level rises as the impedance level rises. At impedance levels below the characteristic noise impedance, (which may typically be in the order of 1 to 10 kΩ) the total output noise is relatively independent of the impedance level, since the noise-voltage source is more significant than the noise-current sources. The second factor involved in the impedance dependence of the output noise is the noise generated by the external resistors themselves (i.e., R_F, R_G, and R_n). By inspection of Fig. 7-48 we see that the noise generated by R_G and R_n will be amplified by the gain of the circuit from the inverting and noninverting input terminals, respectively. The contribution from these two resistors to the total output noise of the circuit will therefore be proportional to the gain, whereas the contribution from R_F will be independent of gain.

The noise voltage generated by a resistor R is given by

$$V_{rms} = \sqrt{4kTRB} \tag{7-235}$$

where

k is Boltzmann's constant, $(1.374 \times 10^{-23} \text{ J/°K} = 8.63 \times 10^{-5} \text{ eV/°K})$,

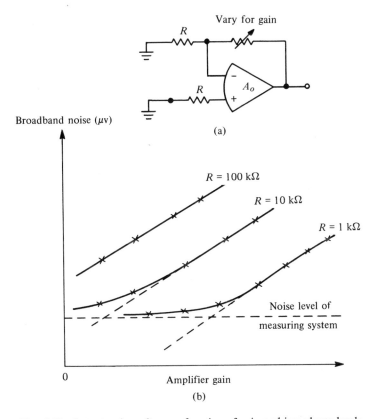

FIG. 7-50. Output noise voltage as function of gain and impedance levels.

T is the absolute temperature (°K),
R is the resistance (Ω), and
B is the bandwidth (Hz).

At room temperature, and with a noise bandwidth of, say, 1 MHz, the rms noise voltages measured with resistor values varying from 1 to 100 kΩ vary from 4 to 40 μV.

One form of noise, power-supply ripple appearing at the output of an amplifier, warrants some special attention of its own. This noise form depends on the ability of an amplifier to reject changes in power-supply voltage. The *power-supply rejection ratio*[21] of an operational amplifier is defined as the ratio of the change in input *offset* voltage to the change in power-supply voltage producing it. If V_{os} is the change in input offset voltage, then the change in output voltage in the absence of feedback will be $A_o V_{os}$, where A_o is the open-loop gain of the operational amplifier. Consequently, for the

21. Often also referred to as *supply-voltage rejection ratio*, it is specified by the manufacturer.

circuit shown in Fig. 7-48, (7-230) is valid if V_N is replaced by $A_o V_{os}$. If the power-supply rejection ratio is denoted by r and the power-supply ripple by V_{ps} then, since $A_o V_{os} = A_o r V_{ps}$ we obtain

$$\frac{V_{out}}{V_{ps}} = A_o r \frac{(R_n + R_i)(R_F + R_G) + R_F R_G}{(R_n + R_i)(R_F + R_G + R_o) + R_G(R_F + R_o) + A_o R_i R_G} \qquad [7\text{-}236]$$

If the amplifier is ideal, this becomes

$$\frac{V_{out}}{V_{ps}} = r\left(1 + \frac{R_F}{R_G}\right) \qquad [7\text{-}237]$$

Suppose, for example, that $V_{ps} = 150$ mV, $r = 200$ μV/V, $R_F/R_G = 1$; then V_{out}, the rms ripple appearing at the output, is 60 μV.

It should be noted that the power-supply rejection ratio of an operational amplifier is a function of the external impedance level. Data sheets usually provide the value only for low impedance values. On the other hand, the rejection ratio may more than double as the impedance level is increased by an order of magnitude, starting from a low impedance value (e.g., 1 kΩ), and may increase significantly beyond that as the impedance levels are increased still further (e.g., between 10 kΩ and 100 kΩ). It should also be noted that the rejection ratio may be different with respect to ripple on the positive or the negative power-supply lead, and is generally frequency-dependent.

We shall now briefly discuss the *noise figure F* of an amplifier. This is defined as[22]

$$F[\text{dB}] = 10 \log \frac{\text{total output noise}}{\text{source resistor noise}} \qquad (7\text{-}238)$$

F provides a measure of the noise contributed by the amplifier in addition to that of the source resistance. With (7-235) and the voltage and current noise sources V_N and I_N, (7-238) becomes

$$F[\text{dB}] = 10 \log \frac{V_N^2 + I_N^2 R_s^2 + 4kTR_s B}{4kTR_s B} \qquad (7\text{-}239)$$

where R_s is the source resistance. The noise figure F is given in decibels; it is zero for an ideal noiseless amplifier. For any practical amplifier F is minimum if

$$(R_s)_{opt} = \frac{V_N}{I_N} \qquad (7\text{-}240)$$

When R_s satisfies (7-240) it is known as the *optimum noise resistance*. Since F depends on bandwidth, $(R_s)_{opt}$ may vary with the frequency band in

22. L. Smith and D. H. Sheingold, Noise and operational amplifier circuits, *Analog Dialogue*, Vol. 3, No. 1, March, 1969. Analog Devices, Inc., Cambridge, Mass. 02142.

question. For this reason (7-238) may be easier to evaluate than (7-239), since the former refers to the actual closed-loop configuration and the measured or computed rms noise.

As a final observation, we should point out that most of the error sources discussed in this section can be controlled, to some degree at least, by keeping the external resistor values of the feedback network, as well as those of other DC paths leading directly from either input terminal to ground, as small as possible. Naturally, this may become difficult at low frequencies, when the resistors in the DC return path of either input terminal are part of a frequency-determining RC network. However, at low frequencies, amplifiers with extremely low biasing, and consequently offset, currents are available; with such amplifiers larger external resistors can generally be tolerated. At higher frequencies, input currents are generally higher, but, at the same time, high-valued resistors are less likely to be required.

7.3 THE GYRATOR

In Chapter 5 the gyrator was discussed in some detail, both as an ideal and a nonideal network element. In the following discussion we shall briefly consider methods of realizing gyrators in silicon integrated form. As we shall see, and would expect, the most promising realizations utilize the building blocks for silicon integrated circuits discussed in Section 7.1.[23]

7.3.1 Expanded Three-Transistor Gyrators

It was shown in Section 5.4.3 that a gyrator can be derived from the following decomposition of its two-port admittance matrix:

$$[y] = \begin{bmatrix} 0 & g_1 \\ -g_2 & 0 \end{bmatrix} = \begin{bmatrix} 0 & g_1 \\ 0 & 0 \end{bmatrix} + \begin{bmatrix} 0 & 0 \\ -g_2 & 0 \end{bmatrix} \qquad (7\text{-}241)$$

This decomposition corresponds to the parallel connection of two VCSs, as shown in Fig. 7-51a, or of two transconductance amplifiers, as shown in Fig. 7-51b; the transconductance constant is g_m for one amplifier, $-g_m$ for the other. While we found in Chapter 5 that other decompositions of the gyrator matrix, e.g., its z matrix, lead to different equivalent circuits (e.g., series connection of two CVSs) the y matrix and the resulting parallel connection of two VCSs has been found to be the most useful in practice.

By considering the nullator–norator equivalents of VCSs, it was shown in Chapter 5 how a variety of three-transistor gyrator circuits çould be

---

23. Remember that, by definition, the ideal gyrator is a passive device (the balance of ipower entering and leaving it is never negative). The fact that active devices are needed for ts implementation all the same is because the gyrator is nonreciprocal.

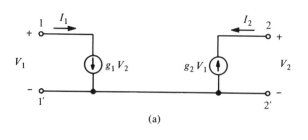

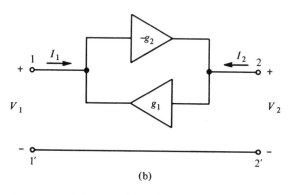

FIG. 7-51. Gyrators as parallel connection of (a) two VCS. (b) two transconductance amplifiers.

derived (see Figs 5-41 and 5-44). We did not pursue the practicability of these circuits in Chapter 5, but it goes without saying that some are more suitable for implementation in silicon integrated form than others. Some of the main practical considerations that must be taken into account are DC biasing, frequency characteristics, and compatibility with silicon integrated circuit technology.

One three-transistor configuration that has received serious attention is that shown in Fig. 5-41b. The VCS in the forward direction is shown in Fig. 7-52a. It consists of a common-emitter stage with current feedback through R_2. The (N_1, n_1) designation corresponds to the designation in Fig. 5-41b and to the nullator–norator equivalent in Fig. 5-41a. The idealized equation of this stage is

$$I_2 = \frac{1}{R_2} V_1 = g_2 V_1 \tag{7-242}$$

It characterizes a VCS with inversion.

The VCS in the reverse direction (see Fig. 7-52b) consists of a common-base stage connected through R_1 with a common-collector stage. This circuit is characterized by

$$I_1 = -\frac{1}{R_1} V_2 = -g_1 V_2 \tag{7-243}$$

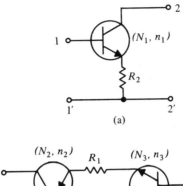

(a)

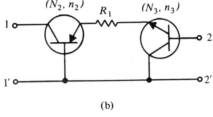

(b)

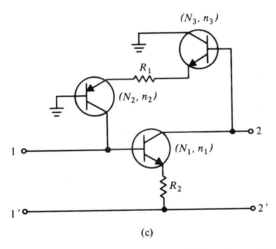

(c)

FIG. 7-52. Decomposition of three-transistor gyrator (c) into forward (a) and reverse (b) VCS.

Connecting the two stages in parallel we obtain the configuration shown in Fig. 7-52c. Assuming that $g_1 = g_2 = g$ and that the three transistors have identical current gains β, the y matrix of the resulting gyrator can be approximated by[24]

$$[y] \approx \begin{bmatrix} g/\beta & -g \\ g & g/\beta \end{bmatrix} \approx \begin{bmatrix} 0 & -g \\ g & 0 \end{bmatrix} \qquad (7\text{-}244)$$

24. Notice that in this configuration y_{12} is negative, y_{21} positive, whereas the opposite has been true in our text so far. This is quite legitimate for a gyrator (see, e.g., Table 5-5, Chapter 5), since the transfer admittances y_{12} and y_{21} are merely required to have opposite polarity.

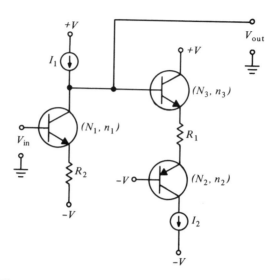

FIG. 7-53. Three-transistor gyrator of Fig. 7-52 with current sources for biasing.

Clearly, the transistor circuit resulting from this derivation cannot be used directly. In this respect it does not differ from any of the other realizations that were derived in Chapter 5. They all evolved in a systematic but undifferentiated manner that took no cognizance of such matters as voltage biasing and current flow. Shenoi[25] modified the structure of Fig. 7-52c into a usable circuit by adding load resistors and base biasing resistors. Later,[26] an improved DC-coupled version was described in which the base resistors were omitted and current sources substituted for the load resistors, as shown in Fig. 7-53.

The current source I_2 used in Fig. 7-53 can conveniently consist of an n-p-n transistor, as described in Section 7.1.2 (Fig. 7-4) and shown again in Fig. 7-54a. To use the same configuration for I_1, we require a p-n-p transistor, as we do for the transistor (N_2, n_2) in Fig. 7-53. The current gain of a lateral p-n-p, the most commonly used kind, is much lower than that of an n-p-n transistor, and its performance will be correspondingly inferior. To help us in this respect, the compound n-p-n–p-n-p configuration shown in Fig. 7-54b is often used. A simple analysis shows that the equivalent current gain of this compound transistor is given by

$$\beta_{n\text{-}p\text{-}n-p\text{-}n\text{-}p} = \beta_{p\text{-}n\text{-}p}(\beta_{n\text{-}p\text{-}n} + 1) \tag{7-245}$$

25. B. A. Shenoi, A practical realization of a gyrator circuit and RC-gyrator filters, *IEEE Trans. Circuit Theory*, **CT-12**, 374–380 (1965).
26. W. H. Holmes, S. Gruetzmann, and W. E. Heinlein, High-performance, direct-coupled gyrators, *Electron. Lett.* **3** (2), 45–46 (1967).

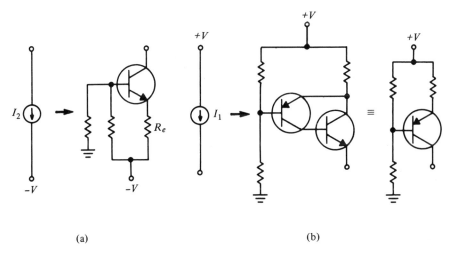

(a)   (b)

FIG. 7-54. Transistor current sources (a) *n-p-n*, (b) *p-n-p* (i.e., compound *n-p-n–p-n-p*).

This can be made at least as high as the current gain of an *n-p-n* transistor alone.

Another three-transistor circuit that has been expanded for actual implementation in silicon integrated circuit form is the circuit that was shown in Fig. 5-44b, Chapter 5, and shown again in Fig. 7-55. It consists of two common-emitter stages, with current feedback in each emitter in the forward

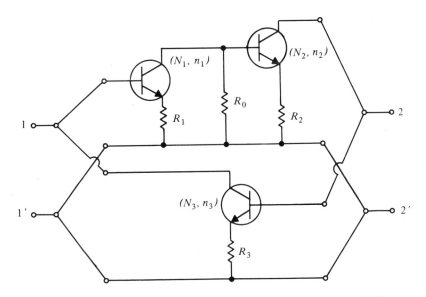

FIG. 7-55. Three-transistor gyrator using common-emitter stages as VCS.

direction and a single common-emitter stage in the reverse direction. These two VCS stages then give us

$$I_1 = \frac{1}{R_3} V_2 = g_1 V_2 \qquad (7\text{-}246a)$$

and, assuming $R_0 = R_1$,

$$I_2 = -\left(\frac{R_0}{R_1 R_2}\right) V_1 \Big|_{R_0 = R_1} = -\frac{1}{R_2} V_1 = -g_2 V_1 \qquad (7\text{-}246b)$$

A practical, and reportedly integrable, realization of Fig. 7-55 is shown in Fig. 7-56a. The circuit consists of n-p-n transistors and compound p-n-p-n-p-n transistors of the kind shown in Fig. 7-54b. The function of the individual transistors becomes clear if we consider the simplified equivalent circuit

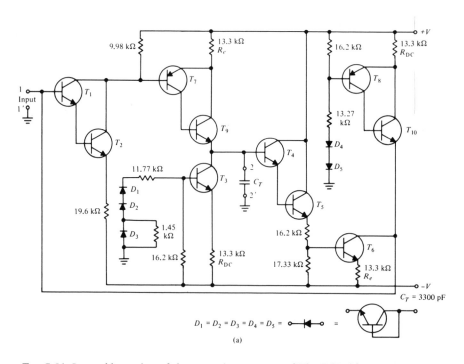

(a)

FIG. 7-56. Integrable version of three-transistor gyrator of Fig. 7-55. (a) Complete circuit diagram (after: J. J. Golembeski and T. N. Rao, Principles and design of gyrator filters, Proc. 20th Electron. Components Conf., Washington, D.C., 1970 pp. 294-300).

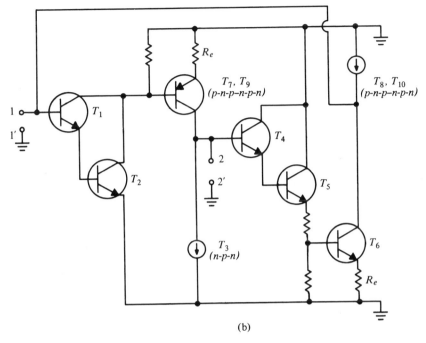

(b)

FIG. 7-56. (b) Simplified circuit diagram.

shown in Fig. 7-56b. Notice that high input impedances are obtained by using Darlington pairs for the input transistor stages T_1, T_2 and T_4, T_5, respectively. High output impedances are obtained by using transistors T_3, and T_8 with T_{10}, as current sources of the kinds shown in Figs. 7-54a and 7-54b, respectively. In order to DC-couple the entire circuit, the second gain stage in the forward direction must use a *p-n-p* transistor and, in order to achieve a high current gain, a compound *p-n-p-n-p-n* structure is used. Straightforward analysis of this network shows that the gyrator conductance turns out to be $g = 1/2R_e$, where R_e is shown in Fig. 7-56a. If the current gains β of the transistors and compound transistors are high enough, the gyrator conductance is virtually independent of β and is determined only by the quality of the resistors R_e. Thus, if a silicon-integrated gyrator is envisaged, R_e, and with it the gyrator conductance, will have an initial tolerance of $\pm 15\%$ and a temperature coefficient, typically, of 2000 ppm/°C. If a hybrid integrated gyrator is to be used, R_e can be made of thin-film material with the attendant advantages of initial accuracy, long-term stability, and low temperature coefficient.

Numerous design features have been added to the basic three-transistor circuit of Fig. 7-55 to obtain the integrable gyrator circuit of Fig. 7-56a. Beside the precautions taken to obtain high input and output impedances (a prerequisite for good gyrator performance), others have been taken to

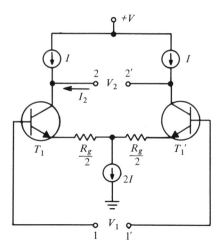

FIG. 7-57. Differential amplifier as a floating VCS.

guarantee low and temperature-stable offset voltages. The diodes D_1 to D_5 are included for this purpose; three are required to stabilize the *n-p-n* current source, two for the compound *p-n-p–n-p-n* current source. In spite of all these design precautions, it may be apparent to the reader that the resulting circuit does not utilize the building blocks for silicon-integrated circuit design discussed in Section 7.1 to the same extent as, say, the operational amplifiers discussed in Section 7.2.1. Furthermore, by connecting a capacitor to the output terminals of this circuit, only grounded inductors can be simulated with it. To obtain a floating inductor, two gyrators are required, as discussed in Chapter 5. A class of gyrators that does not have these disadvantages will be discussed next.

7.3.2 Differential Gyrators

The many advantages of the differential amplifier as a building block in silicon integrated circuit design were discussed in Section 7.1.1. They were recognized by numerous workers in the field and resulted in a variety of novel gyrator designs. One of these is the wideband gyrator designed by Rao[27] that will be discussed in what follows. As will be seen, it makes considerable use of the many features peculiar to silicon integrated circuit technology.

Consider the differential amplifier shown in Fig. 7-57. The output current I_2 at the floating terminals 2, 2' is determined by the input voltage V_1 at the floating input terminals 1, 1':

$$I_2 = \frac{1}{R_g} V_1 = g V_1 \qquad (7\text{-}247)$$

27. T. N. Rao, Readily biased wideband gyrator circuit for floating or earthed inductors, *Electron. Lett.*, **5**, 309–310 (1969); Stability of wideband gyrators, *IEEE Solid-State Circuits*, **SC-5**, 129–132. (1970).

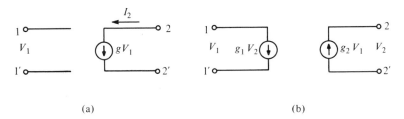

(a)   (b)

FIG. 7-58. (a) Floating VCS. (b) Equivalent VCS diagram of floating gyrator.

Thus, this circuit can be considered a floating VCS, as shown in Fig. 7-58a. By connecting two differential amplifiers in parallel, as shown in Fig. 7-59, and cross connecting one pair of output terminals as shown, to obtain the required inversion, a floating gyrator is obtained. The equivalent VCS diagram is shown in Fig. 7-58b; the gyrator conductance g is $1/R_g$.

The scheme shown in Fig. 7-59 has numerous advantages. For one thing, using differential amplifiers, full advantage can be taken of the matching of transistor parameters on a single chip. Furthermore, a minimum *and equal* number of transistors is involved in the signal path in either direction, resulting in stable operation over a very wide band. Because of the floating terminals, floating inductors can be simulated directly, simply by connecting a capacitor between, say, terminals 2 and 2'. If, on the other hand, a grounded inductor is required, one of the output terminals can be grounded, provided the circuit is properly biased.

The biasing and matching of collector currents of the signal-transmitting transistors (e.g., T_1, T_1') and of the transistors providing the current sources I in the gyrator of Fig. 7-59 is critical. Figure 7-60 may be instructive in regard to the DC-biasing mechanism of a gyrator. High-impedance DC paths are required to close the current paths of the two current sources $-gV_1$ and gV_2. These are designated R_s and R_L in Fig. 7-60. Consider now the case in which the current gV_1 increases. This increases the voltage drop $V_2 = I_L R_L$, which in turn increases the current $I_1 = gV_2$. As a result the input voltage V_1 decreases because of the negative increment caused by the increase in I_1, exerting an opposing and therefore stabilizing effect on the current source gV_1, whose current increase started the cycle. Naturally this stabilizing effect (or negative feedback) is all the more pronounced, the higher the terminating resistors R_s and R_L can be maintained, or, in other words, the more nearly ideal the gyrator itself can be made. With respect to DC biasing, the gyrator, based on a parallel connection of transadmittances, is therefore *open-circuit stable*.

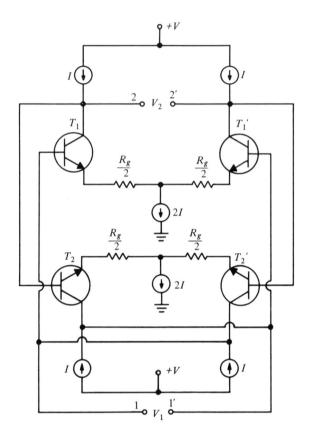

FIG. 7-59. Floating gyrator using differential amplifiers.

Whereas the foregoing discussion explains the general feedback mechanism by which gyrators can be DC-biased it does not explain the specific method by which, e.g., the differential scheme of Fig. 7-59 can be DC-stabilized. For this we must turn our attention to the complete circuit shown in Fig. 7-61.

To begin with, we note that, in order to DC-couple the transistor pairs of the two differential amplifiers of Fig. 7-59, one pair, i.e., T_2 and T_2', are p-n-p-type transistors, while the other pair are n-p-n. Other than that, the signal transmission paths of the circuit are the same as those of Fig. 7-59. The transistor pairs T_1, T_1' and T_2, T_2' comprise the two differential voltage-to-current converters whose transconductances are equal to $1/R_g$. The collectors of T_1, T_1', T_2, T_2' are connected to suitable constant current sources, obtained from p-n-p or n-p-n transistors as required, and these provide the DC currents required for operation. When a signal voltage is applied between the bases of T_1 and T_1', an equivalent differential current source is seen between the terminals 2 and 2'. Similarly, when a signal voltage is applied between the bases of T_2 and T_2', an equivalent differential current source is

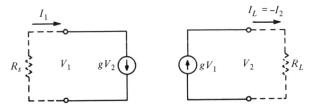

FIG. 7-60. DC stabilization of gyrators.

seen between the terminals 1 and $1'$. The latter is of the required polarity; thus terminals $1,1'$ and $2,2'$ form the two ports of the gyrator. A capacitor C_L connected at the port $2,2'$ appears like an inductor at the port $1,1'$.

The biasing of the circuit is arranged in such a way that only p-n-p transistors are associated with T_2 and T_2' and n-p-n transistors with T_1 and T_1'. Thus, it is advisable to realize the gyrator by two chips, one containing all the n-p-n transistors and the other containing all the p-n-p transistors. In this way, high-quality vertical p-n-p transistors with frequency characteristics similar to those of n-p-n transistors can be used and high-frequency operation can be guaranteed for both differential amplifiers in the circuit. Furthermore, a close match is required only between transistors of the same type, since each is on a separate chip; in other words, it is not necessary that the characteristics of the p-n-p transistors closely match those of the n-p-n transistors.

The current sources used in the circuit are of the type discussed in Section 7.1.2 and shown in Fig. 7-5; in the gyrator (Fig. 7-61) they are intended to operate as follows: Assuming that the base currents of the various transistors are negligible in comparison with their collector currents, the collector currents of transistors T_3 and T_8 are equal. The current I is therefore determined by the resistors R and R_B as well as by the V_{BE} drops of T_3 and T_8. Since the base of T_3, T_4, T_5, T_6, and T_7 have the same potential and their emitters are all returned to the negative supply voltage through equal resistors of value R, each of these transistors has the same collector current I. Similarly, the collector currents of T_8, T_9, T_{10}, T_{11}, and T_{12} are all equal to I. Since the base currents are negligible in comparison with the collector currents and there is a close match between transistors of a kind, the collector currents of T_1 and T_1', T_2 and T_2' are also equal to I. Thus, every transistor in the circuit has the same collector current I, which, for a given R, can be adjusted to any desired value by adjusting the resistor R_B. Thus, R_B, which controls the collector currents of the various transistors of the current sources, has a function similar to that of resistor R_1 in Fig. 7-5. Likewise, transistors T_3 and T_8, which are connected as diodes, play the part of transistor T_1 in Fig. 7-5. The use of T_3, T_8, and R_B in the present circuit enables one to control the collector currents of several transistors of different types simultaneously, where only a match between transistors of the same kind (i.e., n-p-n or p-n-p) need exist. As a result, no variable resistors are necessary to compensate for

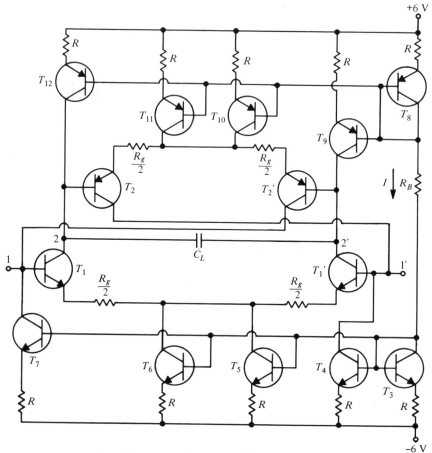

FIG. 7-61. Circuit diagram of differential gyrator.

lack of matching between *p-n-p* and *n-p-n* transistor parameters. Finally, it should be noted that the transistor pairs T_5 and T_6 as well as T_{10} and T_{11} can be combined into single transistors dimensioned to draw twice the current of each individual transistor (i.e., possessing twice the emitter area). The emitter resistor is then, of course, $R/2$.

The gyrator circuit of Fig. 7-61 appears to have some obvious advantages. For one thing, its frequency capabilities should be superior to those of an operational amplifier, the latter being restricted by its gain–bandwidth product. The circuit also makes excellent use of the building blocks most suitable for silicon integrated circuit design and has the advantage of placing relatively few transistors in the actual transmission paths. Furthermore, the gyration resistance can be changed by changing the value of R_g appropriately. On the other hand, the circuit does require two separate chips, one of which consists exclusively of *p-n-p* transistors; this involves a nonstandard technology. Also, the power requirements of this circuit, as of most gyrator circuits, is generally higher than that obtainable with operational amplifiers. The reason for this is that the output stage of an operational amplifier can be designed for Class B or even Class C operation; it thereby dissipates little power while providing high signal capabilities. No comparable method of operation has been found for gyrators. Another reason for high power dissipation becomes apparent when we consider high-frequency operation of the gyrator. In terms of the gyrator resistance R_g and the capacitor C_L the inductor L has the value $R_g^2 C_L$. Assume, for example, that we require an inductor of one 1 μH and that the capacitor C_L is equal to 100 pF. The gyrator resistance R_g must then equal 100 Ω. For such low values we must take the emitter resistance r_e of the transistors of the differential amplifier into account. We therefore have an equivalent gyrator resistance $R_g' = R_g + 2r_e$, where

$$r_e[\Omega] = \frac{kT}{qI} = \frac{25 \text{ mV}}{I[\text{mA}]} \qquad (7\text{-}248)$$

at room temperature. Clearly, r_e is temperature-dependent and should be maintained as small as possible in comparison to R_g. If the DC current I is 1 mA then $2r_e$ is equal to 50 Ω, which is half as large as R_g itself. Thus, the current I must be made very much larger in order to make the effect of r_e negligible with respect to R_g and, as a consequence, the gyrator must be designed to dissipate more power.

Another gyrator based on the parallel connection of differential amplifiers is shown in Fig. 7-62. Notice, however, that an additional *p-n-p* transistor is included in each signal path; it provides the DC biasing that permits DC coupling of the two stages. By appropriate interconnection, the correct, i.e., opposite, signal polarity is obtained in each direction. Clearly a grounded inductor can easily be simulated by connecting a capacitor to terminals 2, 2′. On the other hand the inclusion of a *p-n-p* transistor, presumably a lateral *p-n-p* in order not to complicate the processing, will limit the frequency range obtainable with a silicon integrated version of this gyrator considerably. This problem was avoided in the circuit of Fig. 7-61 at the cost of having to manufacture two separate silicon chips per gyrator circuit.

7.3.3 Practical Considerations in the Design of Integrated Gyrators

The preceding discussion on the design and performance of monolithic silicon integrated gyrator circuits is based largely on conjecture, since very few gyrators have become available commercially in monolithic silicon integrated form. There may be numerous reasons for this commercial scarcity. For one thing the availability of mass-produced monolithic silicon integrated

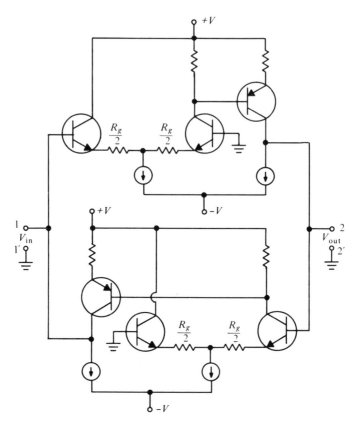

FIG. 7-62. Differential gyrator. (after: W. H. Holmes et al., op. cit.)

operational amplifiers has taken the edge off any need for gyrators. A large variety of high-quality monolithic operational amplifiers are available at such low prices that they provide a strong incentive for the development of active networks incorporating them. Furthermore, it seems unlikely that any other linear integrated devices, such as gyrators, will ever attain the cost advantage of operational amplifiers, since the latter can be used in so many other applications beside active networks. One must only recall that the cost of operational amplifiers was competitively low long before they were incorporated into active networks on any large scale. By contrast, the demand for integrated gyrators is limited to linear network applications—a much more restricted market.

Frequency Limitations   Whereas the aforementioned reasons for the absence of commercially available gyrators are primarily economic in nature, technical reasons are by no means lacking. The most frequently suggested

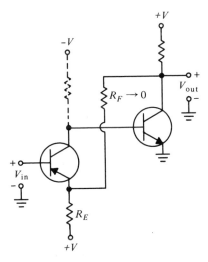

FIG. 7-63. Compound *p-n-p–n-p-n* transistor, redrawn as voltage-feedback pair.

monolithic gyrator designs either require matching *p-n-p* and *n-p-n* transistor pairs with high current gains, or combinations of *n-p-n* transistors and FETs. Both of these methods are limited to low-frequency applications. To operate at high frequencies, the gyrators must be designed entirely with *n-p-n* transistors, which results in biasing problems, or with two separate chips, one of which contains *n-p-n*, the other *p-n-p* transistors. This is uneconomical. Another alternative is to use the compound transistor shown in Fig. 7-54b. However, this consists essentially of a transistor feedback pair, as redrawn in Fig. 7-63; it is generally known as a voltage (or series-shunt) feedback pair.[28] As such it is apt to have stability problems in addition to being limited to low frequencies by the (lateral) *p-n-p* transistor, as well as by the band-width of the loop gain.

Obtaining High-Q Simulated Inductors We recall from the gyrator model described in Chapter 5, that the capacitor-loaded gyrator behaves like a discrete inductor whose Q increases linearly with frequency up to the radian frequency

$$\omega_{max} = \frac{1}{C_L} \sqrt{y_{22}\,\Delta y/y_{11}} \qquad (7\text{-}249)$$

(see (5-87)), where

$$Q_{max} = \frac{g_1 g_2}{2\sqrt{y_{11} y_{22}\,\Delta y}} \qquad (7\text{-}250)$$

28. E. M. Cherry, An engineering approach to the design of transistor feedback amplifiers, *J. Brit. Inst. Radio Engrs.*, **25**, 127–144 (1963); **27**, 349–352 (1964).

(see (5-88)). Here y_{11} and y_{22} are the driving point admittances and $y_{12} = g_1$, $y_{21} = -g_2$ the transfer admittances of the gyrator. Thus $\Delta y = y_{11}y_{22} + g_1g_2$. Assuming that the gyrator conductance $g_1 = g_2 = 1/R_g$, that the gyrator terminal admittances are equal and small, i.e., $y_{11} = y_{22} = 1/R$, and that $y_{11}y_{22} \ll g_1g_2$ or $R \gg R_g$, then (7-249) becomes

$$\omega_{\max} = \frac{1}{R_g C_L} \tag{7-251}$$

and

$$Q_{\max} = \frac{R}{2R_g} \tag{7-252}$$

The simulated inductor has the value

$$L_{eq} = R_g^2 C_L \tag{7-253}$$

and its Q decreases linearly for $\omega > \omega_{\max}$.

It follows from (7-252) that the ratio R/R_g must be all the higher, the higher $Q_{\max}$ is required to be. For example, for a $Q_{\max}$ of 500, R must be 1000 times larger than R_g, i.e., the input and output impedances must be 1000 times larger than the gyrator impedance. This in itself implies practical problems. Either R must be made very large or R_g very small. The second alternative increases the power dissipated by the gyrator (as discussed in the preceding section) and takes up a lot of chip area in a monolithic design. The first implies a high input impedance. This can be obtained, for example, by Darlington transistor pairs or input FETs; however, both cause offset problems and limit the usable frequency range.

Beside the problem of *obtaining* a high Q, there is the second problem of *maintaining* a high Q within certain tolerances. In general Q is proportional to either β or β^2, where β is the common-emitter current gain of the transistors used. If we consider a differential gyrator, for example, such as the one shown in Fig. 7-59, it follows that the input impedance R at terminals 1, 1' is approximately equal to βR_g. With (7-252) it therefore follows that $Q_{\max} \approx \beta/2$. Since β is neither accurately predictable nor constant with temperature, the Q stability of the capacitor-loaded gyrator will suffer accordingly. Furthermore, if R_g is a monolithic resistor it will vary with temperature, typically by 2000 ppm/°C or more. One way of circumventing this large variation is to incorporate the resistors which determine the gyrator impedance (e.g., R_g), as well as the loading capacitor which determines the inductor value, in thin-film form. This allows for the simulation of a wide range of inductor values with the same basic silicon chip. For low to medium Q requirements, the Q of the simulated inductor, which should be high to start with, can be degraded by shunt or series film resistors; thus, the Q variations due to temperature variations of β can be partially swamped out.

A disadvantage of using film capacitors to obtain gyrator-simulated inductors is their high losses compared to discrete capacitors. It will be remembered (see (5-99) in Chapter 5) that the Q of an inductor simulated by an ideal gyrator is equal to the Q of the capacitor C_L loading the gyrator, i.e., $Q_{L_{eq}} = Q_{C_L}$. The losses of film capacitors rapidly increase at high frequencies, so that many of the advantages of using a high-frequency gyrator are lost if a film capacitor is used for C_L.

In this context it may be useful to recall the definition of the dissipation factor D of a resonant system and its relationship to Q:

$$D = \frac{1}{Q} = \frac{1}{2\pi} \frac{\text{energy dissipated in one cycle}}{\text{energy stored}} \qquad (7\text{-}254)$$

This states that, in a resonating system, each dissipative element absorbs a fixed fraction of the total energy during the period of each cycle. Thus, if a resonant network has n nonreactive or ohmic elements, then the total dissipation factor D is[29]

$$D = D_1 + D_2 + \cdots + D_n \qquad (7\text{-}255)$$

i.e., it is equal to the sum of the dissipation factors of each nonreactive element in the resonant network.

The Q of the overall system results from the relationship $Q = 1/D$. As an example of the use of (7-255) consider the problem of calculating the Q of a resonant tank consisting of a capacitor C and a nonideal gyrator loaded by a lossy film capacitor C_L as shown in Fig. 7-64a. Considering only the driving point admittance $y_{in}(s)$ seen by the resonating capacitor C, we can transform the losses in the resonant circuit by following the steps shown in Figs. 7-64b and 7-64c. We obtain the equivalent resonant circuit shown in Fig. 7-64d. Using (7-255) the total dissipation factor D can now easily be obtained by inspection, if we add the sum of the dissipation factors of the individual resistors of the resonant circuit. The dissipation factor of a resistor R_s in series with the inductor has the form $D_s = R_s/\omega L_{eq}$, that of a resistor R_p in parallel with the inductor, the form $D_p = \omega L_{eq}/R_p$. Combining the gyrator output impedance R with the leakage resistance R_C of the capacitor C_L into the resistor R', where $R' = R \| R_C$, we have

$$D = \frac{\omega L_{eq}}{R} + \frac{1}{\omega L_{eq}} \cdot \frac{R_g^2}{R'} \qquad (7\text{-}256)$$

The overall Q is the inverse of this expression.

It should be noted that (7-255) is, strictly speaking, only valid for high-Q systems, i.e., for Q values higher than 5 or even 10. This is because it ignores the effects of the dissipative elements on the "energy stored" of (7-254),

---

29. J. T. May and J. F. Pierce, unpublished notes, Bell Telephone Laboratories.

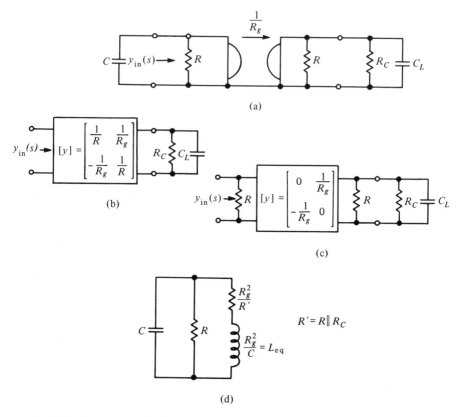

FIG. 7-64. Resonant tank circuit consisting of ideal capacitor C and nonideal gyrator, loaded by lossy capacitor C_L.

i.e., it considers that this energy is stored in a system consisting only of ideal reactive elements. Actually somewhat less energy is stored because of the effect of the dissipative elements. For high-Q systems this error is negligibly small.

One method of adjusting the Q of a simulated inductor is to modify the phase shift of the gyrator loop gain. In this way parasitic effects can be compensated. However, the phase shift of the voltage-controlled current sources making up a gyrator are generally not predictable, so that the phase-shift adjustment must be made individually on each gyrator.

The Dynamic Range   Another practical aspect in the design of monolithic integrated gyrators that must be considered is their dynamic range.[30] The highest allowable signal levels inside any active circuit are limited by the DC

30. This discussion is based on unpublished notes by T. N. Rao, Bell Telephone Laboratories.

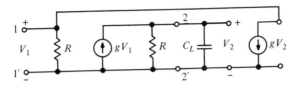

FIG. 7-65. Capacitor-loaded nonideal gyrator.

supply voltages; the lowest by internally generated noise. These dynamic-range limitations are generally more severe in capacitor-loaded gyrators than in the inductors they replace; they may also affect the bandwidths of the filters in which they are used.

To calculate the dynamic range of a gyrator, consider the equivalent diagram of a capacitor-loaded gyrator with finite input and output impedance R (shown in Fig. 7-65). Note the feedback representation of the two VCSs making up the gyrator. This representation is frequently more useful than the commonly used parallel connection of two VCSs. Assuming that the VCSs are realized by transistor circuits operating with a given supply voltage, there is a maximum peak-to-peak voltage V_{max} that is permissible at port 1,1' or port 2,2'. If $v_1(t)$ is given by $V_1 \sin \omega t$ then, neglecting the effects of the finite input and output resistances R, the magnitude of the voltage V_2 at port 2,2' is given by

$$V_2 = \frac{gV_1}{\omega C_L} \qquad (7\text{-}257)$$

This expression is directly proportional to V_1 and inversely proportional to ω for a given ratio g/C_L. In order to guarantee a minimum signal-to-noise ratio, V_1 and V_2 must each be greater than the value V_{min} determined by the required signal-to-noise ratio. For given values of g and C_L, and assuming that the gyrator has a sufficiently high frequency limit, the maximum possible frequency of operation is limited by noise in the second VCS. This maximum frequency ω_h can be obtained by setting $V_1 = V_{max}$ and $V_2 = V_{min}$ in (7-257). It is therefore given by

$$\omega_h = \frac{gV_{max}}{C_L V_{min}} \qquad (7\text{-}258)$$

Thus, for a given signal-to-noise ratio V_{max}/V_{min}, the ratio g/C_L must be as large as possible to obtain a high ω_h.

The lowest possible frequency of operation ω_l is limited by noise in the first voltage-controlled current source and the maximum possible signal level at port 2,2'; ω_l is therefore given by (7-257) when we set $V_1 = V_{min}$ and $V_2 = V_{max}$. Thus

$$\omega_l = \frac{gV_{min}}{C_L V_{max}} \tag{7-259}$$

To obtain a low value for ω_l for a given signal-to-noise ratio V_{max}/V_{min}, we see that the ratio g/C_L should, in this case, be as small as possible. Comparing (7-258) and (7-259) we see that even though ω_h and ω_l are dependent on g/C_L, their ratio depends only on V_{max}/V_{min}:

$$\frac{\omega_h}{\omega_l} = \left(\frac{V_{max}}{V_{min}}\right)^2 \tag{7-260}$$

Thus, if we have a signal-to-noise ratio $V_{max}/V_{min} = 10^3$, then the ratio ω_h/ω_l is on the order of 10^6. In the case of a bandpass filter the dynamic range of the gyrator is therefore not likely to cause problems if the filter covers less than about six decades. In the case of low-pass filters, where, ideally, the ratio ω_h/ω_l is infinite, we must include the finite input and output resistances in our calculations to obtain a realistic measure of the achievable dynamic range. In so doing (7-257) is modified as follows:

$$V_2 = \frac{gV_1}{[(1/R)^2 + \omega^2 C_L^2]^{1/2}} \tag{7-261}$$

Notice that at sufficiently low frequencies the term $1/R^2$ dominates in the denominator and the capacitor-loaded gyrator behaves more like a resistor than like an inductor. This is not necessarily objectionable in a low-pass filter, where the Q of the simulated coils counts most at the band edge. Hence, for satisfactory operation in low-pass filters, the gyrator–capacitor combination and the input voltage must be chosen such that $V_{max} > V_2 = gRV_1$ and $V_1 > V_{min}$, or

$$\frac{V_{max}}{V_{min}} > gR \tag{7-262}$$

For the capacitor-loaded gyrator shown in Fig. 7-65, gR is equal to $2Q_{max}$, where Q_{max} is the maximum Q of the simulated inductor as discussed earlier. Hence, (7-262) can be rewritten as

$$\frac{V_{max}}{V_{min}} > 2Q_{max} \tag{7-263}$$

If this inequality is satisfied, a low-pass filter can be satisfactorily operated up to a radian frequency of $\omega_h = gV_{max}/C_L V_{min}$. Clearly, in high-$Q$ low-pass filters it may be difficult to realize (7-263) because of the dynamic-range limitations discussed above.

From the topics discussed above it will be clear that various practical problems will inevitably be encountered when one attempts to design economically feasible high-performance integrated gyrators. Beside being quite expensive, commercially available gyrators, whether they be integrated or not, may suffer from such shortcomings as frequency dependence of the gyration conductance and Q enhancement at higher frequencies (e.g., when the gyrator is used in bandpass filter configurations). Unfortunately the stimulus to overcome these problems has been impaired by the adequacy of active networks using high-performance operational amplifiers. The only unique uses for gyrators in the voice-frequency range seem to involve applications requiring features peculiar to the gyrator (e.g., voltage-controllability, bilateral capabilities, capacitance transformations). At higher frequencies, strong competition is provided by passive filters in almost all respects (size, cost, stability, etc.) At very low frequencies the gyrator may offer certain advantages, but, as shown in *Linear Integrated Networks: Design*, Chapter 7, an operational-amplifier realization of the gyrator is then generally adequate.

CHAPTER

8

PASSIVE *RC* CIRCUITS FOR LINEAR ACTIVE NETWORK DESIGN

INTRODUCTION

In the preceding chapter, we discussed active integrated devices such as gyrators, operational amplifiers, and the like. To obtain filter networks, these silicon integrated devices must be combined with passive components, i.e., resistors and capacitors. In many cases, the active devices may be combined with individual resistors and capacitors rather than with self-contained *RC* networks; it may then be equally advantageous, from a cost standpoint, to use discrete or integrated *RC* components. In other cases, however, it will be useful to combine *RC* subfunctions—in the form of building blocks— with the active devices, to obtain a required overall network response. It is with these *RC* building blocks that we shall be concerned in this chapter.

8.1 *RC* BUILDING BLOCKS

Useful *RC* building blocks are, for example, bridge networks as well as parallel-*T* and bridged-*T* networks, each of which individually realizes a useful network function. Whereas the poles of these networks are restricted to the negative real axis (see Chapter 1) the zeros can be placed anywhere in the left and, within certain limits, in the right half *s* plane. The latter property is a very useful one, particularly if the *RC* networks are realized by highly stable, temperature-compensated, closely tracking resistors and capacitors, e.g., as may be realized, in thin-film form. The resulting transmission zeros will be considerably more stable and accurate (since their exact location can be precisely adjusted) than would be possible if active devices were involved in their realization as well.

In the following discussion we shall examine some basic *RC* subfunctions, or building blocks, and shall, in particular, discuss in detail one very useful *RC* network, the twin-*T*.

8.1.1 The General Six–Element *RC* Bridge Network[1]

We shall consider here a family of *RC* networks having six arms only, each arm comprising a single resistance or capacitance. The chief property of interest of these networks is the infinite attenuation, or transmission zero, obtainable at some frequency f_z.

Consider the eight-arm mesh shown in Fig. 8-1a. Elements Z_1, Z_3, Z_5, and Z_7 are called the *chords*, elements Z_2, Z_4, Z_6, and Z_8 the *branches* of the mesh. The dual of the mesh, is shown in Fig. 8-1b. Notice that the impedances of Fig. 8-1a have been replaced by admittances in Fig. 8-1b and that the subscripts of the branch impedances in Fig. 8-1a are those of the chord admittances in Fig. 8-1b and vice versa. As we shall see, this inversion brings the equations for the two dual circuits into similar form thereby rendering it necessary to solve only one set of equations.

If the opposite branches Y_1 and Y_5 of Figure 8-1b are terminating impedances, the parallel-*T* configuration of Fig. 8-2 results. If, however, opposite chords are made terminating impedances, e.g., Z_1 and Z_5 in Fig. 8-1a, the configuration of Fig. 8-3 is obtained. These two networks are dual, as

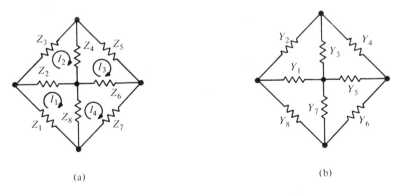

(a) (b)

FIG. 8-1. Eight-arm mesh (a) and its dual (b).

1. G. R. Harris, Bridged reactance–resistance networks, *Proc. IRE*, **33**, 882–887 (1949).

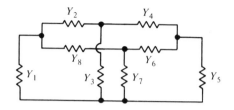

FIG. 8-2. Parallel-T network.

the series and shunt arms of either one appear as the correspondingly numbered shunt and series arms, respectively, of the other.[2] The configuration of Fig. 8-3 might well be called a series-π network by analogy with the parallel T of Fig. 8-2.

Two other symmetrical configurations may be obtained by making adjacent chords of Fig. 8-1b and adjacent branches of Fig. 8-1a the terminating impedances. If Y_2 and Y_8 are so chosen in the first of these cases, the configuration of Fig. 8-4 results. This may be called a shunt-ladder network, since it is, in fact, a ladder in series with a single shunt arm. If Z_2 and Z_8 are chosen as terminating impedances, the configuration of Fig. 8-5 appears, which may be called a bridged-ladder.

The networks of Figs. 8-4 and 8-5 are dual. This may be shown if the configuration of Fig. 8-4, for example, is drawn in solid lines, as in Fig. 8-6. A node is now placed in each of the four meshes of this configuration, as well as one outside, corresponding to the mesh circumscribing the configuration. Branches are then drawn from node to node, so that each node is the junction of as many branches as its corresponding mesh has chords, and the dotted line configuration of Fig. 8-6 is obtained. The dotted-line configuration is the dual of the solid-line configuration, and is seen to correspond to Fig. 8-5.

The four networks developed above may be considered symmetrical about a vertical center line; therefore they have lattice equivalents, which may be derived through Bartlett's bisection theorem (Chapter 3, Section 3-5). It is, therefore, convenient and not inaccurate to speak of the conditions for infinite attenuation as the "balance" conditions of the networks.

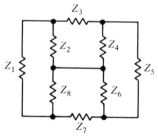

FIG. 8-3. Series-π network.

2. B. D. H. Tellegen, Geometrical configurations and duality of electrical networks, *Philips Tech. Rev.*, **5**, 324–330 (1940).

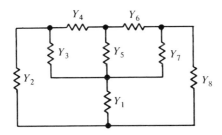

FIG. 8-4. Shunt-ladder network.

The balance conditions may be determined by setting up the general circuit equations in terms of V, I, and either Y or Z, and equating either V or I to zero. As an example, we shall derive the balance conditions for the configuration in Fig. 8-3. Since it is given in terms of impedances, it will be convenient to determine the balance conditions using mesh equations. The order of the meshes is obvious from Fig. 8-1a when V, the applied voltage, is considered to be in series with Z_1. Assuming clockwise currents in each mesh we obtain

$$
\begin{aligned}
V &= I_1(Z_1 + Z_2 + Z_8) - I_2 Z_2 & & - I_4 Z_8 \\
0 &= -I_1 Z_2 & + I_2(Z_2 + Z_3 + Z_4) - I_3 Z_4 & \\
0 &= & - I_2 Z_4 & + I_3(Z_4 + Z_5 + Z_6) - I_4 Z_6 \\
0 &= -I_1 Z_8 & - I_3 Z_6 & + I_4(Z_6 + Z_7 + Z_8)
\end{aligned}
$$

$$(8\text{-}1)$$

Solving for the current I_3 we have

$$
I_3 = V \frac{\begin{vmatrix} -Z_2 & Z_2 + Z_3 + Z_4 & 0 \\ 0 & -Z_4 & -Z_6 \\ -Z_8 & 0 & Z_6 + Z_7 + Z_8 \end{vmatrix}}{\Delta}
\qquad (8\text{-}2)
$$

where Δ is the determinant of (8-1). The requirement for balance is that $I_3 = 0$. If this substitution is made in the above expression, the minor is easily expanded to

$$
0 = Z_2 Z_4 (Z_6 + Z_7 + Z_8) + Z_6 Z_8 (Z_2 + Z_3 + Z_4) \qquad (8\text{-}3)
$$

This expresses the conditions of balance for the network in Fig. 8-3. If the Z's in (8-3) are replaced by Y's with the same subscripts, the resulting equation in Y's expresses the conditions of balance for the parallel-T network of Fig. 8-2.

The balance equation for the networks of Figs. 8-4 and 8-5 is derived in the same manner. For the shunt-ladder configuration of Fig. 8-4 it is found to be

$$
0 = Y_3 Y_7 (Y_4 + Y_5 + Y_6) + Y_5 (Y_3 Y_6 + Y_4 Y_7)
$$

$$
+ Y_4 Y_6 (Y_1 + Y_3 + Y_5 + Y_7). \qquad (8\text{-}4)
$$

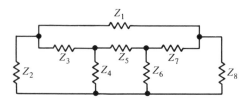

FIG. 8-5. Bridged-ladder network.

This equation in Z's applies to the bridged-ladder configuration in Fig. 8-5.

As mentioned above, we are confining ourselves here to six-arm structures of six elements only. More specifically, we shall assume three resistors and three capacitors per structure. Since each network then has three reactive elements, the general transfer functions for any one of the networks will be of third degree, i.e., of the form

$$T(s) = \frac{b_3 s^3 + b_2 s^2 + b_1 s + 1}{a_3 s^3 + a_2 s^2 + a_1 s + 1} \tag{8-5}$$

Since we are specifically interested in the transfer functions with transmission zeros at a frequency ω_N, (8-5) can be rewritten as

$$T(s) = \frac{(s - z)(s^2 + \omega_N^2)}{(s - p_1)(s - p_2)(s - p_3)} \tag{8-6}$$

It can be shown that when any one of the third-order networks is balanced to have a perfect null or transmission zero at ω_N, the negative real zero z and one of the negative real poles in (8-6) will cancel out, leaving a second-order

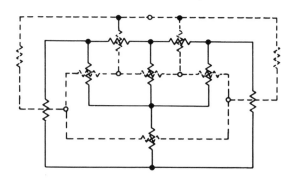

FIG. 8-6. Duality between the shunt-ladder and bridged-ladder networks.

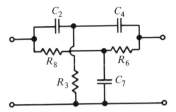

FIG. 8-7. A parallel-*T RC* network.

numerator and denominator. Assuming, then, that z and p_3 cancel in (8-6), we are left with the following second-order transfer function:

$$T(s) = \frac{s^2 + \omega_N^2}{(s - p_1)(s - p_2)} \tag{8-7}$$

To obtain the conditions for transmission zeros at ω_N (and thereby for pole–zero cancellation in (8-6)), we proceed as follows.

Consider the *RC* twin-T^3 network shown in Fig. 8-7. Its balance conditions may be expressed in general terms by the Y form of (8-3):

$$0 = Y_2 Y_4 (Y_6 + Y_7 + Y_8) + Y_6 Y_8 (Y_2 + Y_3 + Y_4) \tag{8-8}$$

Substituting g_3, g_6, g_8, jb_2, jb_4, and jb_7 for the corresponding Y's gives

$$0 = -b_2 b_4 (g_6 + g_8 + jb_7) + g_6 g_8 [g_3 + j(b_2 + b_4)] \tag{8-9}$$

Equating the real and imaginary terms of (8-9) to zero we obtain

$$\text{Real:} \qquad g_3 g_6 g_8 = b_2 b_4 (g_6 + g_8) \tag{8-10}$$

and

$$\text{Imaginary:} \qquad b_2 b_4 b_7 = (b_2 + b_4) g_6 g_8 . \tag{8-11}$$

At this point, it is convenient to replace the g's and b's by the R's and C's of Fig. 8-7, and we obtain for (8-10) and (8-11), respectively,

$$\omega_N^2 C_2 C_4 \left(\frac{1}{R_6} + \frac{1}{R_8} \right) = \frac{1}{R_3 R_6 R_8} \tag{8-12}$$

and

$$\omega_N^3 C_2 C_4 C_7 = \frac{\omega_N (C_2 + C_4)}{R_6 R_8} \tag{8-13}$$

Solving these two equations for ω_N we obtain

$$\omega_N^2 = \frac{1}{C_2 C_4 R_3 (R_6 + R_8)} \tag{8-14}$$

$$\omega_N^2 = \frac{C_2 + C_4}{C_2 C_4 C_7 R_6 R_8} \tag{8-15}$$

3. We are adhering to the convention here of designating a parallel-*T* network, consisting of resistors and capacitors in the configuration shown in Fig. 8-7, as a "twin-*T*" network.

Equating (8-14) to (8-15) we have

$$\frac{C_2 + C_4}{C_7} = \frac{R_6 R_8}{R_3(R_6 + R_8)} \qquad \text{[8-16]}$$

The balance conditions for the structure of Fig. 8-7 may therefore be expressed by two equations, either (8-14) and (8-16) or (8-15) and (8-16), the last-named equation being independent of frequency. These conditions for the twin-T of Fig. 8-7, as well as those for the other networks resulting from the general mesh structures of Fig. 8-1 are listed in Table 8-1. Notice that the balance equations for the series π network (Fig. 8-3) need not be newly derived: they may be obtained from the twin-T conditions merely by replacing all C's with R's of corresponding subscripts, and vice versa.

The parallel T and series π, as well as the shunt–ladder and bridged ladder networks, are dual to one another. In addition, the shunt ladder type A, and the shunt ladder type B, are related to one another by RC duality, as are the bridged ladder types A and B. The reader should take care here not to confuse general network duality with RC duality as defined in Chapter 3, Section 3.4.

The balance conditions for the shunt–ladder and bridged ladder networks in Table 8-1 are derived in precisely the same way as those for the parallel-T derived above. Observe that condition 1 in the table gives the frequency at which a transmission zero is to occur. Condition 2 gives a relation which the network elements must satisfy in order for the third-order networks to have second-order transfer functions. A special case has been added for each of the networks in Table 8-1 by duplicating elements of the structures in such a way as to obtain *symmetrical* networks. The balance conditions are thereby simplified significantly. It is interesting to note that the frequency-determining equations for the shunt-ladder and bridged-ladder network types contain a term which is a ratio of two like elements. This makes it possible to multiply or divide, within certain limits, the frequency determined by this element ratio. Consider, for example, the balance conditions for the symmetrical shunt-ladder type A in Table 8-1. If $C_3/C_5 = 1$, these become

$$\omega_N = \frac{\sqrt{3}}{C_3 R_4} \qquad (8-17)$$

and

$$\frac{R_4}{R_1} = 12 \qquad (8-18)$$

Suppose, however, that $C_3/C_5 = 2$; then

$$\omega_N = \frac{\sqrt{5}}{C_3 R_4} \qquad (8-19)$$

TABLE 8-1. BALANCE CONDITIONS FOR THE GENERAL SIX-ELEMENT *RC* BRIDGE NETWORKS

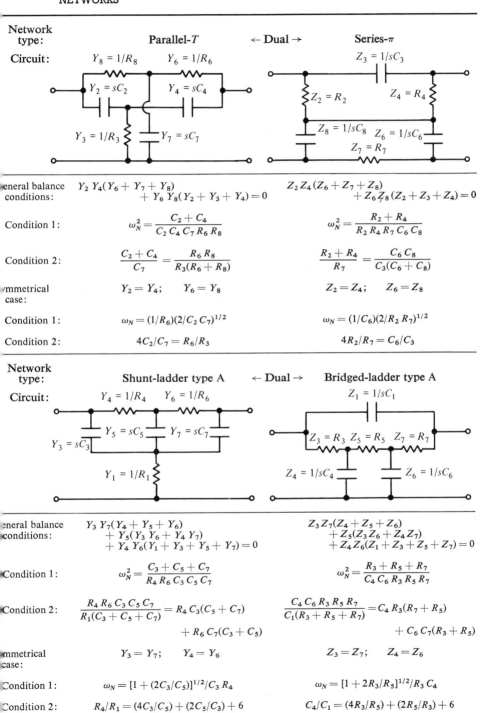

Network type:	Parallel-T	← Dual →	Series-π

General balance conditions:

$$Y_2\,Y_4(Y_6 + Y_7 + Y_8) + Y_6\,Y_8(Y_2 + Y_3 + Y_4) = 0$$

$$Z_2\,Z_4(Z_6 + Z_7 + Z_8) + Z_6\,Z_8(Z_2 + Z_3 + Z_4) = 0$$

Condition 1:

$$\omega_N^2 = \frac{C_2 + C_4}{C_2\,C_4\,C_7\,R_6\,R_8} \qquad \omega_N^2 = \frac{R_2 + R_4}{R_2\,R_4\,R_7\,C_6\,C_8}$$

Condition 2:

$$\frac{C_2 + C_4}{C_7} = \frac{R_6\,R_8}{R_3(R_6 + R_8)} \qquad \frac{R_2 + R_4}{R_7} = \frac{C_6\,C_8}{C_3(C_6 + C_8)}$$

Symmetrical case:

$$Y_2 = Y_4; \qquad Y_6 = Y_8 \qquad\qquad Z_2 = Z_4; \qquad Z_6 = Z_8$$

Condition 1:

$$\omega_N = (1/R_6)(2/C_2\,C_7)^{1/2} \qquad \omega_N = (1/C_6)(2/R_2\,R_7)^{1/2}$$

Condition 2:

$$4C_2/C_7 = R_6/R_3 \qquad\qquad 4R_2/R_7 = C_6/C_3$$

Network type:	Shunt-ladder type A	← Dual →	Bridged-ladder type A

General balance conditions:

$$Y_3\,Y_7(Y_4 + Y_5 + Y_6) + Y_5(Y_3\,Y_6 + Y_4\,Y_7) + Y_4\,Y_6(Y_1 + Y_3 + Y_5 + Y_7) = 0$$

$$Z_3\,Z_7(Z_4 + Z_5 + Z_6) + Z_5(Z_3\,Z_6 + Z_4\,Z_7) + Z_4\,Z_6(Z_1 + Z_3 + Z_5 + Z_7) = 0$$

Condition 1:

$$\omega_N^2 = \frac{C_3 + C_5 + C_7}{R_4\,R_6\,C_3\,C_5\,C_7} \qquad \omega_N^2 = \frac{R_3 + R_5 + R_7}{C_4\,C_6\,R_3\,R_5\,R_7}$$

Condition 2:

$$\frac{R_4\,R_6\,C_3\,C_5\,C_7}{R_1(C_3 + C_5 + C_7)} = R_4\,C_3(C_5 + C_7) + R_6\,C_7(C_3 + C_5)$$

$$\frac{C_4\,C_6\,R_3\,R_5\,R_7}{C_1(R_3 + R_5 + R_7)} = C_4\,R_3(R_7 + R_5) + C_6\,C_7(R_3 + R_5)$$

Symmetrical case:

$$Y_3 = Y_7; \qquad Y_4 = Y_6 \qquad\qquad Z_3 = Z_7; \qquad Z_4 = Z_6$$

Condition 1:

$$\omega_N = [1 + (2C_3/C_5)]^{1/2}/C_3\,R_4 \qquad \omega_N = [1 + 2R_3/R_5]^{1/2}/R_3\,C_4$$

Condition 2:

$$R_4/R_1 = (4C_3/C_5) + (2C_5/C_3) + 6 \qquad C_4/C_1 = (4R_3/R_5) + (2R_5/R_3) + 6$$

(continued)

Table 8-1(*continued*)

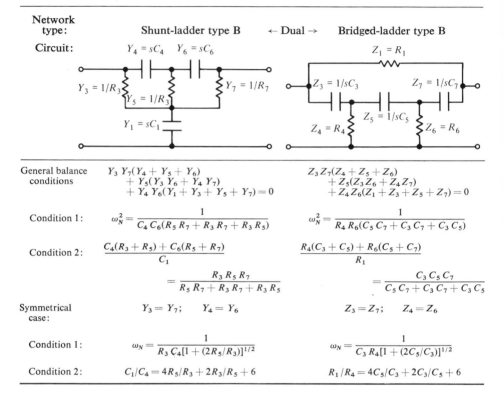

Network type:	Shunt-ladder type B	← Dual →	Bridged-ladder type B

General balance conditions

$$Y_3\,Y_7(Y_4 + Y_5 + Y_6) + Y_5(Y_3\,Y_6 + Y_4\,Y_7) + Y_4\,Y_6(Y_1 + Y_3 + Y_5 + Y_7) = 0$$

$$Z_3\,Z_7(Z_4 + Z_5 + Z_6) + Z_5(Z_3\,Z_6 + Z_4\,Z_7) + Z_4\,Z_6(Z_1 + Z_3 + Z_5 + Z_7) = 0$$

Condition 1:

$$\omega_N^2 = \frac{1}{C_4\,C_6(R_5\,R_7 + R_3\,R_7 + R_3\,R_5)}$$

$$\omega_N^2 = \frac{1}{R_4\,R_6(C_5\,C_7 + C_3\,C_7 + C_3\,C_5)}$$

Condition 2:

$$\frac{C_4(R_3 + R_5) + C_6(R_5 + R_7)}{C_1} = \frac{R_3\,R_5\,R_7}{R_5\,R_7 + R_3\,R_7 + R_3\,R_5}$$

$$\frac{R_4(C_3 + C_5) + R_6(C_5 + C_7)}{R_1} = \frac{C_3\,C_5\,C_7}{C_5\,C_7 + C_3\,C_7 + C_3\,C_5}$$

Symmetrical case:

$$Y_3 = Y_7; \qquad Y_4 = Y_6$$

$$Z_3 = Z_7; \qquad Z_4 = Z_6$$

Condition 1:

$$\omega_N = \frac{1}{R_3\,C_4[1 + (2R_5/R_3)]^{1/2}}$$

$$\omega_N = \frac{1}{C_3\,R_4[1 + (2C_5/C_3)]^{1/2}}$$

Condition 2:

$$C_1/C_4 = 4R_5/R_3 + 2R_3/R_5 + 6$$

$$R_1/R_4 = 4C_5/C_3 + 2C_3/C_5 + 6$$

and

$$\frac{R_4}{R_1} = 15 \tag{8-20}$$

A more interesting property appears from further consideration of these equations. Suppose that $C_3/C_5 = \frac{1}{2}$; then

$$\omega_N = \frac{\sqrt{2}}{C_3\,R_4} \tag{8-21}$$

and

$$\frac{R_4}{R_1} = 12 \tag{8-22}$$

Thus, for given values of C_3, R_4, and R_1 there are two independent frequencies of balance, (8-17) and (8-21), corresponding to two particular values of C_5. This comes about because of the form of condition 2 for the symmetrical shunt-ladder and bridged-ladder networks. For the four network types of this group shown in Table 8-1, condition 2 has the form

$$Y = 4X + (2/X) + 6 \tag{8-23}$$

where Y and X are the capacitor or resistor ratios shown in the table. This is a quadratic expression in X and therefore has the same value for two values of X, each of which is larger than some value Y_{min}. (Thus, for example, we saw above that Y equals 12 for two different values of X.) To find Y_{min} we set

$$\frac{dY}{dX} = 4 - \frac{2}{X^2} = 0 \tag{8-24}$$

Solving for X and Y we obtain

$$X_{min} = 1/\sqrt{2} = 0.707 \tag{8-25}$$

and

$$Y_{min} = (4/\sqrt{2}) + 2\sqrt{2} + 6 = 11.65 \tag{8-26}$$

Values of Y less than Y_{min} cannot be chosen if real results are to be obtained for conditions 1 and 2 in the symmetrical cases shown in Table 8-1.

8.1.2 The Twin-*T* Networks

Of the *RC* building blocks discussed in the previous section, all of which provide conjugate complex, or imaginary, transmission zeros, the twin-*T* network has received the most attention and has been used the most frequently, even long before the advent of hybrid integrated circuit technology. However, because it has been used extensively in hybrid integrated circuit technology as well (for which it has proven to be particularly suitable), we shall discuss it in some detail in what follows.

Invented in 1934 by H. W. Augustadt,[4] the twin-*T* has, since that time, been mainly used as a feedback network around a gain stage to obtain highly selective amplifiers, stable oscillators, and notch filters. The frequency ranges covered were generally low (e.g., voiceband) or very low (e.g., from a fraction of a Hertz to 60 Hz). Recently, with the advent of linear integrated circuits, numerous methods of active *RC* filter synthesis have been developed that rely on the basic frequency characteristics of a twin-*T* network, or modifications thereof, to provide the required filtering properties. These methods depend, for their frequency stability, on the stability of the twin-*T* network. To ensure a high degree of stability, the twin-*T* can be realized by film components and then combined with silicon integrated active circuits to produce hybrid integrated filter networks. It is in particular with respect to its usage in applications of this kind[5] that the twin-*T* network will be examined here.

4. H. W. Augustadt, unpublished memorandum, Bell Telephone Laboratories.
5. G. S. Moschytz, A general approach to twin-*T* design and its application to hybrid integrated linear active networks, *Bell. Syst. Tech. J.*, **49**, 1105–1149 (1970).

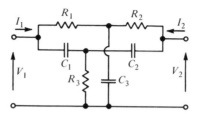

FIG. 8-8. The general twin-T. Note that the component notation used here is more convenient than that used in Fig. 8-7, which was based on that of the general bridge network of Fig. 8-1b.

The General Twin-T with an Infinite Null

The general twin-T network shown in Fig. 8-8 has the following short-circuit admittance matrix

$$[y] = \begin{bmatrix} \dfrac{as^3 + (b+f)s^2 + s(c+g) + 1}{[R_1 R_2 C_3 s + R_1 + R_2]} \times [R_3(C_1 + C_2)s + 1] & -\dfrac{as^3 + bs^2 + cs + 1}{[R_1 R_2 C_3 s + R_1 + R_2]} \times [R_3(C_1 + C_2)s + 1] \\[4ex] -\dfrac{as^3 + bs^2 + cs + 1}{[R_1 R_2 C_3 s + R_1 + R_2]} \times [R_3(C_1 + C_2)s + 1] & \dfrac{as^3 + (b+d)s^2 + (c+e)s + 1}{R_1 R_2 C_3 s + R_1 + R_2]} \times [R_3(C_1 + C_2)s + 1] \end{bmatrix}$$

$$(8\text{-}27)$$

where

$$a = R_1 R_2 R_3 C_1 C_2 C_3 \qquad\qquad e = R_1 C_3 + (R_1 + R_2)$$
$$b = (R_1 + R_2)R_3 C_1 C_2 \qquad\qquad f = R_2 C_3[R_1 C_1 + R_3(C_1 + C_2)]$$
$$c = R_3(C_1 + C_2) \qquad\qquad g = (R_1 + R_2)C_1 + R_2 C_3$$
$$d = R_1 C_3[R_2 C_2 + R_3(C_1 + C_2)] \qquad\qquad (8\text{-}28)$$

The voltage transfer function of the unloaded twin-T follows:

$$T(s) = \left.\frac{V_2}{V_1}\right|_{I_2=0} = \frac{N(s)}{D(s)} = -\frac{y_{21}}{y_{22}} = \frac{as^3 + bs^2 + cs + 1}{as^3 + (b+d)s^2 + (c+e)s + 1} \qquad (8\text{-}29)$$

The conditions for an infinite null, i.e., for transmission zeros of $N(s)$ on the imaginary axis were given by (8-14) or (8-15) together with (8-16) (see also Table 8-1). Rewriting these conditions for the component designations of Fig. 8-8 and introducing the substitutions

$$C_s = \frac{C_1 C_2}{C_1 + C_2} \qquad\qquad (8\text{-}30a)$$

$$C_p = C_1 + C_2 \qquad\qquad (8\text{-}30b)$$

$$R_s = R_1 + R_2 \qquad\qquad (8\text{-}30c)$$

$$R_p = \frac{R_1 R_2}{R_1 + R_2} \qquad\qquad (8\text{-}30d)$$

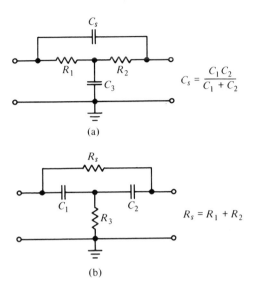

$$C_s = \frac{C_1 C_2}{C_1 + C_2}$$

(a)

$$R_s = R_1 + R_2$$

(b)

FIG. 8-9. Bridged-*T* networks derived from the general twin-*T*. (a) R_3 disconnected. (b) C_3 disconnected.

the following two conditions for the twin-*T* null frequency result:

$$\omega_N^2 = \frac{1}{R_1 R_2 C_s C_3} \qquad\qquad [8\text{-}31]$$

and

$$\frac{C_3}{R_3} = \frac{C_p}{R_p} \qquad\qquad [8\text{-}32]$$

These can be combined as follows:

$$\omega_N^2 = \frac{1}{R_1 R_2 C_s C_3} = \frac{1}{R_s R_3 C_1 C_2}. \qquad\qquad [8\text{-}33]$$

Thus, for a perfect null (i.e., roots z_1 and z_2 of $N(s)$ given by $z_{1,2} = \pm j\omega_N$) the null frequency of each of the two bridged-*T*'s obtained by disconnecting R_3 and C_3, respectively, of the twin-*T* (see Fig. 8-9) must be equal. This fact can be used to develop a two-step tuning method for the twin-*T*[6]. Substituting

6. G. S. Moschytz, Two-step precision tuning of a twin-*T* notch filter, *Proc. IEEE*, **54** 811–812 (1966).

(8-31) and (8-32) into (8-29) gives the transfer-function polynomials of the nulled twin-T:

$$N(s) = \frac{R_3 C_p}{\omega_N^2} (s + \omega_1)(s^2 + \omega_N^2) \qquad (8\text{-}34)$$

and

$$D(s) = \frac{R_3 C_p}{\omega_N^2} (s + \omega_1)\left[s^2 + \left(\omega_N^2 R_1 C_3 + \frac{1}{C_1 R_3}\right)s + \omega_N^2\right] \qquad (8\text{-}35)$$

where

$$\omega_1 = \frac{1}{R_3 C_p} \qquad (8\text{-}36)$$

The voltage-transfer function follows as

$$T(s) = \frac{s^2 + \omega_N^2}{s^2 + (\omega_N/q_N)s + \omega_N^2} \qquad [8\text{-}37]$$

where ω_N is given by (8-33), the pole Q is given by

$$q_N = \frac{\omega_N}{2\sigma_N} = \frac{1}{\omega_N(R_1 C_3 + R_s C_2)} \qquad [8\text{-}38]$$

and

$$2\sigma_N = \omega_N^2 R_1 C_3 + \frac{1}{R_3 C_1} = \frac{1}{R_2 C_s} + \frac{1}{R_3 C_1} \qquad [8\text{-}39]$$

As mentioned previously the degree of the two third-order polynomials (8-34) and (8-35) is reduced by one because of pole–zero cancellation, which occurs when the conditions for a perfect null are satisfied. This corresponds to the pole–zero diagram shown in Fig. 8-10. Inserting (8-32) into (8-27) (this

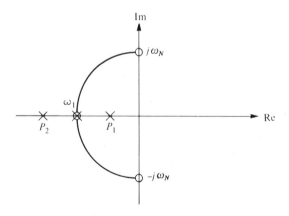

FIG. 8-10. Pole–zero diagram of balanced twin-T.

results in the two first-order factors in the denominators of the y parameters being equal), we obtain the y parameters for the balanced general twin-T:

$$[y] = \begin{bmatrix} \dfrac{s^2 R_1 R_2 C_s C_3 + s(R_s C_1 + R_2 C_3) + 1}{R_s(sR_3 C_p + 1)} & -\dfrac{s^2 R_1 R_2 C_s C_3 + 1}{R_s(sR_3 C_p + 1)} \\[4mm] -\dfrac{s^2 R_1 R_2 C_s C_3 + 1}{R_s(sR_3 C_p + 1)} & \dfrac{s^2 R_1 R_2 C_s C_3 + s(R_s C_2 + R_1 C_3) + 1}{R_s(sR_3 C_p + 1)} \end{bmatrix}$$

[8-40]

Notice that the y parameters are second-order, since pole–zero cancellation has taken place.

The most frequently used twin-T is structurally (and electrically) symmetrical. For this case (see Fig. 8-11a) $R_1 = R_2 = R$, $R_3 = R/2$, $C_1 = C_2 = C$, $C_3 = 2C$, and the coefficients of (8-37) are $\omega_N = 1/RC$, $2\sigma_N = 4/RC$, and $q_N = \frac{1}{4}$. Another commonly used version of the twin-T is the potentially symmetrical configuration. This is obtained by impedance-scaling one-half the symmetrical twin-T by some factor ρ. The resulting twin-T elements are (see Figure 8-11b) $R_1 = R$, $R_2 = \rho R$, $R_3 = [\rho/(1 + \rho)]R$, $C_1 = C$, $C_2 = C/\rho$, and $C_3 = [(1 + \rho)/\rho]C$. The coefficients of (8-37) for this case are $\omega_N = 1/RC$, $2\sigma_N = (2/RC)[(\rho + 1)/\rho]$ and $q_N = \frac{1}{2}\rho/(1 + \rho)$. Notice that for the extreme asymmetrical case for which $\rho \gg 1$, q_N takes on its maximum value of 0.5.

Sometimes it is useful to make all the resistors of the twin-T equal. This enables one to gang three variable resistors in order to vary the null frequency. The twin-T elements are then (see Fig. 8-11c) $R_1 = R_2 = R_3 = R$, $C_1 = C$, $C_2 = C/2$, and $C_3 = 3C$, and the coefficients of the transfer function are the same as those of the symmetrical twin-T. Similarly, if the three capacitors are to be made equal for ganging or other purposes, the twin-T elements are (see Fig. 8-11d) $R_1 = R$, $R_2 = 2R$, $R_3 = R/3$. Here again the coefficients of the transfer function are the same as those of the symmetrical twin-T.

The General Twin-T with a Finite Null   It was shown that the transfer function of a general twin-T is simplified by one degree, due to pole–zero cancellation on the negative real axis at $\omega_1 = 1/R_3 C_p$ (see Fig. 8-10), when the conditions for a perfect null (see (8-31) and (8-32)) are satisfied. It can be shown[7] that this pole–zero cancellation can be maintained for differentially small perturbations of any element of a balanced twin-T configuration so long as the time constants of the series elements remain unchanged. Referring to Fig. 8-8 this implies that

$$R_1 C_1 = R_2 C_2 \tag{8-41}$$

7. G. S. Moschytz, op. cit., pp. 1145–1147 (see footnote 5 of this chapter).

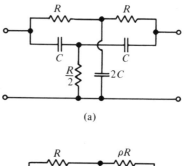

(a)

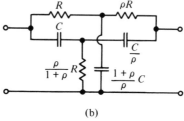

(b)

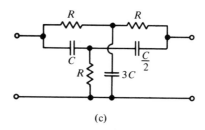

(c)

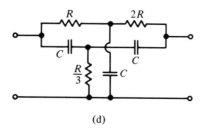

(d)

FIG. 8-11. Frequently used twin-T configurations. (a) Symmetrical. (b) Potentially symmetrical. (c) Equal resistors. (d) Equal capacitors.

For positive element changes (i.e., increasing values) the pole–zero dipole will decrease, i.e., move in the direction of the *s*-plane origin. Twin-*T* networks satisfying (8-41) include all symmetrical configurations in which the series elements are identical, as well as potentially symmetrical configurations, in which the series elements are characterized by relations of the type

$$R_2 = aR_1 \tag{8-42}$$
$$C_2 = C_1/a$$

By using Euclid's algorithm (see Appendix) to find a common divider in the numerator and denominator of (8-29), Lueder has shown[8] that this pole–zero cancellation can be maintained throughout the *s* plane (within the inherent bounds of a third-order network; see Chapter 1, Section 1.5) if the following general condition, expressed in terms of the coefficients given in (8-28), is satisfied:

$$d(d^2 + e^2b) - e(e^2a + cd^2) = 0 \tag{8-43}$$

Inserting the expressions given in (8-28) we find (8-43) to constitute two separate conditions. The first coincides with (8-32) and, as we already know, implies zeros on the imaginary axis. The second is new:

$$(d/e)^2 = R_1 R_3 C_1 C_3 \tag{8-44}$$

or, with the expressions in (8-28)

$$\left(\frac{R_2 C_2 + R_3 C_p}{R_1 C_3 + R_s C_2} \right)^2 = \frac{R_3 C_1}{R_1 C_3} \tag{8-45}$$

If this condition is satisfied, the twin-*T* provides a second-order network whose transfer function, in terms of the coefficients in (8-28), is given by

$$T(s) = \frac{s^2 + [(b/a) - (e/d)]s + (d/ea)}{s^2 + [(b/a) - (e/d) + (d/a)]s + (d/ea)} \tag{8-46}$$

Depending on the polarity of the term $[(b/a) - (e/d)]$, the complex zeros of $T(s)$ may have positive or negative real terms, i.e., lie in the left or right half *s* plane. The poles, of course, are confined to the negative real axis. Nevertheless, assuming that condition (8-45) is fulfilled, we can now generate functions with the passive *RC* twin-*T* network, of the form[9]

$$\hat{T}(s) = \frac{s^2 + (\omega_z/q_z)s + \omega_z^2}{s^2 + (\omega_p/\hat{q}_p)s + \omega_p^2} \tag{8-47}$$

---

8. E. Lueder, The general second order twin-*T* and its application to frequency emphasizing networks, *Bell System Tech. J.*, **51**, 301–316 (1972).
9. We use the general coefficients ω_z, q_z, ω_p, and q_p here to relate the general second-order twin-*T* transfer function to the second-order functions used in preceding, and following chapters; likewise, as elsewhere in this book, the overscript "^" implies a passive *RC* network.

where

$$\omega_z = \omega_p = \omega_N \tag{8-48a}$$

$$0 < q_z|_{LHP} \leq \infty \tag{8-48b}$$

$$1 < |q_z|_{RHP} \leq \infty \tag{8-48c}$$

and

$$0 \leq \hat{q}_p = q_N \leq 0.5 \tag{8-48d}$$

This capability substantially increases the versatility of the twin-T as a zero-placing circuit in active network design.

8.2 Sensitivity of Twin-T Null Characteristics to Component Variations

The null or zero sensitivity of the twin-T to variations of any component x gives a measure of the degree of change of the transmission characteristics in the vicinity of the twin-T null frequency as a result of variations of a component x, or, in other words, of the zero displacement due to an incremental change in the value of the component x. The zero displacement dz is therefore given by

$$dz = \mathcal{S}_x^z \frac{dx}{x} \tag{8-49}$$

where z is the complex null frequency of the network and $\mathcal{S}_x^z$ is the corresponding zero sensitivity.

The displacement dz has a real and an imaginary component; it defines a vector or the direction in which a zero travels from its initial location with incremental changes of a component x. Since x and therefore dx/x must be real, the zero sensitivity defines a vector in the same direction as the zero displacement dz. Herein lies the importance of knowing the root (i.e., pole or zero) sensitivity of a network, since it provides insight into the stability of a system with respect to any component x. It also provides information relevant to network tuning, since it relates adjustments of a component x to its effect on the roots of the transfer function. Conversely, a network can be designed to provide a prescribed sensitivity between some parameter of the transfer function, e.g., the displacement of a specific transmission zero, and the variation of some preselected component x. The choice of sensitivity may be such as to result in a network that responds to a simplified tuning strategy, or whose characteristics may be adjusted in a desirable way.

To calculate the zero sensitivities with respect to the six elements of the balanced twin-T, we proceed as follows. Referring to the concepts of root sensitivity discussed in Chapter 4, Section 4.2.1, we first express the numerator

$N(s)$ of the twin-T transfer function in its bilinear form:

$$N(s) = A_x(s) + xB_x(s) \tag{8-50}$$

The null return difference $F_x^0(s)$ with respect to x is given by

$$F_x^0(s) = \frac{N(s)}{A_x(s)} = 1 + x\frac{B_x(s)}{A_x(s)} \tag{8-51}$$

With (8-28) and (8-29) the null return difference of the twin-T with respect to its six components can be calculated directly.

To obtain the null return difference of the *nulled* twin-T, (8-34) can be substituted into (8-51), i.e.,

$$F_x^0(s) = \frac{(s + \omega_1)(s^2 + \omega_N^2)}{\omega_1 \omega_N^2 A_x(s)} \tag{8-52}$$

The corresponding zero sensitivity then results:

$$\mathscr{S}_x^{j\omega_N} = \frac{s - j\omega_N}{F_x^0(s)}\bigg|_{s=j\omega_N} = \frac{(s - j\omega_N)\omega_1\omega_N^2 A_x(s)}{(s + \omega_1)(s^2 + \omega_N^2)}\bigg|_{s=j\omega_N} \tag{8-53}$$

and simplifies to:

$$\mathscr{S}_x^{j\omega_N} = -\frac{\alpha(1 + j\alpha)}{2(1 + \alpha^2)}\omega_N \cdot A_x(j\omega_N) \tag{8-54}$$

where $\alpha = \omega_1/\omega_N$. The individual $A_x(j\omega_N)$ functions follow directly from (8-28) and (8-29). Substituting these into (8-54) the zero sensitivity of the nulled twin-T with respect to its six components is obtained. These are listed in

TABLE 8-2. ZERO-SENSITIVITY FUNCTIONS

$$\mathscr{S}_{R_1}^{j\omega_N} = \frac{\alpha\omega_N}{2(1 + \alpha^2)}\left[1 - \lambda - j\left(\frac{1}{\alpha} + \alpha\lambda\right)\right]$$

$$\mathscr{S}_{C_1}^{j\omega_N} = -\frac{\alpha\omega_N}{2(1 + \alpha^2)}\left[1 - \eta + j\left(\alpha + \frac{\eta}{\alpha}\right)\right]$$

$$\mathscr{S}_{R_2}^{j\omega_N} = \frac{\alpha\omega_N}{2(1 + \alpha^2)}\left[\lambda - j\left(\frac{1}{\alpha} + \alpha(1 - \lambda)\right)\right]$$

$$\mathscr{S}_{C_2}^{j\omega_N} = -\frac{\alpha\omega_N}{2(1 + \alpha^2)}\left[\eta + j\left(\alpha + \frac{(1 - \eta)}{\alpha}\right)\right]$$

$$\mathscr{S}_{R_3}^{j\omega_N} = -\frac{\alpha\omega_N}{2(1 + \alpha^2)}(1 + j\alpha)$$

$$\mathscr{S}_{C_3}^{j\omega_N} = \frac{\alpha\omega_N}{2(1 + \alpha^2)}\left(1 - \frac{j}{\alpha}\right)$$

$$\omega_N^2 = \frac{1}{R_1 R_2 C_s C_3} = \frac{1}{R_s R_3 C_1 C_2}; \qquad \omega_1 = \frac{1}{R_3 C_p} = \frac{1}{R_p C_3}$$

$$R_s = R_1 + R_2; \qquad R_p = \frac{R_1 R_2}{R_1 + R_2}$$

$$C_s = \frac{C_1 C_2}{C_1 + C_2}; \qquad C_p = C_1 + C_2$$

$$\alpha = \frac{\omega_1}{\omega_N}; \qquad \lambda = \frac{R_1}{R_1 + R_2}; \qquad \eta = \frac{C_2}{C_1 + C_2}$$

Table 8-2. It has thereby been found useful to characterize the general twin-T by the following parameters:

$$\lambda = \frac{R_1}{R_1 + R_2} \tag{8-55}$$

$$\eta = \frac{C_2}{C_1 + C_2} \tag{8-56}$$

and

$$\alpha = \frac{\omega_1}{\omega_N} = \left(\frac{R_s C_s}{R_3 C_p}\right)^{1/2} \tag{8-57}$$

The quantities λ and η give a measure of the degree of symmetry of the series elements of the twin-T; α relates the series elements to the shunt elements.

As an example let us consider the zero sensitivity of the symmetrical balanced twin-T shown in Fig. 8-11a. For this case we have $\omega_N = 1/RC$, $\alpha = 1$, and $\lambda = \eta = 0.5$. The resulting zero-sensitivity functions (for this and the other commonly used twin-T configurations shown in Fig. 8-11) are listed in

TABLE 8-3. ZERO–SENSITIVITY FUNCTIONS

1. Symmetrical twin-T

$$R_1 = R_2 = 2R_3 = R; \quad C_1 = C_2 = C_3/2 = C \quad \omega_N = 1/RC; \quad q_N = \tfrac{1}{4}; \quad \alpha = 1; \quad \lambda = \eta = \tfrac{1}{2}; \quad \gamma = \tfrac{1}{4}$$

$$\mathscr{S}_{R_1}^{j\omega_N} = \mathscr{S}_{R_2}^{j\omega_N} = \frac{\omega_N}{4}\left(\frac{1}{2} - \frac{3}{2}j\right) = 0.395\omega_N \angle -71°30' \qquad \mathscr{S}_{C_1}^{j\omega_N} = \mathscr{S}_{C_2}^{j\omega_N} = -\frac{\omega_N}{4}\left(\frac{1}{2} + \frac{3}{2}j\right) = -0.395 \angle 71°30'$$

$$\mathscr{S}_{R_3}^{j\omega_N} = -\frac{\omega_N}{4}(1+j) = -0.354 \angle 45° \qquad \mathscr{S}_{C_3}^{j\omega_N} = \frac{\omega_N}{4}(1-j) + 0.354 \angle -45°$$

2. Potentially symmetrical twin-T

$$R_1 = R_2/\rho = R_3(1+\rho)/\rho = R; \quad C_1 = \rho C_2 = C_3 \rho/(1+\rho) = C$$

$$\omega_N = \frac{1}{RC}; \quad q_N = \frac{1}{2}\frac{\rho}{1+\rho}; \quad \alpha = 1; \quad \lambda = \eta = \frac{1}{1+\rho}; \quad \gamma = \frac{\rho}{(1+\rho)^2}$$

$$\mathscr{S}_{R_1}^{j\omega_N} = \frac{\omega_N}{4}\left(\frac{\rho}{1+\rho} - j\frac{2+\rho}{1+\rho}\right) \qquad \mathscr{S}_{C_1}^{j\omega_N} = -\frac{\omega_N}{4}\left(\frac{\rho}{1+\rho} + j\frac{2+\rho}{1+\rho}\right)$$

$$\mathscr{S}_{R_2}^{j\omega_N} = \frac{\omega_N}{4}\left(\frac{1}{1+\rho} - j\frac{1+2\rho}{1+\rho}\right) \qquad \mathscr{S}_{C_2}^{j\omega_N} = -\frac{\omega_N}{4}\left(\frac{1}{1+\rho} + j\frac{1+2\rho}{1+\rho}\right)$$

$$\mathscr{S}_{R_3}^{j\omega_N} = -\frac{\omega_N}{4}(1+j) \qquad \mathscr{S}_{C_3}^{j\omega_N} = \frac{\omega_N}{4}(1-j)$$

(continued)

Table 8-3 (*continued*)

2A. Potentially symmetrical twin-T

$$\rho \gg 1; \qquad q_N \to \tfrac{1}{2}; \qquad \alpha = 1; \qquad \lambda = \eta \to 0; \qquad \gamma \to 0$$

$$\mathscr{S}_{R_1}^{j\omega_N} \approx \frac{\omega_N}{4}(1-j) = 0.354\omega_N \angle -45° \qquad\qquad \mathscr{S}_{C_1}^{j\omega_N} \approx -\frac{\omega_N}{4}(1+j) = -0.354 \angle 45°$$

$$\mathscr{S}_{R_2}^{j\omega_N} \approx -j\frac{\omega_N}{2} = 0.5\omega_N \angle -90° \qquad\qquad \mathscr{S}_{C_2}^{j\omega_N} \approx -j\frac{\omega_N}{2} = 0.5\omega_N \angle -90°$$

$$\mathscr{S}_{R_3}^{j\omega_N} = -\frac{\omega_N}{4}(1+j) = -0.354 \angle 45° \qquad\qquad \mathscr{S}_{C_3}^{j\omega_N} = \frac{\omega_N}{4}(1-j) = 0.354 \angle -45°$$

2B. Potentially symmetrical twin-T

$$\rho \ll 1; \qquad q_N \to 0; \qquad \alpha = 1; \qquad \lambda = \eta \to 1; \qquad \gamma \to 0$$

$$\mathscr{S}_{R_1}^{j\omega_N} \approx -j\frac{\omega_N}{2} = 0.5\omega_N \angle -90° \qquad\qquad \mathscr{S}_{C_1}^{j\omega_N} \approx -j\frac{\omega_N}{2} = 0.5\omega_N \angle -90°$$

$$\mathscr{S}_{R_2}^{j\omega_N} \approx \frac{\omega_N}{4}(1-j) = 0.354 \angle -45° \qquad\qquad \mathscr{S}_{C_2}^{j\omega_N} \approx -\frac{\omega_N}{4}(1+j) = -0.354 \angle 45°$$

$$\mathscr{S}_{R_3}^{j\omega_N} = -\frac{\omega_N}{4}(1+j) = -0.354 \angle 45° \qquad\qquad \mathscr{S}_{C_3}^{j\omega_N} = \frac{\omega_N}{4}(1-j) = 0.354 \angle -45°$$

3. Twin-T with equal resistors

$$R_1 = R_2 = R_3 = R; \qquad C_1 = 2C_2 = C_3/3 = C$$
$$\omega_N = 1/RC; \qquad q_N = \tfrac{1}{4}; \qquad \alpha = \tfrac{2}{3}; \qquad \lambda = \tfrac{1}{2}; \qquad \eta = \tfrac{1}{3}; \qquad \gamma = \tfrac{1}{9}$$

$$\mathscr{S}_{R_1}^{j\omega_N} = \mathscr{S}_{R_2}^{j\omega_N} = \frac{\omega_N}{13}(1.5 - 5.5j) = 0.44\omega_N \angle -75° \qquad\qquad \mathscr{S}_{C_1}^{j\omega_N} = -\frac{\omega_N}{13}(2 + 3.5j) = -0.276\omega_N \angle 60°$$

$$\mathscr{S}_{R_3}^{j\omega_N} = -\frac{\omega_N}{13}(3 + 2j) = -0.278\omega_N \angle 34° \qquad\qquad \mathscr{S}_{C_2}^{j\omega_N} = -\frac{\omega_N}{13}(1 + 5j) = -0.392\omega_N \angle 79°$$

$$\mathscr{S}_{C_3}^{j\omega_N} = \frac{\omega_N}{13}(3 - 4.5j) = 0.416\omega_N \angle -56°$$

4. Twin-T with equal capacitors

$$R_1 = R_2/2 = 3R_3 = R; \qquad C_1 = C_2 = C_3 = C$$
$$\omega_N = 1/RC, \qquad q_N = \tfrac{1}{4}; \qquad \alpha = \tfrac{3}{2}; \qquad \lambda = \tfrac{1}{3}; \qquad \eta = \tfrac{1}{2}; \qquad \gamma = \tfrac{1}{2}$$

$$\mathscr{S}_{R_1}^{j\omega_N} = \frac{\omega_N}{13}(2 - 3.5j) = 0.276\omega_N \angle -60° \qquad\qquad \mathscr{S}_{C_1}^{j\omega_N} = \mathscr{S}_{C_2}^{j\omega_N} = -\frac{\omega_N}{13}(1.5 + 5.5j) = -0.44\omega_N \angle 75°$$

$$\mathscr{S}_{R_2}^{j\omega_N} = \frac{\omega_N}{13}(1 - 5j) = 0.392\omega_N \angle -79° \qquad\qquad \mathscr{S}_{C_3}^{j\omega_N} = \frac{\omega_N}{13}(3 - 2j) = 0.278\omega_N \angle -34°$$

$$\mathscr{S}_{R_3}^{j\omega_N} = -\frac{\omega_N}{13}(3 + 4.5j) = -0.416\omega_N \angle 56°$$

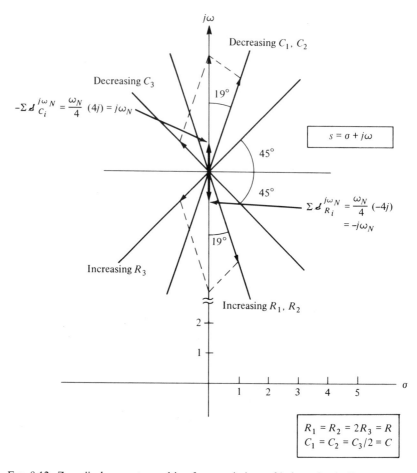

FIG. 8-12. Zero displacements resulting from variations of balanced twin-T components.

Table 8-3 and the corresponding zero displacements shown graphically in the complex frequency plane in Fig. 8-12.[10] As expected we find that

$$\sum \mathscr{S}_{R_i}^z = \sum \mathscr{S}_{C_j}^z = -z$$

Therefore, by assuming tracking and equal but opposite temperature coefficients of the resistors and capacitors, temperature drift of the null frequency can theoretically be eliminated completely. How accurately it can be eliminated in practice will be discussed shortly. First, however, we shall consider some unsymmetrical twin-T configurations that provide features that may be useful in active network design.

10. As pointed out earlier, the root sensitivity function defines a differential vector in the complex s plane. It can be shown that this vector lies on the branch of the root locus with respect to a component x that starts at z, or, in other words, that the root displacement dz and the root sensitivity have the same argument.

8.3 Twin-*T* Networks with Prescribed Tuning Characteristics

The requirement that the six components of a twin-*T* network provide a perfect null, i.e., a pair of imaginary zeros, at a particular frequency, imposes only two design constraints on the network. A third constraint results from the impedance scaling factor chosen for the network. This leaves three parameters to be chosen by whatever criteria seem most important for a given application. Most often circuit simplicity dominates this choice, resulting in the symmetrical twin-*T*. In other instances, practical considerations requiring that either all the resistors or all the capacitors be equal may determine the choice. In the following discussion, the three unconstrained design parameters will be selected with a view to modifying the dependence of the null characteristics to adjustments of certain twin-*T* components. The resulting dependence may provide useful tuning and adjustment strategies or even motivate new applications for the twin-*T* itself.

The dependence of twin-*T* transmission characteristics in the vicinity of the null frequency on variations of any component x are essentially determined by the zero-sensitivity functions listed in Table 8-2. Design equations for twin-*T* networks providing a specified dependence of null characteristics on the adjustment of a particular component x may, therefore, be found by setting corresponding constraints on the zero-sensitivity functions and solving the resulting equations for the twin-*T* components. Instead of designing a twin-*T* for a specified dependence of transmission characteristics on variations of a given component x in the vicinity of the null frequency it therefore suffices for us to design a twin-*T* to a specified *zero sensitivity* with respect to x.

The expressions listed in Table 8-2 show that, after the null frequency ω_N has been specified, there are basically three remaining parameters that must be determined in order to design a twin-*T* to a prescribed zero sensitivity. These are the frequency ratio α, the resistor ratio λ, and the capacitor ratio η. A parameter that sometimes provides clearer physical insight into the design of a twin-*T* than the frequency ratio α is the ratio γ of series to shunt capacitors:

$$\gamma = \frac{C_s}{C_3} = \eta \frac{C_1}{C_3} \qquad (8\text{-}58)$$

For the balanced twin-*T* with $j\omega$-axis zeros, γ and α are related by (8-57), which can be rewritten as

$$\alpha = \left[\frac{\gamma}{\lambda(1-\lambda)} \right]^{1/2} \qquad (8\text{-}59)$$

This expression has been plotted in Fig. 8-13 as a function of λ with the parameter γ. The values of γ for the common twin-*T* configurations are included in Table 8-3. From the defining equations, the limits on the four twin-*T* parameters are

$$0 < \alpha < \infty \qquad (8\text{-}60a)$$

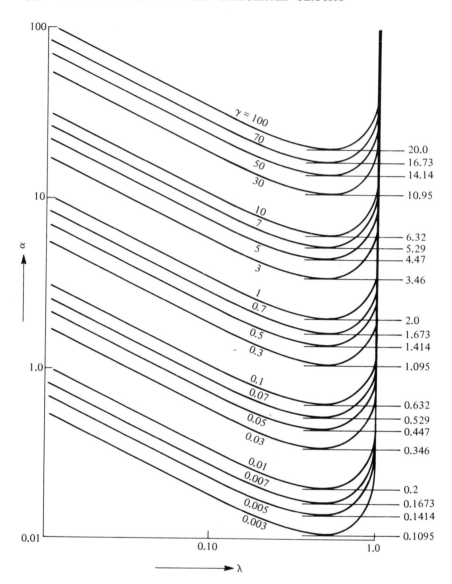

FIG. 8-13. Frequency ratio α as a function of resistor and capacitor ratios λ and γ.

$$0 < \gamma < \infty \qquad\qquad (8\text{-}60\text{b})$$

$$0 < \lambda < 1 \qquad\qquad (8\text{-}60\text{c})$$

$$0 < \eta < 1 \qquad\qquad (8\text{-}60\text{d})$$

A fundamental characteristic of the twin-T is its ability to reject a narrow frequency band, centered at the null frequency f_N, and to pass, with less

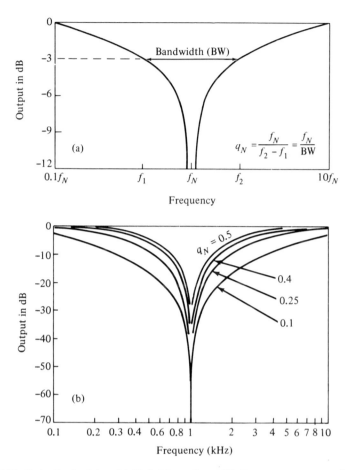

FIG. 8-14. Twin-*T* selectivity. (a) Definition of q_N. (b) Frequency response for various q_N values.

attenuation, the frequencies on either side of this band. A useful parameter characterizing the selectivity of frequency rejection is the pole Q, q_N. Physically, q_N is the ratio of the center frequency f_N divided by the bandwidth at which 3-dB attenuation occurs[11] (see Fig. 8-14a).

It is important, when examining the effects of the parameters listed in (8-60) on the zero sensitivity of the twin-*T* to keep an eye on their effect on the twin-*T* selectivity as expressed by q_N. Obviously, poor selectivity might be too high a price to pay for any set of controlled sensitivity functions. On the other hand, because the twin-*T* is a passive *RC* network, the selectivity

11. This definition is only accurate for nonloaded twin-*T* networks such as those being considered here. For the case of a loaded twin-*T* with an unsymmetrical frequency response it has been found more useful to define selectivity as the slope of the phase ϕ at the null frequency, i.e., by $\tau(\omega_N) = d\phi/d\omega|_{\omega=\omega_N}$.

factor (or pole Q) q_N is limited to values less than 0.5 anyway. The value 0.25 is realized most frequently, with the symmetrical as well as with the equal-resistor and equal-capacitor twin-T's. However, even within the limited possible range of q_N, the difference in actual frequency selectivity can be quite significant. This is illustrated in Fig. 8-14b, where twin-T frequency-response curves have been ploted for various q_N values between 0.1 and 0.5.

Expressing q_N by the same parameters as are used for the zero-sensitivity functions in Table 8-2 we have for the general twin-T

$$q_N = \frac{\alpha(1 - \eta)(1 - \lambda)}{\alpha^2(1 - \lambda) + (1 - \eta)} = \frac{(1 - \eta)[\gamma\lambda(1 - \lambda)]^{1/2}}{\gamma + \lambda(1 - \eta)} \qquad [8\text{-}61]$$

With (8-61) we can now observe the effects on selectivity that any sensitivity constraints on the parameters λ, η, α, or γ may have.

Let us now examine various criteria according to which it would be useful to control the zero-sensitivity functions given in Table 8-2. Due to the RC self-duality of the twin-T we need consider only the resistor or the capacitor functions. Since both discrete and hybrid integrated RC networks are generally tuned by variable or trimmable resistors, the zero sensitivity functions with respect to these will be examined.

8.3.1 Frequency Tuning with One Component

By making the real part of any one of the three sensitivity functions go to zero, it is possible to shift the null frequency accurately over a limited frequency range by varying only the one corresponding resistor (rather than two, as would be necessary in general). Referring to Fig. 8-8 consider, for example, the case of frequency tuning with R_2. We require that

$$\text{Re } \mathscr{S}_{R_2}^{j\omega_N} \to 0 \qquad (8\text{-}62)$$

This condition can be approached if $\lambda \to 0$, or $R_2 \gg R_1$. We then obtain

$$\mathscr{S}_{R_1}^{j\omega_N} \approx \frac{\alpha\omega_N}{2(1 + \alpha^2)}\left(1 - \frac{j}{\alpha}\right) \qquad (8\text{-}63a)$$

$$\mathscr{S}_{R_2}^{j\omega_N} \approx -\frac{j\omega_N}{2} \qquad (8\text{-}63b)$$

$$\mathscr{S}_{R_3}^{j\omega_N} \approx \frac{\alpha\omega_N}{2(1 + \alpha^2)}(1 + j\alpha) \qquad (8\text{-}63c)$$

To see how useful this method of tuning is, we must examine what effect letting λ approach zero has on the selectivity, i.e., on q_N. From (8-61) we find

$$q_N|_{\lambda \to 0} = \frac{\alpha(1 - \eta)}{\alpha^2 + 1 - \eta}. \qquad (8\text{-}64)$$

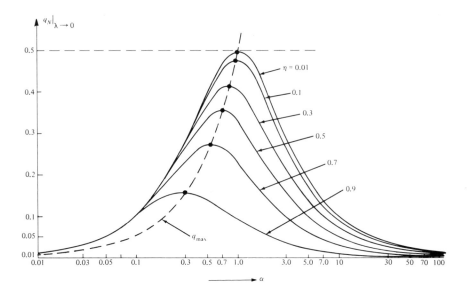

FIG. 8-15. Selectivity q_N as function of α and η for special case that $\lambda \to 0$.

This is a very convenient relationship, for, depending on the choice of α and η, q_N can actually be made to approach its maximum value of 0.5. For example, the potentially symmetrical configuration (see Fig. 8-11b), for which $\rho \gg 1$ represents a special case of (8-62), for which $\alpha = 1$ and η approaches zero at the same rate as λ. This is one of a variety of possible cases for which (8-64) approaches 0.5. Other combinations of α and η are best obtained by plotting (8-64) on semilog paper, as in Fig. 8-15. By setting the derivative of (8-64) with respect to α equal to zero we obtain

$$\alpha_{max} = (1 - \eta)^{1/2} \tag{8-65}$$

and

$$q_{max} = \frac{(1 - \eta)^{1/2}}{2} \tag{8-66}$$

Expression (8-66) is shown in Fig. 8-15 by the broken curve. Clearly, there is a wide practical range of twin-T networks, with good to excellent selectivity, that satisfy condition (8-62) and therefore provide simple frequency tuning over a restricted frequency range.

One of the disadvantages of the twin-T configurations described here is that R_2, the frequency-tuning resistor, is "floating," i.e., it does not have any of its terminals connected to ground. From this point of view it would clearly be advantageous to tune the twin-T null frequency with resistor R_3. It can be shown, however, that to do so entails a deterioration of selectivity since the requirement

$$\text{Re } \mathscr{S}_{R_3}^{j\omega_N} \to 0$$

brings a low value of q_N with it. The same is true of tuning with R_1. The three cases are summarized in Table 8-4. Thus, to tune the null frequency with one resistor while maintaining arbitrarily high selectivity (within the bounds of a passive RC network, i.e., $q_N < 0.5$) resistor R_2 should be used as the tuning resistor and the corresponding design equations listed in Table 8-4 used. If a grounded resistor is required as the tuning element, the design equations in Table 8-4 for R_3 must be used, but a decrease in selectivity will inevitably result.

8.3.2 Null–Depth Tuning with One Component

In some applications it may be desirable to make adjustments in the null depth of a twin-T after it has been initially tuned for a perfect null. This can be achieved with a single component (i.e., resistor) if the imaginary part of the sensitivity function with respect to that component can be made equal to zero. If we consider the general sensitivity functions given in Table 8-2 (and restrict ourselves to variable resistors for practical reasons), three cases are again possible: to tune with R_1, with R_2, or with R_3. Let us, for example, consider the case of tuning the null depth with resistor R_1. This means that

$$\text{Im } \mathscr{S}_{R_1}^{j\omega_N} \to 0 \tag{8-67}$$

TABLE 8-4. TWIN-T DESIGN EQUATIONS FOR CONTROLLED ZERO SENSITIVITY

Design equations for controlled sensitivity and sensitivity functions	Design equations for maximum selectivity	Remarks
1A. Re $\mathscr{S}_{R_1}^{j\omega_N} \to 0$: $\lambda \to 1$	$q_N \approx \dfrac{\alpha(1-\lambda)}{\alpha^2(1-\lambda)+1}$	$q_N \ll 0.5$
$\mathscr{S}_{R_1}^{j\omega_N} \approx -j\,\omega_N/2$	for $\eta \to 0$;	orthogonality between $\mathscr{S}_{R_2}^{j\omega_N}$ and $\mathscr{S}_{R_3}^{j\omega_N}$
$\mathscr{S}_{R_2}^{j\omega_N} \approx \dfrac{\alpha\omega_N}{2(1+\alpha^2)}\left(1-\dfrac{j}{\alpha}\right)$	$q_{Nmax} \approx \sqrt{1-\lambda}\,/2$	
$\mathscr{S}_{R_2}^{j\omega_N} = -\dfrac{\alpha\omega_N}{2(1+\alpha^2)}\left(1-\dfrac{j}{\alpha}\right)$	for $\alpha = 1/\sqrt{1-\lambda}$	

Table 8-4 (*continued*)

Design equations for controlled sensitivity and sensitivity functions	Design equations for maximum selectivity	Remarks
1B. $\mathrm{Re}\ \mathscr{S}_{R_2}^{j\omega_N} \to 0:\ \lambda \to 0$	$q_N \approx \dfrac{\alpha(1-\eta)}{\alpha^2 + 1 - \eta}$	$0 < q_N < 0.5$
$\mathscr{S}_{R_1}^{j\omega_N} \approx \dfrac{\alpha\omega_N}{2(1+\alpha^2)}\left(1 - \dfrac{j}{\alpha}\right)$	$q_{N\max} \approx \sqrt{1-\eta}\,/2$	orthogonality between $\mathscr{S}_{R_1}^{j\omega_N}$ and $\mathscr{S}_{R_3}^{j\omega_N}$
$\mathscr{S}_{R_2}^{j\omega_N} \approx -j\omega_N/2$	for $\alpha_{\max} = \sqrt{1-\eta}$	
$\mathscr{S}_{R_3}^{j\omega_N} = -\dfrac{\alpha\omega_N}{2(1+\alpha^2)}\,(1+j\alpha)$		
1C. $\mathrm{Re}\ \mathscr{S}_{R_3}^{j\omega_N} \to 0:\ \alpha \gg 1$	$q_N \approx (1-\eta)/\alpha$	$q_N \ll 0.5$
$\mathscr{S}_{R_1}^{j\omega_N} = \dfrac{\alpha\omega_N}{2(1+\alpha^2)}\left[1 - \lambda - j\left(\dfrac{1}{\alpha} + \alpha\lambda\right)\right]$	$q_{N\max} \approx 1/\alpha$ for $\eta \to 0$	variable resistor (R_3) tied to ground
$\mathscr{S}_{R_2}^{j\omega_N} = \dfrac{\alpha\omega_N}{2(1+\alpha^2)}\left[\lambda - j\left\{\dfrac{1}{\alpha} + \alpha(1-\lambda)\right\}\right]$		
$\mathscr{S}_{R_3}^{j\omega_N} \approx -j\,\dfrac{\omega_N}{2}$		
2A. $\mathrm{Im}\ \mathscr{S}_{R_1}^{j\omega_N} \to 0:\ \begin{array}{l}\alpha \gg 1 \\ \lambda = 1/\alpha^2 \ll 1\end{array}$	$q_N \approx \dfrac{(\alpha^2-1)(1-\eta)}{\alpha(\alpha^2-\eta)}$	$q_N \ll 0.5$
$\mathscr{S}_{R_1}^{j\omega_N} \approx \omega_N/2\alpha$	$q_{N\max} \approx 1/\alpha$ for $\eta \to 0$	orthogonality between $\mathscr{S}_{R_1}^{j\omega_N}$ and $\mathscr{S}_{R_2}^{j\omega_N} = \mathscr{S}_{R_3}^{j\omega_N}$
$\mathscr{S}_{R_2}^{j\omega_N} \approx -j\,\omega_N/2$		
$\mathscr{S}_{R_3}^{j\omega_N} \approx -j\,\omega_N/2$		
2B. $\mathrm{Im}\ \mathscr{S}_{R_2}^{j\omega_N} \to 0:\ \begin{array}{l}\alpha \gg 1 \\ \lambda = (1 - 1/\alpha^2) \to 1\end{array}$	$q_N \approx (1-\eta)/\alpha(2-\eta)$	$q_N \ll 0.5$
$\mathscr{S}_{R_1}^{j\omega_N} \approx -j\omega_N/2$	$q_{N\max} \approx 1/2\alpha$ for $\eta \to 0$	orthogonality between $\mathscr{S}_{R_2}^{j\omega_N}$ and $\mathscr{S}_{R_1}^{j\omega_N} = \mathscr{S}_{R_3}^{j\omega_N}$
$\mathscr{S}_{R_2}^{j\omega_N} \approx \omega_N/2\alpha$		
$\mathscr{S}_{R_3}^{j\omega_N} \approx -j\omega_N/2$		
2C. $\mathrm{Im}\ \mathscr{S}_{R_3}^{j\omega_N} \to 0:\ \begin{array}{l}\alpha \to 0 \\ \lambda = 0.5\end{array}$	$q_N \approx \dfrac{\alpha(1-\eta)}{\alpha^2 + 2(1-\eta)}$	$q_N \ll 0.5$
$\mathscr{S}_{R_1}^{j\omega_N} \approx \dfrac{\omega_N}{2}\left(\dfrac{\alpha}{2} - j\right)$	$q_{N\max} \approx \alpha/2$ for $\eta \to 0$	variable resistor (R_3) tied to ground
$\mathscr{S}_{R_2}^{j\omega_N} \approx \dfrac{\omega_N}{2}\left(\dfrac{\alpha}{2} - j\right)$		
$\mathscr{S}_{R_3}^{j\omega_N} \approx -\alpha\omega_N/2$		

The minimum of the imaginary part of $\mathscr{S}^{j\omega_N}_{R_1}$ occurs when $\lambda = 1/\alpha^2$, in which case the sensitivity functions become

$$\mathscr{S}^{j\omega_N}_{R_1} = \frac{\alpha\omega_N}{2(1+\alpha^2)}\left[1 - \frac{1}{\alpha^2} - j\frac{2}{\alpha}\right]\Bigg|_{\alpha \gg 1} \approx \frac{\omega_N}{2\alpha} \tag{8-68a}$$

$$\mathscr{S}^{j\omega_N}_{R_2} = \frac{\alpha\omega_N}{2(1+\alpha^2)}\left[\frac{1}{\alpha^2} - j\alpha\right]\Bigg|_{\alpha \gg 1} \approx -j\frac{\omega_N}{2} \tag{8-68b}$$

$$\mathscr{S}^{j\omega_N}_{R_3} = -\frac{\alpha\omega_N}{2(1+\alpha^2)}(1+j\alpha)\Bigg|_{\alpha \gg 1} \approx -j\frac{\omega_N}{2} \tag{8-68c}$$

Condition (8-67) can be realized by (8-68a) if $\alpha \gg 1$. Furthermore, the other two sensitivity functions turn out to be orthogonal to (8-68a), permitting independent null-frequency and null-depth control by two individual resistors (e.g., R_1 and R_2 or R_1 and R_3).

Whereas any sensitivity functions requiring large values of α may very likely be dismissed as impractical in the case of frequency tuning, because of the existence of one case (tuning with R_2) in which no decrease in selectivity need be suffered, no such freedom exists here. It can be shown that any configuration allowing null-depth tuning by one component invariably results in selectivity deterioration. Practical implementation therefore requires a compromise between the realization of any one of the three possible sensitivity functions and selectivity. Nevertheless, not all the cases discussed are equally disadvantageous with respect to this compromise; null-depth tuning with R_1 results in q_N values twice as large as in the other two cases.

Substituting $\lambda = 1/\alpha^2$ into (8-61) we obtain

$$q_N = \frac{(\alpha^2 - 1)(1 - \eta)}{\alpha(\alpha^2 - \eta)}. \tag{8-69}$$

To obtain as large a value as possible for q_N, we let $\eta \to 0$, in which case $q_N \approx 1/\alpha$. Thus the more accurately we realize (8-68a), the smaller the selectivity will become. This as well as the remaining two cases are summarized in Table 8-4 together with the design equations that must apply. Here again the advantage of using R_3 for the null-depth adjustment may be that it has one terminal grounded. This will be useful, for example, if the null depth is to be varied by a switch or a voltage-dependent resistor (e.g., FET).

8.3.3 Orthogonal Tuning With Two Components

Orthogonality between two zero-sensitivity functions simplifies null adjustments in the vicinity of a perfect null, particularly if the two functions are parallel with the real and imaginary axes. Some of the configurations described in the previous sections provided the latter type of orthogonality, but only at the cost of selectivity. General orthogonality, which is discussed here, may be of interest for a variety of reasons; e.g., the 90° phase reference required here for tuning purposes may be easier to generate than any arbitrary phase reference.

Two vectors $\bar{q} = u + jv$ and $\bar{p} = w + jz$ are orthogonal if

$$uw + vz = 0 \tag{8-70}$$

Thus, to obtain orthogonality between pairs of the functions listed in Table 8-2, we must investigate if they can be made to satisfy this condition. Let us, for example, examine the possibility of orthogonal tuning between R_1 and R_3. This requires that

$$-(1 - \lambda) + \alpha\left(\frac{1}{\alpha} + \alpha\lambda\right) = 0 \tag{8-71}$$

Solving for α results in two nonrealizable (i.e., imaginary) roots for α and one additional solution for which $\lambda = 0$. The latter condition can only be approximated (see (8-60c)) and has been dealt with in the two previous cases (see 1B and 2A in Table 8-4); as expected, $\mathscr{S}_{R_1}^{j\omega_N}$ is orthogonal to $\mathscr{S}_{R_3}^{j\omega_N}$. Similarly, it can be shown that orthogonal tuning between R_2 and R_3 is only possible in the circumstances given by cases 1A and 2B in Table 8-4, and orthogonality between R_1 and R_2 cannot be achieved at all.

8.3.4 Two Design Examples

Using the design equations listed in Table 8-4, the detailed procedure for the design of two twin-*T* networks with prescribed tuning characteristics follow.

Twin-*T* With Null Frequency Tunable by R_2   To satisfy condition (8-62) we find from Table 8-4, part 1B, that $\lambda \ll 1$, and therefore we select $\lambda = 0.01$. Furthermore, specifying that $q_N = 0.25$, $\omega_N = 2\pi(1 \ \text{kHz})$, and $R_1 = 1 \ \text{k}\Omega$, we find $R_2 = 99 \ \text{k}\Omega$. From Fig. 8-15 we find that the $q_{\max}$ curve passes through the value 0.25 when $\alpha = 0.5$. However, with $\lambda = 0.01$, γ takes on a simpler value for $\alpha = 0.55$ (see Fig. 8-13), namely,

$$\gamma = \alpha^2\lambda(1 - \lambda) = 0.003$$

Solving (8-64) for η we obtain

$$\eta = 1 - \frac{\alpha^2}{4\alpha - 1} = 0.75$$

and from (8-58)

$$C_1 = \frac{\gamma}{\eta} \cdot C_3 = 0.004C_3$$

From (8-31) and (8-58) we obtain

$$C_3 = \frac{1}{\omega_N \sqrt{\gamma R_1 R_2}} = 0.292 \ \mu F$$

and

$$C_1 = \frac{\gamma}{\eta} C_3 = 1.168 \ nF$$

Finally, with (8-56), we obtain

$$C_2 = \frac{\eta}{1 - \eta} C_1 = 3.504 \ nF$$

and, from (8-32),

$$R_3 = \frac{C_3}{C_p} \cdot R_p = 62 \ k\Omega,$$

The corresponding sensitivity functions can be calculated directly by substituting the values obtained above into the expressions listed in Table 8-2. Considering only the relative values of the resistor sensitivity functions, we obtain

$$\mathscr{S}_{R_1}^{j\omega_N} = \frac{\alpha \omega_N}{2(1 + \alpha^2)} (0.99 - 1.83j)$$

$$\mathscr{S}_{R_2}^{j\omega_N} = \frac{\alpha \omega_N}{2(1 + \alpha^2)} (0.01 - 2.365j)$$

$$\mathscr{S}_{R_3}^{j\omega_N} = \frac{\alpha \omega_N}{2(1 + \alpha^2)} (1 + 0.55j)$$

The twin-T network resulting from the above calculations, as well as the zero-displacement vectors given above, are shown in Fig. 8-16. Thus the condition for frequency tuning (see (8-62)) can be realized very closely, while high selectivity is maintained; this, however, at the expense of correspondingly large spreads in resistor and capacitor values.

Twin-T With Null Depth Tunable by R_3   To satisfy the condition that Im $\mathscr{S}_{R_3}^{j\omega_N} \to 0$, we find from Table 8-4, part 2C, that $\alpha \ll 1$ and select $\alpha = 0.1$. Furthermore, with $\lambda = 0.5$ and $\eta = 0.01$, for maximum selectivity, we find from (8-61) that $q_N = 0.048$.

As in the previous example, we specify that $\omega_N = 2\pi$ (1 kHz) and, because $\lambda = 0.5$, we select $R_1 = R_2 = 10 \ k\Omega$. From (8-59) we find $\gamma = 0.0025$ and, in

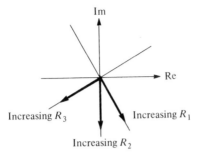

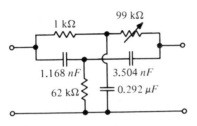

FIG. 8-16. Zero displacement and twin-T for $f_N = 1$ kHz, $q_N = 0.25$, $\lambda = 0.01$, $\eta = 0.75$, $\alpha = 0.55$.

precisely the same way as in the preceding example, $C_3 = 0.318$ μF, $C_1 = 79.5$ nF, $C_2 = 0.804$ nF, and $R_3 = 19.8$ kΩ. The corresponding zero-sensitivity functions follow directly from Table 8-4. The relative values of the resistor sensitivity functions are

$$\mathscr{S}_{R_1}^{j\omega_N} = \frac{\alpha\omega_N}{2(1 + \alpha^2)}(0.5 - 10.05j)$$

$$\mathscr{S}_{R_2}^{j\omega_N} = \frac{\alpha\omega_N}{2(1 + \alpha^2)}(0.5 - 10.05j)$$

$$\mathscr{S}_{R_3}^{j\omega_N} = -\frac{\alpha\omega_N}{2(1 + \alpha^2)}(1 + 0.1j)$$

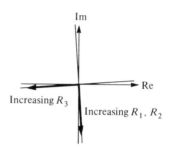

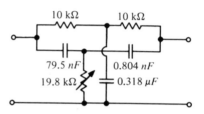

FIG. 8-17. Zero displacement and twin-T for $f_N = 1$ kHz, $q_N = 0.048$, $\lambda = 0.5$, $\eta = 0.01$, $\alpha = 0.1$.

The resulting twin-T network and the zero-displacement vectors are shown in Fig. 8-17. Notice that orthogonality between frequency (adjusting R_1 or R_2) and null depth (adjusting R_3) is almost accurately obtained. This may be of considerable advantage in some networks, but at the cost of low selectivity ($q_N = 0.048$) and a relatively large spread in capacitor values.

It may be timely here to remind the reader that the preceding discussion on sensitivity with a view to tuning and selectivity (and, in fact, the following discussion on null stability) is by no means restricted to the twin-T network. Indeed, similar reasoning applies to any RC bridge network (including, of course, those discussed in Section 8.1.1). In all cases, by considering the balance conditions of the network and deriving the null sensitivity with respect to any component, design equations for the networks can be obtained that will modify the corresponding null-sensitivity expressions in such a way as to provide convenient or otherwise desirable adjustment conditions in the vicinity of the null.

8.4 TWIN-T NULL STABILITY USING THIN–FILM COMPONENTS

The twin-T is frequently used to provide stable zeros in the design of hybrid integrated linear active networks. If a high degree of stability is required, thin-film components should be used for the twin-T network. The degree of

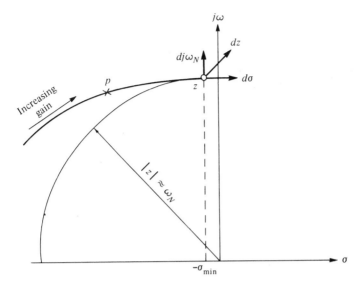

FIG. 8-18. Typical root locus of feedback network using a twin-*T* to provide the open-loop zeros; $z =$ open-loop zero, $p =$ closed-loop pole.

null stability that can be expected with thin-film components whose temperature coefficients and aging characteristics are known follows directly from the sensitivity functions discussed in the previous sections. This will be demonstrated in the following discussion.

A displacement dz of the transmission zero z of a twin-*T* network, caused by variations in its six components, can be expressed in terms of the zero sensitivity as follows:

$$dz = \sum_{i=1}^{3} \mathscr{S}_{R_i}^z \frac{dR_i}{R_i} + \sum_{i=1}^{3} \mathscr{S}_{C_i}^z \frac{dC_i}{C_i} \qquad (8\text{-}72)$$

As shown in Fig. 8-18, if the twin-*T* transmission zero is close to the $j\omega$ axis it can be considered purely imaginary i.e.,

$$\mathscr{S}_x^z \approx \mathscr{S}_x^{j\omega_N} \qquad (8\text{-}73)$$

Furthermore, the zero displacement is given as

$$dz = d\sigma + dj\omega_N \qquad (8\text{-}74)$$

Equating (8-72) and (8-74) and substituting the expressions of Table 8-2 in (8-72), it is possible to solve for $d\sigma$ and $dj\omega_N$.

First, however, some characteristics peculiar to thin-film integrated circuitry must be considered. As discussed in Chapter 6, variations of thin-film components tend to track very closely on a given glass or ceramic substrate, and these variations can be precisely predicted and controlled. Component tracking simplifies the following calculations considerably. Thus, we can write

$$\Delta R_i / R_i = (\delta_r \pm \varepsilon_r) \Delta T + \kappa_r = \Delta R / R \qquad (8\text{-}75)$$

and

$$\Delta C_i / C_i = (\delta_c \pm \varepsilon_c) \Delta T + \kappa_c = \Delta C / C \qquad (8\text{-}76)$$

The temperature coefficients of the resistors and capacitors are δ_r and δ_c, respectively; ε_r and ε_c are the tracking ratios among the three components of a kind; ΔT is the temperature range under consideration; and κ_r and κ_c are the percentage resistor and capacitor aging, respectively. Substituting (8-75) and (8-60) as well as the expressions in Table 8-2 into (8-72) we obtain for the real zero displacement

$$\frac{\Delta \sigma}{\omega_N} = \frac{\alpha}{2(1 + \alpha^2)} [\varepsilon_c(1 - \eta) - \varepsilon_r(1 - \lambda)] \Delta T \qquad [8\text{-}77]$$

Notice that the zero displacement parallel to the real axis depends only on the amount of mistracking between components and not on the absolute drift of the individual components. In other words, if all components of a kind drift by the same percentage, the null depth of a tuned twin-T will not change. This is to be expected since equal component drift corresponds to a frequency-scaling process.

The imaginary zero displacement is obtained in the same way as the real displacement:

$$-\frac{\Delta \omega_N}{\omega_N} = (\delta_r + \delta_c) \Delta T + \kappa_r + \kappa_c + \frac{\varepsilon_r \, \Delta T}{2} \frac{1 + \alpha^2(2 - \lambda)}{1 + \alpha^2} + \frac{\varepsilon_c \, \Delta T}{2} \frac{\alpha^2 + 2 - \eta}{1 + \alpha^2}$$

$$[8\text{-}78]$$

Thus, the zero displacement along the $j\omega$ axis depends on the actual drift of the individual components. Clearly, if the temperature and aging coefficients of the resistors can be made equal in value but opposite in sign to those of the capacitors, drift along the $j\omega$ axis can be eliminated.

In various active filter schemes the network poles are tied closely to the transmission zeros generated by a twin-T. In high-Q networks, uncontrolled drift of the twin-T zero into the right half s plane could therefore pull the poles over with it, causing oscillation. Similarly, drift of the twin-T zero along the $j\omega$ axis would cause a frequency shift in the active-filter frequency response.

To prevent oscillation due to drift into the right half plane, the transmission zero of the twin-T must be located left of the $j\omega$ axis by some distance σ_{min} such that, under worst-case component drift, it will not travel across the $j\omega$ axis. Referring to Fig. 8-18, this implies that

$$\sigma_{min} \geq \text{Re}\,(\Delta z_{max}) = \Delta \sigma_{max} \qquad (8\text{-}79)$$

This condition, in turn, implies that the twin-*T* null depth may not exceed a certain maximum attenuation $T_{N\,max}$, which can now be calculated directly.

It follows from Section 8.1.2 that the transfer function of a twin-*T* with an imperfect null will have the form

$$T_N(s) \approx \frac{s^2 + 2\sigma s + \omega_N^2}{s^2 + (\omega_N/q_N)s + \omega_N^2} \tag{8-80}$$

With (8-79) the maximum null attenuation for left-half-plane transmission zeros is then

$$T_N \max \bigg|_{s=j\omega_N} = 2\frac{q_N}{\omega_N}\sigma_{min} = 2\frac{q_N}{\omega_N}\Delta\sigma_{max} \tag{8-81}$$

With (8-61) and (8-77) this becomes

$$T_{N\,max} = \frac{\alpha^2}{1+\alpha^2}\left(\frac{(1-\eta)(1-\lambda)}{\alpha^2(1-\lambda)+(1-\eta)}\right)[\varepsilon_r(1-\lambda)+\varepsilon_c(1-\eta)]\,\Delta T \tag{8-82}$$

In active-filter applications, in which the twin-*T* transmission zero *z* represents the open-loop zero of the root locus of a pole *p* with respect to gain (see Fig. 8-18), the highest attainable *Q* of the network is all the more limited, the larger σ_{min} has to be chosen for stability. In the limit, as the loop gain approaches infinity, the closed-loop pole *p* coincides with *z*. The upper limit on *Q* is therefore given by

$$Q_{max} < \frac{\omega_N}{2\sigma_{min}} \tag{8-83}$$

or, with (8-77)

$$Q_{max} < \frac{(1+\alpha^2)}{\alpha\,\Delta T}\frac{1}{\varepsilon_r(1-\lambda)+\varepsilon_c(1-\eta)} \tag{8-84}$$

Thus, with the type of active-network design represented by the root locus in Fig. 8-18, both network stability and maximum *Q* may ultimately depend on the stability of the twin-*T* network.

Consider, for example, the stability of a symmetrical twin-*T* network fabricated with tantalum thin-film resistors and capacitors. The required ambient temperature range is assumed to be from 0°C to 60°C. From (8-82) we obtain

$$T_N \max \bigg|_{\substack{\lambda=\eta=0.5 \\ \alpha=1}} = \frac{1}{16}(\varepsilon_r + \varepsilon_c)\,\Delta T \tag{8-85}$$

Typically, for tantalum thin-film resistors and capacitors $\varepsilon_r = \pm 5$ ppm/°C and $\varepsilon_c = \pm 15$ ppm/°C, in which case $T_{N\,max} = 7.5 \times 10^{-5} = -83$ dB.

The frequency drift for a symmetrical twin-*T* results from (8-78):

$$-\Delta\omega_N/\omega_N = [(\delta_r + \delta_c) + \tfrac{5}{8}(\varepsilon_r + \varepsilon_c)]\,\Delta T + \kappa_r + \kappa_c \tag{8-86}$$

Typically, for tantalum thin-film components $\kappa_r = \kappa_c = 0.1\%$ and for tantalum thin-film capacitors $\delta_c = 200$ ppm/°C. The TC of tantalum thin-film resistors can be controlled by oxygen doping during the sputtering process (oxynitride). It may therefore be of interest to solve (8-86) for the required TCR (i.e., δ_r) when a maximum acceptable frequency drift is specified. Assuming that $(\Delta\omega_N/\omega_N)_{max} \leq 0.5\%$ we obtain $\delta_r = -215 \pm 50$ ppm/°C.

APPENDIX

To Find the Largest Common Divider of Two Polynomials: Euclid's Algorithm

Consider two polynomials $D_0(s)$ and $D_1(s)$ where

$$\text{degree } D_1(s) \leq \text{degree } D_0(s) \tag{1}$$

We wish to find the largest common divider of these two polynomials. Using Euclid's algorithm we proceed as follows:*

First step: Form the ratio

$$\frac{D_0(s)}{D_1(s)} = P_1(s) + \frac{D_2(s)}{D_1(s)} \tag{2}$$

where the degree of $D_2(s)$ is smaller than the degree of $D_1(s)$. Now, if $D_0(s)$ and $D_1(s)$ have a common root, p_0, then this root must also be contained in $D_2(s)$. This follows directly from (2) since:

$$D_0(p_0) = P_1(p_0)D_1(p_0) + D_2(p_0) = 0 \tag{3}$$

Obviously, if p_0 were *not* a root of $D_2(s)$ then (3) would not hold and p_0 would not be a root of $D_0(s)$. Thus we continue with the *second step* forming the ratio:

$$\frac{D_1(s)}{D_2(s)} = P_2(s) + \frac{D_3(s)}{D_2(s)} \tag{4}$$

Reasoning as we did above, we can state that p_0, the common root of $D_1(s)$ and $D_2(s)$, must also be a root of $D_3(s)$ and that the degree of $D_3(s)$ is smaller than the degree of $D_2(s)$. Continuing in this manner we arrive at the jth *step* and obtain the general form:

$$\frac{D_{j-1}(s)}{D_j(s)} = P_j(s) + \frac{D_{j+1}(s)}{D_j(s)} \tag{5}$$

Now, *if $D_{j+1}(s) \equiv 0$, then $D_j(s)$ must be the largest common divider of $\cdot D_0(s)$ and $D_1(s)$.* This is so because we already know that $D_{j-1}(s)$ and $D_j(s)$ have the same common divider as $D_0(s)$ and $D_1(s)$. If in addition $D_{j+1}(s) \equiv 0$ then we have from (5):

$$D_{j-1}(s) = P_j(s)D_j(s) \tag{6}$$

which tells us that $D_j(s)$ is also contained in $D_{j-1}(s)$; thus $D_j(s)$ is the sought-after largest common divider of $D_0(s)$ and $D_1(s)$.

---

* E. Lueder: "The General Second-Order Twin-T and Its Application to Frequency Emphasizing Networks," BSTJ, Vol. 51, No. 1, January 1972, pp. 301–316.

If, now, we are looking for a common divider of first degree in s (as in the case of the twin-T in Chapter 8, section 8-1.2) then we continue dividing as above until a $D_j(s)$ of first degree is found. The rest, $D_{j+1}(s)$ then has zero degree. Setting this rest equal to zero, we obtain the condition for which our two polynomials $D_0(s)$ and $D_1(s)$ have a common divider of first degree, that is a common single root.

ADDITIONAL REFERENCES AND SUGGESTED READING

Chapter 1
Angelo Jr., E. J. and Papoulis, A.,: *Pole-Zero Patterns in the Analysis and Design of Low-Order Systems*, McGraw-Hill Book Co., New York, 1964.
Balabanian, N., *Network Synthesis*, Prentice-Hall, Englewood Cliffs, N. J., 1958.
Balabanian N. and Bickart T. A., *Electrical Network Theory*, John Wiley & Sons, New York, 1969.
Guillemin, E. A., *Synthesis of Passive Networks*, John Wiley & Sons, New York, 1957.
Karni S., *Network Theory: Analysis and Synthesis*, Allyn and Bacon, Boston, 1966.
Kuh, E. S. and Rohrer, R. A., *Theory of Linear Active Networks*, Holden-Day, San Francisco. 1967.
Truxal, J. G., *Control System Synthesis*, McGraw-Hill Book Co., New York, 1955.
Van Valkenburg, M. E., *Introduction to Modern Network Synthesis*, John Wiley & Sons, New York, 1960.
Weinberg, L., *Network Analysis and Synthesis*, McGraw-Hill Book Co., New York, 1962.

Chapter 2
Bode, H. W., *Network Analysis and Feedback Amplifier Design*, Van Nostrand Reinhold Co., New York, 1945.
Chow, Y. and Cassignol, E., *Linear Signal-Flow Graphs and Applications*, John Wiley & Sons, New York, 1962.
DePian, L., *Linear Active Network Theory*, Prentice Hall, Englewood Cliffs, N. J., 1962.
Haykin, S. S., *Active Network Theory*, Addison-Wesley Publishing Co., Reading, Mass., 1961.
Horowitz, I. M., *Synthesis of Feedback Systems*, Academic Press, New York, 1963.
Lorens, C. S., *Flowgraphs for the Modeling and Analysis of Linear Systems*, McGraw-Hill Book Co., New York, 1964.
Mason, S. J., and Zimmerman, H. J., *Electronic Circuits, Signals, and Systems*, John Wiley & Sons, New York, 1960.
Sheshu, S. and Reed, M. B., *Linear Graphs and Electrical Networks*, Addison-Wesley Publishing Co., Reading, Massachusetts, 1961.
Shu-Park Chan, *Introductory Topological Analysis of Electrical Networks*, Holt, Rinehart and Winston, Inc., New York, 1969.
Truxal, J. G., *Control System Synthesis*, McGraw-Hill Book Co., New York, 1955.

Chapter 3
Balabanian, N. and Bickart, T. A., *Electrical Network Theory*, John Wiley & Sons, New York, 1969.

Braae, R., *Matrix Algebra for Electrical Engineers*, Addison-Wesley Publishing Co., New York, 1963.

DePian, L., *Linear Active Network Theory*, Prentice-Hall, Inc., Englewood Cliffs, N. J., 1962.

Desoer, C. A. and Kuh, E. S., *Basic Circuit Theory*, McGraw-Hill Book Co., New York, 1969.

Ghausi, M. S., *Principles and Design of Linear Active Circuits*, McGraw-Hill Book Co., New York, 1965.

Haykin, S. S., *Active Network Theory*, Addison-Wesley Publishing Company, Reading, Massachusetts, 1970.

Hlawiczka, P., *Matrix Algebra for Electronic Engineers*, Haydn Book Co., New York, 1965.

Huelsman, L. P., *Circuits, Matrices, and Linear Vector Spaces*, McGraw-Hill Book Co., New York, 1963.

Linvill, J. G., and Gibbons, J. F., *Transistors and Active Circuits*, McGraw-Hill Book Co., New York, 1961.

Mitra, S. K., *Analysis and Synthesis of Linear Active Networks*, John Wiley & Sons, New York, 1969.

Muir, T. and Metzler, W. H., *A Treatise on the Theory of Determinants*, Dover Publications, New York, 1960.

Van Valkenburg, M. E., *Network Analysis*, Prentice-Hall, Englewood Cliffs, N. J., 1964.

Chapter 4

Bode, H. W., *Network Analysis and Feedback Amplifier Design*, Van Nostrand Reinhold Co., New York, 1945.

Blostein, M. L., *Sensitivity Considerations in RLC Networks*, Doctoral Dissertation. Univ, of Illinois, 1963.

Calahan, D. A., *Modern Network Theory*, Hayden Book Co., New York, 1964.

Huelsman, L. P., *Theory and Design of Active RC Circuits*, McGraw-Hill Book Co., New York, 1968.

Mitra, S. K., *Analysis and Synthesis of Linear Active Networks*, John Wiley & Sons, New York, 1969.

Spence, R., *Linear Active Networks*, John Wiley & Sons, London, 1970.

Chapter 5

Burr-Brown Research Corp., *Handbook of Operational Amplifier Applications*, Tucson, Arizona, 1963. Also: *Handbook of Operational Amplifier Active RC Networks*, 1966.

Haykin, S. S., *Synthesis of RC Active Filter Networks*, McGraw-Hill Book Co., London, 1969

Huelsman, L. P., *Theory and Design of Active RC Circuits*, McGraw-Hill Book Co., New York, 1968.

Mitra, S. K., *Analysis and Synthesis of Linear Active Networks*, John Wiley & Sons, New York, 1969.

Newcomb, R. W., *Active Integrated Circuit Synthesis*, Prentice-Hall, Englewood Cliffs, N. J., 1968.

Spence, R., *Linear Active Networks*, John Wiley & Sons, London, 1970.

Su, K. L., *Active Network Synthesis*, McGraw-Hill Book Co., New York, 1965.

Chapter 6

Berry, R. W., Hall, P. M. and Harris, M. T., *Thin Film Technology*, Van Nostrand Reinhold, Co., New York, 1968.

Hamer, D. W., *Thick Film Hybrid Microcircuit Technology*, John Wiley & Sons, New York, 1972.

Hnatek, E. R., *A User's Handbook of Integrated Circuits*, John Wiley & Sons, New York, 1973.

Fogiel, M., *Microelectronics*, Research and Education Assoc., New York, 1968.

Charschan, S. S., *Lasers in Industry*, Van Nostrand Reinhold Co., New York, 1972.

Madland, G. R. *et al.*, *Integrated Circuit Engineering: Basic Technology*, Boston Technical Publishers, Cambridge, Massachusetts, 1966.

Stern, L., *Fundamentals of Integrated Circuits*, Hayden Book Co., New York, 1968.

Topfer, M. L., *Thick-Film Microelectronics: Fabrication, Design and Applications*, Van Nostrand Reinhold Co., New York, 1971.

Warner Jr., R. M., and Fordemwalt, *Integrated Circuits, Design Principles and Fabrication*, McGraw-Hill Book Co., New York, 1965.

Chapter 7

Angelo, E. J., *Electronics: BJTs, FETs, and Microcircuits*, McGraw-Hill Book Co., New York, 1969.

Burr-Brown, *Operational Amplifiers, Design and Applications*, McGraw-Hill Book Co., New York, 1971.

Camenzind, H. R., *Electronic Integrated Systems Design*, Van Nostrand Reinhold Co., New York, 1972.

Chirlian, P. M., *Integrated and Active Network Analysis and Synthesis*, Prentice-Hall, Inc., Englewood Cliffs, N. J., 1967.

Clayton, C. B., *Operational Amplifiers*, Butterworth & Co., London, 1971.

Eimbinder, J., *Linear Integrated Circuits: Theory and Applications*, John Wiley & Sons, New York, 1968.

Eimbinder, J., *Application Considerations for Linear Integrated Circuits*, John Wiley & Sons, New York, 1970.

Fitchen, F. C., *Electronic Integrated Circuits and Systems*, Van Nostrand Reinhold Co., New York, 1970.

Ghausi, M. S., *Electronic Circuits*, Van Nostrand Reinhold Co., New York, 1971.

Grebene, A. B., *Analog Integrated Circuit Design*, Van Nostrand Reinhold Co., New York, 1972.

Huelsman, L. P., *Active Filters: Lumped, Distributed, Integrated, Digital and Parametric*, McGraw-Hill Book Co., New York, 1970.

Lynn, D. K. *et al.*, *Analysis and Design of Integrated Circuits*, McGraw-Hill Book Co., New York, 1967.

Millman, J. and Halkias, C. C., *Integrated Electronics: Analog and Digital Circuits and Systems*, McGraw-Hill Book Co., New York, 1972.

Newcomb, R. W., *Active Integrated Circuit Synthesis*, Prentice-Hall, Englewood Cliffs, N. J., 1968.

RCA: *Linear Integrated Circuits*, Technical Series 1C-41, RCA, Harrison, N. J., 1967.

Chapter 8

Balabanian, N., *Network Synthesis*, Prentice-Hall, Englewood Cliffs, N. J., 1958.

Ghausi, M. S. and Kelly, J. J., *Introduction to Distributed-Parameter Networks With Application to Integrated Circuits*, Holt, Rineholt and Winston, New York, 1968.

Guillemin, E. A., *Synthesis of Passive Networks*, John Wiley & Sons, New York, 1969.

Kuo, F. F., *Network Analysis and Synthesis*, John Wiley & Sons, New York, 1966.

Sun, H. H., *Synthesis of RC Networks*, Hayden Book Co., New York, 1967.

Wohlers, M. R., *Lumped and Distributed Passive Networks*, Academic Press, New York, 1969.

INDEX